中国生物多样性红色名录：脊椎动物
China's Red List of Biodiversity: Vertebrates

主编 蒋志刚
Chief Editor: Zhigang Jiang

第二卷 鸟 类
Volume II, Birds

主编 张雁云 郑光美
Chief Editors: Yanyun Zhang Guangmei Zheng

科学出版社
北 京

内 容 简 介

为全面评估中国鸟类受威胁状况，我们利用IUCN红色名录划分标准，对《中国鸟类分类与分布名录》（第三版）中收录的1,445种鸟类进行了评估，其中3种鸟类被列为区域灭绝（RE），还有极危（CR）18种、濒危（EN）48种、易危（VU）78种、近危（NT）223种、无危（LC）969种，数据缺乏（DD）105种，1种为无效种。本书还给出了受威胁物种或野外灭绝物种的分类地位、评估信息、地理分布、种群现状以及发展趋势、栖息地、威胁因子、保护行动等内容。

本工作是对中国鸟类受威胁状况进行的一次全面评估和系统分析，为我国鸟类研究、保护、管理工作者提供借鉴，对我国鸟类保护和履行国际公约等工作有重要价值。

审图号：GS(2020)3061号

图书在版编目（CIP）数据

中国生物多样性红色名录. 脊椎动物. 第二卷，鸟类 = China's Red List of Biodiversity: Vertebrates, Volume II, Birds: 汉英对照 / 蒋志刚主编；张雁云，郑光美分册主编. —北京：科学出版社，2021.3
国家出版基金项目
ISBN 978-7-03-067491-3

Ⅰ. ①中… Ⅱ. ①蒋… ②张… ③郑… Ⅲ. ①珍稀动物-中国-名录-汉、英 ②珍稀植物-中国-名录-汉、英 ③鸟类-中国-名录-汉、英 Ⅳ. ①Q958.52-62 ②Q948.52-62 ③Q959.7

中国版本图书馆CIP数据核字（2020）第256690号

责任编辑：马　俊　李　迪　孙　青 / 责任校对：严　娜
责任印制：肖　兴 / 排版设计：北京鑫诚文化传播有限公司

科学出版社 出版
北京东黄城根北街16号
邮政编码：100717
http://www.sciencep.com

中国科学院印刷厂 印刷
科学出版社发行　各地新华书店经销

*

2021年3月第　一　版　开本：889×1194　1/16
2021年3月第一次印刷　印张：31 1/2
字数：907 000

定价：528.00元
（如有印装质量问题，我社负责调换）

China's Red List of Biodiversity: Vertebrates, Volume II, Birds

Chief Editors: Yanyun Zhang   Guangmei Zheng

Abstract

To get comprehensive knowledge of the current threatened status of birds in China, we assessed 1,445 bird species listed in the book entitled with *A checklist on the classification and distribution of the birds of China* using IUCN Red List criteria. 3 species were identified as Regionally Extinct (RE); as well as 18 species Critically Endangered (CR), 48 species Endangered (EN), 78 species Vulnerable (VU), 223 species Near Threatened (NT), 969 species Least Concern (LC), 105 species Data Deficient (DD), and one species as invalid. For each threatened species (including CR, EN and VU) and RE species, the taxonomic status, assessment information, geographical distribution, population, habitats, threatened factors and conservation actions were present.

This project aims to make a comprehensive assessment and systematic analysis of the threatened status of birds in China. It provides a reference to the researchers, conservationists, and managers of birds in China. It is of great value to the conservation of birds in China and the implementation of international conventions.

ISBN 978-7-03-067491-3

Copyright© 2021, Science Press (Beijing)
All rights reserved. No Part of this publication may be reproduced, stored in a retrieval system, or transmitted in any form or by any means, mechanical, photocopying, recording or otherwise, without the prior written permission of the copyright owner.

《中国生物多样性红色名录：脊椎动物》编委会

顾问
陈宜瑜　郑光美　张亚平　金鉴明　马建章　曹文宣

主编
蒋志刚

副主编
江建平　王跃招　张　鹗　张雁云

编委会成员
蔡　波　曹　亮　车　静　陈小勇　陈晓虹　丁　平
董　路　胡慧建　胡军华　计　翔　江建平　蒋学龙
蒋志刚　李　成　李春旺　李家堂　李丕鹏　梁　伟
刘　阳　刘少英　卢　欣　马　鸣　马　勇　马志军
饶定齐　史海涛　王　斌　王剑伟　王英永　王跃招
吴　华　吴　毅　吴孝兵　谢　锋　杨道德　杨晓君
曾晓茂　张　鹗　张　洁　张保卫　张雁云　张正旺
赵亚辉　周　放　周开亚

责任编辑
马　俊　李　迪　郝晨扬　孙　青

Editorial Committee of China's Red List of Biodiversity: Vertebrates

Consultants of the Editorial Committee

Yiyu Chen   Guangmei Zheng   Yaping Zhang   Jianming Jin   Jianzhang Ma
Wenxuan Cao

Chief Editor of the Editorial Committee

Zhigang Jiang

Vice-Chief Editors of the Editorial Committee

Jianping Jiang   Yuezhao Wang   E Zhang   Yanyun Zhang

Members of the Editorial Committee

Bo Cai   Liang Cao   Jing Che   Xiaoyong Chen   Xiaohong Chen   Ping Ding
Lu Dong   Huijian Hu   Junhua Hu   Xiang Ji   Jianping Jiang   Xuelong Jiang
Zhigang Jiang   Cheng Li   Chunwang Li   Jiatang Li   Pipeng Li   Wei Liang
Yang Liu   Shaoying Liu   Xin Lu   Ming Ma   Yong Ma   Zhijun Ma
Dingqi Rao   Haitao Shi   Bin Wang   Jianwei Wang   Yingyong Wang
Yuezhao Wang   Hua Wu   Yi Wu   Xiaobing Wu   Feng Xie   Daode Yang
Xiaojun Yang   Xiaomao Zeng   E Zhang   Jie Zhang   Baowei Zhang
Yanyun Zhang   Zhengwang Zhang   Yahui Zhao   Fang Zhou   Kaiya Zhou

Responsible Editors

Jun Ma   Di Li   Chenyang Hao   Qing Sun

本书编委会

主　编　张雁云　郑光美

编写人员

北京林业大学：丁长青
北京师范大学：郑光美　张正旺　张雁云　邓文洪　董　路
东北师范大学：王海涛
复　旦　大　学：马志军
广　西　大　学：周　放
海南师范大学：梁　伟　杨灿朝
内蒙古大学：邢莲莲　杨贵生
武　汉　大　学：卢　欣
浙　江　大　学：丁　平
浙江自然博物馆：陈水华　范忠勇
中国科学院动物研究所：孙悦华　朱　磊
中国科学院昆明动物研究所：杨晓君
中国科学院生态环境研究中心：曹　垒
中国科学院新疆生态与地理研究所：马　鸣
中国林业科学研究院：马　强
中　山　大　学：刘　阳

英文翻译

北京林业大学：曾　晴

Editorial Committee of The Book

Chief Editors Yanyun Zhang, Guangmei Zheng

Compilers

Beijing Forestry University: Changqing Ding

Beijing Normal University: Guangmei Zheng, Zhengwang Zhang, Yanyun Zhang, Wenhong Deng, Lu Dong

Northeast Normal University: Haitao Wang

Fudan University: Zhijun Ma

Guangxi University: Fang Zhou

Hainan Normal University: Wei Liang, Canchao Yang

Inner Mongolia University: Lianlian Xing, Guisheng Yang

Wuhan University: Xin Lu

Zhejiang University: Ping Ding

Zhejiang Museum of Natural History: Shuihua Chen, Zhongyong Fan

Institute of Zoology, Chinese Academy of Sciences: Yuehua Sun, Lei Zhu

Kunming Institute of Zoology, Chinese Academy of Sciences: Xiaojun Yang

Research Center for Eco-Environmental Sciences, Chinese Academy of Sciences: Lei Cao

Xinjiang Institute of Ecology and Geography, Chinese Academy of Sciences: Ming Ma

Chinese Academy of Forestry: Qiang Ma

Sun Yat-sen University: Yang Liu

Translation

Beijing Forestry University: Qing Zeng

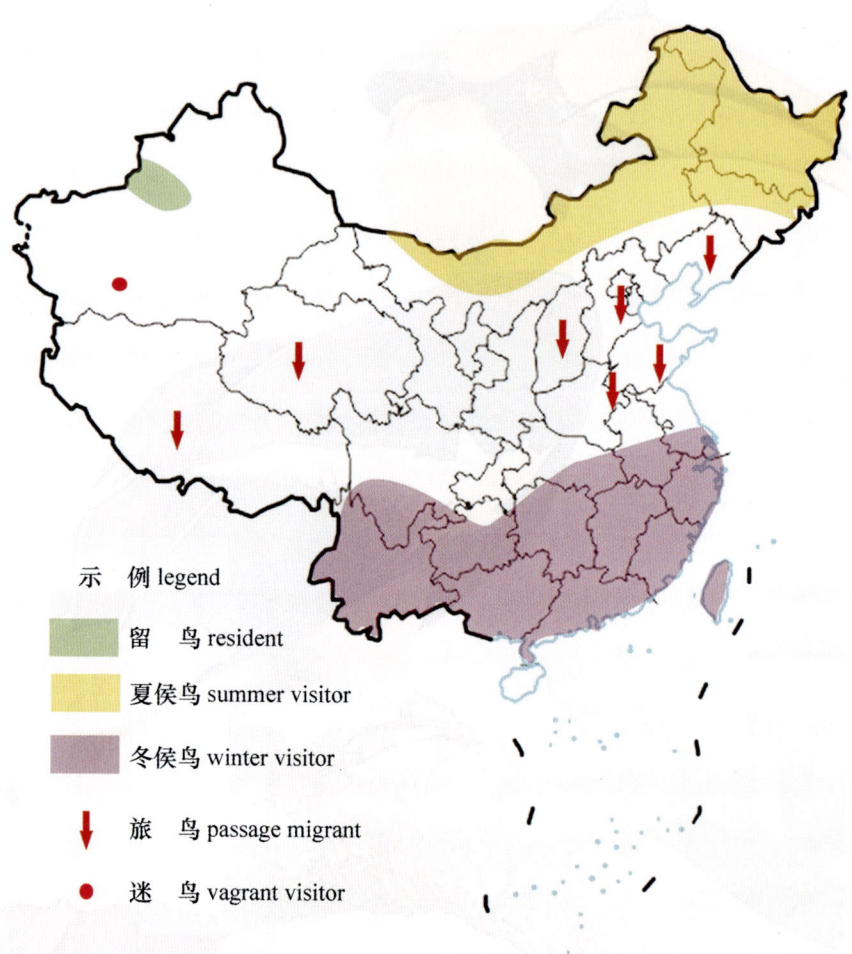

地理分布图示例
Example of geographical distribution

序 一

地球进入了一个崭新的地质纪元——人类世（Anthropocene），而地球上的人口仍呈指数增长。人类社会进入了全球化、信息化时代。人类的生态足迹日益扩大，人类对自然资源的消耗、对环境的污染达到了一个前所未有的水平，人类的影响已经遍及地球各个角落，导致了全球变化，影响了地球生物圈的结构与功能，危及了许多野生动植物的生存，造成全球范围的生物多样性危机，影响了人类社会的可持续发展。

中国是一个生物多样性大国，是地球上生物多样性最丰富的国家之一。根据《中国生物物种名录》（2019），中国已经记载生物达 106,509 个物种及种下单元，其中物种 94,260 个，种下单元 12,249 个。中国南北纬度跨度大，海拔跨度也大。中国还有多种气候类型、多样生境类型，栖息着丰富的高等生物。这些生物物种是国家重要的战略资源，是社会经济可持续发展中不可替代的物质基础。

濒危物种红色名录已经成为重要的生物多样性保护研究工具。目前，《世界自然保护联盟受威胁物种红色名录》（《IUCN 受威胁物种红色名录》）评估了 98,500 多个物种，发现其中 32,000 多个面临灭绝威胁，包括 41% 的两栖动物、34% 的针叶树、33% 的造礁珊瑚、26% 的哺乳动物和 14% 的鸟类。然而，《IUCN 受威胁物种红色名录》对物种的生存状况的评估是基于全球资料所做的，并不代表物种在各个分布国的生存状况。国家是生物多样性保护的主体，各国须开展自己的濒危物种红色名录研究，对其生物物种的生存状况进行评估。在某种程度上，可以说濒危物种红色名录研究反映了一个国家生物多样性综合研究的能力。

早在 20 世纪 90 年代，中国即引入 IUCN 受威胁物种红色名录等级标准开展了物种濒危状况评估工作，如 1991 年，我国学者发表了《中国植物红皮书》。1998 年，国家环境保护总局联合国家濒危物种科学委员会发表了《中国濒危动物红皮书·鱼类》《中国濒危动物红皮书·两栖类和爬行类》《中国濒危动物红皮书·鸟类》《中国濒危动物红皮书·兽类》等著作。另外，2004 年和 2009 年，相关领域专家开展了不同类群物种的濒危状况评估工作，先后发表了《中国物种红色名录

（第一卷）红色名录》和《中国物种红色名录（第二卷）脊椎动物》。

物种生存状况是变化的，于是，IUCN 每年定期更新 IUCN 红色名录。IUCN 红色名录并不反映一个跨越国家分布的物种在一个分布国家的生存状况。国家是濒危物种的管理主体，各国需要应用国际标准进行红色名录评估。鉴于此，为全面评估中国野生脊椎动物濒危状况，中国研究人员于 2013 年启动了"中国生物多样性红色名录——脊椎动物卷"的物种评估和报告编制工作。这次评估组织全国鱼类、两栖类、爬行类、鸟类与哺乳类专家收集数据，采用综合分析和专家评估相结合的方法，依据中国鱼类、两栖类、爬行类、鸟类和哺乳类野生种群与生境现状，利用 IUCN 红色名录标准第 3.1 版，编制了"中国生物多样性红色名录——脊椎动物卷"。该卷红色名录于 2015 年 5 月 6 日通过环境保护部和中国科学院的联合验收，并于 5 月 22 日以环境保护部、中国科学院 2015 年第 32 号公告形式正式发布。

2016 年以来，中国脊椎动物红色名录工作组组织中国研究人员再次厘定了中国脊椎动物多样性，重新评估了中国脊椎动物的生存状况。完成了《中国生物多样性红色名录：脊椎动物》(2021)。本次脊椎动物红色名录评估发现中国脊椎动物生存状况严峻，中国脊椎动物的灭绝风险高于世界平均水平。中国脊椎动物哺乳类、鸟类、两栖类、爬行类和鱼类等各个类群都发现了野外灭绝或区域灭绝的物种，有许多物种处于极危、濒危和易危的受威胁状态。

中国正处于发展期，人口众多，地貌复杂，区域发展程度差异大。如何拯救这些濒危物种是中国生物多样性保护面临的一项艰巨任务。中国政府十分重视生物多样性保护，缔结了《生物多样性公约》《濒危野生动植物种国际贸易公约》《关于特别是作为水禽栖息地的国际重要湿地公约》（简称《湿地公约》）等国际公约，并积极主动履约。中国大力开展了以自然保护区为主体、以国家公园为龙头的保护地建设。目前，我国建立的各类保护地已达 1.18 万处，面积占国土面积的 18% 以上。其中有 474 个国家级自然保护区。自然保护区保护了 90.5% 的陆地生态系统类型、85% 的野生动植物种类、65% 的高等植物群落。中国的森林覆盖率逐年增加，为濒危物种的种群与栖息地恢复奠定了基础。

生物多样性研究既是一项综合性研究，也是一项组合型研究。生物物种编目与受威胁状态评估需要不同学科的联合研究。为了生物多样性科学研究与保护，来自不同学科的学者走到一起，完成中国生物多样性红色名录研究。这项研究是中国整体生物多样性研究的重要组成部分。《中国生物多样性红色名录：脊椎动物》各卷在《生物多样性公约》第 15 次缔约方大会即将在中国召开之前出版发行，为中国生物多样性保护提供了基础数据，为监测中国生物多样性现状、开展阶段性 IUCN 红色名录指数研究积累了参数，也是中国保护生物学研究成果的展示。

中国科学院院士
国家自然科学基金委员会前主任
国家濒危物种科学委员会主任
2020 年 6 月 26 日

Foreword I

The earth has entered a new geological epoch, the Anthropocene, while the human population on the earth is still increasing exponentially. Human society has entered the era of globalization and information. The human ecological footprint is enlarging while the human consumption of natural resources increases; environmental pollution created by human being has reached an unprecedented level. Impact of human reaches throughout all corners of the earth, causes the global change and affects the structure and function of the earth's biosphere, threatens the survival of many wild animals and plants, causing a global biodiversity crisis, consequently, influences the sustainable development of human society.

China is a country with great biodiversity and one of the countries with the richest biodiversity on the earth. According to the *Species Catalogue of China* (2019), China has already recorded 106,509 species and subspecies taxa, including 94,260 species and 12,249 subspecies. The territory of China spans a large latitude and has huge elevation differences. China also has a variety of climate types, diverse habitat types, rich niches of higher organisms. These biological species are important strategic resources of the country and irreplaceable material basis for sustainable social and economic development.

The red list of endangered species has become an important tool for biodiversity conservation research. Currently, the *IUCN Red List of Threatened Species* has assessed more than 98,500 species and found that more than 32,000 of them are threatened with extinction, including 41% of amphibians, 34% of conifers, 33% of reef-building corals, 26% of mammals and 14% of birds. However, the *IUCN Red List of Threatened Species* is based on global data and does necessarily not represent the status of species in each range country. A country is the main sovereign body of biodiversity conservation, and each country needs to carry out the red list study of its endangered species

to assess the survival status of its biological species. To some extent, red list study reflects the comprehensive research capacity of biodiversity in a country.

The red list of endangered species has become an important tool for biodiversity conservation research. As early as in the 1990s, China introduced the IUCN red list criteria for threatened species to carry out the assessment of status of endangered species. For example, in 1991, Chinese scholars published the *Red Data Book of Chinese Plants*. In 1998, Environmental Protection Administration, together with the National Scientific Committee on Endangered Species, published such books as *China Red Data Book of Endangered Animals*: *Pisces*; *China Red Data Book of Endangered Animals*: *Amphibia & Reptilia*; *China Red Data Book of Endangered Animals*: *Aves*; *China Red Data Book of Endangered Animals*: *Mammalia*. In addition, in 2004 and 2009, experts in related fields carried out the assessment of the endangered status of different taxa of species, and published the *China Species Red List Vol I Red List* and the *China Species Red List Vol II Vertebrates*.

Status of species is changing, therefore the IUCN updates its red list every year. However, the IUCN red list does not reflect the status of particular species in a range country if a species distributes in multi-countries. Countries are the main management bodies of endangered species; thus, each country needs to apply international standards for red list assessment. Therefore, in order to comprehensively assess the endangered status of wild vertebrates in China, Chinese researchers launched the compilation of the "China's Red List of Biodiversity: Volume of Vertebrates" in 2013. During that evaluation, by adopting the combination of comprehensive analysis and expert evaluation method, the experts were coordinated to collect data for assessing wild population and habitat status of China's fishes, amphibians, reptiles, birds and mammals and compiled the *China's Red List of Biodiversity: Volume of Vertebrates*. The red list was approved by Ministry of Environmental Protection and the Chinese Academy of Sciences on May 6, 2015, and was officially released on May 22 in the form of Announcement No. 32 of 2015 by the Ministry of Environmental Protection and Chinese Academy of Sciences.

Since 2016, the China's vertebrate red list working group has organized Chinese researchers to reassess the diversity and the status of Chinese vertebrates. The working group completed the *China's Red List of Biodiversity: Vertebrates* (2021). The assessment found that China's vertebrate survival situation is still grim, and the risk of extinction of Chinese vertebrates is higher than the world average. Various groups of vertebrates, including mammals, birds, amphibians, reptiles and fishes in China have found species of Extinct in the Wild or Regionally Extinct, and many species are in a state of Critically Endangered, Endangered and Vulnerable.

China is in the process of rapid development. China has the largest human population, complex landforms and huge differences in regional development levels. How to save these endangered species is a difficult task for China's biodiversity conservation. The Chinese government attaches great importance to the protection of biological diversity, and has signed and actively implemented international conventions such as the *Convention on Biological Diversity*, the *Convention on International Trade in Endangered Species of Wild Fauna and Flora*, and the *Convention on Wetlands of International Importance, Especially as Waterfowl Habitats* (*Convention on Wetlands* for short). China has vigorously established protected areas with the nature reserves as the main body and national parks as the leading part. At present, China has set up 11,800 protected areas of various types, accounting for more than 18% of the country's land area. There are 474 national nature reserves. The nature reserves protect 90.5% of terrestrial ecosystem types, 85% of wildlife species, and 65% of higher plant communities. On the other hand, China's forest coverage is increasing year by year, laying a sound foundation for the population and habitat restoration of endangered species.

Biodiversity research is not only a comprehensive research, but also a combined study. The inventory of biological species and the assessment of threatened status require joint efforts of different disciplines. For the scientific research and conservation of biodiversity, scholars from different disciplines came together to complete the red list of China's biodiversity. The study is an important part of China's overall biodiversity research. All volumes of *China's Red List of Biodiversity: Vertebrates* are published before the fifteenth meeting of the Conference of the Parties to the *Convention on Biological Diversity* which will be held in Kunming, China. The set of books will provide basic data for the biodiversity conservation in China, for monitoring the current situation of biological diversity, for accumulating parameters to conduct periodic IUCN red list index research, and is also an outcome of the Chinese conservation biology research.

Yiyu Chen

Member of the Chinese Academy of Sciences

Former Director of National Natural Science Foundation of China

Director of Endangered Species Scientific Commission, P. R. China

June 26, 2020

序二

1948年，在法国枫丹白露举行的一次由23个政府、126个国家组织和8个国际组织参与的国际会议上，世界自然保护联盟（International Union for the Protection of Nature，IUPN）成立了。当时，这个组织没有财政来源、没有长期预算，甚至没有永久雇员，但是IUPN成为世界政府与非政府组织（Governmental and Nongovernmental Organization，GONGO）的发端。世界自然保护联盟成立后的第一个重大举措是在1950年建立了"生存服务（Survival Service）机构"。"生存服务机构"利用当时筹集到的2,500美元，召集全球的科学家、志愿者为全球濒危物种撰写评估报告，要求各国政府保护其境内的濒危物种。1964年，世界自然保护联盟正式发布《濒危物种红皮书》（Endangered Species Red Book）。今天，世界自然保护联盟已经完全改变了它自己，包括其名称也改变为International Union for Conservation of Nature（IUCN）。IUCN已经成为联合国的观察员、世界范围内主要保护组织。IUCN"生存服务机构"已经演化为物种存续委员会（Species Survival Commission，SSC），IUCN《濒危物种红皮书》也演化为《IUCN受威胁物种红色名录》。现在，IUCN物种存续委员会每年发布《IUCN受威胁物种红色名录》。《IUCN受威胁物种红色名录》已发展成为世界上关于动物、植物和真菌物种全球保护状况最全面的信息源。

世界各国是生物多样性保护的主体。一个物种的IUCN红色名录等级并不一定等同于其在一个国家红色名录中的等级，除非这一物种是该国特有的物种。于是，世界各国也在制定各自的濒危物种红色名录。通过濒危物种红色名录的研究，各国对其境内分布的植物与动物物种的分布、生存状况和保护状况进行调查，然后，对物种的生存状况进行全面评估。因此，濒危物种红色名录是一份物种及其分布的清单，是对物种生存状况、保护状况的客观评估，是生物多样性健康状况的一个重要指标。

濒危物种红色名录被各国政府、自然保护地与野生动植物主管部门、与保护有关的非政府组织、研究人员、自然资源规划人员、教育机构使用。红色名录为生物多样性保护和政策变化提供了信息和促进行动的有力工具，对保护我们赖以生存的自然资源至关重要。通过网络应用，濒危物种红色名录也成为濒危物种信息库，成为保护工作者与研究人员的工具。

中国是 IUCN 的成员，中国也是联合国《生物多样性公约》的最早缔约方之一。中国一直走在生物多样性保护的前沿。中国还是世界上生物多样性最丰富的国家之一，有 7,300 余种脊椎动物，约占全球脊椎动物总数的 11%。中国动物区系组成复杂，空间分布格局差异显著，起源古老，拥有生物演化系统中的各种类群，如有"活化石"之称的大熊猫（*Ailuropoda melanoleuca*）、白鱀豚（*Lipotes vexillifer*）和扬子鳄（*Alligator sinensis*）等。此外，中国还是许多家养动物的起源中心。中国也是生物多样性受威胁最严重的国家之一。人类活动造成的资源过度利用、生境丧失与退化、环境污染及气候变化等因素导致脊椎动物多样性受到严重的威胁。

近年来，党中央和国务院高度重视生物多样性保护工作，将生物多样性保护上升为国家战略，发布了《中国生物多样性保护战略与行动计划（2011－2030 年）》，建立了生物物种资源保护部际联席会议制度，成立了中国生物多样性保护国家委员会，制定和实施了一系列生物多样性保护规划和计划，取得了积极进展。然而，中国生物多样性下降的总体趋势尚未得到有效遏制，保护形势依然严峻，特别是由于目前对中国物种受威胁状况缺乏全面的了解，影响了生物多样性的有效保护。因此，评估物种的受威胁状况，制定红色名录，从而提出针对性的保护策略，对于推动实施《中国生物多样性保护战略与行动计划（2011－2030 年）》和生态文明建设具有重要意义。

到目前，在中国科学家的努力下，中国哺乳动物、鸟类、两栖动物、爬行动物、淡水鱼类都得到了全面的评估。除了评估新发现的物种，中国濒危脊椎动物红色名录还重新评估了一些现存物种的状况，如大熊猫、藏羚（*Pantholops hodgsonii*）等物种，由于中国的保护努力，这些物种的中国濒危脊椎动物红色名录濒危等级下降。然而，中国生物多样性濒危局面仍然严峻。

尽管中国受威胁物种的比例很高，但中国政府正在加强生态环境保护，加强自然保护区、国家公园、世界遗产地及其他类型的保护地建设，加强荒漠化治理、湿地恢复、植树造林，努力扭转或至少制止生物多样性的下降。《生物多样性公约》第 15 次缔约方大会即将在中国昆明召开之际，中国科学家发表最新版《中国生物多样性红色名录：脊椎动物 第一卷 哺乳动物》、《中国生物多样性红色名录：脊椎动物 第二卷 鸟类》、《中国生物多样性红色名录：脊椎动物 第三卷 爬行动物》、《中国生物多样性红色名录：脊椎动物 第四卷 两栖动物》和《中国生物多样性红色名录：脊椎动物 第五卷 淡水鱼类》，全面更新了中国脊椎动物生存状况与种群和栖息地保护状况，从而为确定哪些物种须有针对性地努力恢复，为确定须保护的关键种群和栖息地提供了依据，有助于鉴别未来的濒危脊椎动物保护重点。这套图书的出版是中国自然保护史上的一件大事。

IUCN 主席
2020 年 6 月 26 日

Foreword II

The International Union for the Protection of Nature (IUPN) was established in Fontainebleau of France in 1948 at an international conference that 23 governments, 126 national organizations and 8 international organizations participated in. At that time, the organization had no financial resources, no long-term budget or not even a permanent employee, but it marked the born of the first world Governmental and Nongovernmental Organization (GONGO). Its first move was to establish the "Survival Service" in 1950. Using the $2,500 it raised at the time, the Survival Services called on scientists and volunteers from all around the world to write assessments of the world's endangered species, asking governments to protect those species within their borders. In 1964, the IUCN officially released the *Endangered Species Red Book*. Today, the International Union for Conservation of Nature (IUCN) has completely changed itself, including its name. The IUCN has become an observer of the United Nations and a leading conservation organization worldwide. IUCN Survival Service has evolved into the Species Survival Commission (SSC), the IUCN *Endangered Species Red Book* has been expanded into the website of *IUCN Red List of Threatened Species*, which is now renewed annually by the IUCN Species Survival Committee. The *IUCN Red List of Threatened Species* is the world's most comprehensive source of information on the status of global conservation of animal, plant and fungal species.

Sovereignty countries in the world are the main body of biodiversity protection. The status of a species in IUCN red list is not the affirmatively same in a country's red list except that the species is an endemic species in that country; countries around the world are also developing their own red lists of endangered species. Through the study of the red list of endangered species, countries conduct surveys on the distribution, survival and conservation status of plant and animal species in their territory and then conduct a comprehensive assessment of the survival status of species. Therefore, the red list of endangered species is a list of species and their distribution, an objective assessment of the survival and conservation status of species, and an important indicator of the health status of biodiversity. The red list is used by governments, natural protected areas and wildlife authorities, conservation NGOs, researchers, natural resource planners and educational institutions. It provides information and powerful tools for promoting action on biodiversity conservation and policy formation and is critical to safely guarding the natural resources on which we depend. Through the internet, the red list of endangered species has also become an information base of endangered species and a tool for conservation workers and researchers.

China is a member of the IUCN and one of the earliest parties to the UN *Convention on Biological Diversity*. China has been at the forefront of biodiversity conservation. China is also one of the most biodiverse countries in the world, with more than 7,300 vertebrate species, accounting for about 11% of the total number of vertebrates in the world. China has complex fauna composition, significant differences in spatial distribution pattern, ancient origins and various groups in the biological evolution system, such as Giant Panda (*Ailuropoda melanoleuca*), Baiji (*Lipotes vexillifer*) and Yangtze Alligator (*Alligator sinensis*). In addition, China is the origin center of many domestic animals and plants. China is also one of the countries where biodiversity is most threatened. Due to the overuse of resources, habitat loss and degradation caused by human activities, environmental pollution, climate change and other factors, vertebrate diversity is seriously threatened.

During recent years, the CPC Central Committee and the State Council attach great importance to the protection of biodiversity. Biodiversity conservation is announced as the national strategy, *China's Biodiversity Conservation Strategy and Action Plan (2011-2030)* is issued, the Joint Inter-Ministerial Meeting for Biological Species Resources Protection is regularly held, the China National Committee for Biodiversity Conservation has been set up, a series of biological diversity protection programs and plans have been formulated and implemented, and positive progress in the field has been made. However, the overall trend of biodiversity decline in China has not been effectively stopped, and the conservation situation is still pressing, especially, lack of comprehensive understanding of threatened species in China has hindered the effective conservation of biodiversity. Therefore, it is of great significance to assess the threatened status of species and to formulate the red list of endangered species, and to propose targeted conservation strategies for promoting the implementation of *China's Biodiversity Conservation Strategy and Action Plan (2011-2030)* and the construction of ecological civilization.

Now, the status of Chinese mammals, birds, reptiles, amphibians and freshwater fishes have all been comprehensively assessed by Chinese scientists. In addition to assessing newly discovered species, China's red list of endangered vertebrates has also reassessed the status of some species, including the Giant Panda and Tibetan Antelope (*Pantholops hodgsonii*), whose status on the red list of endangered vertebrates in China has been downgraded due to conservation efforts. However, the overall situation of endangered biodiversity in China is still serious.

Despite the high proportion of threatened species in China, the Chinese government is strengthening ecological protection, stepping up the construction of nature reserves, national parks, World Heritage Sites and other protected areas, working on desertification control, wetland restoration and afforestation, and trying to reverse or at least stop the trend of biodiversity decline. On the occasion that the fifteenth meeting of the Conference of the Parties to the *Convention on Biological Diversity*, which will be held in Kunming, China, Chinese scientists published the latest edition of the "*China's Red List of Biodiversity: Vertebrates, Volume I, Mammals*", "*China's Red List of Biodiversity: Vertebrates, Volume II, Birds*", "*China's Red List of Biodiversity: Vertebrates, Volume III, Reptiles*", "*China's Red List of Biodiversity: Vertebrates, Volume IV, Amphibians*" and "*China's Red List of Biodiversity: Vertebrates, Volume V, Freshwater Fishes*". This set of books comprehensively update the survival status, population and habitat protection of vertebrate, determine the recovery efforts needed for the targeted species, identify key populations and habitats that need to be protected, thus provide a basis for identifying future priorities for endangered vertebrates conservation. Publication of these books is an important event in the history of Chinese nature conservation.

Xinsheng Zhang
President of IUCN,
the International Union for Conservation of Nature
June 26, 2020

总前言

物种的濒危现状和濒危机制是保护生物学的核心研究内容，其研究目标是评估人类对生物多样性的影响，提出防止物种灭绝及保护的策略，通过保护生物物种的种群和栖息地，避免物种受到灭绝的威胁。保护生物学研究既关注全球性问题，又具有鲜明的地域特色。中国具有世界上最多的人口，国土面积为世界第三，监测和评估其生物多样性、保护濒危物种，将为中国实现可持续发展提供科学支撑。中国研究人员通过濒危物种红色名录研究，量化了物种灭绝风险，预警了潜在的生态危机，为中国履行《生物多样性公约》等提供科技支撑。

世界自然保护联盟（International Union for Conservation of Nature，IUCN）成立后的第一个重大举措是在1950年建立了"生存服务（Survival Service）机构"。"生存服务机构"利用当时募集的2,500美元，召集科学家、志愿者评估全球濒危物种灭绝风险，发表有灭绝风险的物种研究报告，呼吁各国政府保护其境内的濒危物种，这是《IUCN受威胁物种红色名录》的发端。直到1964年，世界自然保护联盟才正式发布《濒危物种红皮书》。今天，世界自然保护联盟完全改变了它自己，"生存服务机构"已经演化成为物种存续委员会（Species Survival Commission，SSC）。IUCN《濒危物种红皮书》已经演变为网络版的《IUCN受威胁物种红色名录》。物种存续委员会从不定期发布IUCN濒危物种红皮书发展到现在每年发布更新的《IUCN受威胁物种红色名录》。

《IUCN受威胁物种红色名录》是世界生物多样性健康状况的重要指标。它是世界上最全面的一份动物、植物和真菌物种濒危状况清单。濒危物种红色名录基于物种种群数量、种群数量下降速率、生境破碎程度、生境面积及下降速率、预测灭绝概率等指标估测物种灭绝概率。《IUCN受威胁物种红色名录》（2020-2）发现地球上的32,000多个物种面临着灭绝的风险，占所有被评估物种的27%。其中，41%的两栖类、26%的哺

乳类、34%的针叶树、14%的鸟类、30%的鲨鱼、33%的造礁珊瑚，以及28%的特定甲壳类物种面临灭绝风险。《IUCN受威胁物种红色名录》为保护我们赖以生存的自然资源提供了至关重要的信息。

项目背景

自1980年以来，中国经济步入高速发展期。目前，中国已经成为世界第二大经济体。在人口增长、经济发展、全球变化的背景下，中国的生物多样性正面临着前所未有的城镇化、乡村和社会基础设施建设及全球变化的压力，野生生物的生存受到威胁。许多证据显示地球上的生物正面临生物进化中的第六次大灭绝。保护濒危物种是生物多样性保护的核心问题。评估物种濒危等级是生物多样性监测与保护的迫切需要。

虽然《IUCN受威胁物种红色名录》没有国际法和国家法律的效力，但它是专家对全部物种生存状况的评估，它不仅限于评估濒危物种和明星物种，而是最大限度地涵盖了已知的物种，它不仅仅指导世界范围的濒危物种保护，也是指导生物多样性研究的有用工具。《IUCN受威胁物种红色名录》对于政府间组织和非政府组织的保护决策及各国自然与自然保护法律法规的制定都产生了重要影响。

综上所述，濒危物种红色名录是物种灭绝风险的测度，IUCN定期更新其濒危物种红色名录，预警全球物种的生存危机。同时，各国也开展了本国濒危物种红色名录研究。那么，既然已经有《IUCN受威胁物种红色名录》，为什么还要开展国家濒危物种红色名录研究？

《IUCN受威胁物种红色名录》与国家濒危物种红色名录都是物种灭绝风险的测度，前者是全球性评估，后者则是依国别的研究，两者的研究空间尺度不同。《IUCN受威胁物种红色名录》预警了全球物种的濒危状况，为全球生物多样性研究提供了大数据；各国红色名录则确定了各国物种受威胁状况，填补了前者的知识空缺，两份红色名录互为补充。

基于如下原因，应当重视依国别的濒危物种红色名录。①国家是濒危物种保护的行为主体，物种在一个国家的生存状况是确定其保护级别、开展濒危物种保育的依据。②对于仅分布于一个国家的特有物种来说，其按国别的濒危物种红色名录等级即是其全球濒危等级。③《IUCN受威胁物种红色名录》只提供了全球范围的物种濒危信息，并没有评估每一个国家所有物种的生存状况，特别是一些特有物种和跨越国境分布的物种。一些物种跨越国界分布，全球的生存状况并不反映其在个别国家的生存状况，一些全球无危的物种在其边缘分布区的国家里却是极度濒危的或受威胁的物种。世界各国的物种濒危状况有待各国科学家的研究。对于跨国境分布的物种来说，依国别的濒危物种红色名录等级则确定了该物种在本国的生存状况。④结合《IUCN受威胁物种红色名录》，依国别的濒危物种红色名录为建立跨国保护地、保护迁徙物种的栖息地与跨国迁徙洄游通道提供依据。⑤依国别的濒危物种红色名录所特有的"区域灭绝"等级，反映了一个物种边缘种群在该国的区域灭绝，对于一个国家来说，事关重大；恢复"区域灭绝"物种是该物种原分布国家重新引入的相关保育工作的重点。⑥物种濒危状况是不断变化的。近年来，新种、新记录不断被发现。随着人们对生命世界认识的深入，脊椎动物分类系统也发生了变化。依国别的濒危物种红色名录提供了该国物种编目、分类、分布和生存状况的最新信息（蒋志刚等，2020）。

国家红色名录的重要性在许多情况下被忽视了。在研究报告和科普作品中，对国家濒危物种红色名录重视不够。论及物种濒危属性时，作者通常言必《IUCN 受威胁物种红色名录》濒危等级而不提其国家级的红色名录濒危等级。目前正值全球新型冠状病毒肺炎大流行，人们正在重新审视人与野生动物的关系。我国将修订有关野生动物保护与防疫法规和法律、重点保护野生物种名录，防控新的人与野生动物共患疾病再次暴发。对于确定《国家重点保护野生动物名录》而言，物种受威胁程度是物种列为国家重点保护野生物种的特征之一。重视依国别的红色名录有特别的意义。于是，生态环境部（原环境保护部）与中国科学院联合开展了中国生物多样性红色名录研究。

中国动物学家掌握了中国动物分布和生存状况的第一手资料，有必要组织全国淡水鱼类、两栖类、爬行类、鸟类与哺乳类专家及时更新中国脊椎动物分类系统，提供中国脊椎动物多样性的全面、完整的信息；有必要应用统一的国际物种濒危等级标准评估物种生存状况。在国家层面，定期组织全国淡水鱼类、两栖类、爬行类、鸟类与哺乳类专家应用 IUCN 受威胁物种红色名录等级标准和 IUCN 区域受威胁物种红色名录标准，全面评估更新的中国脊椎动物生物多样性红色名录，提供与国际红色名录研究可对比的结果，为红色名录指数的研究积累数据。

经过系统评审制定的中国生物多样性红色名录，由国家权威机构发布。中国生物多样性红色名录淡水鱼类、两栖类、爬行类、鸟类与哺乳类各卷将为监测中国生物多样性现状、为开展阶段性 IUCN 红色名录指数研究和履行《生物多样性公约》提供数据。

中国在 1998 年首次出版了《中国濒危动物红皮书》，2004 年，又出版了《中国物种红色名录》，2009 年，环境保护部组织开展了"中国陆栖脊椎动物物种濒危等级评估"。时隔多年，有必要重新全面评估中国生物多样性的濒危状况。于是，环境保护部委托中国科学院组织有关专家开展了"中国生物多样性红色名录——脊椎动物卷"的研究。在环境保护部和中国科学院的领导下，我们依据 IUCN 受威胁物种红色名录等级标准和 IUCN 区域受威胁物种红色名录标准，全面评估了中国哺乳动物生存状况。

2015 年，环境保护部与中国科学院联合发布了"中国生物多样性红色名录——脊椎动物卷"。现在，历时 6 年，我们全面编研、更新、丰富了此名录，形成了此 2021 版的《中国生物多样性红色名录：脊椎动物》。

项目目标

通过脊椎动物各类群的研究，收集整理中国脊椎动物现有物种种群、生境研究数据、资源监测数据，充实数据库；组织专家，采用综合分析和专家评估相结合的方法，依据中国脊椎动物野生种群与生境现状，利用 IUCN 受威胁物种红色名录等级标准第 3.1 版和 IUCN 区域受威胁物种红色名录标准第 4.0 版综合评价中国脊椎动物濒危状况，编制 2021 版《中国生物多样性红色名录：脊椎动物》。全面评价中国脊椎动物的灭绝风险，对中国濒危物种保护及时提供基础信息。

编研过程

2013 年 5 月 16 日，在中国科学院动物研究所召开了研究启动会。项目聘请陈宜瑜院士、郑光美院士、张亚平院士、金鉴明院士、马建章院士、曹文宣院士为咨询专家，并成立了哺乳类、鸟类、爬行类、两栖类、淡水鱼类课题组。各课题组就评估程序和规范展开了研讨，对典型物种进行了评估并听取了专家委员会的意见。会后总结了专家意见，完善了中国脊椎动物红色名录评估程序和规范。

针对哺乳类、鸟类、爬行类、两栖类和淡水鱼类分别建立了工作组、核心专家组和咨询专家组。工作组负责按照预定的红色名录判定规程开展工作，工作包括资料收集与整理、红色名录初步评定、与通讯评审专家联络及通讯评估结果汇总。核心专家组对红色名录评估的方法、标准使用、数据来源等重要科学问题进行界定，讨论审核有关物种的受威胁等级。工作组在全国范围遴选咨询专家，建立咨询专家库。咨询专家参加了红色名录的通讯评审和会议评审。评审结束后，工作组按照统一格式，整理每个物种包含的信息，形成最终的物种评估说明书。物种评估说明书的内容包括物种的学名、中文名、评估受威胁等级及 IUCN 红色名录等级。"中国生物多样性红色名录——脊椎动物卷"于 2015 年 5 月 6 日通过环境保护部和中国科学院的联合验收，并于 5 月 22 日以环境保护部、中国科学院 2015 年第 32 号公告形式发布。

红色名录评估的信息来源主要有研究积累、标本数据、文献数据和专家咨询。项目各课题组相关研究团队是工作在中国淡水鱼类、两栖类、爬行类、鸟类和哺乳类研究一线的研究团队，在数十年的研究中积累了大量的科学数据，各分卷主持人还是国家濒危物种科学机构，以及淡水鱼类、两栖类、爬行类、鸟类和哺乳类学术团体的骨干，所在单位是有关动物物种分类、标本收藏、研究的信息交换所，并各自建立了数据库。各分卷主持人还主持或参与了国家有关物种资源本底调查、科学评估、自然保护区生物多样性考察及相关的保护政策制定。

实践意义

《中国生物多样性红色名录：脊椎动物》的出版是一项重大的系统工程。这次生物多样性红色名录评估是迄今评估对象最广、涉及信息最全、参与专家人数最多的一次评估。通过 2015 版红色名录研究，我们更新了中国脊椎动物编目。中国有 2,854 种陆生脊椎动物，其中，有 407 种两栖类，402 种爬行类，1,372 种鸟类，673 种哺乳类。在 2021 版《中国生物多样性红色名录：脊椎动物》的编研中，我们再次更新了中国脊椎动物分类系统和编目。中国有 3,147 种陆生脊椎动物，其中，有 475 种两栖类，527 种爬行类，1,445 种鸟类，700 种哺乳类，比 2015 年的统计数据增加了 293 种。我们发现，中国是全球哺乳动物物种数最多的国家。中国陆生脊椎动物中，特有种超过 20%。我们还分析了中国脊椎动物的分布格局和特有类群，探讨了其濒危种类的空间分布规律。

中国濒危脊椎物种濒危模式与分布格局在我国生物多样性和生态系统保护中具有指导意义，这将为我国重点保护物种确定、国土空间开发和生态功能区的划分及各类保护地规划设计提供重要参考依据。也是确定中国物种多样性保护热点的依据之一。我们发现，中国脊椎动物生存危机依然严重，中国濒危脊椎物种的分布格局不均衡。物种的空间分布是一种立体格局，除了水平纬度上的物种分布格局，我们也需要物种多样性和濒危种类的垂直分布格局，这些格局对物种多样性保护具有重要参考价值。我们发现，高海拔地区受威胁哺乳动物的比例比低海拔地区高，高海拔地区的濒危物种应受到更多的关注。

展望与致谢

《中国生物多样性红色名录》的编制和发布为生物多样性保护政策和规划的制定提供了科学依据，发挥了中国科学家作为中国《生物多样性公约》履约"智库"的功能，同时，为开展生物多样性科学基础研究积累了基础数据，更新了脊椎动物分类系统与编目，为公众参与生物多样性保护创造了必要条件。《中国生物多样性红色名录》的编制是贯彻实施《中国生物多样性保护战略与行动计划（2011—2030 年）》和积极履行《生物多样性公约》的具体行动。通过《中国生物多样性红色名录》的编制，中国在生物多样性评价方面已经在全球先行一步，使我国在履行《生物多样性公约》方面走在世界的前列。

本项目得到了生态环境部（原环境保护部）、中国科学院、国家林业与草原局（原国家林业局）、中国科学院大学、科学出版社的关怀、指导和大力支持；得到了国家出版基金的大力支持。课题组还得到了如下项目的资助：中国科学院战略性先导科技专项（A 类）"地球大数据科学工程"（项目编号：XDA19050204）、国家重点研发计划项目（项目编号：2016YFC0503303）、国家科技基础性工作专项（项目编号：2013FY110300）的资助。在此谨致感谢！

蒋志刚
中国科学院动物研究所研究员
中国科学院大学岗位教授
国家濒危物种科学委员会前常务副主任
2020 年 6 月 6 日

Series' Foreword

The current status and threats to species are the key issues in conservation biology. The primary goal of putting forward an endangered species red list is to evaluate human impact on biodiversity, identifying key threats and preventing species extinction by protecting the populations and habitats of threatened species. Conservation research not only pays attention to global issues, but also must focus attention on regional and national problems. China has the largest human population and the third largest terrestrial area in the world. Monitoring and evaluating the country's biodiversity and protecting endangered species will provide scientific support for China's sustainable development. Therefore, Chinese researchers are working to quantify the risk of extinctions through studies related to the red list of threatened species, providing early warning about potential ecological hazards, thus offering scientific support for implementation of the *Convention on Biological Diversity* in China.

The first major move for the International Union for Conservation of Nature (IUCN) after its establishment was to launch the Survival Service in 1950. Using the $2,500 raised at that time, the Survival Service called on scientists and volunteers to dedicate their expertise and time to assess the extinction risk of globally threatened species. The Survival Service then publicized their research reports on species at risk of extinction and called on governments to protect endangered species within their borders. Such an act marked the beginning of the *IUCN Red List of Threatened Species*. However, it was not until 1964 that IUCN officially published its first *Endangered Species Red Book*. Today, IUCN has completely changed itself, the Survival Service has been renamed as the Species Survival Commission (SSC). The IUCN

Endangered Species Red Book has evolved into the online version of *IUCN Red List of Threatened Species*. The Species Survival Commission refreshes and revises the *IUCN Red List of Threatened Species* periodically, and updates the *IUCN Red List of Threatened Species* annually.

The *IUCN Red List of Threatened Species* is an important indicator of the health of the world's biodiversity. It is the world's most comprehensive list of rare and threatened animal, plant and fungal species. The *IUCN Red List of Threatened Species* estimates the extinction probability of species based on population size, population decline rate, degree of habitat fragmentation, rate of decline of habitat area and other indicators. The *IUCN Red List of Threatened Species* (2020-2) estimated that more than 32,000 species on the earth are at risk of extinction, accounting for 27% of all assessed species globally. 41% of amphibians, 26% of mammals, 34% of conifers, 14% of birds, 30% of sharks, 33% of corals, and 28% of certain crustaceans are presently at risk of extinction. The *IUCN Red List of Threatened Species* provides vital information for protecting the biodiversity and natural resources on which we all collectively depend.

The Background

Since the 1980s, China has embarked on a fast track of socioeconomic development. China has become the world's second largest economy. Against a backdrop of population growth, rapid economic development and many global changes, China's biodiversity is under unprecedented pressure from urbanization, infrastructure development and a wide range of other factors, and the survival of wildlife is under threat. Ample evidence shows that life on the earth is facing its Sixth Mass Extinction in its long evolutionary history. Protecting endangered species is the core issue for biodiversity conservation. Thus, assessing the endangerment level of species is the primary and most urgent need that biodiversity monitoring and protection measures seek to address.

Though the *IUCN Red List of Threatened Species* does not possess the power of international or national laws, it is an expert assessment of the survival status of all species, not only endangered or charismatic species, but all known species to the greatest extent possible. It thus serves as a most useful tool not only to guide worldwide protection of endangered species but also for the study of biodiversity. The *IUCN Red List of Threatened Species* has significant impact on the conservation decisions of intergovernmental and non-governmental organizations as well as for the formulation of national laws and regulations regarding wildlife and nature conservation.

As stated above, the red list of threatened species provides a measure of the risk of extinction of species. IUCN regularly updates its global red list of endangered species in order to raise public awareness of the global status of wildlife and the species survival crisis. At the same time, countries also conduct national-level studies on the status of endangered species. However, since there is already an *IUCN Red List of Threatened Species*, the question may arise, why bother to conduct research at country level to produce national red lists of endangered species?

Both the *IUCN Red List of Threatened Species* and country red lists of threatened species assess species' risk of extinction, with the former being global in scope while the latter are regional assessments. The

Series' Preface

IUCN Red List of Threatened Species alerts the world to the status of endangered species, and also serves as a database of global biodiversity. Country red lists, on the other hand, ascertain the status of species in particular countries, filling knowledge gaps in the former. The two lists are thus complementary to each other.

Country-level red lists should be given greater attention for at least the following reasons: (i) A sovereign country is the main authority for taking conservation action in regard to wildlife species within its boundaries, based on the level of endangerment (conservation status) of the species; (ii) For endemic species in a country, the country red list status constitutes its global status; (iii) The *IUCN Red List of Threatened Species* provides only the information on species at risk worldwide and does not assess the status of all species in each country, especially endemic species and those species with transboundary distribution. Some species are distributed across national boundaries and the global conservation status does not entirely reflect the survival status of the species in any particular country. Some global non-threatened species are critically endangered or threatened in the countries where they have peripheral ranges. The endangered status of species in different countries of the world thus remains to be studied by scientists in relation to specific countries. For species whose ranges cross national borders, the country's red list status reflects the survival status of the species in the country; (iv) Combined with the global *IUCN Red List of Threatened Species*, country red lists provide a basis from which to consider the establishment of transnational protected areas, the protection of important habitats for migratory species, and the protection of international migration corridors; (v) The category "Regionally Extinct" is unique to country (regional) red lists of endangered species as it refers only to a subset of the broader geographic distribution of the species, yet the national status is still indicative of the species' overall risk of extinction, this matters a lot for a country, and the restoration of "regionally extinct" species is the focus of conservation efforts for reintroduction in countries where the species originated; (vi) Country red lists provide updated information about endangered species with national inventories as well as with national reviews of classification, geographic distribution, and status of species at national level, which are also relevant for global species descriptions and assessments (Jiang *et al.*, 2020).

Despite these benefits, the significance of country-level red lists is often overlooked. Following onset of the global COVID-19 pandemic, however, people's outlook has been changing in regard to the relationship between people and wildlife. Consequently, China is amending its national laws on wildlife protection, epidemic prevention, and the list of state key protected wild species, in order to better prevent and control emerging zoonoses. The status of wildlife species included in China's red list of threatened species should be one of the defining elements for identifying and updating species on the *List of State Key Protected Wild Animal Species* in China. It is therefore critical to duly recognize the significance of the country red list at this special moment in time. For this purpose, the Ministry of Ecology and Environment (former Ministry of Environmental Protection) and the Chinese Academy of Sciences have jointly launched China's biodiversity red list.

Chinese zoologists have obtained first-hand information on the distribution and living status of animals in China. It is necessary to organize national experts on freshwater fishes, amphibians, reptiles, birds and mammals to update the taxonomy of vertebrates in China in a timely manner and to provide systematic and comprehensive information on the diversity of vertebrates in China. It is necessary to apply standard international criteria for threatened species to assess the status of species. At the national level, it is necessary to coordinate national experts on freshwater fishes, amphibians, reptiles, birds and mammals to apply the IUCN red list criteria for threatened species and the IUCN regional red list criteria for threatened species, and through this process also to comprehensively update the *China's Red List of Biodiversity: Volume of Vertebrates* and thus to provide a country red list that is comparable to international red lists, and to enable index studies of red lists.

The red list of China's biodiversity, which has been systematically reviewed and formulated, shall be issued by the state authorities. The volumes of freshwater fishes, amphibians, reptiles, birds and mammals of the red list of china's biodiversity will provide data for the implementation of the *Convention on Biological Diversity*, for monitoring the state of biodiversity in China, as well as for conducting periodic IUCN red list index studies in the country.

China firstly published its *China Red Data Book of Endangered Animals* in 1998, followed by the *China Species Red List* in 2004 and in 2009, the Ministry of Environmental Protection coordinated the assessment and publishing of the "Assessment of the Red List of Endangered Species of Terrestrial Vertebrates in China", which is a multi-year project for the comprehensive re-assessment of the threatened status of China's biodiversity. Therefore, the Ministry of Environmental Protection entrusts the Chinese Academy of Sciences to organize experts to carry out research on the *China's Red List of Biodiversity: Volume of Vertebrates*. Under the leadership of the Ministry of Environmental Protection and the Chinese Academy of Sciences, we have conducted a comprehensive assessment of the living status of the vertebrates in China based on the IUCN red list criteria for threatened species and the IUCN regional red list criteria for endangered species.

In 2015, the Ministry of Environmental Protection and the Chinese Academy of Sciences jointly released the *China's Red List of Biodiversity: Volume of Vertebrates*. Now, six years on, we have thoroughly updated and compiled the series of books of *China's Red List of Biodiversity: Vertebrates* (2021).

The Goal

Through the study and preparation for each volume of vertebrates in China's biodiversity red list, we collected and sorted existing information on the population and habitat status of vertebrates in China and completed the database. We systematically and comprehensively evaluated the status of vertebrates in China, using the *IUCN Red List Categories and Criteria* (version 3.1) and the *Guidelines for Application of IUCN Red List Criteria at Regional and National Levels* (version 4.0) based on the status of the species' wild population and habitat. Combined with the empirical analysis and expert evaluation, we compiled the *China's Red List of Biodiversity: Vertebrates* (2021), which is a comprehensive assessment of extinction risk of China's vertebrates, providing the basic information pertinent for the protection of endangered species over the coming years.

The Assessment

The project launch meeting was held at the Institute of Zoology, Chinese Academy of Sciences on May 16, 2013. Academicians Yiyu Chen, Guangmei Zheng, Yaping Zhang, Jianming Jin, Jianzhang Ma and Wenxuan Cao were invited to participate in the meeting as consulting experts. Mammals, birds, reptiles, amphibians and freshwater fishes research groups were formed at the meeting. Each research group held a discussion on evaluation procedures and norms, assessed the typical species of their own taxonomic group, and consulted the opinion of the expert committee. After the meeting, all experts' opinions were summarized and the evaluation procedures and norms of the red list of vertebrates in China were finalized.

Working groups, core expert groups and communication expert groups were formed for each research group, focused respectively on mammals, birds, reptiles, amphibians and fresh water fishes. The working groups were responsible for carrying out work in accordance with the red list category assessment procedures, including data collection and classification, preliminary red list category evaluation, liaison with experts by correspondence, and providing summaries and communicating evaluation results. The core expert group defined the methods, standards, data sources and other important scientific issues of the red list assessment, discussed and reviewed the status of species. The working group selected consulting experts nationwide and established a database of consulting experts. Each consulting expert participated in the red list evaluation and conference review. After the review, the information about every species was sorted and summarized in a unified format to provide the final species evaluation specifications. The species description and assessment includes scientific name, Chinese name, threat level assessment and IUCN red list category criteria. The "*China's Red List of Biodiversity: Volume of Vertebrates*" was jointly approved by the Ministry of Environmental Protection and the Chinese Academy of Sciences on May 6, 2015, and officially released on May 22, 2015, through the Announcement No. 32 of the Ministry of Environmental Protection and the Chinese Academy of Sciences, on International Biodiversity Day of 2015.

The information sources for the evaluation of vertebrates in the red list of China's biodiversity included published and unpublished literature, specimen data, and expert consultation. Experts from the red list working groups for freshwater fishes, amphibians, reptiles, birds and mammals are experts who have accumulated a large amount of scientific data over decades. The coordinators of working groups are people from state endangered species scientific authorities and established academics from scientific communities focused on freshwater fishes, amphibians, reptiles, birds and mammals from across the country. The research institutions are the centers of taxonomy, specimen collections, and databases for animal species. The principal scientists of freshwater fishes, amphibians, reptiles, birds and mammals also often coordinated or participated in background investigations on national species resources, scientific assessments, biodiversity investigations of nature reserves, and the formulation of relevant government conservation policies.

The significance

The publishing of the *China's Red List of Biodiversity: Vertebrates* (2021) is a major systematic project. The biodiversity red list assessment covered the widest range of subjects, providing the most complete information and involving the largest number of experts so far in the country. During the process of developing the 2015 edition of the red list, we updated the Chinese vertebrate inventory. On this basis, it was found that China has 2,854 terrestrial vertebrates, of which 407 are amphibians, 402 are reptiles, 1,372 are birds and 673 are mammals. In the preparation and research of the 2021 edition, we have once again updated the classification system and produced an updated inventory of vertebrates in China. Altogether there are 3,147 terrestrial vertebrates in China, among which there are 475 species of amphibians, 527 reptiles, 1,445 birds and 700 mammals. A further 293 vertebrate species were assessed for preparing the 2021 edition. China is now found to have the largest number of mammal species in the world. Among land vertebrates in China, more than 20% are endemic. We also analyzed the distribution pattern and endemic groups of vertebrates in China and discussed the spatial distribution pattern of endangered species.

The conservation status and distribution patterns of threatened vertebrates in China that are shared in this book provide an important reference for identification of key protected vertebrate species, planning and development of national strategic land use blueprints, the design of ecological functional zones and various protected sites, which are of great significance for biodiversity and ecosystem conservation in China. This information is also one of the criteria for determining hotspot locations for species diversity in China. We have found that the survival crisis of vertebrates is still present in China and the distribution pattern of endangered vertebrates remains uneven. The spatial distribution of species presents a three-dimensional pattern, including their geographic distribution (two dimensions) as well as vertical or elevational dimension where species including threatened species are generally situated. In particular, we found that the proportions of threatened mammals in different families and orders were greater at higher altitudes than those found at lower altitudes, and additionally we found that endangered species that live at higher altitudes often should receive more attention from the public as well as from the government.

The Outlook and Appreciation

The compiling and publishing of *China's Red List of Biodiversity* provides a scientific basis for biodiversity conservation planning and policy in China, based on the long-standing work of Chinese scientists, who constitute a *de facto* "think-tank" for research and implementation of the *Convention on Biological Diversity* in China. At the same time, the red list study has updated the vertebrate taxonomy and inventory in China, accumulated data for basic zoological research in biodiversity science both nationally and globally, and created the necessary conditions for public participation in biodiversity conservation. Producing the *China's Red List of Biodiversity* has been a concrete action in the implementation of *China's Biodiversity Conservation Strategy and Action Plan (2011-2030)* and has also been a key step in implementing the *Convention on Biological Diversity* in its territory. Through the compilation of the *China's Red List of Biodiversity*, China has demonstrated its leading role in the global assessment of the current status of biodiversity.

The project that enabled development and publication of this red list book received guidance and support from the Ministry of Ecology and Environment (former Ministry of Environmental Protection), the Chinese

Academy of Sciences, the National Forestry and Grassland Administration (former National Forestry Administration), the University of Chinese Academy of Sciences, and the Science Press (Beijing). The project received funding from the National Publication Foundation, and each individual research group also received support through the following projects: "Earth Big-Data Scientific Project" (XDA19050204) of the Strategic Leading Science and Technology Project, Chinese Academy of Sciences (Category A); National Key Research and Development Project (2016YFC0503303); Basic Science Special Project of the Ministry of Science and Technology of China (2013FY110300). We express our most sincere gratitude to all of these governmental bodies, institutions and funding agencies for their many different forms of support.

Zhigang Jiang, Ph.D.

Professor of Institute of Zoology, Chinese Academy of Sciences

Professor of University of Chinese Academy of Sciences

Former Executive Director of the Endangered Species Scientific Commission, P. R. China

前 言

中国跨越寒温带、温带、亚热带和热带，包括东洋界和古北界两个具有不同特色的动物地理界；海岸线纵长，有上万个岛屿，为各种海鸟提供了优越的栖息条件。长江、黄河、鄱阳湖、洞庭湖等众多内陆江河湖泊、湿地，是许多鸟类繁殖、迁徙和越冬期间的重要栖息地。中国还地处"东亚–澳大利亚"、"东非–西亚"和"中亚"3条候鸟迁徙路线上，数以万计的、往返于南北越冬地和繁殖地的鸟类途经于此，因而中国的鸟类区系和生物多样性极为复杂和丰富，是世界上鸟类多样性最为丰富的国家之一，郑光美（2017）记录了中国鸟类达1,445种，隶属于26目109科497属，包括93种特有种。

长期以来，资源过度利用、栖息地丧失和片断化、环境污染等因素，致使我国的鸟类多样性保护面临严峻的挑战。大面积、持续的围海造地，导致适宜鸟类栖息的滩涂锐减，水鸟种类和数量下降明显；部分地区食用野生鸟类和鸟蛋的现象屡禁不止，严重威胁鸟类的生存；投毒捕猎造成大批水禽死亡的现象时有发生，非法猎捕导致黄胸鹀（*Emberiza aureola*）等鸟类种群数量急剧下降，红嘴相思鸟（*Leiothrix lutea*）、画眉（*Garrulax canorus*）等鸟类资源在一些地区几乎枯竭；猎隼（*Falco cherrug*）等猛禽的买卖和非法走私现象尽管受到政府的严厉打击，依然猖獗。此外，基因污染也对野生鸟类资源产生了影响，如红原鸡（*Gallus gallus*）分布区内家鸡散放所引起的自然杂交，已使现今很难觅得纯种红原鸡；在绿孔雀（*Pavo muticus*）自然分布区中人工饲养蓝孔雀（*Pavo cristatus*）的现象，也让人担忧这一极危物种的基因受到污染。

为保护我国野生动物资源，1989年公布《国家重点保护野生动物名录》中，将短尾信天翁（*Phoebastria albatrus*）等41种鸟类（或类群）列为国家Ⅰ级重点保护野生鸟类，角䴙䴘（*Podiceps auritus*）等184种鸟类（或类群）列为国家Ⅱ级重点保护野生鸟类。10年后出版的《中国濒危动物红皮书·鸟类》（郑光美和王岐山，1998）依据世界自然保护联盟（International Union for Conservation of Nature，IUCN）的标准，并结合中国鸟类的实际情况，将中国183种受威胁鸟类划分为野生绝迹（Extinct，Ex）、国内绝迹（Extirpated，Et）、濒危（Endangered，E）、易危（Vulnerable，V）、稀有（Rare，R）、未定（Indeterminate，I）等类别。多年来，《国家重点保护野生动物名录》和《中国濒危动物红皮书·鸟类》被国内外广为引用，影响深远。《中国物种红色名录》（第一、第二卷）（汪松和解焱，2004，2009），依据IUCN修订的标准厘定了中国受威胁鸟类名录，并给出了受威胁等级和所依据的标准及理由。以上工作均受到我国政府和民间的高度关注，对鸟类及其栖息地的保护工作起到了重要的促进和推动作用。

随着我国鸟类学研究的不断深入和观鸟活动的蓬勃发展，对鸟类的数量、分布、种群动态等信息的了解也不断增加。为了全面评估当前中国野生鸟类的受威胁状况，启动了本次《中国生物多样性红色名录：脊椎动物 第二卷 鸟类》的编制工作。依照IUCN的标准，集国内20余位鸟类学工作者的力量，最初以《中国鸟类分类与分布名录》（第二版）（郑光美，2011）所列的鸟类为评估对象进行了讨论，并依照后续出版的《中国鸟类分类与分布名录》（第三版）（郑光美，2017）的新分类系统和物种地位变动进行了调整。经过充分评估分析之后提出了中国鸟类红色名录和评估等级，完成了包括分类地位、评估信息、地理分布、种群现状以及发展趋势、栖息地、威胁因子、保护行动等内容的受威胁鸟类的评估书等。所评估的1,445种鸟类中，白鹳（*Ciconia ciconia*）、镰翅鸡（*Falcipennis falcipennis*）和赤颈鹤（*Grus antigone*）3种（和亚种）被列为区域灭绝（RE）；还有极危（CR）18种、濒危（EN）48种、易危（VU）78种、近危（NT）223种、无危（LC）969种，中亚夜鹰（*Caprimulgus centralasicus*）为无效种。此外尚有105种鸟类信息非常少，无法对其评估，受威胁状况不明，在本次评估中均列为数据缺乏（DD）。本工作是依照IUCN的新修订标准，对中国鸟类的受威胁状况进行的一次全面评估和系统分析，对于我国鸟类保护和履行国际公约等工作均有重要意义。

参加本书编写工作的有：北京师范大学郑光美教授、张正旺教授、张雁云教授、邓文洪教授、董路教授，北京林业大学丁长青教授，东北师范大学王海涛教授，复旦大学马志军教授，广西大学周放教授，海南师范大学梁伟教授、杨灿朝教授，内蒙古大学邢莲莲教授、杨贵生教授，武汉大学卢欣教授，浙江大学丁平教授，浙江自然博物馆陈水华研究员、范忠勇研究员，中国科学院动物研究所孙悦华研究员、朱磊博士，中国科学院昆明动物研究所杨晓君研究员，中国科学院生态环境研究中心曹垒研究员，中国科学院新疆生态与地理研究所马鸣研究员，中国林业科学研究院马强副研究员，中山大学刘阳教授。

参加名录审定工作的有：安徽大学周立志教授、广东省科学院动物研究所邹发生研究员、兰州大学刘迺发教授、中国科学院动物研究所雷富民研究员。

北京师范大学王宁博士协助核查了有关鸟类照片，北京师范大学田怀玉博士依照底图重新绘制了分布图，夏灿玮博士，研究生林玉英、邓竹青、刘金、薛泊宁、赵凯参加了部分分布区数据核查、参考文献整理和部分翻译工作，鸟网的王肖阳先生帮助征集了部分鸟类照片。感谢诸多专家学者和观鸟爱好者提供照片，在此一并致谢。

Preface

China stretches from the sub-frigid, temperate, to sub-tropical and tropical zones, and covers two zoogeographic regions, the palearctic realm and oriental realm which embrace different characteristics. China has a long coastline and more than ten thousands islands (National Oceanic Administration, People's Republic of China, 2012) provide ideal habitats for various seabirds. Many inland rivers, lakes and wetlands, including the Changjiang River, the Yellow River, the Poyang Lake, and the Dongting Lake, are important habitats for many birds during breeding, migration and wintering. Moreover, China is located on three bird migration routes including the East Asia-Australia Flyway, the East Africa-West Asia Flyway and the Central Asia Flyway, where thousands of birds travel between the breeding and wintering ranges in the north and south. Therefore, China is a country with the extremely complex bird fauna and rich biodiversity, and is one of the countries with the most abundant bird diversity among the world. Zheng (2017) recorded 1,445 species of Chinese birds, which belong to 26 orders, 109 families, and 497 genera, including 93 endemic species.

Due to the long-term excessive exploitation of resources, habitat loss and fragmentation, environmental pollution, the conservation of bird diversity in China is facing severe challenges: the large-scale and continuous coastal reclamation projects have led to a sharp decline of the tidal flats, which has significantly reduced the number of species and population size of waterbirds; eating wild birds and eggs in some areas fails to prohibit, which seriously threatens the survival of birds; poison often causes the death of a large number of waterfowls, and illegal hunting has also led to a sharp decline in bird population (*e.g.*, Yellow-breasted Bunting *Emberiza aureola*); other bird

resources like Red-billed Leiothrix (*Leiothrix lutea*) and Hwamei (*Garrulax canorus*), are almost exhausted in some areas. Despite the government's severe crackdown on illegal smuggling, raptor trading (*e.g.*, Saker Falcon *Falco cherrug*), remains rampant. Furthermore, gene contamination is also a threat to wild bird resources. The natural cross-breeding caused by free-range chicken in the distribution area of Red Junglefowl (*Gallus gallus*) has made it difficult to find a purebred Red Junglefowl today; and artificial breeding of Indian Peafowl (*Pavo cristatus*) in the natural distribution areas of Green Peafowl (*Pavo muticus*) has also raised concerns about the genes contamination of the critically endangered species.

In the aim of protecting the wildlife resources in China, 41 species (Short-tailed Albatross *Phoebastria albatrus* etc.) were listed as the first class and 184 species (Horned Grebe *Podiceps auritus* etc.) were listed as the second class in "*National Key Protected Wild Animal List*" in the Law of the People's Republic of China on the Protection of Wildlife, issued in 1989 with the approval of the State Council. The *China Red Data Book of Endangered Animals: Aves* (Zheng and Wang, 1998) was published ten years later, identified 183 threatened species as Extinct (Ex), Extirpated (Et), Endangered (E), Vulnerable (V), Rare (R), and Indeterminate (I), based on the actual situation of Chinese birds and referred to the International Union for Conservation of Nature (IUCN) criteria. The "*National Key Protected Wildlife List*" and the *China Red Data Book of Endangered Animals: Aves* have been widely cited over years. *China Species Red List* (*Vol. I-II*) (Wang and Xie, 2004, 2009) list all the threatened birds in China based on the IUCN criteria, with category and justification for each species. The above work has been highly concerned by the Chinese government and the public, and has made the significant contributions in conservation of birds and their habitats.

With the improvement of ornithological researches and the promotion of public bird watching activities, the knowledge of population size, distribution, and population dynamics has increased and accumulated. In order to comprehensively assess the status of threatened birds in China, the *China's Red List of Biodiversity: Vertebrates, Volume II, Birds* was initiated. Refer to the criteria of IUCN red list, more than 20 ornithologists researchers were engaged in the assessment, and initially discussed assessment objects based on the birds listed in the *A Checklist on the Classification and Distribution of the Birds of China* (Second Edition) (Zheng, 2011); moreover, some modifications have also made in accordance with changes in the new taxonomic system and the status of the species in the following edition (Third Edition) of *A Checklist on the Classification and Distribution of the Birds of China* (Zheng, 2017). After sufficient assessment and analysis, the Red List and Categories of Chinese birds were proposed. The completed assessment report included: taxonomic status, assessment information, geographical distribution, population status and trends, habitats, threats and conservation actions *etc*. Among the assessed 1,445 species, three species (White Stork *Ciconia ciconia*, Siberian Spruce Grouse *Falcipennis falcipennis*, and Sarus Crane *Grus antigone*) were identified as Regionally Extinct (RE), and 18 species Critically Endangered (CR), 48 species Endangered (EN), 78 species Vulnerable (VU), 223 species Near Threatened (NT), and 969 species Least Concern (LC). Vaurie's Nightjar (*Caprimulgus centralasicus*) was identified as invalid species. In addition, there were 105 species with very little information for assessment, and their status were unclear, and therefore they were categorized as Data Deficient (DD) in this assessment. This work is a comprehensive and systematic assessment of the threatened status of Chinese birds in accordance with the newly revised criteria of IUCN, which is of great significance for China's bird conservation and implementation of international conventions.

Scholars engaged in the compiling of this book included: Professor Guangmei Zheng, Professor Zhengwang Zhang, Professor Yanyun Zhang, Professor Wenhong Deng, Professor Lu Dong of Beijing Normal University, Professor Changqing Ding of Beijing Forestry University, Professor Haitao Wang of Northeast Normal University, Professor Zhijun Ma of Fudan University, Professor Fang Zhou of Guangxi University, Professor Wei Liang and Professor Canchao Yang of Hainan Normal University, Professor Lianlian Xing and Guisheng Yang of Inner Mongolia University, Professor Xin Lu of Wuhan University, Professor Ping Ding of Zhejiang University, Professor Shuihua Chen and Professor Zhongyong Fan of Zhejiang Museum of Natural History, Professor Yuehua Sun and Dr. Lei Zhu of Institute of Zoology, Chinese Academy of Sciences, Professor Xiaojun Yang of Kunming Institute of Zoology, Chinese Academy of Sciences, Professor Lei Cao of Research Center for Eco-Environmental Sciences, Chinese Academy of Sciences, Professor Ming Ma of Xinjiang Institute of Ecology and Geography, Chinese Academy of Sciences, Associate Professor Qiang Ma of Chinese Academy of Forestry, Professor Yang Liu of Sun Yat-sen University.

Scholars engaged in the review of the list include: Professor Lizhi Zhou of Anhui University, Professor Fasheng Zou of Institute of Zoology, Guangdong Academy of Science, Professor Naifa Liu of Lanzhou University, and Professor Fumin Lei of Institute of Zoology, Chinese Academy of Sciences.

Dr. Ning Wang from Beijing Normal University assisted in checking photos of birds. Dr. Huaiyu Tian re-drew the distribution map according to the new base map. Dr. Canwei Xia, and graduate students Yuying Lin, Zhuqing Deng, Jin Liu, Boning Xue, Kai Zhao participated in part of the distribution area verification, translation and literature work. Mr. Xiaoyang Wang from Birdnet helped collecting a part of photos. We are also grateful to the experts, scholars and bird-watchers for providing photos of birds.

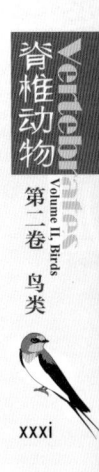

目 录 Contents

序一	i	Foreword I	iii
序二	vii	Foreword II	ix
总前言	xiii	Series' Foreword	xix
前言	xxvii	Preface	xxix

总 论 General Introduction

1	鸟类的多样性	2	1	Diversity of Birds	3
2	中国鸟类多样性及地理分布	4	2	The Diversity and Geographical Distribution of Birds in China	5
3	中国鸟类系统分类	10	3	Taxonomy of Birds in China	11
4	中国鸟类红色名录评估对象	12	4	Species assessed in the China's Red List of Birds	13
5	中国鸟类保护评估现状	12	5	Assessment on Birds Conservation Status in China	13
6	鸟类红色名录评估过程	12	6	The Assessment Process of China's Red List of Birds	13
7	鸟类红色名录评估标准	14	7	The Assessment Criteria of China's Red List of Birds	15
8	建立数据库	20	8	Setting up the Database	21
9	初步评定	20	9	A Preliminary Assessment	21
10	书面评估和会议评审	20	10	Written Evaluation and Meeting Review	21
11	评估修订	22	11	Checks and Revision	23
12	评估结果	22	12	Evaluation Results	23
13	与IUCN红色名录的比较	32	13	Comparison with IUCN Red List	33
14	灭绝物种分析	34	14	Analysis of Extinct Species	35
15	中国鸟类主要受威胁因素	34	15	Major Threats to Bird in China	35
16	保护成效	38	16	Conservation Effectiveness	39
17	结束语	44	17	Concluding Remarks	45

各 论 Species Monograph

区域灭绝 | RE

中文名	页码	学名	页码
镰翅鸡	48	Falcipennis falcipennis	48
赤颈鹤	50	Grus antigone	50
白鹳	52	Ciconia ciconia	52

极危 | CR

中文名	页码	学名	页码
海南孔雀雉	54	Polyplectron katsumatae	54
绿孔雀	56	Pavo muticus	56
青头潜鸭	58	Aythya baeri	58
长尾鸭	60	Clangula hyemalis	60
白头硬尾鸭	62	Oxyura leucocephala	62
爪哇金丝燕	64	Aerodramus fuciphagus	64
白鹤	66	Grus leucogeranus	66
勺嘴鹬	68	Calidris pygmeus	68
中华凤头燕鸥	70	Thalasseus bernsteini	70
黑头白鹮	72	Threskiornis melanocephalus	72
黑兀鹫	74	Sarcogyps calvus	74
毛腿雕鸮	76	Bubo blakistoni	76
冠斑犀鸟	78	Anthracoceros albirostris	78
双角犀鸟	80	Buceros bicornis	80
棕颈犀鸟	82	Aceros nipalensis	82
蓝冠噪鹛	84	Garrulax courtoisi	84
黑冠薮鹛	86	Liocichla bugunorum	86
黄胸鹀	88	Emberiza aureola	88

濒危 | EN

中文名	页码	学名	页码
四川山鹧鸪	90	Arborophila rufipectus	90
海南山鹧鸪	92	Arborophila ardens	92
松鸡	94	Tetrao urogallus	94
黑嘴松鸡	96	Tetrao urogalloides	96
黄腹角雉	98	Tragopan caboti	98
白尾梢虹雉	100	Lophophorus sclateri	100
绿尾虹雉	102	Lophophorus lhuysii	102
白冠长尾雉	104	Syrmaticus reevesii	104
灰孔雀雉	106	Polyplectron bicalcaratum	106
棉凫	108	Nettapus coromandelianus	108
中华秋沙鸭	110	Mergus squamatus	110
紫林鸽	112	Columba punicea	112

中文名	页码	学名	页码
绿皇鸠	114	Ducula aenea	114
大鸨	116	Otis tarda	116
波斑鸨	118	Chlamydotis macqueenii	118
白枕鹤	120	Grus vipio	120
丹顶鹤	122	Grus japonensis	122
白头鹤	124	Grus monacha	124
小青脚鹬	126	Tringa guttifer	126
遗鸥	128	Ichthyaetus relictus	128
东方白鹳	130	Ciconia boyciana	130
朱鹮	132	Nipponia nippon	132
黑脸琵鹭	134	Platalea minor	134
海南鳽	136	Gorsachius magnificus	136
白鹈鹕	138	Pelecanus onocrotalus	138
斑嘴鹈鹕	140	Pelecanus philippensis	140
卷羽鹈鹕	142	Pelecanus crispus	142
乌雕	144	Clanga clanga	144
白肩雕	146	Aquila heliaca	146
玉带海雕	148	Haliaeetus leucoryphus	148
虎头海雕	150	Haliaeetus pelagicus	150
褐渔鸮	152	Ketupa zeylonensis	152
黄腿渔鸮	154	Ketupa flavipes	154
花冠皱盔犀鸟	156	Rhyticeros undulatus	156
大黄冠啄木鸟	158	Chrysophlegma flavinucha	158
猎隼	160	Falco cherrug	160
蓝背八色鸫	162	Pitta soror	162
鹊鹂	164	Oriolus mellianus	164
细纹苇莺	166	Acrocephalus sorghophilus	166
灰冠鸦雀	168	Sinosuthora przewalskii	168
弄岗穗鹛	170	Stachyris nonggangensis	170
巨䴓	172	Sitta magna	172
丽䴓	174	Sitta formosa	174
黑喉歌鸲	176	Calliope obscura	176
棕头歌鸲	178	Larvivora ruficeps	178
贺兰山红尾鸲	180	Phoenicurus alaschanicus	180
白喉石䳭	182	Saxicola insignis	182
栗斑腹鹀	184	Emberiza jankowskii	184

易危 / VU

中文名	页码	学名	页码
白眉山鹧鸪	186	Arborophila gingica	186
红胸山鹧鸪	188	Arborophila mandellii	188
柳雷鸟	190	Lagopus lagopus	190

中文名	页码	学名	页码
红喉雉鹑	192	Tetraophasis obscurus	192
黄喉雉鹑	194	Tetraophasis szechenyii	194
阿尔泰雪鸡	196	Tetraogallus altaicus	196
红胸角雉	198	Tragopan satyra	198
褐马鸡	200	Crossoptilon mantchuricum	200
白颈长尾雉	204	Syrmaticus ellioti	204
黑颈长尾雉	206	Syrmaticus humiae	206
栗树鸭	208	Dendrocygna javanica	208
鸿雁	210	Anser cygnoid	210
小白额雁	212	Anser erythropus	212
红顶绿鸠	214	Treron formosae	214
花田鸡	216	Coturnicops exquisitus	216
白喉斑秧鸡	218	Rallina eurizonoides	218
长脚秧鸡	220	Crex crex	220
斑胁田鸡	222	Zapornia paykullii	222
紫水鸡	224	Porphyrio porphyrio	224
黑颈鹤	226	Grus nigricollis	226
林沙锥	228	Gallinago nemoricola	228
大杓鹬	230	Numenius madagascariensis	230
大滨鹬	232	Calidris tenuirostris	232
红腹滨鹬	234	Calidris canutus	234
黑嘴鸥	236	Saundersilarus saundersi	236
短尾信天翁	238	Phoebastria albatrus	238
黑鹳	240	Ciconia nigra	240
黄嘴白鹭	242	Egretta eulophotes	242
林雕	244	Ictinaetus malalensis	244
靴隼雕	246	Hieraaetus pennatus	246
草原雕	248	Aquila nipalensis	248
金雕	250	Aquila chrysaetos	250
白腹隼雕	252	Aquila fasciata	252
栗鸢	254	Haliastur indus	254
白腹海雕	256	Haliaeetus leucogaster	256
白尾海雕	258	Haliaeetus albicilla	258
大鵟	260	Buteo hemilasius	260
四川林鸮	262	Strix davidi	262
鬼鸮	264	Aegolius funereus	264
白喉犀鸟	266	Anorrhinus austeni	266
蓝须蜂虎	268	Nyctyornis athertoni	268
斑头大翠鸟	270	Alcedo hercules	270
白腿小隼	272	Microhierax melanoleucos	272
黄爪隼	274	Falco naumanni	274

中文名	页码	学名	页码
大紫胸鹦鹉	278	Psittacula derbiana	278
绯胸鹦鹉	280	Psittacula alexandri	280
双辫八色鸫	282	Pitta phayrei	282
蓝枕八色鸫	284	Pitta nipalensis	284
栗头八色鸫	286	Pitta oatesi	286
绿胸八色鸫	288	Pitta sordida	288
仙八色鸫	290	Pitta nympha	290
大盘尾	292	Dicrurus paradiseus	292
黑头噪鸦	294	Perisoreus internigrans	294
黑尾地鸦	296	Podoces hendersoni	296
白尾地鸦	298	Podoces biddulphi	298
歌百灵	300	Mirafra javanica	300
蒙古百灵	302	Melanocorypha mongolica	302
远东苇莺	304	Acrocephalus tangorum	304
东亚蝗莺	306	Locustella pleskei	306
台湾鹎	308	Pycnonotus taivanus	308
海南柳莺	310	Phylloscopus hainanus	310
暗色鸦雀	312	Sinosuthora zappeyi	312
金额雀鹛	314	Schoeniparus variegaticeps	314
黑额山噪鹛	316	Garrulax sukatschewi	316
白点噪鹛	318	Garrulax bieti	318
灰胸薮鹛	320	Liocichla omeiensis	320
四川旋木雀	322	Certhia tianquanensis	322
滇䴓	324	Sitta yunnanensis	324
淡紫䴓	326	Sitta solangiae	326
鹩哥	328	Gracula religiosa	328
褐头鸫	330	Turdus feae	330
金胸歌鸲	332	Calliope pectardens	332
白喉林鹟	334	Cyornis brunneatus	334
贺兰山岩鹨	336	Prunella koslowi	336
禾雀	338	Lonchura oryzivora	338
藏雀	340	Carpodacus roborowskii	340
藏鹀	342	Emberiza koslowi	342
硫黄鹀	344	Emberiza sulphurata	344

参考文献 ·········346
附录　中国鸟类濒危等级评估名录 ·········372

物种中文名索引 ·········427
物种学名索引 ·········443

References ·········346
Appendix　Assessment List of Endangered Levels of Chinese Birds ·········372
Index of Chinese Names of Species ·········427
Index of Scientific Names of Species ·········443

总论

General Introduction

1 鸟类的多样性

鸟类是陆生脊椎动物中种类最多、分布最广、研究最为深入的类群。鸟类起源于蜥臀目兽脚类恐龙,中生代侏罗纪(距今约 1.5 亿年前)的始祖鸟是迄今发现的最古老鸟类。鸟类在中生代晚侏罗纪已经发生了广泛的辐射演化,在新生代经历了快速的辐射演化,形成现在的鸟类多样性格局。基于多基因分子标记和基因组学技术构建的鸟类系统分类,对传统的鸟类分类系统进行了全面梳理和修订,形成了新的、已经被广为接受的世界鸟类分类系统。在新的分类系统中,出现了许多新的目、科和属,以及各分类阶元的重新归类,原来分类系统中的平胸总目 (Ratitae)、楔翼总目 (Implennes) 和突胸总目 (Carinatae) 3 个总目,被归入新分类系统中的古颚总目 (Paleognathae) 和今颚总目 (Neognathae)。

现存鸟类有大约 11,000 种,其中雀形目占 60% 以上。Gill 等 (2020) 认为世界有 10,770 种 (20,005 种及亚种) 现生鸟类和 158 个已灭绝的物种,隶属于 40 目 250 科;*Handbook of the Birds of the World and Birdlife International* (2019) 的世界鸟类名录则包括了 11,147 种现生鸟类和 159 种已灭绝的鸟类,隶属于 36 目 243 科。

鸟类也是脊椎动物中分布最广的类群,从北极苔原到白雪皑皑的南极大陆(图 1)、从热带雨林到荒漠地带、从海洋孤岛到无垠草原(图 2),

图 1　南极恩克斯堡岛的阿德利企鹅 (*Pygoscelis adeliae*) 繁殖群(张雁云摄)

Figure 1　The breeding colony of Adelie Penguin in Inexpressible Island, Antarctica (Photographed by Yanyun Zhang)

1 Diversity of Birds

Birds are the most diverse, the most widely distributed, and the most thoroughly studied taxon of terrestrial vertebrates.

Birds originate from theropod dinosaurs of Saurischia, and *Archaeopteryx* living in the Jurassic period of Mesozoic Era (about 150 million years ago) is the oldest bird ever found. Birds undergone extensive radiation evolution in Late Jurassic, and experienced rapid radiation evolution in the Cenozoic Era, forming the current pattern of avian diversity. The ornithological classification system based on multi-gene molecular markers and genomics techniques has comprehensively revised the traditional taxonomy system, forming a new and widely accepted bird taxonomy system of world. The new taxonomy system includes many new orders, families and genera, as well as a reclassification of taxonomic categories. Ratitae, Implennes and Carinatae in the original taxonomy system have been categorized into Paleognathae and Neognathae in the new system.

There are about 11,000 bird species in existence, of which passerine accounts for more than 60%. Gill *et al.* (2020) proposed that there were 10,770 (20,005 species and subspecies) living bird species and 158 known extinct species in the world, belonging to 250 families of 40 orders. While, the checklists of birds from Handbook of the Birds of the World and Birdlife International (2019) include 11,147 living bird species and 159 known extinct bird species, belonging to 243 families of 36 orders.

Birds are also the most widely distributed group of vertebrates. Breeding habitats of birds range from the Arctic Tundra to the snow-capped Antarctic (**Figure 1**), from tropical rainforests to desert areas, from solitary islands in the ocean to expansive grasslands (**Figure 2**). Most birds have strong abilities on flight and dispersal, and their ability to reach somewhere is unmatched by other organisms. For example, Hawkes *et al.* (2013) have found Bar-headed Geese can reach an altitude of 7,290 m when migrating over Qinghai-Tibet

图2 栖息于印度热带林区的红原鸡 (*Gallus gallus*)(张雁云摄)

Figure 2 Red Junglefowl (*Gallus gallus*)(inhabiting tropical forests in India)(Photographed by Yanyun Zhang)

都有鸟类的繁殖地。大多数鸟类飞行、扩散能力强，其到达能力更是其他生物无法比拟的，如 Hawkes 等 (2013) 监测到斑头雁迁徙飞越青藏高原时的飞行高度达 7,290 m，Fijn 等 (2013) 则发现北极燕鸥每年迁徙往返南北极的飞行距离达到 90,000 km。

鸟类的鸣声响亮、多变，有多彩的羽饰，是动物界中最容易被观察到、研究最深入的类群。许多生物学、行为学和生态学的理论，如物种概念、性选择、栖息地利用、觅食策略等最先在鸟类中发现并提出，然后在其他类群中得到验证和支持。动物各类群中，对鸟类整体的认识也是最深入的，如 IUCN 对动物、植物和微生物三大生物类群的 21 个亚类评估结果显示，只有鸟纲和肢口纲 (仅包括 4 个物种) 100% 的物种完成了评估，其他脊椎动物中，哺乳类、爬行类、两栖类、鱼类评估的物种数分别占已描述物种数的 90%、71%、84% 和 54%，无脊椎动物除肢口纲外的其他类群仅对 0.31%～40% 的物种进行了评估 (IUCN，2019)。

2 中国鸟类多样性及地理分布

中国目前已记录鸟类 1,445 种 (2,344 种及亚种)，隶属于 26 目 109 科 497 属 (郑光美，2017)。最近的研究认为中亚夜鹰 (*Caprimulgus centralasicus*) 是欧夜鹰 (*C. europaeus*) 的同物异名 (Schweizer *et al.*，2020)，减少一种。中国鸟类物种数居世界各国前列，这与中国丰富、独特的自然环境有关。中国是幅员辽阔、地理位置适中、地形和气候十分复杂、自然景观和生态系统多样性非常丰富的国家。沿西藏南部的喜马拉雅山向西，经横断山脉—秦岭—淮河一线，将中国大陆分为南、北两个具有不同特色的动物地理界，即东洋界和古北界。南方的东洋界位于我国东部季风区的南部，包括热带和亚热带动物群；北方的古北界包括我国东部季风区的北部以及蒙新区和青藏高原，包括寒温带、温带以及高原寒漠、草原及荒漠等动物群 (图 3)；辽阔的海岸线以及数以千计的从寒温带到热带的岛屿，为各种海鸟提供了优越的栖息条件；2,500 多万 hm² 的湿地 (包括河流、沼泽、湖泊、滩涂、盐沼等)，是水鸟繁殖、迁徙和越冬期间的重要栖息地 (图 4 和图 5)。许多繁殖于寒带和亚寒带国家的鸟类，在迁徙时途经我国，大多到我国长江流域以南越冬，在每年春秋季节还有大批旅鸟途经我国，全球 9 个主要的候鸟迁飞区中，有 3 个经过我国。上述因素使我国的鸟类区系复杂和生物多样性丰富。

世界 15 种鹤类中，曾有 9 种分布于中国，其中黑颈鹤 (*Grus nigricollis*) 是青藏高原的特有种。在我国的 1,444 种鸟类中，有 92 种特有鸟类。64 种雉科鸟类中有 22 种为我国特有鸟类，其中的虹雉属 (*Lophophorus*)、马鸡属 (*Crossoptilon*)、长尾雉属 (*Syrmaticus*)、锦鸡属 (*Chrysolophus*) 等均是中国的特有属或主要分布区在中国；我国有 68 种噪鹛科 (Leiothrichidae) 鸟类，其中 19 种为我国特有鸟类。雉类和画眉类的大多数不具迁徙习性，是"永久居民"，因而中国素有"雉鸡王国"和"画眉乐园"的美称。

中国陆地区域依地形、温度和降水的显著差异，可划分为东部的

(Xizang) Plateau. Fijn *et al.* (2013) have reported that Arctic Terns migrate up to 90,000 km each year between the North and South Poles.

Birds have loud songs and colorful plumages. They are the most easily observed and deeply studied group compared to other animals. Many biological, behavioral, and ecological theories and hypotheses have been first proposed in birds, such as the concept of species, sexual selection, habitat use, and foraging strategies, which have been verified and supported in other groups of species. Among the various animal groups, the overall knowledge of birds is also the most profound. For example, IUCN has assessed 21 subgroups of three major biological groups of animals, plants and microorganisms, 100% of the species of Aves and Merostomata (only four species included) have been evaluated. However, for other vertebrates, such as mammals, reptiles, amphibians, and fish, about 90%, 71%, 84%, and 54% have been evaluated, respectively. As to other groups of invertebrates, there are only 0.31%~40% of species have been assessed (IUCN, 2019).

2 The Diversity and Geographical Distribution of Birds in China

China has currently recorded 1,445 bird species (2,344 including subspecies), belonging to 497 genera of 109 families of 26 orders (Zheng, 2017), and recent studies suggest that the Vaurie's Nightjar (*Caprimulgus centralasicus*) is a synonym of European Nightjar (*C. europaeus*) (Schweizer *et al.*, 2020). The number of bird species ranks among the top in the world, which is related to the rich and unique natural environment of China, with vast territory, moderate geographic location, complex topography and climate, as well as rich and diverse natural landscapes and ecosystems. Along to the Himalayas in southern Tibet (Xizang), through the Hengduan Mountains-Qinling Mountains and Huaihe River, the mainland of China is divided into two geographic realms with different characteristics, namely the Oriental Realm and the Palearctic realm. The Oriental Realm includes the southern part of the eastern monsoon region of China, containing subtropical and tropical fauna (**Figure 3**); the Palearctic realm includes the northern part of the eastern monsoon region of China, as well as the Inner Mongolia-Xinjiang region and the Tibetan Plateau, containing the cold temperate fauna, the temperate fauna, the highland cold deserts fauna, grasslands fauna and desert fauna. The vast coastline and thousands of islands ranging from the temperate zone to tropical zone, provide suitable habitats for various seabirds; moreover, approximately more than 25 million hectares of wetlands (including rivers, swamps, lakes, tidal flats, salt marsh, *etc.*) are important habitats for breeding, migration and overwintering to waterbirds (**Figure 4** and **Figure 5**). Many birds breed in frigid zone and sub-frigid zone, pass through or overwinter in the south of the Changjiang River in China. In addition, three of nine world main bird flyways cover China and there are large number of migratory birds in spring and autumn. All these contribute to complex and abundant bird biodiversity in China.

There are fifteen crane species in the world and nine occur in China, of which the Black-necked Crane (*Grus nigricollis*) is endemic in the Qinghai-Tibet (Xizang) Plateau. Among the 1,444 birds in China, 92 species are

图 3　栖息于内蒙古草原的大鸨 (*Otis tarda*)（张雁云摄）

Figure 3　Great Bustard (*Otis tarda*) inhabiting the grasslands of Inner Mongolia (Nei Mongol)(Photographed by Yanyun Zhang)

图 4　辽宁湿地鸟类（张雁云摄）

Figure 4　Waterbirds in wetland, Liaoning (Photographed by Yanyun Zhang)

季风区、西北部的干旱区和西南部的青藏高原区 3 个大的自然地理区。根据郑作新和张荣祖（1956）、张荣祖和赵肯堂（1978）、张荣祖（1999）的观点，可将中国划分为 7 个鸟类大区和 19 个亚区，其中 10 个亚区属于古北界，9 个亚区属于东洋界（表 1）。

endemic. Of the 64 species of pheasant birds, 22 species are endemic to China, and *Lophophorus, Crossoptilon, Syrmaticus* and *Chrysolophus* are native to China or mainly distributed in China. There are 68 species of Leiothrichidae in China, 19 of which are endemic to China. Most of the pheasants and thrushes are "permanent residents" without migration behavior, and therefore China has been known as "Kingdom of Pheasant" and "Laughingthrush Paradise of Laughingthrush".

China's terrestrial region can be divided into three large natural geographic regions based on the significant differences in topography, temperature and precipitation, including the monsoon region in the east, the arid region in the northwest, and the Qinghai-Tibet (Xizang) Plateau region in the southwest. China can be divided into 7 bird regions and 19 subregions, of which 10 sub-regions belong to the Oriental Realm and nine subregions belong to the Palearctic region (**Table 1**) (Zheng and Zhang, 1956; Zhang and Zhao, 1978; Zhang, 1999).

Figure 5 Common Crane (*Grus grus*) overwintering in wetland of Wild Duck Lake, Beijing (Photographed by Yanyun Zhang)

表 1 中国鸟类地理分区情况

动物地理界	自然区	地理区	地理亚区	鸟类群落	代表鸟类
古北界	季风区北部	东北区	大兴安岭亚区（含阿尔泰山）	寒温带针叶林鸟类	黑嘴松鸡 (Tetrao urogalloides)、雪鸮 (Bubo scandiacus)、黑啄木鸟 (Dryocopus martius)、北噪鸦 (Perisoreus infaustus)、红交嘴雀 (Loxia curvirostra)、普通䴓 (Sitta europaea) 等
			长白山山地亚区 松辽平原亚区	中温带森林、森林草原、农田鸟类	北领角鸮 (Otus semitorques)、丹顶鹤 (Grus japonensis)、松鸦 (Garrulus glandarius)、牛头伯劳 (Lanius bucephalus)、黑头䴓 (Sitta villosa) 等
		华北区	黄淮平原亚区 黄土高原亚区	暖温带森林草原、农田鸟类	石鸡 (Alectoris chukar)、斑翅山鹑 (Perdix dauurica)、大杜鹃 (Cuculus canorus)、大斑啄木鸟 (Dendrocopos major)、灰喜鹊 (Cyanopica cyanus)、红嘴山鸦 (Pyrrhocorax pyrrhocorax)、灰鹡鸰 (Motacilla cinerea)、灰眉岩鹀 (Emberiza godlewskii) 等
	西部干旱区	蒙新区	东部草原亚区	温带草原鸟类	大鸨 (Otis tarda)、毛腿沙鸡 (Syrrhaptes paradoxus)、草原雕 (Aquila nipalensis)、角百灵 (Eremophila alpestris)、蒙古百灵 (Melanocorypha mongolica)、荒漠伯劳 (Lanius isabellinus) 等
			西部荒漠亚区	温带荒漠、半荒漠鸟类	沙䳭 (Oenanthe isabellina)、白顶䳭 (Oenanthe pleschanka)、短趾百灵 (Alaudala cheleensis)、黑顶麻雀 (Passer ammodendri)、白尾地鸦 (Podoces biddulphi) 等
			天山山地亚区	山地森林草原、荒漠鸟类	暗腹雪鸡 (Tetraogallus himalayensis)、红背伯劳 (Lanius collurio)、灰蓝山雀 (Cyanistes cyanus)、欧亚旋木雀 (Certhia familiaris)、花彩雀莺 (Leptopoecile sophiae)、金额丝雀 (Serinus pusillus) 等
	青藏高寒区	青藏区	羌塘高原亚区	高原寒漠鸟类	藏雪鸡 (Tetraogallus tibetanus)、西藏毛腿沙鸡 (Syrrhaptes tibetanus)、棕头鸥 (Chroicocephalus brunnicephalus)、斑头雁 (Anser indicus)、黑颈鹤 (Grus nigricollis)、地山雀 (Pseudopodoces humilis)、棕背雪雀 (Pyrgilauda blanfordi) 等
			青海藏南亚区	高原草原、草甸鸟类	红喉雉鹑 (Tetraophasis obscurus)、白马鸡 (Crossoptilon crossoptilon)、血雉 (Ithaginis cruentus)、藏雀 (Carpodacus roborowskii)、藏鹀 (Emberiza koslowi)、朱鹀 (Urocynchramus pylzowi)、黑头金翅雀 (Chloris ambigua) 等
东洋界	季风区南部	西南区	西南山地亚区	南方亚高山森林草原、草甸鸟类	红腹角雉 (Tragopan temminckii)、绿尾虹雉 (Lophophorus lhuysii)、白腹锦鸡 (Chrysolophus amherstiae)、灰胸薮鹛 (Liocichla omeiensis)、斑背噪鹛 (Garrulax lunulatus)、绿翅短脚鹎 (Ixos mcclellandii) 等
			喜马拉雅亚区	亚热带山地森林鸟类	红胸角雉 (Tragopan satyra)、火尾太阳鸟 (Aethopyga ignicauda)、绿背山雀 (Parus monticolus)、杂色噪鹛 (Trochalopteron variegatum)、红眉朱雀 (Carpodacus pulcherrimus) 等
		华中区	东部丘陵平原亚区 西部山地高原亚区	亚热带森林灌丛、草地、农田鸟类	黄腹角雉 (Tragopan caboti)、红腹锦鸡 (Chrysolophus pictus)、灰胸竹鸡 (Bambusicola thoracicus)、乌鸫 (Turdus mandarinus)、画眉 (Garrulax canorus)、黄臀鹎 (Pycnonotus xanthorrhous)、红头长尾山雀 (Aegithalos concinnus)、灰头鸦雀 (Psittiparus gularis) 等
		华南区	闽广沿海亚区 滇南山地亚区 海南岛亚区 台湾亚区 南海诸岛亚区	热带森林、森林灌丛、草地、农田鸟类	红原鸡 (Gallus gallus)、绿孔雀 (Pavo muticus)、海南山鹧鸪 (Arborophila ardens)、黑长尾雉 (Syrmaticus mikado)、鲣鸟 (Sula spp.)、红嘴鹲 (Phaethon aethereus)、白斑军舰鸟 (Fregata ariel)、长尾阔嘴鸟 (Psarisomus dalhousiae)、八色鸫 (Pitta spp.)、双角犀鸟 (Buceros bicornis)、海南柳莺 (Phylloscopus hainanus)、橙腹叶鹎 (Chloropsis hardwickii)、鸦鹃 (Centropus spp.)、台湾蓝鹊 (Urocissa caerulea)、褐头凤鹛 (Yuhina brunneiceps)、台湾黄山雀 (Machlolophus holsti) 等

Table 1 Geographical Division of Birds in China

Fauna realm	Natural region	Geographic Region	Geographic Sub-region	Avian Community	Typical Avian Species
Palearctic realm	North of Monsoon Region	Northeast China Region	Da Hinggan Ling Mountains Sub-region (Altay Shan affiliated)	Cold temperate taiga avian community	*Tetrao urogalloides, Bubo scandiacus, Dryocopus martius, Perisoreus infaustus, Loxia curvirostra, Sitta europaea,* etc.
			Changbai Shan Mountain Sub-region, Songliao Plain Sub-region	Medium temperate forest, forest steppe, farmland avian community	*Otus semitorques, Grus japonensis, Garrulus glandarius, Lanius bucephalus, Sitta villosa,* etc.
		North China Region	Huang-huai Plain Sub-region, Loess Plateau Sub-region	Warm temperate forest steppe, farmland avian community	*Alectoris chukar, Perdix dauurica, Cuculus canorus, Dendrocopos major, Cyanopica cyanus, Pyrrhocorax pyrrhocorax, Motacilla cinerea, Emberiza godlewskii,* etc.
	Western Drought Region	Inner Mongolia Region	Eastern Steppe Sub-region	Temperate steppe avian community	*Otis tarda, Syrrhaptes paradoxus, Aquila nipalensis, Eremophila alpestris, Melanocorypha mongolica, Lanius isabellinus,* etc.
			Western Desert Sub-region	Temperate desert, semi-desert avian community	*Oenanthe isabellina, Oenanthe pleschanka, Alaudala cheleensis, Passer ammodendri, Podoces biddulphi,* etc.
			Tianshan Mountains Sub-region	Alpine forest steppe, desert avian community	*Tetraogallus himalayensis, Lanius collurio, Cyanistes cyanus, Certhia familiaris, Leptopoecile sophiae, Serinus pusillus,* etc.
	High and Cold Region	Qinghai-Tibbet Region	Qiangtang Plateau Sub-region	Plateau cold desert avian community	*Tetraogallus tibetanus, Syrrhaptes tibetanus, Chroicocephalus brunnicephalus, Anser indicus, Grus nigricollis, Pseudopodoces humilis, Pyrgilauda blanfordi,* etc.
			Qinghai south-Tibet Sub-region	Plateau steppe, meadow avian community	*Tetraophasis obscurus, Crossoptilon crossoptilon, Ithaginis cruentus, Carpodacus roborowskii, Emberiza koslowi, Urocynchramus pylzowi, Chloris ambigua,* etc.
Oriental realm	South of Monsoon Region	Southwest China Region	Southwest Mountains Sub-region	Southern sub-alpine forest steppe, meadow avian community	*Tragopan temminckii, Lophophorus lhuysii, Chrysolophus amherstiae, Liocichla omeiensis, Garrulax lunulatus, Ixos mcclellandii,* etc.
			Himalaya Sub-region	Subtropical mountain forest avian community	*Tragopan satyra, Aethopyga ignicauda, Parus monticolus, Trochalopteron variegatum, Carpodacus pulcherrimus,* etc.
		Central China Reion	Eastern Hilly-plain Sub-region Western Mountain-plateau Sub-region	Subtropical forest-scrub, grassland, farmland avian community	*Tragopan caboti, Chrysolophus pictus, Bambusicola thoracicus, Turdus mandarinus, Garrulax canorus, Pycnonotus xanthorrhous, Aegithalos concinnus, Psittiparus gularis,* etc.
		South China Region	Fujian-Guangdong Coastal Sub-region Southern Yunnan Mountain Sub-region Hainan Island Sub-region Taiwan Sub-region South China Sea Islands Sub-region	Tropical forest, forest-scrub, grassland, farmland avian community	*Gallus gallus, Pavo muticus, Arborophila ardens, Syrmaticus mikado, Sula* spp., *Phaethon aethereus, Fregata ariel, Psarisomus dalhousiae, Pitta* spp., *Buceros bicornis, Phylloscopus hainanus, Chloropsis hardwickii, Centropus* spp., *Urocissa caerulea, Yuhina brunneiceps, Machlolophus holsti,* etc.

3 中国鸟类系统分类

我国鸟类系统分类研究起步于国外的学者或传教士进入中国后，收集鸟类标本并整理和发表了一些关于中国鸟类分类和区系的报告和专著，如 Swinhoe (1863) 发表了包括 454 种鸟类的中国鸟类名录；Gee 等 (1926) 发表了《中国鸟类目录试编》(A Tentative List of Chinese Birds)，并于 1931 年对此名录进行了修订，共记录鸟类 1,093 种。

从 20 世纪 40 年代开始，我国鸟类学主要奠基人郑作新院士对中国鸟类分类系统开展了全面、系统的研究和梳理。他在 1947 年发表的 Checklist of Chinese Birds 中，记录中国鸟类 1,087 种，是中国鸟类学者对中国鸟类分类系统进行深入研究的标志，此后 A Synopsis of the Avifauna of China (1987) 记录鸟类 1,186 种，《中国鸟类种和亚种分类名录大全》第一版 (1994 年) 和第二版 (2000 年) 分别收录鸟类 1,244 种和 1,253 种。

马敬能等 (2000) 发表的《中国鸟类野外手册》，收录中国鸟类 1,329 种。2005 年，由郑光美院士组织编写、发表的《中国鸟类分类与分布名录》，共收录中国鸟类 1,332 种 (2,261 种及亚种)，隶属于 24 目 101 科；2011 年修订出版的第二版收录了鸟类 1,371 种 (2,304 种及亚种)，隶属于 24 目 101 科；2017 年修订出版的第三版收录了鸟类 1,445 种 (2,344 种及亚种)，隶属于 26 目 109 科 (图 6)。

近年来，我国鸟类种数增加较多，一方面是随着分子生物学技术、鸟鸣录音与声谱分析技术等的不断发展，为研究鸟类的系统发育以及种上和种下分类提供了重要的支撑。另一方面归因于我国蓬勃发展的群众性和专业性鸟类调查和观鸟活动，不断发现中国鸟类新分布记录。

3 Taxonomy of Birds in China

Research on taxonomy of bird in China started with foreign scholars or missionaries. They collected bird specimens and published some reports and monographs on the classification and fauna of Chinese birds. For example, Swinhoe (1863) has proposed a checklist of Chinese birds which contains 454 species. Gee *et al.* (1926) published *A Tentative List of Chinese Birds* in 1926, and revised it in 1931, which records 1,093 bird species.

Since the 1940s, Prof. Zuoxin Zheng (Tso-Hsin Cheng), Academician of Chinese Academy of Science (CAS), the main founder of ornithology in China, had carried out a comprehensive and systematic research on the classification system of birds in China. 1,087 species of birds were recorded in the *Checklist of Chinese Birds* (Zheng, 1947), as a sign that Chinese ornithologists had conducted in-depth research on the classification system of the birds in China. Later, 1,186 species of birds were recorded in *A Synopsis of the Avifauna of China* (1987); 1,244 and 1,253 species were included in the first edition (1994) and the second edition (2000) of the *A Complete Checklist of Species and Subspecies of the Chinese Birds*, respectively.

A Field Guide to the Birds of China published by Mackinnon *et al.* (2000), included 1,329 species of Chinese birds. In 2005, Prof. Guang-mei Zheng, Academician of Chinese Academy of Science (CAS), organized and published *A Checklists on the Classification and Distribution of the Birds of China*. 1,332 species (2,261 including subspecies) of birds were included, affiliated with 101 families of 24 orders. 1,371 species (2,304 including subspecies) with 101 families of 24 orders were included in the second edition in 2011. And 1,445 species (2,344 including subspecies) with 109 families of 26 orders were included in the third edition in 2017 (**Figure 6**).

In recent years, the number of bird species has increased greatly in China, with the continuous development of molecular biology technology, and technologies on sound recording, sound spectrum analysis , which provides an important basis for the study of phylogeny and taxonomy of birds. Furthermore, the vigorous development of bird surveys and bird watching activities by experts as well as the public in China, has contributed greatly to the new records of bird in China.

图 6 中国鸟类分类专著

Figure 6 Chinese monographs on checklist of the classification of the birds

4 中国鸟类红色名录评估对象

评估工作 2015 年展开，当时以《中国鸟类分类与分布名录》(第二版)（郑光美，2011）所记录的种类为基础。该名录包含 1,371 种鸟类的分类和分布信息，是当时国内已出版的文献中使用最广泛、最翔实的分类与分布名录。

在中国鸟类红色名录编制后期，由国内多位知名学者历时一年编写的《中国鸟类分类与分布名录》(第三版) 已经定稿，该书采纳最新鸟类分类学研究成果，分类系统有很大变化：如增补了小美洲黑雁 (*Branta hutchinsii*)、白腰滨鹬 (*Calidris fuscicollis*) 等中国鸟类新记录，同时有多个亚种被提升为种；过去国内通称的印度八色鸫 (*Pitta brachyura*)、褐头山雀 (*Parus songarus*)、黄胸山雀 (*Parus flavipectus*)、西域山雀 (*Parus bokharensis*)、灰腹灰雀 (*Pyrrhula griseiventris*) 等所隶属的种属关系有较大调整；银鸥 (*Larus argentatus*) 广泛分布于欧洲而不见于我国。

本书共对 1,445 种鸟类进行了评估，鸟类名称、分类系统、中国特有鸟类也依照《中国鸟类分类与分布名录》(第三版)。

5 中国鸟类保护评估现状

IUCN 红色名录和 CITES 附录物种历经多次修订，对全球的受威胁物种进行了科学评估，为鸟类保护提供重要的支撑。

1989 年，国家颁布了《国家重点保护野生动物名录》，是我国野生动物保护法制化的重要标志，其中 41 种鸟类被列为国家 I 级重点保护野生动物、184 种鸟类被列入国家 II 级重点保护野生动物。为了进一步贯彻落实《中华人民共和国野生动物保护法》，加强对我国国家和地方重点保护野生动物以外的陆生野生动物资源的保护和管理，国家野生动物主管部门于 2000 年发布了《国家保护的有益的或者有重要经济、科学研究价值的陆生野生动物名录》，其中鸟纲包括 18 目 61 科 707 种。《中国濒危动物红皮书·鸟类》（郑光美和王岐山，1998）借鉴 IUCN 早期的评估标准，并结合中国鸟类的实际情况，将中国 183 种受威胁鸟类划分为野生绝迹 (Extinct, Ex)、国内绝迹 (Extirpated, Et)、濒危 (Endangered, E)、易危 (Vulnerable, V)、稀有 (Rare, R)、未定 (Indeterminate, I) 等类别。汪松和解焱 (2004, 2009) 依据 IUCN 修订的标准，厘定了中国受威胁鸟类名录，编制发表了《中国物种红色名录》（第一、第二卷），并给出了物种的受威胁等级和所依据的标准。这些工作均受到我国政府和民间的高度关注，对鸟类及其栖息地的保护工作起到了重要的促进和推动作用。

6 鸟类红色名录评估过程

鸟类红色名录评估过程包括确定评估对象、数据收集、初评、初核、复审、定稿等步骤。首先由来自北京师范大学、复旦大学、北京林业大学、浙江大学等单位的学者组成工作组，对红色名录评估的方法、标准使用、数据来源等进行界定。然后，由北京师范大学牵头，邀请来自安徽大学、北京林业大学、东北师范大学、复旦大学、广西大学、海南师范大学、

4 Species assessed in the China's Red List of Birds

The assessment began in 2015, based on *A Checklist on the Classification and Distribution of the Birds of China* (2nd edition) (Zheng, 2011), which is the most widely used checklist in China, with background information on classification and distribution of 1,372 species.

In the late stage of the process of China's Red List of Birds, *A Checklist on the Classification and Distribution of the Birds of China* (3rd edition) was finalized with the updated research result of bird taxonomy, in which the taxonomy system and status of species had changed a lot: some new records of birds have been added, *e.g.* Cackling Goose (*Branta hutchinsii*), White-rumped Sandpiper (*Calidris fuscicollis*), *etc.*; some subspecies have been upgraded as species; the genera of some species have been reclassified, *e.g.* Indian Pitta (*Pitta brachyura*), Songar Tit (*Parus songarus*), Yellow-breasted Tit (*Parus flavipectus*), Turkestan Tit (*Parus bokharensis*), Oriental Bullfinch (*Pyrrhula griseiventris*), *etc.*; and some species, *e.g.* Herring Gull (*Larus argentatus*), have been confirmed they do not distribute in China.

The species name, taxonomy system and endemic species were based on *A Checklist on the Classification and Distribution of the Birds of China* (3rd edition), and 1,445 species were assessed in total.

5 Assessment on Birds Conservation Status in China

The IUCN Red List and the CITES Appendix have been revised several times with scientific assessment on threatened species, and played an important role in bird conservation in the world. In 1989, the *National Key Protected Wild Animal List* in China was published, which is an important symbol of the legalization of wildlife protection in China. There were 41 species of birds listed as first-class key protected wild animals and 184 species listed as second-class. For further implementation of the *Wildlife Protection Law of the People's Republic of China* and the reinforce of the protection and management of terrestrial wildlife resources other than the national and local key protected wildlife, in 2000, the State Forestry Administration has issued the *National Protected List of Terrestrial Wild Animals with Good Benefits or Important Economic or Scientific Values*, which included 18 orders, 61 families and 707 species in Aves. *China Red Data Book of Endangered Animals: Aves* (Zheng and Wang, 1998) referred to the early assessment criteria of IUCN and combines the actual situation of birds in China to categorize 183 threatened species into Extinct (Ex), Extirpated (Et), Endangered (E), Vulnerable (V), Rare (R), and Indeterminate (I). According to the revised standard of IUCN, Wang and Xie (2004, 2009) assessed the birds of China with IUCN Red List criteria and published the *China Species Red List* (volumes I-II). These work got high concern by the government and public, and played an important role in promoting the conservation of birds and their habitats.

6 The Assessment Process of China's Red List of Birds

The process of China's Red List of Birds includes determining the species list, collecting data, preliminary assessment, initial check, rechecking, finalization, *etc.* Firstly, the working group of scholars from Beijing Normal University, Fudan University, Beijing Forestry University, Zhejiang University *etc.* was set up to discuss and determine the methodology, standard and data sources of the assessment. Afterwards, Beijing Normal University invited 27 experts

华南濒危动物研究所、兰州大学、内蒙古大学、武汉大学、浙江大学、浙江自然博物馆、中国科学院动物研究所、中国科学院昆明动物研究所、中国科学院新疆生态与地理研究所、中国科学院生态环境研究中心、中国林业科学研究院、中山大学等高校和科研院所的 27 位专家组成红色名录专家组（图 7），仔细研究和厘定有关物种的受威胁等级及致危因素。

7 鸟类红色名录评估标准

依据 IUCN 物种红色名录濒危等级和标准（3.1 版）、IUCN 物种红色名录标准在区域和国家的应用指南（4.0 版），对中国鸟类进行评估。

本次评估使用了以下 IUCN 等级：灭绝（Extinct, EX）、野外灭绝（Extinct in the Wild, EW）、区域灭绝（Regionally Extinct, RE）、极危（Critically Endangered, CR）、濒危（Endangered, EN）、易危（Vulnerable, VU）、近危（Near Threatened, NT）、无危（Least Concern, LC）、数据缺乏（Data Deficient, DD）。

各等级的含义和评估标准如下所述（图 8，表 2）。

灭绝 (EX)。一个物种的最后一个个体已经死亡，则该种灭绝。

野外灭绝 (EW)。一个物种的所有个体只生活在人工养殖状态下，则该种野外灭绝。

区域灭绝 (RE)。一个物种在某个区域内的最后一个个体已经死亡，则该物种已经区域灭绝。

Figure 7 Ornithologists discussed Red List (Photographed by Lu Dong)

from Anhui University, Beijing Forestry University, Northeast Normal University, Fudan University, Guangxi University, Hainan Normal University, Guangdong Institute of Applied Biological Resources, Lanzhou University, Inner Mongolia University, Wuhan University, Zhejiang University, Zhejiang Museum of Natural History, Institute of Zoology (CAS), Kunming Institute of Zoology (CAS), Xinjiang Institute of Ecology and Geography (CAS), Research Center for Eco-Environmental Sciences (CAS), Chinese Academy of Forestry, and Sun Yat-Sen University, *etc*. to form the experts group (**Figure 7**), in order to study and define the threat level and major threats of species concerned.

7 The Assessment Criteria of China's Red List of Birds

Birds in China were evaluated according to the IUCN Red List Categories and Criteria (Version 3.1) and Guidelines for Application of IUCN Red List Criteria at Regional and National Levels (Version 4.0).

The IUCN Red List category were used in the assessment: Extinct (EX), Extinct in the Wild (EW), Regionally Extinct (RE), Critically Endangered (CR), Endangered (EN), Vulnerable (VU), Near Threatened (NT), Least Concern (LC) and Data Deficient (DD).

The definition and evaluation criteria of each category are as follows (**Figure 8, Table 2**).

Extinct (EX). A taxon is Extinct when there is no reasonable doubt that the last individual has died.

Extinct in the Wild (EW). A taxon is Extinct in the Wild when it is known only to survive in cultivation and in captivity.

Regionally Extinct (RE). A taxon is Regionally Extinct when the last individual of a species in the region has died.

图 8 IUCN 受威胁物种红色名录评估等级 (IUCN Standards and Petitions Subcommittee, 2017)

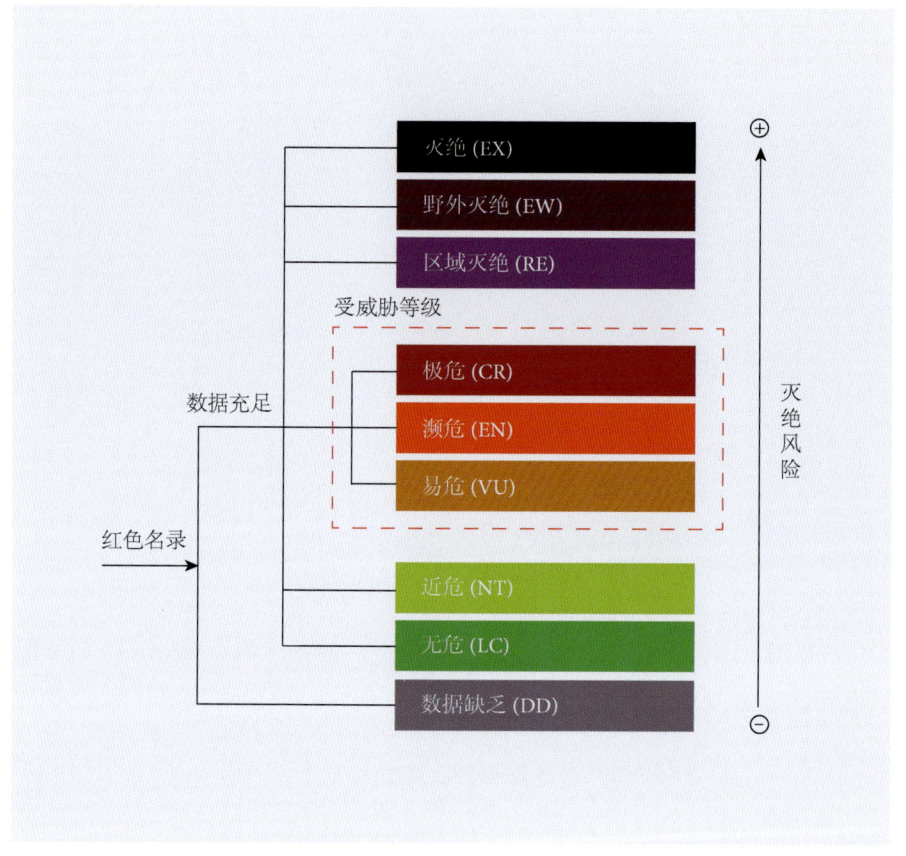

极危 (CR)、濒危 (EN) 和易危 (VU)。这三个等级统称为受威胁等级 (Threatened Categories)，灭绝的风险由高到低。当某一物种符合**表 2** 中 A～E 任一标准时，该种被列为相应的受威胁等级。如果根据不同标准评定的受威胁等级不同，则该种应被归于风险最高的受威胁等级。

近危 (NT)。当一物种未达到极危、濒危或易危标准，但在未来一段时间内，接近符合或可能符合受威胁等级，则该种为"近危"。

无危 (LC)。当某一物种评估为未达到极危、濒危、易危或近危标准，则该种为无危。广泛分布和个体数量多的物种都属于该等级。

数据缺乏 (DD)。缺乏足够的信息对某一物种的灭绝风险进行评估时，则该物种属于数据缺乏。

Critically Endangered (CR), **Endangered (EN)** and **Vulnerable (VU)**, these three categories are referred to as Threatened Categories, with high to low risk of extinction. When a species meets any one of the criteria A to E in **Table 2**, it is categorized as threatened. If an threatened species can be categorized differently with different criteria, the highest category should be used.

Near Threatened (NT). A taxon is Near Threatened when it has been evaluated against the criteria but does not qualify for Critically Endangered, Endangered or Vulnerable now, but is close to qualifying for or is likely to qualify for a threatened category in the near future.

Least Concern (LC). A taxon is Least Concern when it has been evaluated against the criteria and does not qualify for Critically Endangered, Endangered, Vulnerable or Near Threatened. Widespread and abundant taxa are included in this category.

Data Deficient (DD). A taxon is Data Deficient when there is inadequate information to make a assessment of its risk of extinction.

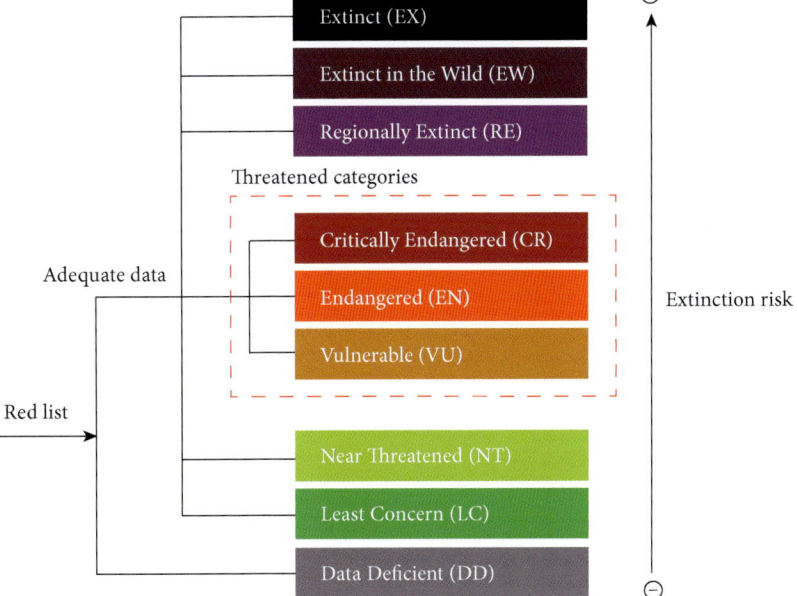

Figure 8 Levels criteria of *IUCN threatened species red List* (IUCN Standards and Petitions Subcommittee, 2017)

表 2　IUCN 红色名录濒危等级标准中 A ~ E 的生物学指标与数量阈值（自 IUCN，2012）

A. 种群数量减少 . 基于任意 A1~A4 的种群下降（测算时间超过 10 年或 3 个世代）			
	极危 CR	濒危 EN	易危 VU
A1	≥ 90%	≥ 70%	≥ 50%
A2，A3 & A4	≥ 80%	≥ 50%	≥ 30%
A1 过去 10 年或 3 个世代内种群减少的比例，其减少的原因是可逆转的且被理解和已经停止的 A2 观察、估计、推断或猜测到已经发生种群下降，这些种群下降的原因可能不会停止，或不被理解，或不可逆 A3 预期、推断或猜测到未来将会发生的种群下降（时间上限为 100 年）[易危一列的 "(a)" 不适用于此条] A4 观察、估计、推断、预测或怀疑的种群减少，其时间周期必须包括过去和未来（未来时间上限 100 年），并且这些种群下降的原因可能不会停止，或不被理解，或不可逆	基于以下任意方面：		(a) 直接观察 (A3 除外) (b) 适合该分类单元的丰富度指数 (c) 占有面积 (AOO) 减少，分布范围 (EOO) 减少和 (或) 栖息地质量下降 (d) 实际的或潜在的开发水平 (e) 外来物种、杂交、病原体、污染物、竞争者或寄生物的影响

B. 以分布范围 (B1) 和 (或) 占有面积 (B2) 体现的地理范围			
	极危 CR	濒危 EN	易危 VU
B1. 分布范围 (EOO)	< 100km²	< 5,000km²	< 20,000km²
B2. 占有面积 (AOO)	< 10km²	< 500km²	< 2,000km²
以及以下 3 个条件中的至少 2 个：			
(a) 严重片段化或分布地点数	=1	≤ 5	≤ 10
(b) 在以下方面观察、估计、推断或预期持续下降：(i) 分布范围；(ii) 占有面积；(iii) 占有面积、分布范围和 (或) 栖息地质量；(iv) 分布地点或亚种群数；(v) 成熟个体数			
(c) 以下任何方面的极度波动：(i) 分布范围；(ii) 占有面积；(iii) 分布地点或亚种群数；(iv) 成熟个体数			

C. 小种群的规模和下降情况			
	极危 CR	濒危 EN	易危 VU
成熟个体数	< 250	< 2,500	< 10,000
和至少 C1 或 C2 其一			
C1. 观察、估计或预期的持续下降的最小比例（未来时间上限 100 年）	3 年或 1 个世代内 25%（以较长时间为准）	5 年或 2 个世代内 20%（以较长时间为准）	10 年或 3 个世代内 10%（以较长时间为准）
C2. 观察、估计或预期的持续下降和至少以下 3 个条件之一			
(a)　(I) 每个亚种群中的成熟个体数	≤ 50	≤ 250	≤ 1,000
(II) 亚种群中的成熟个体数的比例 (%)	90%~100%	95%~100%	100%
(b) 成熟个体数量极度波动			

D. 种群数量极小或分布范围局限			
	极危 CR	濒危 EN	易危 VU
D. 成熟个体数	< 50	< 250	D1. < 1,000
D2 仅适用于易危等级 占有有限区域的面积或分布点数目，并在未来很短时间内有一个可信的、可能驱动该分类单元走向极危或灭绝的威胁	—	—	D2. 一般情况下： AOO < 20km² 或 分布点数目 ≤ 5

E. 定量分析			
	极危 CR	濒危 EN	易危 VU
使用定量模型评估的野外灭绝率：	未来 10 年或 3 代内 ≥ 50%（以较长时间为准，上限为 100 年）	未来 20 年或 5 代内 ≥ 20%（以较长时间为准，上限为 100 年）	未来 100 年内 ≥ 10%

Table 2 Summary of the five criteria (A~E) used to evaluate the status of a taxon (IUCN, 2012)

A. Population size reduction. Population reduction (measured over the longer of 10 years or 3 generations) based on any of A1 to A4			
	Critically Endangered	**Endangered**	**Vulnerable**
A1	≥ 90%	≥ 70%	≥ 50%
A2, A3 & A4	≥ 80%	≥ 50%	≥ 30%
A1 Population reduction observed, estimated, inferred, or suspected in the past where the causes of the reduction are clearly reversible AND understood AND have ceased. A2 Population reduction observed, estimated, inferred, or suspected in the past where the causes of reduction may not have ceased OR may not be understood OR may not be reversible. A3 Population reduction projected, inferred or suspected to be met in the future (up to a maximum of 100 years) [*(a) cannot be used for A3*] A4 An observed, estimated, inferred, projected or suspected population reduction where the time period must include both the past and the future (up to a max. of 100 years in future), and where the causes of reduction may not have ceased OR may not be understood OR may not be reversible.	*Based on any of the following:*		(a) direct observation (*except A3*) (b) an index of abundance appropriate to the taxon (c) a decline in area of occupancy (AOO), extent of occurrence (EOO) and/or habitat quality (d) actual or potential levels of exploitation (e) effects of introduced taxa, hybridization, pathogens, pollutants, competitors or parasites
B. Geographic range in the form of either B1 (extent of occurrence) AND/OR B2 (area of occupancy)			
	Critically Endangered	**Endangered**	**Vulnerable**
B1. Extent of occurrence (EOO)	< 100km^2	< 5,000km^2	< 20,000km^2
B2. Area of occupancy (AOO)	< 10km^2	< 500km^2	< 2,000km^2
AND at least 2 of the following 3 conditions:			
(a) Severely fragmented OR Number of locations	=1	≤ 5	≤ 10
(b) Continuing decline observed, estimated, inferred, or projected in any of: (i) extent of occurrence; (ii) area of occupancy; (iii) area, extent and/or quality of habitat; (iv) number of locations or subpopulations; (v) number of mature individuals			
(c) Extreme fluctuations in any of: (i) extent of occurrence; (ii) area of occupancy; (iii) number of locations or subpopulations; (iv) numbers of mature individuals			
C. Small population size and decline			
	Critically Endangered	**Endangered**	**Vulnerable**
Number of mature individuals	< 250	< 2,500	< 10,000
AND at least one of C1 or C2			
C1. An observed, estimated or projected continuing decline of at least (up to a max of 100 years in future)	25% in 3 years or 1 generation (whichever is longer)	20% in 5 years or 2 generations (whichever is longer)	10% in 10 years or 3 generations (whichever is longer)
C2. An observed, estimated, projected or inferred continuing decline AND at least 1 of the following 3 conditions:			
(a) (I) Number of mature individuals in each subpopulation	≤ 50	≤ 250	≤ 1,000
(a) (II) % of mature individuals in one subpopulation =	90%~100%	95%~100%	100%
(b) Extreme fluctuations in the number of mature individuals			
D. Very small or restricted population			
	Critically Endangered	**Endangered**	**Vulnerable**
D. Number of mature individuals	< 50	< 250	D1. < 1,000
D2 *Only applies to the VU category* Restricted area of occupancy or number of locations with a plausible future threat that could drive the taxon to CR or EX in a very short time	—	—	D2. Typically: AOO < 20km^2 or number of locations ≤ 5
E. Quantitative Analysis			
	Critically Endangered	Endangered	Vulnerable
Indicating the probability of extinction in the wild to be:	≥ 50% in 10 years or 3 generations, whichever is longer (100 years max)	≥ 20% in 20 years or 5 generations, whichever is longer (100 years max)	≥ 10% in 100 years

8 建立数据库

在中国鸟类名录的基础上,汇总国家重点保护鸟类名录、《中国濒危动物红皮书·鸟类》(郑光美和王岐山,1998)、《中国物种红色名录》(第一、第二卷)(汪松和解焱,2009)、《IUCN受威胁物种红色名录》(http://www.iucnredlist.org)、CITES公约附录鸟类名录(CITES Appendix: http://www.cites.org.cn/database/)、*Threatened Birds of Asia: The BirdLife International Red Data Book* (BirdLife International,2001)等(图9),列出了各受威胁鸟种评估的主要评判依据,建立中国鸟类评估数据库。

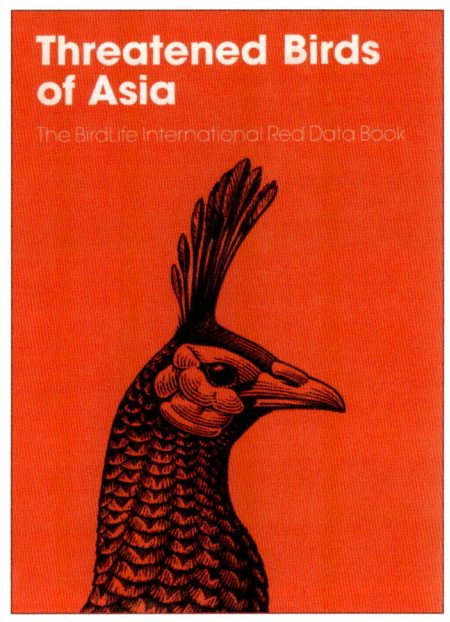

图9 鸟类红色名录评估的重要参考书

Figure 9 Some key reference books for evaluation of the Red List of Birds

9 初步评定

评估工作组利用上述数据库,结合文献和专家掌握的信息,辅以观鸟记录,依据IUCN红色名录评定标准,提出中国鸟类红色名录初步名单。

10 书面评估和会议评审

专家组召开讨论会,依据IUCN标准,对工作组提出的中国受威胁鸟类初步名单进行审议,对210种受威胁鸟类进行书面评估,并建议对情况不明、潜在受威胁的另外68个鸟种,在查阅文献和观鸟报告等资料之后进行书面评估。

专家组成员根据熟悉的区系或类群选择评估种类,将210个物种分配给鸟类学专家进行书面评估。各个专家通过查阅文献和有关数据库,并结合多年的研究积累,按照IUCN红色名录评定标准对每一个物种进行定量评估并编写评估说明书。对专家组提出的68种潜在受威胁鸟类,由熟悉国内观鸟活动的学者核查文献、比对观鸟数据,进行书面评估。每一个物种的评估书内容包括分类地位、评估信息、地理分布、种群、栖息地、威胁因子、保护行动、参考文献等。

总论 General Introduction

8 Setting up the Database

On the basis of the checklist of birds in China, combined with the list of national key protected birds, *China Red Data Book of Endangered Animals* (Zheng and Wang, 1998), *China Species Red List* (volumes I-II) (Wang and Xie, 2009), *IUCN Red List of Threatened Species* (http://www.iucnredlist.org), CITES Appendix (http://www.cites.org.cn/database/), and the *Threatened Birds of Asia: The Bird Life International Red Data Book* (BirdLife International, 2001) (**Figure 9**), the database of threatened birds in China was built.

9 A Preliminary Assessment

According to the above database, and combined the information from literatures and experts, as well as bird watching records, the working group proposed a preliminary red list of the birds in China on the basis of the IUCN Red List criteria.

10 Written Evaluation and Meeting Review

Based on the IUCN criteria, the expert group held a meeting to review the preliminary list, submitted by the working group, proposed an assessment of 210 species of threatened birds in writing, and suggested to conduct an assessment in writing of another 68 species with unclear status and potentially threatened status after reviewing literatures and birdwatching reports.

Members of experts group picked up species based on their familiar fauna or taxa and assigned 210 species to ornithologists for written assessment. By reviewing literatures and related databases and combing with accumulation of years of research, every ornithologist assessed the species quantitatively and compiled assessment reports according to the criteria of IUCN Red List. Ornithologists who are familiar with the domestic bird watching activities assessed the 68 potential threatened species through literature reviewing

再将专家书面评估意见发给熟悉相关物种的其他鸟类学专家进行复核。依据评估意见，整理出中国鸟类红色名录，提交中国脊椎动物红色名录中期验收专家组审议。

中国脊椎动物红色名录中期验收专家组在充分肯定我们工作的同时，给出如下建议：进一步加强对中国特有鸟类的关注，参照其他脊椎动物类群，鸟类保护等级可以适当提升。专家组通过会议讨论的方式，对中国鸟类红色名录逐一做了最后的审核和修订，形成了中国鸟类红色名录终稿。

11 评估修订

在中国鸟类红色名录编制后期，由国内多位知名学者历时一年编写的《中国鸟类分类与分布名录》(第三版)已经定稿，该书采纳最新鸟类分类学研究成果，分类系统有很大变化。本评估的物种名称和分类系统即基于该书，共对 1,445 种鸟类进行了评估，中国特有鸟类的评估也依照《中国鸟类分类与分布名录》(第三版)。

12 评估结果

12.1 总体情况

对 1,445 种鸟类的评估结果表明：白鹳 (*Ciconia ciconia*)、镰翅鸡 (*Falcipennis falcipennis*)、赤颈鹤 (*Grus antigone*)3 种区域灭绝 (RE)；极危 (CR) 18 种，濒危 (EN) 48 种，易危 (VU)78 种，近危 (NT) 223 种，无危 (LC) 969 种，数据缺乏 (DD) 105 种，无效种 1 种。依照 IUCN 红色名录标准，列入极危、濒危、易危等级的种类为受威胁鸟类合计 144 种，受威胁比例占全国鸟类种数的 10.0% (图 10)，低于 IUCN 红色名录评估的全球鸟

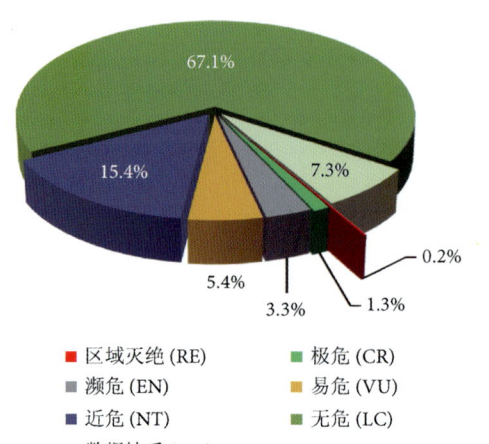

图 10 各评估等级的鸟种占中国鸟类种数的比例

and comparing with bird watching data. The assessment report of each species includes taxonomic status, assessment information, geographical distribution, population, habitat, threats, conservation actions, references, *etc*.

The expert's assessment report has been sent to other ornithologists to recheck. According to the evaluation opinion of recheck, the working group compiled the China bird species red list, and submitted to the midterm acceptance panel of experts of the China vertebrate species red list for reviewing.

The midterm acceptance panel of experts of the China vertebrate species red list highly appreciated our work, at the meantime, they also recommended to promote the attention of endemic bird species, and to upgrade the protection class referring to other vertebrate. Based on this, ornithologists discussed, reviewed and revised the list one by one, and finalized the China bird species red list.

11 Checks and Revision

In the late stage of compiling, the third edition of *A Checklist on the Classification and Distribution of the Birds of China* was finalized with the updated research result of bird taxonomy, in which the classification system has greatly changed. The species name and classification system of the assessment was based on the checklist and 1,445 species have been assessed in total, and the assessment of endemic species is also based on it.

12 Evaluation Results

12.1 General situation

The assessment of 1,445 bird species shows: three species such as, White Stork (*Ciconia ciconia*), Siberian Grouse (*Falcipennis falcipennis*), Sarus Crane (*Grus antigone*) are Regionally Extinct (RE); 18 species are Critically Endangered (CR), 48 species are Endangered (EN), 78 species are Vulnerable (VU), 223 species are Near Threatened (NT), 969 species are Least Concern (LC), and 105 species are Data Deficient (DD), and one invalid species. Based on the criteria of IUCN Red List, 144 species have been listed as CR, EN, or VU. The national threatened rate is 10.0% (**Figure 10**), which is lower than the global

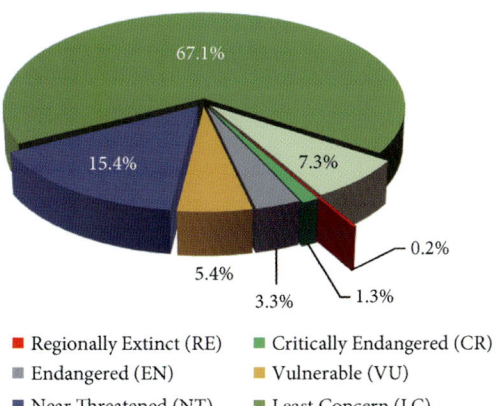

Figure 10 The proportion of bird species in each assessment category in China

类受威胁比例 (13.7%)，但高于 2009 年出版的《中国物种红色名录》(第 2 卷 脊椎动物，下册)(汪松和解焱，2009) 中受威胁鸟类比例 (7.5%)。

本次评估结果显示，我国受威胁鸟类为 144 种 (不包括 3 种区域灭绝鸟类)，占全国鸟类种数的 10.0%。实际上，列为近危的 223 种鸟类也处于受威胁边缘；另有 105 种鸟类列为数据缺乏，这些鸟类中由于数量、分布等信息不详而未列入"中国鸟类红色名录"的受威胁等级 (极危、濒危和易危)，其实际生存状况可能很差，需引起高度关注，即所评估的中国 1,445 种鸟类中，有 475 种 (32.9%) 需要关注。

12.2 不同类群受威胁状况

中国鸟类中各目鸟类受威胁状况见**表 3**，受威胁种类最多的是鸡形目，有 21 种受威胁鸟类。受威胁物种比例最高的前 10 个科见**表 4**，受威胁程度最高的是鹈鹕科 (Pelecanidae) 和犀鸟科 (Bucerotidae)，这 2 个科所有的物种均处于受威胁状态。

表 3　中国鸟类各目受威胁物种数

目	物种数	受威胁物种数	受威胁物种占比 /%
鸡形目	64	21	33
雁形目	54	8	15
䴙䴘目	5	0	0
红鹳目	1	0	0
鸽形目	31	3	10
沙鸡目	3	0	0
夜鹰目	21	1	5
鹃形目	20	0	0
鸨形目	3	2	67
鹤形目	29	10	34
鸻形目	134	10	7
鹲形目	3	0	0
潜鸟目	4	0	0
鹱形目	16	1	6
鹳形目	7	2	29
鲣鸟目	11	0	0
鹈形目	35	8	23
鹰形目	55	14	25
鸮形目	32	5	16
咬鹃目	3	0	0
犀鸟目	6	5	83
佛法僧目	23	2	9
啄木鸟目	43	1	2
隼形目	12	3	25
鹦鹉目	9	2	22
雀形目	820	48	6

bird threatened rate (13.7%), but is higher than the rate (7.5%) in the *China Species Red List* (volumes II) (Wang and Xie, 2009).

The result of the assessment shows that there are 144 threatened species in China (three RE species excluded), accounting for 10.0% of the bird species in China. In fact, 223 species listed as NT are also on the edge of being threatened. And 105 species listed as DD are not included due to data deficiency of population size and distribution range, while their actual situation might be quite poor and need to be paid more attention. In summary, 32.9% (475 species) of the 1,445 species should be highly concerned.

12.2 Threatened status of different groups

The threatened status of birds of each order in China is shown in **Table 3**. The most threatened order is Galliformes, and there are 21 threatened species. The most threatened ten families are listed in **Table 4**, especially for Bucerotidae and Pelecanidae, and all species in both families are threatened.

Table 3 The threatened species of birds of each order in China

Order	Number of species	Number of threatened species	Percentage of threatened species (%)
Galliformes	64	21	33
Anseriformes	54	8	15
Podicipediformes	5	0	0
Phoenicopteriformes	1	0	0
Columbiformes	31	3	10
Pterocliformes	3	0	0
Caprimulgiformes	21	1	5
Cuculiformes	20	0	0
Otidiformes	3	2	67
Gruiformes	29	10	34
Charadriiformes	134	10	7
Phaethontiformes	3	0	0
Gaviiformes	4	0	0
Procellariiformes	16	1	6
Ciconiiformes	7	2	29
Suliformes	11	0	0
Pelecaniformes	35	8	23
Accipitriformes	55	14	25
Strigiformes	32	5	16
Trogoniformes	3	0	0
Bucerotiformes	6	5	83
Coraciiformes	23	2	9
Piciformes	43	1	2
Falconiformes	12	3	25
Psittaciformes	9	2	22
Passeriformes	820	48	6

表4 中国鸟类受威胁程度最高的前10个科

科	种数	受威胁种数	受威胁比例/%
鹈鹕科	3	3	100
犀鸟科	5	5	100
八色鸫科	8	6	75
鸨科	3	2	67
鹤科	9	5	56
鹦科	6	3	50
信天翁科	3	1	33
鸠科	12	4	33
雉科	64	21	33
鹳科	7	2	29

12.3 中国特有鸟类受威胁状况

92种（不含中亚夜鹰）中国特有鸟类中有29种被列入受威胁等级，占31.5%，远超过全国受威胁鸟类比例（表5）。中国特有鸟类中，蓝冠噪鹛 (*Garrulax courtoisi*) 和海南孔雀雉 (*Polyplectron katsumatae*) 2种列为极危，濒危8种，易危19种，近危21种，无危41种，褐头朱雀 (*Carpodacus sillemi*) 列为数据缺乏。

中国特有鸟类中，蓝冠噪鹛和海南孔雀雉列为极危，蓝冠噪鹛繁殖地面积不足10 km^2，种群数量小于200只，且繁殖点有减少的趋势。海南孔雀雉种群数量少，由于非法捕猎压力和栖息地的丧失，在过去的20年间种群数量下降非常严重，且下降趋势没有得到有效遏制。

8种濒危鸟类中，海南山鹧鸪 (*Arborophila ardens*)、绿尾虹雉 (*Lophophorus lhuysii*)、四川山鹧鸪 (*Arborophila rufipectus*)、黄腹角雉 (*Tragopan caboti*)、白冠长尾雉 (*Syrmaticus reevesii*) 为我国特产鸡形目鸟类；贺兰山红尾鸲 (*Phoenicurus alaschanicus*)、弄岗穗鹛 (*Stachyris nonggangensis*)、灰冠鸦雀 (*Sinosuthora przewalskii*) 3种属于种群数量少、狭域分布的雀形目鸟类。

19种易危鸟类中，包括5种雉科鸟类、4种画眉科鸟类，它们属于对原生栖息地依赖性强、扩散能力差的物种，其他10种分属于9科。其他特有鸟类中，有17种近危，29种无危（其中12种为我国台湾地区特有），褐头朱雀 (*Carpodacus sillemi*) 被列为数据缺乏的鸟种。

Table 4 Top ten Families with the highest proportion of threatened bird species in China

Family	Number of species	Number of threatened species	Percentage of threatened species (%)
Pelecanidae	3	3	100
Bucerotidae	5	5	100
Pittidae	8	6	75
Otididae	3	2	67
Gruidae	9	5	56
Threskiornithidae	6	3	50
Diomedeidae	3	1	33
Sittidae	12	4	33
Phasianidae	64	21	33
Ciconiidae	7	2	29

12.3 Threatened conditions of endemic birds in China

29 out of 92 endemic species (Varie's Nightjar excluded) are threatened, accounting for 31.5%, which is greatly higher than the rate of all threatened bird species in China (**Table 5**). Two species, Blue-crowned Laughingthrush (*Garrulax courtoisi*) and Hainan Peacock Pheasant (*Polyplectron katsumatae*), are Critically Endangered, 8 species are Endangered, 19 species are Vulnerable, 21 species are Near Threatened, 41 species are Least Concern.

In the endemic species in China, Blue-crowned Laughingthrush and Hainan Peacock Pheasant are listed as CR. The breeding area of *G. courtoisi* is less than 10 km^2 and disappearing, and its population is less than 200. The population size of Hainan Peacock Pheasant is small and undergoing a rapid decline in the past 20 years owing to habitat loss and illegal hunting, while the decreasing trend has not been effectively contained.

In 8 EN species, Hainan Hill Partridge (*Arborophila ardens*), Chinese Monal (*Lophophorus lhuysii*), Sichuan Hill Partridge (*Arborophila rufipectus*), Cabot's Tragopan (*Tragopan caboti*), Reeves's Pheasant (*Syrmaticus reevesii*) are endemic Galliformes; Alashan Redstart (*Phoenicurus alaschanicus*), Nonggang Babbler (*Stachyris nonggangensis*), Rusty-throated Parrotbill (*Sinosuthora przewalskii*) are Passeriformes with narrow distribution area and small population size.

In 19 VU species, there are five species from Phasianidae and four species from Timaliidae, which are strongly dependent on the original habitat with poor dispersal ability; and the other 10 species fall into nine families. There are 17 NT species, 29 LC species (12 species were endemic in Taiwan, China) and one DD species Sillem's Rosefinch (*Carpodacus sillemi*).

表5 中国特有鸟类受威胁状况

目	科	物种中文名	物种学名	受威胁等级
鸡形目	雉科	四川山鹧鸪	*Arborophila rufipectus*	EN
		白眉山鹧鸪	*Arborophila gingica*	VU
		台湾山鹧鸪	*Arborophila crudigularis*	NT
		海南山鹧鸪	*Arborophila ardens*	EN
		斑尾榛鸡	*Tetrastes sewerzowi*	NT
		红喉雉鹑	*Tetraophasis obscurus*	VU
		黄喉雉鹑	*Tetraophasis szechenyii*	VU
		大石鸡	*Alectoris magna*	NT
		灰胸竹鸡	*Bambusicola thoracicus*	LC
		台湾竹鸡	*Bambusicola sonorivox*	LC
		黄腹角雉	*Tragopan caboti*	EN
		绿尾虹雉	*Lophophorus lhuysii*	EN
		蓝腹鹇	*Lophura swinhoii*	NT
		白马鸡	*Crossoptilon crossoptilon*	NT
		藏马鸡	*Crossoptilon harmani*	NT
		褐马鸡	*Crossoptilon mantchuricum*	VU
		蓝马鸡	*Crossoptilon auritum*	NT
		白颈长尾雉	*Syrmaticus ellioti*	VU
		黑长尾雉	*Syrmaticus mikado*	NT
		白冠长尾雉	*Syrmaticus reevesii*	EN
		红腹锦鸡	*Chrysolophus pictus*	NT
		海南孔雀雉	*Polyplectron katsumatae*	CR
鸮形目	鸱鸮科	四川林鸮	*Strix davidi*	VU
啄木鸟目	拟啄木鸟科	台湾拟啄木鸟	*Psilopogon nuchalis*	LC
雀形目	鸦科	黑头噪鸦	*Perisoreus internigrans*	VU
		台湾蓝鹊	*Urocissa caerulea*	LC
		白尾地鸦	*Podoces biddulphi*	VU
	山雀科	黄腹山雀	*Pardaliparus venustulus*	LC
		台湾杂色山雀	*Sittiparus castaneoventris*	LC
		白眉山雀	*Poecile superciliosus*	NT
		红腹山雀	*Poecile davidi*	LC
		四川褐头山雀	*Poecile weigoldicus*	NT
		地山雀	*Pseudopodoces humilis*	LC
		台湾黄山雀	*Machlolophus holsti*	LC
	鳞胸鹪鹛科	台湾鹪鹛	*Pnoepyga formosana*	LC
	蝗莺科	台湾短翅蝗莺	*Locustella alishanensis*	LC
		四川短翅蝗莺	*Locustella chengi*	NT
	鹎科	台湾鹎	*Pycnonotus taivanus*	VU

Table 5 Threatened status of endemic birds in China

Order	Family	English Names of Species	Scientific Names of Species	Threatened categories
GALLIFORMES	Phasianidae	Sichuan Hill Partridge	Arborophila rufipectus	EN
		White-necklaced Hill Partridge	Arborophila gingica	VU
		Taiwan Hill Partridge	Arborophila crudigularis	NT
		Hainan Hill Partridge	Arborophila ardens	EN
		Chinese Grouse	Tetrastes sewerzowi	NT
		Chestnut-throated Partridge	Tetraophasis obscurus	VU
		Buff-throated Partridge	Tetraophasis szechenyii	VU
		Rusty-necklaced Partridge	Alectoris magna	NT
		Chinese Bamboo Partridge	Bambusicola thoracicus	LC
		Taiwan Bamboo Partridge	Bambusicola sonorivox	LC
		Cabot's Tragopan	Tragopan caboti	EN
		Chinese Monal	Lophophorus lhuysii	EN
		Swinhoe's Pheasant	Lophura swinhoii	NT
		White Eared Pheasant	Crossoptilon crossoptilon	NT
		Tibetan Eared Pheasant	Crossoptilon harmani	NT
		Brown Eared Pheasant	Crossoptilon mantchuricum	VU
		Blue Eared Pheasant	Crossoptilon auritum	NT
		Elliot's Pheasant	Syrmaticus ellioti	VU
		Mikado Pheasant	Syrmaticus mikado	NT
		Reeves's Pheasant	Syrmaticus reevesii	EN
		Golden Pheasant	Chrysolophus pictus	NT
		Hainan Peacock Pheasant	Polyplectron katsumatae	CR
STRIGIFORMES	Strigidae	Sichuan Wood Owl	Strix davidi	VU
PICIFORMES	Megalaimidae	Taiwan Barbet	Psilopogon nuchalis	LC
PASSERIFORMES	Corvidae	Sichuan Jay	Perisoreus internigrans	VU
		Taiwan Blue Magpie	Urocissa caerulea	LC
		Xinjiang Ground Jay	Podoces biddulphi	VU
	Paridae	Yellow-bellied Tit	Pardaliparus venustulus	LC
		Chestnut-bellied Tit	Sittiparus castaneoventris	LC
		White-browed Tit	Poecile superciliosus	NT
		Rusty-breasted Tit	Poecile davidi	LC
		Sichuan Tit	Poecile weigoldicus	NT
		Ground Tit	Pseudopodoces humilis	LC
		Yellow Tit	Machlolophus holsti	LC
	Pnoepygidae	Taiwan Wren-Babbler	Pnoepyga formosana	LC
	Locustellidae	Taiwan Bush Warbler	Locustella alishanensis	LC
		Sichuan Bush Warbler	Locustella chengi	NT
	Pycnonotidae	Styan's Bulbul	Pycnonotus taivanus	VU

续表

目	科	物种中文名	物种学名	受威胁等级
雀形目	柳莺科	甘肃柳莺	*Phylloscopus kansuensis*	LC
		峨眉柳莺	*Phylloscopus emeiensis*	LC
		海南柳莺	*Phylloscopus hainanus*	VU
	长尾山雀科	银喉长尾山雀	*Aegithalos glaucogularis*	LC
		银脸长尾山雀	*Aegithalos fuliginosus*	LC
		凤头雀莺	*Leptopoecile elegans*	NT
	莺鹛科	宝兴鹛雀	*Moupinia poecilotis*	LC
		中华雀鹛	*Fulvetta striaticollis*	LC
		三趾鸦雀	*Cholornis paradoxus*	NT
		白眶鸦雀	*Sinosuthora conspicillata*	NT
		暗色鸦雀	*Sinosuthora zappeyi*	VU
		灰冠鸦雀	*Sinosuthora przewalskii*	EN
	绣眼鸟科	褐头凤鹛	*Yuhina brunneiceps*	LC
	林鹛科	华南斑胸钩嘴鹛	*Erythrogenys swinhoei*	LC
		台湾斑胸钩嘴鹛	*Erythrogenys erythrocnemis*	LC
		台湾棕颈钩嘴鹛	*Pomatorhinus musicus*	LC
		弄岗穗鹛	*Stachyris nonggangensis*	EN
	幽鹛科	金额雀鹛	*Schoeniparus variegaticeps*	VU
	噪鹛科	棕草鹛	*Babax koslowi*	NT
		海南画眉	*Garrulax owstoni*	NT
		台湾画眉	*Garrulax taewanus*	NT
		黑额山噪鹛	*Garrulax sukatschewi*	VU
		斑背噪鹛	*Garrulax lunulatus*	LC
		白点噪鹛	*Garrulax bieti*	VU
		大噪鹛	*Garrulax maximus*	LC
		台湾白喉噪鹛	*Garrulax ruficeps*	LC
		蓝冠噪鹛	*Garrulax courtoisi*	CR
		山噪鹛	*Garrulax davidi*	LC
		棕噪鹛	*Garrulax berthemyi*	LC
		台湾棕噪鹛	*Garrulax poecilorhynchus*	LC
		橙翅噪鹛	*Trochalopteron elliotii*	LC
		灰腹噪鹛	*Trochalopteron henrici*	LC
		台湾噪鹛	*Trochalopteron morrisonianum*	LC
		灰胸薮鹛	*Liocichla omeiensis*	VU
		黄痣薮鹛	*Liocichla steerii*	LC
		台湾斑翅鹛	*Sibia morrisoniana*	LC
		白耳奇鹛	*Heterophasia auricularis*	LC

continued

Order	Family	English Names of Species	Scientific Names of Species	Threatened categories
Passeriformes	Phylloscopidae	Gansu Leaf Warbler	*Phylloscopus kansuensis*	LC
		Emei Leaf Warbler	*Phylloscopus emeiensis*	LC
		Hainan Leaf Warbler	*Phylloscopus hainanus*	VU
	Aegithalidae	Silver-throated Bushtit	*Aegithalos glaucogularis*	LC
		Sooty Bushtit	*Aegithalos fuliginosus*	LC
		Crested Tit Warbler	*Leptopoecile elegans*	NT
	Sylviidae	Rufous-tailed Babbler	*Moupinia poecilotis*	LC
		Chinese Fulvetta	*Fulvetta striaticollis*	LC
		Three-toed Parrotbill	*Cholornis paradoxus*	NT
		Spectacled Parrotbill	*Sinosuthora conspicillata*	NT
		Grey-hooded Parrotbill	*Sinosuthora zappeyi*	VU
		Rusty-throated Parrotbill	*Sinosuthora przewalskii*	EN
	Zosteropidae	Taiwan Yuhina	*Yuhina brunneiceps*	LC
	Timaliidae	Grey-sided Scimitar-Babbler	*Erythrogenys swinhoei*	LC
		Black-necked Scimitar-Babbler	*Erythrogenys erythrocnemis*	LC
		Taiwan Scimitar Babbler	*Pomatorhinus musicus*	LC
		Nonggang Babbler	*Stachyris nonggangensis*	EN
	Pellorneidae	Golden-fronted Fulvetta	*Schoeniparus variegaticeps*	VU
	Leiothrichidae	Tibetan Babax	*Babax koslowi*	NT
		Hainan Hwamei	*Garrulax owstoni*	NT
		Taiwan Hwamei	*Garrulax taewanus*	NT
		Snowy-cheeked Laughingthrush	*Garrulax sukatschewi*	VU
		Barred Laughingthrush	*Garrulax lunulatus*	LC
		White-speckled Laughingthrush	*Garrulax bieti*	VU
		Giant Laughingthrush	*Garrulax maximus*	LC
		Rufous-crowned Laughingthrush	*Garrulax ruficeps*	LC
		Blue-crowned Laughingthrush	*Garrulax courtoisi*	CR
		Plain Laughingthrush	*Garrulax davidi*	LC
		Buffy Laughingthrush	*Garrulax berthemyi*	LC
		Rusty Laughingthrush	*Garrulax poecilorhynchus*	LC
		Elliot's Laughingthrush	*Trochalopteron elliotii*	LC
		Brown-cheeked Laughingthrush	*Trochalopteron henrici*	LC
		White-whiskered Laughingthrush	*Trochalopteron morrisonianum*	LC
		Emei Shan Liocichla	*Liocichla omeiensis*	VU
		Steere's Liocichla	*Liocichla steerii*	LC
		Taiwan Barwing	*Sibia morrisoniana*	LC
		White-eared Sibia	*Heterophasia auricularis*	LC

续表

目	科	物种中文名	物种学名	受威胁等级
雀形目	旋木雀科	四川旋木雀	*Certhia tianquanensis*	VU
	鸸科	滇䴓	*Sitta yunnanensis*	VU
	鸫科	乌鸫	*Turdus mandarinus*	LC
		宝兴歌鸫	*Turdus mupinensis*	LC
	鹟科	台湾林鸲	*Tarsiger johnstoniae*	LC
		贺兰山红尾鸲	*Phoenicurus alaschanicus*	EN
		台湾紫啸鸫	*Myophonus insularis*	LC
	戴菊科	台湾戴菊	*Regulus goodfellowi*	LC
	朱鹀科	朱鹀	*Urocynchramus pylzowi*	NT
	雀科	藏雪雀	*Montifringilla henrici*	NT
	燕雀科	褐头朱雀	*Carpodacus sillemi*	DD
		中华朱雀	*Carpodacus davidianus*	LC
		台湾酒红朱雀	*Carpodacus formosanus*	NT
		藏雀	*Carpodacus roborowskii*	VU
		斑翅朱雀	*Carpodacus trifasciatus*	LC
	鹀科	蓝鹀	*Emberiza siemsseni*	LC
		藏鹀	*Emberiza koslowi*	VU

13 与 IUCN 红色名录的比较

IUCN 对全球 11,147 种鸟类进行了评估，包括 EX 159 种、EW 5 种、CR 225 种、EN 461 种、VU 800 种、NT 1,017 种、LC 8,427 种、DD 53 种 (Handbook of the Birds of the World and BirdLife International, 2019)。鸟类受威胁物种占评估物种数量的 13%，而哺乳类和两栖类的受威胁比例约为 25% 和 41%，鱼类和爬行类评估物种比例相对较少，目前评估显示的受威胁物种比例尚不能反映这两个纲的受威胁状况 (IUCN, 2019)。

我国鸟类受威胁比例占全国鸟类种数的 10.0%，低于 IUCN 红色名录评估的全球鸟类受威胁比例。本次评估中列入 DD 的鸟种中，有 24 种鸟类被 IUCN 分别列为 CR (6 种)、EN (2 种) 和 VU (16 种)。

IUCN 评定中国极危鸟类 14 种，本次评估中确定为 18 种，其中有 8 个共同的物种，IUCN 评估确定的其他 6 个极危种，如黄颊麦鸡 (*Vanellus gregarius*)、白肩黑鹮 (*Pseudibis davisoni*)、白腹军舰鸟 (*Fregata andrewsi*) 等，仅分布于我国藏南、云南或沿海等边境地区，资料不足，本次评估均列为 DD。

在 IUCN 评估的 28 种 EN 鸟类中，本次评估中的 15 种属于 EN 等级，4 种属于 CR，3 种因为资料缺乏定为 DD，6 种被评估为 VU 或 NT。

continued

Order	Family	English Names of Species	Scientific Names of Species	Threatened categories
Passeriformes	Certhiidae	Sichuan Treecreeper	*Certhia tianquanensis*	VU
	Sittidae	Yunnan Nuthatch	*Sitta yunnanensis*	VU
	Turdidae	Chinese Blackbird	*Turdus mandarinus*	LC
		Chinese Thrush	*Turdus mupinensis*	LC
	Muscicapidae	Collared Bush Robin	*Tarsiger johnstoniae*	LC
		Ala Shan Redstart	*Phoenicurus alaschanicus*	EN
		Taiwan Whistling Thrush	*Myophonus insularis*	LC
	Regulidae	Flamecrest	*Regulus goodfellowi*	LC
	Urocynchramidae	Pink-tailed Rosefinch	*Urocynchramus pylzowi*	NT
	Passeridae	Henri's Snowfinch	*Montifringilla henrici*	NT
	Fringillidae	Sillem's Mountain Finch	*Carpodacus sillemi*	DD
		Chinese Beautiful Rosefinch	*Carpodacus davidianus*	LC
		Taiwan Rosefinch	*Carpodacus formosanus*	NT
		Tibetan Rosefinch	*Carpodacus roborowskii*	VU
		Three-banded Rosefinch	*Carpodacus trifasciatus*	LC
	Emberizidae	Slaty Bunting	*Emberiza siemsseni*	LC
		Tibetan Bunting	*Emberiza koslowi*	VU

13 Comparison with IUCN Red List

11,147 species of birds have been assessed by IUCN, including 159 EX, 5 EW, 225 CR, 461 EN, 800 VU, 1,017 NT, 8,427 LC, and 53 DD (Handbook of the Birds of the World and BirdLife International, 2019). Threatened species accounted for 13% of the total number of birds, and the proportions of threatened species of mammals and amphibians are approximately 25% and 41%, respectively. As not enough fish and reptile species have been assessed, the proportion of species is insufficient to reflect the threatened status of these two groups (IUCN, 2019).

It is indicated that 10.0% of the bird species assessed are threatened in China, which is lower than that in IUCN Red List, mainly due to that those 24 DD species are listed as CR (6 species), EN (2 species) and VU (16 species) by IUCN respectively.

14 species of birds were listed as CR in China by IUCN, while 18 species were listed as CR in this assessment. 8 species are the same in both lists, while 6 species (*e.g.* Sociable Lapwing *Vanellus gregarious*, White-shouldered Ibis *Pseudibis davisoni*, Christmas Island Frigatebird *Fregata andrewsi*) in IUCN Red List was assessed as DD with only occurance in the border region of China, *e.g.* south of Tibet, Yunnan or coastal areas.

Of the 28 EN species assessed by IUCN Red List, 15 species were assessed as EN, 4 species CR, 3 species DD and 6 species VU or NT in this assessment.

IUCN 评估中，只有褐头朱雀和中亚夜鹰评估为 DD。褐头朱雀是 20 世纪 90 年代初期，根据 1929 年在新疆西南部近喀喇昆仑山口采集到 1 只成鸟和 1 只雏鸟标本命名 (Roselaar, 1992)，2012 年 6 月，法国野生生物和自然摄影师 Yann Muzika 在青海海西的野牛沟中拍摄到了雄鸟和雌鸟，进一步确认了该物种的有效性。中亚夜鹰是 Vaurie(1960) 据新疆皮山县所采 1 只雌鸟标本命名的，Schweizer 等 (2020) 的研究发现中亚夜鹰是欧夜鹰的同物异名，所以本次评估将中亚夜鹰列为无效种，褐头朱雀列入评估范围并定为数据缺乏；有 59 种鸟类缺乏进行准确评估的有效信息，本次评估将它们列为数据缺乏，另有 46 种鸟类在中国偶见，依照 IUCN 标准，将其列为不予评估。生态环境部组织的专家评审中，建议本次评估将不予评估和数据缺乏合称数据缺乏，故本次评估数据缺乏的鸟种有 105 种，其中 63 种为水鸟。

14 灭绝物种分析

冠麻鸭 (*Tadorna cristata*) 自从 1964 年以来再无任何记录，且诸多鸟类名录已不再将冠麻鸭列入现存鸟类中，已是公认的灭绝物种，本次评估未对该物种进行评估。

镰翅鸡 (*Dendragapus falcipennis*) 一直没有在国内收集到标本或照片证据，近年来的历次系统调查也没有发现该物种的踪迹，2001 年黑龙江省野生动物研究所宣布镰翅鸡在中国境内灭绝，该物种仅见于俄罗斯远东地区 (del Hoyo et al., 2014)，本次评估中将镰翅鸡列为区域灭绝。

中国科学院昆明动物研究所动物资源考察队曾于 1959 年和 1960 年在云南西双版纳州勐腊发现过赤颈鹤 (*Grus antigone*)，迄今再无活体记录，仅 1973 年在云南高黎贡山发现 1 具残骸 (杨岚，1987)。另外经过长期监测，马鸣 (2002) 提出白鹳 (*Ciconia ciconia*) 已在我国灭绝。鉴于这 2 个物种栖息于开阔地区，且极易识别，却长期没有记录，故将其列入区域灭绝。

15 中国鸟类主要受威胁因素

栖息地退化和丧失、捕猎和贸易、人类活动、气候变化、自然灾害等是影响我国鸟类的主要威胁因素 (郑光美，2012)。本次评估中对受威胁鸟类致危因子的分析也发现，由于森林砍伐和替代种植经济林、湿地围垦等引起的"栖息地退化和丧失"为鸟类致危因子之首。其次是食用、贸易、笼鸟饲养等从野外捕捉鸟类 (图 11)。例如，2016 年和 2017 年在全国 200 个鸟市开展的调查，记录鸟类 420 种 240,406 只，数量最大的是画眉，2016 年冬季就在鸟市调查到 23,104 只 (数据源自：朱雀会，"还野鸟自由翅膀"全国鸟市调查报告，2017)。此外，建坝、旅游等人类活动以及疫病、自然灾害、气候变化、生物入侵对鸟类造成严重影响。这些都充分表明目前中国鸟类的濒危主要是各种人类活动造成。

Only Sillem's Rosefinch and Vaurie's Nightjar were assessed as DD in IUCN Red List. Sillem's Rosefinch was named in 1929 by two specimens (one adult and one sub-adult) collected in Karakorum Passin southwest Xinjiang (Roselaar, 1992). In June 2012, French Bio-photographer Yann Muzika recorded female and male by camera in Yeniugou, Haixi of Qinghai, further confirming the species. Vaurie's Nightjar was named based on a female specimen collected in Pishan, Xinjiang (Vaurie, 1960), and it was revealed that Vaurie's Nightjar (*Caprimulgus centralasicus*) is a synonym of European Nightjar (*C. europaeus*) (Schweizer *et al.*, 2020). Therefore, Sillem's Rosefinch was evaluated as DD in this assessment. Besides, 59 species were assessed as DD owing to the data deficiency and 46 species were not assessed as rarely seen in China based on the IUCN criteria. According to the experts' opinions on reviewing workshop organized by Ministry of Ecology and Environment, those not evaluated and data deficiency were referred as DD in this assessment, and thus 105 species were assessed as DD, with 63 species of waterbirds.

14 Analysis of Extinct Species

There has been no record of *T. cristata* since 1964, and it was widely regarded as extinct and not included in most lists, therefore it was not evaluated in this assessment.

There was not any evidence of specimen or photographs collected in China, and there has been no records in the recent systematic surveys. It was reported to be extinct in China by Heilongjiang Wildlife Institute in 2001, and only found in Russian Far East area (del Hoyo *et al*, 2014). Therefore, *D. falcipennis* has been evaluated as Regionally Extinct in this assessment.

Grus antigone was recorded in 1959 and 1960 in Mengla, Xishuangbanna, Yunnan Province by Wildlife Resource Investigation Group of Kunming Institute of Zoology (CAS), while there has been no living record till now, except for only one carcass found in Gaoligong Shan in 1973 (Yang, 1987). In addition, after long term monitoring, Ma (2002) proposed that *Ciconia ciconia* had been extinct in China. These two species were listed as Regionally Extinct, considering that they inhabit open areas and are easily recognized, while no record for a long time.

15 Major Threats to Bird in China

Major threats to bird species in China include habitat loss and degradation, hunting and trading, human activities, climate change and natural disasters (Zheng, 2012). In this assessment, the most serious threat was habitat degradation and loss owing to deforestation, economic forests plantation, and wetland reclamation, followed with hunting in the wild for food, trading or as pets (**Figure 11**). *e.g.*, in 2016 and 2017, a total of 240,406 individuals of 420 species were recorded in a survey of 200 bird markets nationwide. The top number was Hwamei (*Garrulax canorus*) with 23,104 individuals recorded in the bird markets in a single winter of 2016 (National Bird Market Survey Report, 2017). In addition, human activities such as dam construction and tourism, as well as diseases, natural disasters, climate change and biological invasion were all among the major threats. The assessment indicates that most of the threats were induced by activities of human.

图 11　被捕猎的白鹇（李立摄）

Figure 11　Hunted Silver Pheasant (Photographed by Li Li)

本次评估的所有受威胁鸟类类群中，受威胁比例最高的为犀鸟目 (Bucerotiformes) 鸟类。其种群密度与栖息地内大树的数量有一定的相关性（周放和周解，2004；James and Kannan，2009；罗益奎等，2013），而犀鸟栖息地同时也适宜种植香蕉、橡胶和甘蔗等热带作物。因此砍伐森林，大面积种植热带作物所造成的生境破坏、退化和片断化，是其主要的威胁因子。由于体型大，飞行时的声响以及鸣声响亮，犀鸟容易被发现并被猎杀，其巨大的喙也经常作为装饰品销售，捕猎和贸易也是犀鸟的重要威胁因子。

除了栖息地丧失等原因外，非法捕杀也是鸡形目鸟类数量迅速下降的直接原因（郑光美，2015）。除了为获取"雉鸡翎"而猎捕外，当地农民为了保护庄稼而在山区小块耕地周边投放毒饵，常造成集群白冠长尾雉 (*Syrmaticus reevesii*) 中毒死亡。一些地区存在的捕猎以及人工散养种群与野生近缘种类的杂交所致的基因污染，仍是严重威胁，如目前已经无法确定哪些红原鸡 (*Gallus gallus*) 种群属于纯系种群。

湿地丧失与退化所导致的栖息地丧失和质量下降是鹤类面临的主要威胁，繁殖地的沼泽湿地和越冬地的滩涂湿地被围垦开发为农田，使其适宜的栖息地消失 (Lu et al.，2007；Kong et al.，2011；杨晓君和常云艳，2014)。滨海湿地的围垦和开发而导致的生境丧失与退化，致使我国丹顶鹤 (*Grus japonensis*) 种群数量持续下降，在其主要越冬地江苏盐城，种群数量由 20 世纪末的近 1,200 只下降到目前的 600 只左右（吕士成，2009），其他威胁还包括冬季觅食地（农田）里农药和杀虫剂使用、非法捕猎活动，以及输电线和风电场对鹤类的伤害。而传统耕作模式的改变及牧民游牧范围的扩展，使得黑颈鹤 (*Grus nigricollis*) 与当地居民冲突越来越严重（马鸣等，2014；杨晓君和常云艳，2014）。水利工程的建设等对大部分越冬个体可能会造成严重的威胁，白鹤 (*Grus leucogeranus*) 种群数量在未来很可能会急剧下降。

鸻形目 (Charadriiformes) 鸟类受威胁因素主要是滩涂环境的改变。例如，勺嘴鹬 (*Calidris pygmeus*) 全球总数量估计不超过 1,000 只 (MacKinnon et al.，2012)。江苏如东滩涂湿地是目前本种已知的南迁时数量最大的迁徙停歇地 (Tong et al.，2012)，在 2011 年秋季曾一次记录到 103 只，2013 年 10 月调查记录到的数量达 143 只，表明该区域对勺嘴鹬迁徙的重要性。滩涂围垦、开发以及外来植物互花米草 (*Spartina alterniflora*) 入侵所造成的栖息地丧失，将会是其面临的主要威胁。

In all the threatened groups, species of Bucerotiformes were most highly threatened. The population densities were related with the number of trees in the habitat (Zhou and Zhou, 2004; James and Kannan, 2009; Luo *et al.*, 2013), while such area is also suitable for tropical crops, *e.g.* banana, rubber and sugar cane. Therefore, the major threats are habitat loss, degradation and fragmentation derived from deforestation and tropical crops plantation. Meanwhile, the large body, loud noise during flight and loud and clear calling songs, have made hornbills easily to found, hunted and killed, and their large beaks are often targeted as adornments. Therefore, hunting and trade are also key threats.

In addition to habitat loss, illegal hunting is the direct cause of the rapid decline of Galliformes populations (Zheng, 2015). Besides caught for their long-tail feathers to be used as a decoration in the Peking Opera costumes, Reeves's Pheasant (*Syrmaticus reevesii*) also was poisoned in mountain areas when local farmers tried to protect crops from birds. The hunting and gene pollution of hybridization between captive population and wild relatives are still the major threat. For example, it is hardly distinguish the pure Red Junglefowl (*Gallus gallus*).

Habitat loss and degradation due to the loss of wetlands are the major threats to cranes. The marshes of breeding grounds and the tidal flats of wintering grounds have been reclaimed and developed to farmland (Lu *et al.*, 2007; Kong *et al.*, 2011; Yang and Chang, 2014). Habitat loss and degradation caused by the reclamation and exploitation of coastal wetlands led to the continuous decline of the population of Red-crowned Crane (*Grus japonensis*) in China. In its major wintering ground Yancheng of Jiangsu Province, the population has been decreased from 1,200 to 600 so far (Lu, 2009). Other threats include the use of pesticides, insecticides in farmlands in winters, illegal hunting, power lines and wind farms. The change of traditional farming and the extension of grazing range has aggravated the conflicts between Black-necked Crane (*Grus nigricollis*) and local residents (Ma *et al.*, 2014; Yang and Chang, 2014). Most of the overwintering cranes might be seriously threatened by the hydraulic engineering construction. The population size of Siberian Crane (*Grus leucogeranus*) is likely to decline sharply in the future.

The major threat of Charadriiformes is the change of coastal environment. The global population of Spoon-billed Sandpiper (*Calidris pygmeus*) is less than 1,000 (MacKinnon *et al.*, 2012). The mudflat at Rudong of Jiangsu is one of the most important habitats and the known largest stopover to support the largest number of Spoon-billed Sandpiper during southbound migration (Tong *et al.*, 2012). 103 individuals was recorded in the autumn of 2011, and 143 individuals were recorded in October, 2013, which shows the importance of the area to the migration of Spoon-billed Sandpiper. While the habitat loss due to reclamation of mudflat, development and invasion by *Spartina alterniflora* will be the major threats.

本次评估是对中国鸟类受威胁状况的一次全面梳理，相信会对中国鸟类保护工作有重要的推动作用。同时也看到，尚有105种鸟类的种群数量、分布区面积、种群波动情况、受威胁信息等资料不清、生存状况不明，期望得到鸟类学研究者和公众的更多关注。

16 保护成效

经过中国政府、学者和公众的不懈努力，中国鸟类保护已经初见成效。迄今全国已建成各种类型、不同级别的自然保护地1.18万处，占国土面积的18%，其中各类国家级自然保护地3766处，这些保护地为鸟类提供了重要的庇护所。鸟类学工作者不断深入的工作为鸟类保护提供了有力支撑，并推动了鸟类保护工作。公众的积极参与，为鸟类保护工作深入开展奠定了扎实的群众基础，并带动形成了鸟类保护的燎原之势。

朱鹮 (*Nipponia nippon*) (图12) 在1981年发现时只有7只，由于种群数量少、分布区狭窄，被IUCN评估为极危物种，经过多年的保护，朱鹮种群数量不断增加，2000年被IUCN列为濒危物种；此后朱鹮种群数量进一步稳定增长、分布区范围不断扩大，2012年野生种群的种群数量为1,100～1,200只（王超等，2012）。多个人工繁育场所形成了

This assessment is a comprehensive review of the threatened status of bird species in China, and it is believed to be important to promote the conservation of birds in China. Meanwhile, the information of population size, distribution area, population trend and threats of 105 species are still far less than clear, expecting more attention from researchers and public.

16 Conservation Effectiveness

With the efforts of the Chinese government, experts and the public, bird conservation in China has already made a great progress. A total of 1.18 million of protected areas of various types and levels have been established nationwide. The land area of the nature reserve is accounts for 18% of the land area of China. There are 3766 national protected areas in China. These protected areas provide important shelters for birds. The continuous work by ornithologists has provided a strong support for bird conservation and also has promoted conservation of birds. The active public engagement has been a solid foundation for bird conservation, and has also promote the conservation.

Only seven individuals of Crested Ibis (*Nipponia nippon*) (**Figure 12**) were found in 1981, and evaluated as CR by IUCN considering the small population and narrow distribution. After several decades of conservation, the population has been increasing, and it was listed as EN by IUCN in 2000; since then, the population has increased steadily, and the range of distribution has been continuously expanded. In 2012, its wild populations was 1,100 ~ 1,200 (Wang *et al.*, 2012). Several stable populations in captivity have been developed, and their reintroduction was successfully carried out in Zhejiang, Henan, *etc*. The conservation of Crested Ibis is a typical case on conservation of threatened bird species.

图 12 朱鹮 (*Nipponia nippon*) 在野外栖息地觅食（丁长青摄）

Figure12 Crested Ibis (*Nipponia nippon*) foraging in the natural habitat (Photographed by Changqing Ding)

图 13 人工招引繁育成功的中华凤头燕鸥 (*Thalasseus bernsteini*)(洪崇航摄)

Figure13 Restoration of the Chinese Crested Terns (*Thalasseus bernsteini*) (Photographed by Chonghang Hong)

规模化的人工繁育种群，并在浙江、河南等地成功开展了再引入，成为我国濒危鸟类保护成效的代表。

中华凤头燕鸥 (*Thalasseus bernsteini*)(图 13) 自 1937 年在青岛采集到标本之后，销声匿迹长达 60 多年，2000 年在马祖列岛意外发现了一个极小的群体 (Liang et al., 2000)。2004 年在浙江韭山列岛发现了中华凤头燕鸥的繁殖群体，该繁殖群体在 2007 年因遭遇人为捡蛋后离开 (Chen et al., 2009，2010)。为了有效保护中华凤头燕鸥繁殖群体，对其繁殖栖息地进行了严格的保护管理，并于 2013 年开始在韭山列岛实施人工招引和恢复项目，2014 年和 2015 年分别成功孵化出 13 只和 16 只雏鸟 (周晓等，2017)，目前该物种的种群数量稳步增长，是我国成功开展海鸟保护的一个典型案例。

在栖息地保护方面，近年来国家海洋局陆续发布了《关于加强滨海湿地管理与保护工作的指导意见》、《关于全面建立实施海洋生态红线制度的意见》等，加强重要自然滨海湿地保护。通过建立海洋自然保护区、海洋特别保护区 (海洋公园) 等形式，将当前亟须保护的重要滨海湿地纳入保护范围，将重要滨海湿地等纳入海洋生态红线。通过严格滨海湿地开发利用管理，禁止在鸟类栖息地进行围填海活动，为水鸟栖息地的保护提供了政策保障，有效遏制了无序开垦、破坏水鸟繁殖、破坏水鸟越冬和中途停歇栖息地的行为。

除水鸟外，在陆生鸟类保护方面，也取得积极进展。中国目前有关于受威胁雉类的各级自然保护区 123 个，其中国家级自然保护区 25 个，面积为 84.8 万 hm^2，为我国雉类资源保护发挥了重要作用。2008 年，太行山、吕梁山和黄龙山三大山系的 8 个褐马鸡 (*Crossoptilon mantchuricum*)(图 14) 自然保护区组建"姐妹保护区"，构建成了布局合理、管理有序的褐马鸡自然保护区网络，促进了褐马鸡及其栖息地的保护与恢复。

There had been no record of Chinese Crested Tern (*Thalasseus bernsteini*) (**Figure 13**) for more than 60 years since specimens were collected in Qingdao in 1937. A small group was occasionally discovered on the Matsu Islands in 2000 (Liang *et al.*, 2000), and a breeding colony was discovered on the Jiushan Islands in Zhejiang in 2004. The breeding group abandoned the islands in 2007 because illegal egg-collection (Chen *et al.*, 2009, 2010). In order to protect the breeding population of Chinese Crested Tern, their breeding habitat has been strictly managed. In 2013, an attraction and restoration project was carried out on Jiushan Islands, and 13 and 16 chicks were successfully hatched in 2014 and 2015, respectively (Zhou *et al.*, 2017). The population of this species has been steadily increasing, which is a typical and successful case of seabird conservation in China.

In terms of the habitat protection of waterbirds, the State Oceanic Administration has successively issued *Guidelines on Strengthening the Management and Protection of Coastal Wetlands* and *Opinions on the Comprehensive Establishment and Implementation of the Marine Ecological Red Line System*, etc., to strengthen the protection of important natural coastal wetlands in recent years. By establishing of marine nature reserves, marine special reserves (marine parks) *etc.*, the important coastal wetlands are under protection, and included in the marine ecological red line. By strictly control of the development and utilization of coastal wetlands and prohibit reclamation in bird habitats, there provided a policy guarantee for the protection of waterbird habitats. Disorderly reclamation as well as destruction activities of breeding grounds, wintering sites and stopovers were effectively stopped under policy umbrella with strict control of the development and utilization of coastal wetlands and prohibition reclamation in bird habitats.

There is also great progress in conservation of terrestrial bird. There are 123 nature reserves for threatened pheasants, including 25 national nature

Figure 14 Brown-eared pheasant inhabiting the Baihua mountain of Beijing (Photographed by Dongsheng Guo)

图 15　在野外安放的人工巢中孵卵的黄腹角雉（*Tragopan caboti*）（张雁云摄）　　**Figure15**　Cabot's Tragopan (*Tragopan caboti*) hatching eggs in artificial nests placed in the natural habitat (Photographed by Yanyun Zhang)

我国学者自 20 世纪 80 年代开始，对黄腹角雉（*Tragopan caboti*）（图 15）栖息地利用、活动区与活动性、繁殖、食性、行为等进行了全面、深入、系统的研究，查明了该物种的濒危机制和生态适应性。通过实施一系列针对性的保护措施，浙江乌岩岭保护区的黄腹角雉种群数量稳步提升。从 80 年代中期开始，陆续建立起实验性的黄腹角雉易地保护种群（图 16），并成功地在浙江乌岩岭保护区实施了再补充，在湖南桃源洞保护区实施了再引入。

近年来，在广大公众和社会组织的积极参与下，我国的鸟类监测和保护工作有了飞速发展和进步。由中国观鸟组织联合行动平台（朱雀会）牵头组织的全国中华秋沙鸭调查，有超过 50 家各地观鸟组织与保护区等机构参与，是迄今针对雁形目某一特定种类最大规模的越冬情况调查。2015～2016 年调查到中华秋沙鸭 634 只，在多地新发现了中华秋沙鸭的分布，同时记录了精确的分布信息。中国沿海水鸟普查组在 2005 年 9 月至 2013 年 12 月，每月对 14 个地点的水鸟进行一次同步调查，并对 18 个地点的水鸟进行不定期调查，发现 26 个地点至少有 1 种水鸟的数

reserves with an area of 848,000 hectares, and playing an important role in the conservation of pheasants in China. Eight nature reserves for Brown Eared Pheasant (*Crossoptilon mantchuricum*) (**Figure 14**) in the three major mountains (Taihang Shan, Lüliang Mountain, and Huanglong Mountain) formed "sister reserves" in 2008, building a network of nature reserves with reasonable patterns and effective management, which promoted the conservation and restoration of Brown Eared Pheasant and their habitats.

Since 1980s, Chinese scholars have conducted comprehensive and systematic studies on habitat use, home range and activities, reproduction, foraging and food, and behaviors of the Cabot's Tragopan (*Tragopan caboti*) (**Figure 15**), and revealed their endangered mechanism and ecological adaptation. The population of Cabot's Tragopan in Wuyanling Nature Reserve in Zhejiang province has been steadily increasing through series of conservation actions. In addition, the *ex situ* conservation of Cabot's Tragopan has been made gradually improvement since mid of 1980s (**Figure 16**). The population has been successfully supplemented in Wuyanling National Nature Reserve in Zhejiang Province and reintroduced in Taoyuandong National Nature Reserve in Hunan province.

In recent years, with the active participation of the public and social organizations, the bird monitoring and conservation has made great development and progress in China. The national survey of Scaly-sided Merganser led by China's Birdwatching Association , involved more than 50 local bird watching organizations and institutes of nature reserves. And it was the largest survey of a specific species in China during winter. In the winter of 2015 , 634 Scaly-sided Merganser were recorded, and many new sites were reported, and the accurate distribution information and threat factors have been recorded as well. From September 2005 to December 2013, the China

图 16 笼养的黄腹角雉幼鸟（张雁云摄）

Figure 16 The chick of Cabot's Tragopan in capativity (Photographed by Yanyun Zhang)

量达到了全球总数量的 1%，有 75 种水鸟在一个调查点或一次调查中的种群数量达到全球种群数量的 1% 水平 (Bai et al., 2015)。《中国观鸟年报》(2003～2007 年) 共记录观鸟记录 30,936 条，包含 17 目 70 科 1,078 种，占全部鸟类的 70% 以上，包括 14 种中国新发现鸟类，还有 109 种鸟类出现在新分布区 (李雪艳等，2012)。个人和观鸟组织队伍日益壮大，成为我国鸟类野外监测和保护的重要力量。

17 结束语

2020 年全国人民代表大会常务委员会出台了《关于全面禁止非法野生动物交易、革除滥食野生动物陋习、切实保障人民群众生命健康安全的决定》，"全面禁止食用国家保护的'有重要生态、科学、社会价值的陆生野生动物'以及其他陆生野生动物，包括人工繁育、人工饲养的陆生野生动物"。全面禁止以食用为目的猎捕、交易、运输在野外环境自然生长繁殖的陆生野生动物。2021 年，国家颁布了调整后的《国家重点保护野生动物名录》，92 种鸟类被列为国家 I 级重点保护野生动物，302 种鸟类被列为国家 II 级重点保护野生动物。随着全民科学素养、野生动物保护意识和管理水平的不断提升，随着我国生态文明建设和国家公园为主体的保护地体系的建设和发展，我国濒危鸟类的保护工作必将进入一个新阶段。

Coastal Waterbird Survey Team conducted census of waterbirds in 14 sites monthly and irregular census of waterfowls in 18 sites. It was found that at least one species reached 1% of its global population in 26 sites, and 75 species of waterbirds reached 1% of its global population in one site or one survey (Bai *et al.*, 2015). *China Bird Watching Annual Report* (2003~2007) collected 30,936 bird watching records, including 1,078 species affiliated with 70 families of 17 orders, accounting for more than 70% of bird species, and 14 new records of China, and new distribution areas for 109 species a (Li *et al.*, 2012). Individuals and birdwatching organizations are growing and developing increasingly, which have become an important support of field monitoring and conservation of birds in China.

17 Concluding Remarks

In 2020, the National People's Congress Standing Committee issued the *Decision on the Comprehensive Ban on Illegal Wildlife Trade, Elimination of Eating Wild Animal Abuse, and Effective Protection of People's Health and Safety*. It is prohibited to eat "Terrestrial wildlife of important ecological, scientific and social value" and other protected terrestrial wildlife, including those that are artificially bred and raised. It is forbidden to hunt, trade and transport for the purpose of food to terrestrial wild animals that grow and reproduce naturally in the wild. In 2021, the revised National Key Protected Wild Animal List in China was published, and there are 92 species of birds listed as first class key protected wild animals and 302 species of birds listed as second class key protected wild animals. With the continuous improvement of the scientific knowledge and the awareness of the public, as well as the development of wildlife management, as well as the construction of ecological civilization and protected area system in China, the conservation of threatened birds in China is bound to enter a new stage.

各 论
Species Monograph

镰翅鸡
Falcipennis falcipennis

区域灭绝 RE

| 数据缺乏 DD | 无危 LC | 近危 NT | 易危 VU | 濒危 EN | 极危 CR | 区域灭绝 RE | 野外灭绝 EW | 灭绝 EX |

分类地位 Taxonomic Status

动物界 Animalia	脊索动物门 Chordata	鸟纲 Aves	鸡形目 Galliformes	雉科 Phasianidae
学名 Scientific Name		*Falcipennis falcipennis*		
命名人 Species Authority		Hartlaub, 1855		
英文名 English Name(s)		Siberian Grouse		
同物异名 Synonym(s)		*Dendragapus falcipennis*		
种下单元评估 Infra-specific Taxa Assessed		无 / None		

评估信息 Assessment Information

评估年份 Year Assessed	2016
评定人 Assessor(s)	张雁云 / Yanyun Zhang
其他贡献人 Other Contributor(s)	无 / None

理由 Justification: 在中国一直未获得标本和野外照片，也未见野外考察正式发表的观察记录。郑作新 (1987) 提出该物种在中国极其少见，可能已经灭绝。因此，列为区域灭绝等级 / No specimen, field photo or formal record of *Falcipennis falcipennis* from field survey. Zheng (1987) perceived this species extremely rare in China, and might be extinct. Thus, it is listed as Regionally Extinct

评估历史 Assessment History: 《国家重点保护野生动物名录》: II (1989, 2021); 《中国濒危动物红皮书·鸟类》(1998): 濒危; 《中国物种红色名录: 第二卷 脊椎动物 下册》(2009): 无危 / *National Key Protected Wild Animal List*: II (1989, 2021); *China Red Data Book of Endangered Animals*: *Aves* (1998): E; *China Species Red List*: *Vol. II Vertebrates Part 2* (2009): LC

地理分布 Geographical Distribution

国内分布 Domestic Distribution

《中国动物志》记载其见于我国小兴安岭及黑龙江下游。近70年从未有野外记录 / It recorded in Xiao Hinggan Ling and lower reach of Heilongjiang River in China (*Fauna Sinica*). There is not any record in recent 70 years

分布标注 Distribution Note

非特有种 / Non-endemic

国内分布图
Map of Domestic Distribution

种群 Population

种群数量 Population Size	对该物种一直缺乏有效监测，Brazil (2009) 估计俄罗斯种群数量为 10,000～100,000 对，但在中国近几十年一直未调查到 / No effective monitoring on the species till now, and no record in China for decades. The population size was estimated to be 10,000～100,000 pairs in Russia (Brazil, 2009)
种群趋势 Population Trend	下降 / Decreasing

生境与生态系统 Habitat(s) and Ecosystem(s)

生境 Habitat(s)	典型的针叶林鸟类，栖息在冷杉、云杉和落叶松林中，林下常密布灌木和草本植物，或栖息于林间空地、草地、田地等的交错地带。以嫩枝、芽、花及植物浆果为食，冬季采食落叶松和云杉的叶子 / *F. falcipennis* is a typical species in coniferous forests, and uses firs, spruces and pines with dense shrubs and grass. It also uses cross areas of the open lands in forests, grasslands, and farmlands. It feeds on twigs, shoots, buds and berries, and leaves of pines and spruces in winter
生态系统 Ecosystem(s)	陆地生态系统 / Terrestrial Ecosystem

威胁 Threat(s)

主要威胁 Major Threat(s)	大规模的森林砍伐、林火和狩猎是该物种面临的主要威胁 / The main threats are deforestation, forest fire and illegal trapping

保护级别与保护行动 Protection Category and Conservation Action(s)

IUCN 红色名录 (2016) IUCN Red List (2016)	近危 / NT
保护行动 Conservation Action(s)	列入国家 II 级重点保护野生动物 / It has been listed as the second class in National Key Protected Wild Animal List

镰翅鸡 *Falcipennis falcipennis*　　　　　　　　　　　　　　　　张春福 摄　By Chunfu Zhang

赤颈鹤
Grus antigone

区域灭绝 RE

| 数据缺乏 DD | 无危 LC | 近危 NT | 易危 VU | 濒危 EN | 极危 CR | 区域灭绝 RE | 野外灭绝 EW | 灭绝 EX |

分类地位 Taxonomic Status

动物界 Animalia	脊索动物门 Chordata	鸟纲 Aves	鹤形目 Gruiformes	鹤科 Gruidae
学名 Scientific Name		*Grus antigone*		
命名人 Species Authority		Linnaeus, 1758		
英文名 English Name(s)		Sarus Crane		
同物异名 Synonym(s)		*Antigone antigone*		
种下单元评估 Infra-specific Taxa Assessed		无 / None		

评估信息 Assessment Information

评估年份 Year Assessed	2016
评定人 Assessor(s)	丁长青 / Changqing Ding
其他贡献人 Other Contributor(s)	无 / None

理由 Justification: 中国仅 1973 年在云南发现一具标本，国内赤颈鹤在 20 世纪七八十年代以来一直无存活个体记录（何晓瑞，1994；杨岚，1987）。因此，列为区域灭绝等级 / Only a specimen of *Grus antigone* was found in Yunnan in 1973, and no individual has been recorded in the field since 1970s (He, 1994; Yang, 1987). Thus, it is listed as Regionally Extinct

评估历史 Assessment History:《国家重点保护野生动物名录》：Ⅰ（1989，2021）；《中国濒危动物红皮书·鸟类》(1998)：稀有；《中国物种红色名录：第二卷 脊椎动物 下册》(2009)：易危 / *National Key Protected Wild Animal List*: Ⅰ (1989, 2021); *China Red Data Book of Endangered Animals*: *Aves* (1998): R; *China Species Red List*: Vol. Ⅱ Vertebrates Part 2: VU

地理分布 Geographical Distribution

国内分布 Domestic Distribution
历史分布于云南 / It occured in Yunnan in history
分布标注 Distribution Note
非特有种 / Non-endemic

国内分布图
Map of Domestic Distribution

种群 Population

种群数量 Population Size	全球种群数量 19,000～21,800 只，其中成熟个体 13,000～15,000 只（IUCN，2012）/ The global population size is 19,000～21,800 individuals, and 13,000～15,000 mature individuals (IUCN, 2012)
种群趋势 Population Trend	下降 / Decreasing

生境与生态系统 Habitat(s) and Ecosystem(s)

生　　境 Habitat(s)	栖息地生境类型广泛，包括农田、草地及湿地和旱地的混合生境 / *G. antigone* uses various habitats, including farmlands, grasslands, and mosaic of wetlands and dry lands
生态系统 Ecosystem(s)	陆地生态系统 / Terrestrial Ecosystem

威胁 Threat(s)

主要威胁 Major Threat(s)	栖息地丧失 / Habitat loss

保护级别与保护行动 Protection Category and Conservation Action(s)

IUCN 红色名录 (2016) IUCN Red List (2016)	易危 / VU A2cde+3cde+4cde
保护行动 Conservation Action(s)	列入国家 I 级重点保护野生动物和 CITES 公约附录 II / It has been listed as the first class in National Key Protected Wild Animal List and the Appendix II of CITES

赤颈鹤 *Grus antigone*　　　　　王易 摄　By Yi Wang

白鹳
Ciconia ciconia
区域灭绝 RE

| 数据缺乏 DD | 无危 LC | 近危 NT | 易危 VU | 濒危 EN | 极危 CR | 区域灭绝 RE | 野外灭绝 EW | 灭绝 EX |

分类地位 Taxonomic Status

动物界 Animalia	脊索动物门 Chordata	鸟纲 Aves	鹳形目 Ciconiiformes	鹳科 Ciconiidae
学　名 Scientific Name		*Ciconia ciconia*		
命 名 人 Species Authority		Linnaeus, 1758		
英 文 名 English Name(s)		White Stork		
同物异名 Synonym(s)		无 / None		
种下单元评估 Infra-specific Taxa Assessed		无 / None		

评估信息 Assessment Information

评估年份 Year Assessed	2016
评定人 Assessor(s)	马鸣 / Ming Ma
其他贡献人 Other Contributor(s)	才代、巴图尔汗、杨小敏 / Dai Cai, Baturhan, Xiaomin Yang

理由 Justification: 在过去的30年里，国内已经没有白鹳的确切记录（马鸣，2001a）。因此，列为区域灭绝等级 / No exact record of *Ciconia ciconia* in China in the past 30 years (Ma, 2001a). Thus, it is listed as Regionally Extinct

评估历史 Assessment History:《国家重点保护野生动物名录》：Ⅰ (2021)；《国家重点保护野生动物名录》(1989) 和《中国濒危动物红皮书·鸟类》(1998) 评估的白鹳主要为东方白鹳 *Ciconia boyciana*；《中国物种红色名录：第二卷 脊椎动物 下册》(2009)：未评估 / *National Key Protected Wild Animal List*: I (2021); The assessed species in *National Key Protected Wild Animal List* (1989) and *China Red Data Book of Endangered Animals*: *Aves* (1998) are mainly *Ciconia boyciana*; *China Species Red List*: *Vol. II Vertebrates Part 2* (2009): NA

地理分布 Geographical Distribution

国内分布 Domestic Distribution

历史上分布区位于新疆西部的皮山、叶城、泽普、莎车、英吉沙、疏勒、喀什地区、天山、伊犁等地，有繁殖记录 (Scully，1876)。偶见于内蒙古（迷鸟）/ It occured in Pishan, Yecheng, Zepu, Shache, Yingjisha, Shule, Kashi, Tianshan and Ili of Xinjiang in history, with breeding records (Scully, 1876), and was seldomly seen in Inner Mongolia (Nei Mongol) (vagrant visitor)

分布标注 Distribution Note

非特有种 / Non-endemic

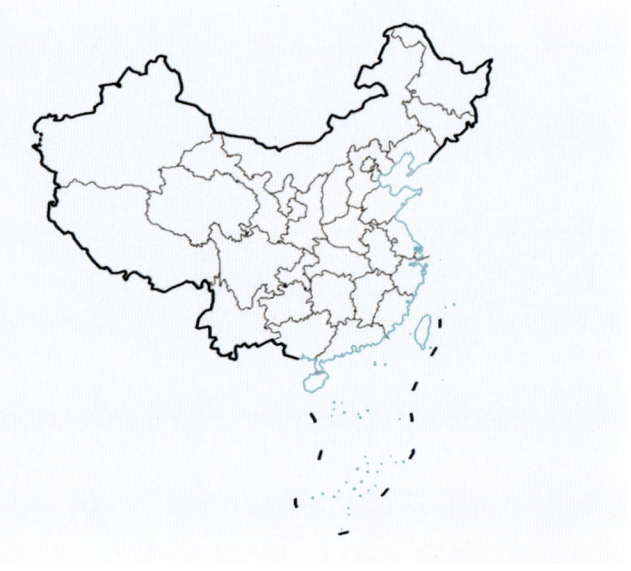

国内分布图
Map of Domestic Distribution

种群 Population

种群数量 Population Size	种群数量 0～10 只（观鸟者有一些记录，尚需核实）/ The population size is only 0～10 individuals based on records of bird watchers, which should be checked
种群趋势 Population Trend	下降 / Decreasing

生境与生态系统 Habitat(s) and Ecosystem(s)

生境 Habitat(s)	生活在荒漠绿洲、内陆湿地，与人类关系密切。营巢于大树上、屋顶、电杆及水塔之上。喜欢活动于池塘、湖泊、水库，海拔 600～1,100 m。食物有鱼类、蛙、蜥蜴、昆虫等。在新疆为夏候鸟，冬季迁徙至南亚 / *C. ciconia* inhabits in oasis and inner wetlands, and nests in big trees, roofs, electrical pole, and water tower, which are closely related to human people. It uses ponds, lakes and reservoirs at 600～1,100 m. The diet includes fish, frogs, lizards, and insects. It bred in Xinjiang and winters in south Asia
生态系统 Ecosystem(s)	陆地生态系统 / Terrestrial Ecosystem

威胁 Threat(s)

主要威胁 Major Threat(s)	近 60 年来的西部环境变化很大，生存空间缩小，白鹳可选择地繁殖地越来越少，遭遇灭顶之灾（马鸣，2001a）。威胁包括过度的砍伐森林（胡杨林）、传统狩猎、毁巢（取卵食用）、栖息地改变、农业开垦、人口增加、自然湖泊退化及被改造成为水库（浅水区的消失）、大河断流、过度捕捞及食物资源缺乏等（马鸣等，1997a；马鸣，2002）/ The environment has largely changed, breeding sites has been largely decreased in the past 60 years (Ma, 2001a). The threats include over deforestation, traditional hunting, egg collection, change of habitats, agricultural reclamation, increase of human population, loss of shallow water area by reduction of natural lakes and construction of dams, cutoff of large rivers, food resource shortage owing to over fishing (Ma *et al.*, 1997a; Ma, 2002)

保护级别与保护行动 Protection Category and Conservation Action(s)

IUCN 红色名录 (2016) IUCN Red List (2016)	无危 / LC
保护行动 Conservation Action(s)	列入国家 I 级重点保护野生动物、迁徙物种保护公约 (CMS) 附录 II。期待着环境改善、人们的保护意识增强，也许有一天这个物种还有可能重新返回新疆。因为白鹳 (*Ciconia ciconia asiatica*) 在邻国还有分布，种群恢复是完全有可能的（马鸣，2002）/ It has been listed as the first class in National Key Protected Wild Animal List, appendix II of the Convention on Migratory Species. It might occur in Xinjiang some day in the future, with the improvement of habitat and conservation conscious of the public. The restoration of the population is feasible on consideration that *Ciconia ciconia asiatica* inhabits neighboring countries (Ma, 2002)

白鹳 *Ciconia ciconia*　　　　　　　　　　　　　　　　By David Blank

中国生物多样性红色名录
海南孔雀雉
Polyplectron katsumatae

极危 CR A2cd+3cd+4cd; C2a(i)

| 数据缺乏 DD | 无危 LC | 近危 NT | 易危 VU | 濒危 EN | 极危 CR | 区域灭绝 RE | 野外灭绝 EW | 灭绝 EX |

分类地位 Taxonomic Status

| 动物界 Animalia | 脊索动物门 Chordata | 鸟纲 Aves | 鸡形目 Galliformes | 雉科 Phasianidae |

学名 Scientific Name	*Polyplectron katsumatae*
命名人 Species Authority	Rothschild, 1906
英文名 English Name(s)	Hainan Peacock Pheasant
同物异名 Synonym(s)	无 / None
种下单元评估 Infra-specific Taxa Assessed	霸王岭国家级自然保护区内海南孔雀雉个体数在15年内下降了50%～79% / The population in Bawangling National Nature Reserve has decreased by 50%～79% in 15 years

评估信息 Assessment Information

评估年份 Year Assessed	2016
评定人 Assessor(s)	丁平 / Ping Ding
其他贡献人 Other Contributor(s)	无 / None

理由 Justification: 海南孔雀雉的分布范围十分狭窄且分散，再加上森林破坏、捕猎活动对其造成的影响，估计其种群数量在300只左右 (马福和张建龙，2009)，正在以极快的速度减少。在霸王岭国家自然保护区对海南孔雀雉的调查发现，其个体数在15年内下降了50%～79%。同时也急需在全岛范围内全面的调查统计 (Liang and Zhang, 2011)。因此，列为极危等级 / The distribution areas of *Polyplectron katsumatae* are small and fragmentated. Negatively affected by deforestation and hunting, the population is estimated to number c. 300 (Ma and Zhang, 2009) and undergoing a rapid decline. The population in Bawangling National Nature Reserve has decreased by 50%～79% in 15 years. Surveys covering the whole Hainan Island is in urgent need (Liang and Zhang, 2011). Thus, it is listed as Critically Endangered

评估历史 Assessment History: 《国家重点保护野生动物名录》：Ⅰ (2021)；《中国物种红色名录：第二卷 脊椎动物 下册》(2009)：极危 / *National Key Protected Wild Animal List*: Ⅰ (2021); *China Species Red List: Vol. II Vertebrates Part 2* (2009): CR

地理分布 Geographical Distribution

| 国内分布 Domestic Distribution |
| 仅分布于海南 / It only occurs in Hainan |
| 分布标注 Distribution Note |
| 特有种 / Endemic |

国内分布图
Map of Domestic Distribution

种群 Population

种群数量 Population Size	1990 年种群数量估计有 2,700 只，2000 年种群数量约为 300 只 (马福和张建龙，2009)。但这个种群数量可能被低估，约 350 ～ 1,500 只。目前急需进一步的调查以评估其真正的种群数量现状 / The population size was estimated to be 2,700 individuals in 1990, and the number was 300 in 2000 (Ma and Zhang, 2009). Possibly it may be underestimated, and the population size is supposed to be 350 ～ 1,500 individuals. Systematic surveys are need to assess the status of population size
种群趋势 Population Trend	下降 / Decreasing

 ## 生境与生态系统 Habitat (s) and Ecosystem (s)

生　　境 Habitat(s)	栖息于海拔 200 ～ 1,300 m 的热带原始林或者成熟的次生林内。倾向于选择具有浓密的林下郁闭度、靠近水源地、草木茂盛、远离人为干扰的区域 / *P. katsumatae* inhabits tropical primitive forests or mature secondary forests at 200 ～ 1,300 m. It prefer areas with dense canopy and rich and varied vegetation, close to water, and away from human disturbance
生态系统 Ecosystem(s)	陆地生态系统 / Terrestrial Ecosystem

 ## 威胁 Threat (s)

主要威胁 Major Threat(s)	从 20 世纪 50 年代开始，由于分布区内栖息地破坏、片断化，以及捕猎的影响使海南孔雀雉的数量急剧减少 (Liang and Zhang，2011)。80 年代末，海南森林的覆盖度只有 10% 左右，其分布区逐渐狭窄而分散 (高育仁等，1990) / The population size has decreased since 1950s due to the habitat loss and fragmentation, as well as hunting (Liang and Zhang, 2011). The forest coverage in Hainan is only 10% till late 1980s and the distribution areas became small and scattered (Gao *et al.*, 1990)

 ## 保护级别与保护行动 Protection Category and Conservation Action (s)

IUCN 红色名录 (2016) IUCN Red List (2016)	濒危 / EN A2cd+3cd+4cd; C2a (i)
保护行动 Conservation Action(s)	列入国家 I 级重点保护野生动物。海南的 3 个国家级自然保护区及 5 个省级自然保护区内有分布，但仍无法有效地降低捕猎对海南孔雀雉的影响 / It has been listed as the first class in National Key Protected Wild Animal List. The distribution areas of *P. katsumatae* covers 3 national nature reserves and 5 provincial nature reserves in Hainan, while the hunting pressure has not been effectively prevented

海南孔雀雉 *Polyplectron katsumatae*　　　　　　　　　　　　　　　　　　　张正旺 摄 By Zhengwang Zhang

中国生物多样性 红色名录
China's Red List of Biodiversity

绿孔雀
Pavo muticus

极危　CR A2cd+3cd+4cd

| 数据缺乏 DD | 无危 LC | 近危 NT | 易危 VU | 濒危 EN | **极危 CR** | 区域灭绝 RE | 野外灭绝 EW | 灭绝 EX |

分类地位 Taxonomic Status

动物界 Animalia	脊索动物门 Chordata	鸟纲 Aves	鸡形目 Galliformes	雉科 Phasianidae
学　名 Scientific Name		*Pavo muticus*		
命名人 Species Authority		Linnaeus, 1766		
英文名 English Name(s)		Green Peafowl		
同物异名 Synonym(s)		无 / None		
种下单元评估 Infra-specific Taxa Assessed		无 / None		

评估信息 Assessment Information

评估年份 Year Assessed	2016
评定人 Assessor(s)	张正旺 / Zhengwang Zhang
其他贡献人 Other Contributor(s)	杨晓君 / Xiaojun Yang

理由 Justification: 由于栖息地的改变和捕猎压力较大，绿孔雀的种群数量快速下降，栖息地片断化严重。而且这种趋势将会持续下去。中国绿孔雀的数量已经十分稀少。如果不加强保护，该物种在我国境内率先灭绝的风险很大。因此，列为极危等级 / The population of *Pavo muticus* is undergoing a rapid decline owing to the habitat change and high pressure of hunting, and the trend might continue. The population size is extremely small in China, and is under high risk of region extinct without effective protection. Thus, it is listed as Critically Endangered

评估历史 Assessment History: 《国家重点保护野生动物名录》：Ⅰ(1989，2021)；《中国濒危动物红皮书·鸟类》(1998)：濒危；《中国物种红色名录：第二卷 脊椎动物 下册》(2009)：濒危 / *National Key Protected Wild Animal List*: I (1989, 2021); *China Red Data Book of Endangered Animals*: *Aves* (1998): E; *China Species Red List*: *Vol. II Vertebrates Part 2* (2009): EN

地理分布 Geographical Distribution

国内分布 Domestic Distribution
仅见于云南西部、中部和南部。20世纪90年代主要见于云南省怒江、德宏、保山、临沧、普洱、红河、西双版纳、玉溪、楚雄9地州34县，2014年仅在云南的22个县调查到绿孔雀 / It only occurs in west, middle and south Yunnan in China, including Nujiang, Dehong, Baoshan, Lincang, Puer, Honghe, Xishuangbanna, Yuxi and Chuxiong in 1990s. However, it was investigated in 22 counties of Yunnan in 2014
分布标注 Distribution Note
非特有种 / Non-endemic

国内分布图
Map of Domestic Distribution

种群 Population

种群数量 Population Size	全球种群数量估计有 15,000 ~ 30,000 只 (BirdLife, 2014)。20 世纪末中国绿孔雀种群数量为 800 ~ 1,100 只 (文贤继等，1995；杨晓君等，1997)。目前绿孔雀的数量持续下降，2014 年调查到的种群数量在 194 ~ 248 只，现存不足 500 只 (Kong *et al.*, 2018) / The global population size is estimated to be 15,000 ~ 30,000 (BirdLife, 2014), and at the end of the 20th century, the population size was approximately 800 ~ 1,100 (Wen *et al.*, 1995; Yang *et al.*, 1997). 194 ~ 248 individuals were found in 2014, and population size in China is less than 500 individuals and in a trend of decreasing (Kong *et al.*, 2018)
种群趋势 Population Trend	由于栖息地的破坏和非法猎捕，无论在东南亚还是在中国，其种群数量呈持续下降趋势，并且在中国的种群下降更为显著 / *P. muticus* is undergoing a decline in southeast Asia and China owing to the habitat loss and illegal hunting, while the situation in China is much more serious

生境与生态系统 Habitat (s) and Ecosystem (s)

生境 Habitat(s)	主要栖息于海拔 2,000 m 以下的热带、亚热带常绿阔叶林和混交林，尤其喜欢在疏林草地、河岸或地边丛林以及林间草地和林中空旷的开阔地带活动。常成 5 ~ 10 只小群边走边觅食，尤以清晨和临近傍晚时觅食活动较为频繁。通常在草丛中寻找种子、浆果，也吃稻谷、嫩芽、禾苗，有时也会在河边捉食昆虫、蜥蜴、青蛙等。主要食蕈类、浆果、谷物种子、草籽等，也兼食昆虫、蛙类、蜥蜴等 / *P. muticus* inhabits below 2,000 m in tropical, subtropical evergreen broad-leaved forests and mixed forests, especially in the grassland with sparse trees, river banks or open areas in forests. It normally forages in dawn and dusk in small flocks of 5 ~ 10 individuals. It searches for seeds and berries in grassland, and also grains, buds and crops; sometimes insects, lizards and frogs at the river side. Its main food includes fungi, berries, seeds of grains and grass, and insects, frogs and lizards are complementary food
生态系统 Ecosystem(s)	陆地生态系统 / Terrestrial Ecosystem

威胁 Threat (s)

主要威胁 Major Threat(s)	由于其赖以生存的次生落叶季雨林和常绿季雨林生态环境遭到破坏，绿孔雀的分布区正在逐步减少和退缩；目前的开矿和建坝活动对楚雄恐龙河保护区的种群产生重要影响 (Wu *et al.*, 2019)。此外，非法捕猎和农药中毒导致该物种在许多分布地点的数量都出现了下降，甚至出现局域性灭绝。在绿孔雀分布区，有些农民饲养蓝孔雀，存在逃逸和基因污染的风险 / The distribution areas of *P. muticus* are declining owing to the destroy of secondary deciduous and evergreen seasonal rainforests. The population in Chuxiong Konglonghe Nature Reserve is negatively affected by mining and damming (Wu *et al.*, 2019). Populations at several sites have declined, or even locally extinct owing to illegal hunting and pesticide. Farmers raise *P. cristatus* in captivity in the distribution area of *P. muticus*, and there is risk of escape and genetic contamination

保护级别与保护行动 Protection Category and Conservation Action (s)

IUCN 红色名录 (2016) IUCN Red List (2016)	濒危 / EN A2c+3c+4cd
保护行动 Conservation Action(s)	列入国家 I 级重点保护野生动物和 CITES 公约附录 II，国家林业局将其列为极小濒危种群拯救工程规划 / It has been listed as the first class in National Key Protected Wild Animal List, the Appendix II of CITES, and the conservation project of endangered small population

绿孔雀
Pavo muticus

吴飞 摄
By Fei Wu

青头潜鸭
Aythya baeri

极危 CR A2cd+3cd+4cd

数据缺乏 DD	无危 LC	近危 NT	易危 VU	濒危 EN	极危 CR	区域灭绝 RE	野外灭绝 EW	灭绝 EX

分类地位 Taxonomic Status

动物界 Animalia	脊索动物门 Chordata	鸟纲 Aves	雁形目 Anseriformes	鸭科 Anatidae

学名 Scientific Name	*Aythya baeri*
命名人 Species Authority	Radde, 1863
英文名 English Name(s)	Baer's Pochard
同物异名 Synonym(s)	*Anas baeri*
种下单元评估 Infra-specific Taxa Assessed	无 / None

评估信息 Assessment Information

评估年份 Year Assessed	2016
评定人 Assessor(s)	王海涛 / Haitao Wang
其他贡献人 Other Contributor(s)	刘宇 / Yu Liu

理由 Justification: 青头潜鸭历史上分布范围广且数量大，曾为我国常见狩猎鸟（高继宏，1992）。目前种群数量不足 1,000 只，繁殖区和越冬地种群数量都急剧下降，已从部分分布区消失。湿地破坏和狩猎是该种数量下降的关键原因。因此，列为极危等级 / *Aythya baeri* was used to be game birds with large population and distribution in history (Gao, 1992). The population is estimated to be less than 1,000 individuals and has sharply decreased in both breeding and wintering sites, and even locally extinct. It is thought that wetland destruction and over-harvesting are the key reasons for its decline. Thus, it is listed as Critically Endangered

评估历史 Assessment History: 《国家重点保护野生动物名录》：Ⅰ (2021)；《中国物种红色名录：第二卷 脊椎动物 下册》(2009)：易危 / *National Key Protected Wild Animal List*: I (2021); *China Species Red List: Vol. II Vertebrates Part 2* (2009): VU

地理分布 Geographical Distribution

国内分布 Domestic Distribution

繁殖于黑龙江、吉林、辽宁、北京、河北、内蒙古、湖北、安徽、河南等，主要在长江以南地区越冬 / It breeds in Heilongjiang, Jilin, Liaoning, Beijing, Hebei, Inner Mongolia (Nei Mongol), Hubei, Anhui, Henan *etc.*, and mainly winters in the south of Changjiang River

分布标注 Distribution Note

非特有种 / Non-endemic

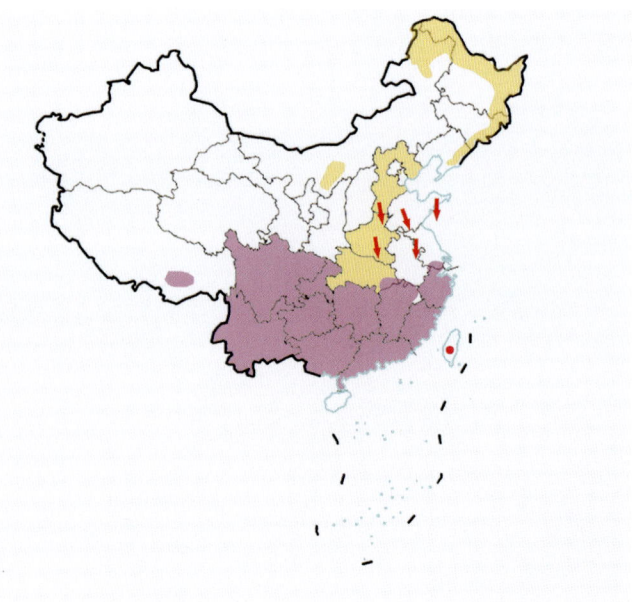

国内分布图
Map of Domestic Distribution

种群 Population

种群数量 Population Size	历史上数量较多，目前中国种群数量不足 1,000 只 (Wang et al., 2012)，东部越冬种群 850 只 (Cao et al., 2008)。黑龙江兴凯湖 2005 年调查到 2,900 只 (王凤坤等，2007)；江西鄱阳湖 2004～2007 年调查，种群数量稳定增长，分别调查到 263 只、285 只、1,213 只、1,916 只 (涂业苟等，2007)。2015 年在黑龙江调查到约 200 只，2016 年 3 月在江西九江附近观察到 150～200 只 (郭玉民等，2016)，2017 年 3 月，在河北衡水湖观测到 308 只，这是近十年记录的最大的青头潜鸭种群 / The population size in China is less than 1,000 (Wang et al., 2012), and there were 850 individuals of wintering populations in eastern China (Cao et al., 2008). Xingkai Lake in Heilongjiang province was investigated to 2,900 in 2005 (Wang et al., 2007), and Poyang Lake in Jiangxi province was surveyed in 2004～2007, and 263, 285, 1213, and 1916 individuals were observed (Tu et al., 2007). About 200 individuals were surveyed in Heilongjiang in 2015 and 150～200 individuals were observed near Jiujiang, Jiangxi province in March 2016 (Guo et al., 2016). In March 2017, 308 individuals were observed in Hengshui Lake, Hebei province, which is the largest population of this species in last ten years
种群趋势 Population Trend	下降 / Decreasing

生境与生态系统 Habitat(s) and Ecosystem(s)

生 境 Habitat(s)	繁殖栖息地为多水生植物的湖泊、河流和池塘，营巢于河湖岸边的草丛中或灌丛下，有时为浮巢。越冬栖息地主要在淡水湖泊和水库 / A. baeri breeds in lakes, rivers and ponds with rich aquatic vegetation, and nests in grasses of shrubs along the banks, sometimes floating. In winter, it occurs on freshwater lakes and reservoirs
生态系统 Ecosystem(s)	湿地生态系统 / Wetland Ecosystem

威胁 Threat(s)

主要威胁 Major Threat(s)	狩猎、栖息地丧失和退化 (Cao et al., 2010)，湿地开垦、鱼类过捕和人为干扰加剧 (王凤坤，2007) / Over-harvesting and habitat loss and degradation (Cao et al., 2010); reclamation of wetlands, over fishing and increased human disturbance (Wang Fengkun, 2007)

保护级别与保护行动 Protection Category and Conservation Action(s)

IUCN 红色名录 (2016) IUCN Red List (2016)	极危 / CR A2cd+3cd+4cd
保护行动 Conservation Action(s)	列入国家 I 级重点保护野生动物和迁徙物种保护公约 (CMS) 附录 I，越冬分布范围和重要越冬地不明确，部分繁殖地和越冬地处于自然保护区内。2018 年，在河北衡水召开的青头潜鸭保护国际研讨会制定了中国青头潜鸭保护行动计划 / It has been listed as the first class in National Key Protected Wild Animal List, Appendix I of the Convention on Migratory Species (CMS). The wintering range and key sites are not clear. Some of the breeding and wintering sites are in nature reserves. In 2018, the International Symposium on the Protection of the Baer's Pochard held in Hengshui, formulated the China Baer's Pochard Protection Action Plan

青头潜鸭 Aythya baeri

韦铭 摄 By Ming Wei

长尾鸭
Clangula hyemalis

极危 CR C2a(i)

| 数据缺乏 DD | 无危 LC | 近危 NT | 易危 VU | 濒危 EN | 极危 CR | 区域灭绝 RE | 野外灭绝 EW | 灭绝 EX |

分类地位 Taxonomic Status

动物界 Animalia	脊索动物门 Chordata	鸟 纲 Aves	雁形目 Anseriformes	鸭 科 Anatidae
学 名 Scientific Name		*Clangula hyemalis*		
命名人 Species Authority		Linnaeus, 1758		
英 文 名 English Name(s)		Long-tailed Duck		
同物异名 Synonym(s)		*Anas hyemalis*		
种下单元评估 Infra-specific Taxa Assessed		无 / None		

评估信息 Assessment Information

评估年份 Year Assessed	2016
评定人 Assessor(s)	陈水华 / Shuihua Chen
其他贡献人 Other Contributor(s)	王思宇 / Siyu Wang

理由 Justification: 长尾鸭在20世纪90年代以前以较大数量在我国东北部沿海地区越冬或迁徙停歇，数量可达2,000只以上，但2000年以后变得罕见，种群规模一般不超过50只，估计国内每年过境的个体数量不超过250只。可能是迁徙路线改变，也可能是因为该物种全球范围内的种群数量下降所致。因此，列为极危等级 / *Clangula hyemalis* wintered or migrated through northeast coast before 1990s, its population size is over 2,000. It became less seen after 2000 with a small population of less than 50 individuals. It is estimated that the population size in China is less than 250 individuals. The cause might be change of migrate route, or the decrease of its global population. Thus, it is listed as Critically Endangered

评估历史 Assessment History: 《中国物种红色名录：第二卷 脊椎动物 下册》(2009)：无危 / *China Species Red List: Vol. II Vertebrates Part 2* (2009): LC

地理分布 Geographical Distribution

国内分布 Domestic Distribution

中国罕见冬候鸟和旅鸟，在黑龙江、吉林、辽宁东部、河北东北部、北京、天津、山东、河南、山西、内蒙古东部、甘肃、新疆、四川中部、重庆、湖南、江苏、浙江、福建、广东有记录 / It is rare winter visitor and passage migrant in China. There were records in Heilongjiang, Jilin, east Liaoning, northeast Hebei, Beijing, Tianjin, Shandong, Henan, Shanxi, northeast Inner Mongolia (Nei Mongol), Gansu, Xinjiang, middle Sichuan, Chongqing, Hunan, Jiangsu, Zhejiang, Fujian, Guangdong

分布标注 Distribution Note

非特有种 / Non-endemic

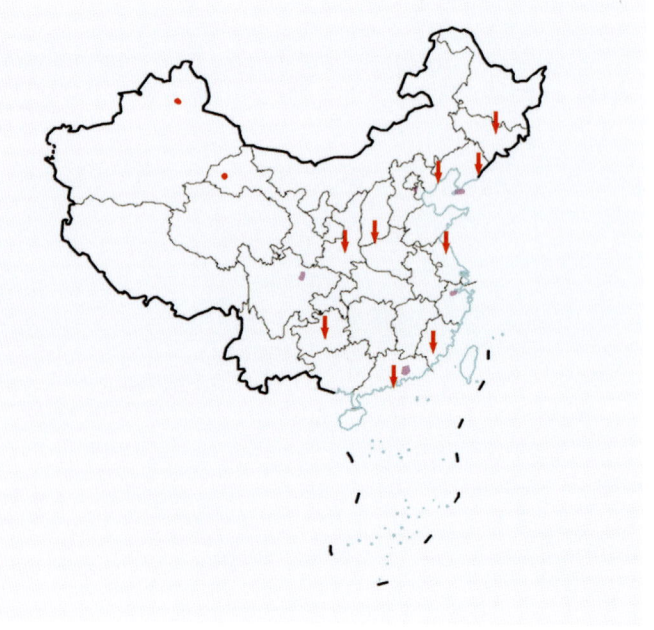

国内分布图
Map of Domestic Distribution

种群 Population

种群数量 Population Size	目前种群数量小于 250 只。1990 年辽宁大王家岛附近海域记录到共计 431 群 2,056 只，分布在开阔海面或海湾（韩晓东等，1994）；2009 年秋天在甘肃敦煌西湖地区党河水库记录到 2 只（邱冠华，2009），2006～2007 年在新疆 3 个湖泊中记录到该物种的 1～2 个个体（马鸣和梅宇，2007；Holt，2008）/ The population size in China is less than 250 individuals. 2,056 individuals in 431 flocks were recorded in open water or gulf around Dawangjia Islands, Liaoning province in 1990 (Han *et al.*, 1994). Two were recorded in Danghe Reservoir in Dunhuang West Lake, Gansu province in autumn of 2009 (Qiu, 2009). 1～2 individuals were recorded in three lakes in Xinjiang (Ma and Mei, 2007; Holt, 2008)
种群趋势 Population Trend	下降 / Decreasing

生境与生态系统 Habitat (s) and Ecosystem (s)

生境 Habitat(s)	夏季出没于草地和矮桦树林，包括苔原植被区。繁殖期主要栖息于北极冻原上的水塘和小型湖泊中，也栖息在流速缓慢的河流及河口地区。非繁殖季节主要栖息在沿海水域、海岛和海湾，偶尔到内陆大的湖泊与江河中 / *C. hyemalis* inhabits grasslands, birch forests and tundra in summer. It winters in coastal area, brackish lagoons, islands and gulf, and inland (very rarely) freshwater rivers or lakes; and breeds in pools and small lakes in Arctic tundra, as well as slow rivers and estuaries
生态系统 Ecosystem(s)	湿地和苔原生态系统 / Wetland and Tundra Ecosystem

威胁 Threat (s)

主要威胁 Major Threat(s)	西部地区近年来的开发活动使环境变化剧烈，筑坝截留引发断流，抽取湖水晒盐，开挖蟹池；捕捞卤虫，捡蛋，偷猎；上游工业污染，滥用农药、毒鼠药等；内陆湖泊湖面锐减，水质下降，食物资源紧张（马鸣等，2007）/ *C. hyemalis* is threatened by environment change owing to the development of the west areas: the cutoff of rivers by dams, pumping water from lakes for salt extraction, reclamation for crab cultivation; egg collection, illegal hunting; industrial pollution from upper reach, abuse of pesticide and raticide; the decline of open waters, deterioration of water quality, and shortage of food resources (Ma *et al.*, 2007)

保护级别与保护行动 Protection Category and Conservation Action (s)

IUCN 红色名录 (2016) IUCN Red List (2016)	易危 / VU A4bce
保护行动 Conservation Action(s)	列入中国《国家保护的有益的或者有重要经济、科学研究价值的陆生野生动物名录》/ It has been listed as *National Protected List of Terrestrial Wild Animals with Good Benefits or Important Economic and Scientific Values*

长尾鸭 *Clangula hyemalis*　　　　　　　　钱斌 摄　By Bin Qian

白头硬尾鸭 Oxyura leucocephala

中国生物多样性红色名录 China's Red List of Biodiversity

极危 CR A2bcde+4bcde; B1ab(i); C2a(i)

| 数据缺乏 DD | 无危 LC | 近危 NT | 易危 VU | 濒危 EN | **极危 CR** | 区域灭绝 RE | 野外灭绝 EW | 灭绝 EX |

分类地位 Taxonomic Status

动物界 Animalia	脊索动物门 Chordata	鸟纲 Aves	雁形目 Anseriformes	鸭科 Anatidae
学 名 Scientific Name		*Oxyura leucocephala*		
命 名 人 Species Authority		Scopoli, 1769		
英 文 名 English Name(s)		White-headed Duck		
同物异名 Synonym(s)		*Anser leucocephala*, *Anas leucocephala*		
种下单元评估 Infra-specific Taxa Assessed		无 / None		

评估信息 Assessment Information

评估年份 Year Assessed	2016
评定人 Assessor(s)	马鸣 / Ming Ma
其他贡献人 Other Contributor(s)	张同、梅宇、赵序茅、苟军、张耀东、邢睿、黄亚慧、王传波、林宣龙、夏咏 / Tong Zhang, Yu Mei, Xumao Zhao, Jun Gou, Yaodong Zhang, Rui Xing, Yahui Huang, Chuanbo Wang, Xuanlong Lin, Yong Xia

理由 Justification: 过去55年国内一直没有白头硬尾鸭的记录,近年才有零星分布(马鸣和梅宇,2007),国内种群数量不足150只。在新疆北部,其分布区域狭窄,见于少数几个封闭的水域,种群数量很不稳定。孤立散布,面临迅速绝灭的危险。因此,列为极危等级 / No *Oxyura leucocephala* was recorded in past 55 years, and scattered individuals were recorded recent years (Ma and Mei, 2007). The population size is less than 150 individuals. The distribution area in Xinjiang is small and only in several closed waters. The population is not stable and isolated, which is under the pressure of extinct. Thus, it is listed as Critically Endangered

评估历史 Assessment History:《国家重点保护野生动物名录》:I(2021);《中国濒危动物红皮书·鸟类》(1998):稀有;《中国物种红色名录:第二卷 脊椎动物 下册》(2009):濒危 / *National Key Protected Wild Animal List*: I (2021); *China Red Data Book of Endangered Animals*: Aves (1998): R; *China Species Red List: Vol. II Vertebrates Part 2* (2009): EN

地理分布 Geographical Distribution

国内分布 Domestic Distribution

分布于新疆的石河子、博乐、精河、奎屯、福海、乌鲁木齐、米泉、五家渠、吉木萨尔等县(马鸣,2011b)。在内蒙古和湖北偶然记录 / It occurs in Shihezi, Bole, Jinghe, Kuitun, Fuhai, Urumqi, Miquan, Wujiaqu, Jimsar etc. (Ma, 2011b). It occasionally occurs in Inner Mongolia (Nei Mongol) and Hubei

分布标注 Distribution Note

非特有种 / Non-endemic

国内分布图 Map of Domestic Distribution

种群 Population

种群数量 Population Size	初步估计新疆有 120 余只（繁殖鸟），2007 年在新疆乌鲁木齐白鸟湖发现 47 只 / The breeding population in Xinjiang is estimated to be more than 120 individuals. In 2007, 47 individuals were found in Bainiao Lake, Urumqi, Xinjiang
种群趋势 Population Trend	波动大，不稳定 / Not stable with large fluctuation

生境与生态系统 Habitat (s) and Ecosystem (s)

生　　　境 Habitat(s)	在新疆的奎屯、乌鲁木齐等地，白头硬尾鸭 4 月迁来繁殖地，9 月迁离。其繁殖期主要栖息于内陆开阔平原地区的淡水湖泊、封闭型的水域，尤其喜欢在紧靠大的湖泊附近的一些浅水小湖和水塘营巢繁殖。夏天活动于具有岸边植物和挺水植物的淡水湖泊、城市废水池，偶尔也到盐水湖附近。冬季主要栖息于开阔的大型湖泊，特别是有水边植物的湖泊，也栖息于盐水湖泊和沿海地带。白头硬尾鸭主要通过潜水觅食，也常在水边浅水处觅食。食物主要以眼子菜、水草等水生植物为主，也吃昆虫幼虫、小鱼、蛙、甲壳类、软体动物、蠕虫等水生动物 / In breeding sites in Kuitun and Urumqi in Xinjiang, *O. leucocephala* arrives at April, and migrate to wintering sites in September and October. It uses freshwater lakes in inland open plain during breeding, especially shallow lakes or pools close to large lakes, normally with emergent plants in lakes or vegetation around. It also occurs in waste waters in cities and occasionally uses salty lakes. It winters in open large lakes in winters, especially those with hydrophyte, sometimes use salty lakes and coastal areas. It forages underwater or at the shallow water. The main food include hydrophyte (pondweeds and weeds) and aquatic animals (larvae, small fish, frog, Crustanceans, molluscs, and worms)
生态系统 Ecosystem(s)	湿地生态系统 / Wetland Ecosystem

威胁 Threat (s)

主要威胁 Major Threat(s)	在新疆，数量下降的主要原因是近几年来 (2007～2014 年) 人类活动的干扰和繁殖地破坏，包括在栖息地附近采石放炮、捡蛋、毁巢、偷猎、城市扩张及建设开发区、西部移民、废物处理及污水排放、过度放牧、农业开垦、水位下降、旅游娱乐与焚烧芦苇等 / The population size of *O. leucocephala* has decreased in recent years (2007～2014) as a result of human disturbance and habitat loss, including quarrying, eggs collection, destroy of nests, illegal hunting, development of city, immigration in the west, industry and domestic waste, over grazing, reclamation, falling of water level, tourism and burning reeds, *etc.*

保护级别与保护行动 Protection Category and Conservation Action (s)

IUCN 红色名录 (2016) IUCN Red List (2016)	濒危 / EN A2bcde+4bcde
保护行动 Conservation Action(s)	列入国家 I 级重点保护野生动物、《国家保护的有益的或者有重要经济、科学研究价值的陆生野生动物名录》、CITES 公约附录 II 和迁徙物种保护公约 (CMS) 附录 I。白头硬尾鸭在中国过去几十年都没有确切记录。只是最近几年才发现几个零星的繁殖地。栖息地丧失非常快，种群数量屈指可数，岌岌可危。尚未受到应有的保护 / It has been listed as the first class in National Key Protected Wild Animal List, *National Protected List of Terrestrial Wild Animals with Good Benefits or Important Economic and Scientific Values*, Appendix II of CITES, and Appendix I of the Convention on Migratory Species (CMS). No *O. leucocephala* was recorded in past tens of years, and only scattered breeding sites were found in recent years. The population size is extremely small and the habitats have been destroyed rapidly. The species has not been under necessary protection

白头硬尾鸭 *Oxyura leucocephala*　　　　　　　　　　　　　　李韬 摄　By Tao Li

爪哇金丝燕
Aerodramus fuciphagus

极危　CR B1b(ii, iii); C1+2a(i)

| 数据缺乏 DD | 无危 LC | 近危 NT | 易危 VU | 濒危 EN | 极危 CR | 区域灭绝 RE | 野外灭绝 EW | 灭绝 EX |

分类地位 Taxonomic Status

动物界 Animalia	脊索动物门 Chordata	鸟纲 Aves	夜鹰目 Caprimulgiformes	雨燕科 Apodidae
学名 Scientific Name		*Aerodramus fuciphagus*		
命名人 Species Authority		Thunberg, 1812		
英文名 English Name(s)		Edible-nest Swiftlet		
同物异名 Synonym(s)		*Collocalia germani, Hirundo fuciphagus*		
种下单元评估 Infra-specific Taxa Assessed		无 / None		

评估信息 Assessment Information

评估年份 Year Assessed	2016
评定人 Assessor(s)	梁伟 / Wei Liang
其他贡献人 Other Contributor(s)	杨灿朝 / Canchao Yang

理由 Justification: 爪哇金丝燕分布区小，数量少，采集鸟巢等人类干扰和破坏严重。因此，列为极危等级 / The distribution range of *Aerodramus fuciphagus* is small. The population size of the isolated subspecies in China is small. It has been heavily threatened by human disturbance, *e.g.* nest collection. Thus, it is listed as Critically Endangered

评估历史 Assessment History: 《国家重点保护野生动物名录》：II (2021)；《中国物种红色名录：第二卷 脊椎动物 下册》(2009)：无危 / *National Key Protected Wild Animal List*: II (2021); *China Species Red List*: *Vol. II Vertebrates Part 2* (2009): LC

地理分布 Geographical Distribution

国内分布 Domestic Distribution
分布于海南大洲岛 / It occurs in Dazhou Island in Hainan
分布标注 Distribution Note
非特有种 / Non-endemic

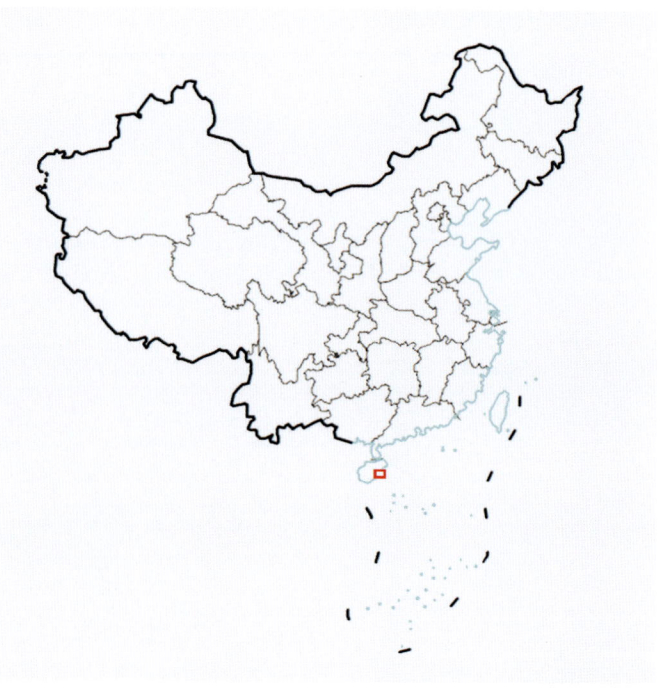

国内分布图
Map of Domestic Distribution

种群 Population

种群数量 Population Size	估计在海南大洲岛的种群数量 60～70 只，但还在持续下降 / It is estimated to be 60~70 individuals in Dazhou Island in Hainan, and the population is undergoing a continuing decline
种群趋势 Population Trend	下降 / Decreasing

生境与生态系统 Habitat (s) and Ecosystem (s)

生　　境 Habitat(s)	热带海岛 / Tropical island
生态系统 Ecosystem(s)	岛屿生态系统 / Island Ecosystem

威胁 Threat (s)

主要威胁 Major Threat(s)	分布区狭小，仅分布于海南的大洲岛。数量少，受人为干扰和破坏 / The distribution range is small and only in Dazhou Island in Hainan. The population is small and threatened by human disturbance

保护级别与保护行动 Protection Category and Conservation Action (s)

IUCN 红色名录 (2016) IUCN Red List (2016)	无危 / LC
保护行动 Conservation Action(s)	列入国家 II 级重点保护野生动物、《国家保护的有益的或者有重要经济、科学研究价值的陆生野生动物名录》。应加大对其繁殖地的保护，禁止采摘燕窝 / It has been listed as the second class in National Key Protected Wild Animal List, *National Protected List of Terrestrial Wild Animals with Good Benefits or Important Economic and Scientific Values*. The protection in breeding sites are suggested to be improved and nest collection should be prohibited

爪哇金丝燕 *Aerodramus fuciphagus*　　　　黄秦 摄　By Qin Huang

白鹤
Grus leucogeranus

极危　CR A3bcd+4bcd; B2b

| 数据缺乏 DD | 无危 LC | 近危 NT | 易危 VU | 濒危 EN | 极危 CR | 区域灭绝 RE | 野外灭绝 EW | 灭绝 EX |

分类地位 Taxonomic Status

动物界 Animalia	脊索动物门 Chordata	鸟纲 Aves	鹤形目 Gruiformes	鹤科 Gruidae
学　名 Scientific Name		*Grus leucogeranus*		
命名人 Species Authority		Pallas, 1773		
英文名 English Name(s)		Siberian Crane		
同物异名 Synonym(s)		*Leucogeranus leucogeranus*		
种下单元评估 Infra-specific Taxa Assessed		无 / None		

评估信息 Assessment Information

评估年份 Year Assessed	2016
评　定　人 Assessor(s)	马志军 / Zhijun Ma
其他贡献人 Other Contributor(s)	李凤山 / Fengshan Li

理由 Justification: 虽然越冬期的数量调查表明，近年来白鹤的种群数量保持稳定甚至略有增加，但这可能与数量调查的区域扩大有关。长江干流和支流的大型水利工程建设已导致白鹤栖息环境的改变，未来栖息地变化可能会进一步加剧，这将给白鹤的生存带来严重威胁。由于白鹤种群数量较小，栖息地的变化可能会导致白鹤种群数量急剧下降。因此，列为极危等级 / The wintering surveys indicated that the population of *Grus leucogeranus* is stable and even slightly increasing, which might be relate to the increase of survey area. The habitat has been changed and keeps changing with the development of the Three Gorges Dam and a large number of other dams on its tributaries, which would threaten the survival of *G. leucogeranus*. Its population is small and might be undergoing a sharply decline owing to the habitat change. Thus, it is listed as Critically Endangered

评估历史 Assessment History: 《国家重点保护野生动物名录》：Ⅰ (1989, 2021);《中国濒危动物红皮书·鸟类》(1998): 濒危;《中国物种红色名录: 第二卷 脊椎动物 下册》(2009): 极危 / National Key Protected Wild Animal List: I (1989, 2021); *China Red Data Book of Endangered Animals: Aves* (1998): E; *China Species Red List: Vol. II Vertebrates Part 2* (2009): CR

地理分布 Geographical Distribution

国内分布 Domestic Distribution

主要分布于东北到长江中下游地区。迁徙时见于河北、内蒙古东部、辽宁、吉林、黑龙江、安徽、山东、河南，越冬地主要在江西鄱阳湖和湖南洞庭湖，其中95%以上的个体在鄱阳湖越冬，越冬期间零星个体见于辽宁、江苏、上海、浙江、山东沿海地区 / It distributed from northeast to middle and lower Changjiang River. It occurs in Hebei, east Inner Mongolia (Nei Mongol), Liaoning, Jilin, Heilongjiang, Anhui, Shandong, Henan. When migrating, and winters mainly in Poyang Lake in Jiangxi and Dongting Lake in Hunan, with the vast majority (95%) of individuals wintering in Poyang Lake. Only a few individuals winters in the coastal of Liaoning, Jiangsu, Shanghai, Zhejiang and Shandong

分布标注 Distribution Note
非特有种 / Non-endemic

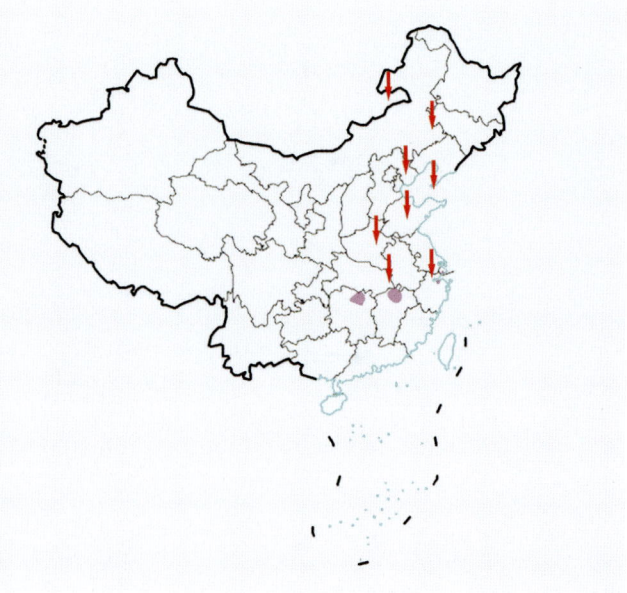

国内分布图
Map of Domestic Distribution

种群 Population

种群数量 Population Size	估计种群数量近 4,000 只 / The population is estimated to be 4,000 individuals
种群趋势 Population Trend	稳定 / Stable

生境与生态系统 Habitat (s) and Ecosystem (s)

生　　境 Habitat(s)	繁殖期于苔原湿地营巢，在湖泊湿地越冬。主要以陆生和水生植物幼嫩的根、茎、叶、种子、果实为食，偶尔也食底栖动物、昆虫、小鱼及其他小型脊椎动物。在越冬地主要摄取苦草等水生植物的冬芽 / G. leucogeranus nests in tundra wetlands and winters in lakes. Its diet consists primarily of roots, tubers, leaves, seeds, and fruits, and occasionally benthonic animals, insects, fish and other small animals. During the non-breeding season it feeds mainly on tubers (especially of sedges) of aquatic plants
生态系统 Ecosystem(s)	湿地生态系统 / Wetland Ecosystem

威胁 Threat (s)

主要威胁 Major Threat(s)	湿地丧失与退化导致的栖息地丧失和质量下降是其面临的主要威胁。三峡大坝修建后使鄱阳湖的水文状况发生了巨大改变，未来有可能在鄱阳湖修建大坝，这使得其栖息地面临着严重威胁 / The key threat is wetland loss and degradation at wintering sites. Construction of the Three Gorges Dam has greatly changed the hydrology and the proposed dam at the outlet to the Poyang lake might threaten the wintering grounds of G. leucogeranus

保护级别与保护行动 Protection Category and Conservation Action (s)

IUCN 红色名录 (2016) IUCN Red List (2016)	极危 / CR A3bcd+4bcd
保护行动 Conservation Action(s)	列入国家 I 级重点保护野生动物、CITES 公约附录 I 和迁徙物种保护公约 (CMS) 附录 I。对其繁殖生态、越冬生态都有过研究，对其迁徙路线开展过卫星跟踪。在主要的繁殖地和越冬地已建立多个自然保护区。在多家动物园有人工饲养的个体或种群 / It has been listed as the first class in National Key Protected Wild Animal List, Appendix I of CITES and Appendix I of the Convention on Migratory Species. There are studies on breeding and wintering ecology, as well as satellite tracking on migration route. Several nature reserves have been set up in its main breeding and wintering sites. There are feeding individual or population in several zoos

白鹤 *Grus leucogeranus*　　　　　　　　　　　　　　　　　　　　　　袁晓 摄　By Xiao Yuan

勺嘴鹬
Calidris pygmeus

中国生物多样性 红色名录 | China's Red List of Biodiversity

极危 CR A2bcd+3bcd+4bcd; C2a(i)

| 数据缺乏 DD | 无危 LC | 近危 NT | 易危 VU | 濒危 EN | **极危 CR** | 区域灭绝 RE | 野外灭绝 EW | 灭绝 EX |

分类地位 Taxonomic Status

动物界 Animalia	脊索动物门 Chordata	鸟纲 Aves	鸻形目 Charadriiformes	鹬科 Scolopacidae
学名 Scientific Name		*Calidris pygmeus*		
命名人 Species Authority		Linnaeus, 1758		
英文名 English Name(s)		Spoon-billed Sandpiper		
同物异名 Synonym(s)		无 / None		
种下单元评估 Infra-specific Taxa Assessed		无 / None		

评估信息 Assessment Information

评估年份 Year Assessed	2016
评定人 Assessor(s)	马志军 / Zhijun Ma
其他贡献人 Other Contributor(s)	中国沿海水鸟同步调查项目组 / China Coastal Waterbird Survey Group

理由 Justification: 长期对勺嘴鹬种群数量缺乏了解。近十余年来开始受到关注，发现其种群数量极少，且数量迅速下降。因此，列为极危等级 / No clear information on *Calidris pygmaea* for a long time and it came into attention in recent ten years. The population size is extremely small and it is undergoing a rapid decline. Thus, it is listed as Critically Endangered

评估历史 Assessment History: 《国家重点保护野生动物名录》: Ⅰ (2021); 《中国物种红色名录: 第二卷 脊椎动物 下册》(2009): 易危 / *National Key Protected Wild Animal List*: Ⅰ (2021); *China Species Red List: Vol. Ⅱ Vertebrates Part 2* (2009): VU

地理分布 Geographical Distribution

国内分布 Domestic Distribution

在我国主要为旅鸟，春季于4～5月、秋季于9～10月迁经中国，沿海各省均有分布记录，少量个体在福建、香港、广东、广西等华南沿海地区越冬 / It migrates across China in April to May in spring and September to October in autumn. There are records in coastal provinces and only a few individual winters in South China coast, *e.g.* Fujian, Hong Kong, Guangdong, and Guangxi

分布标注 Distribution Note

非特有种 / Non-endemic

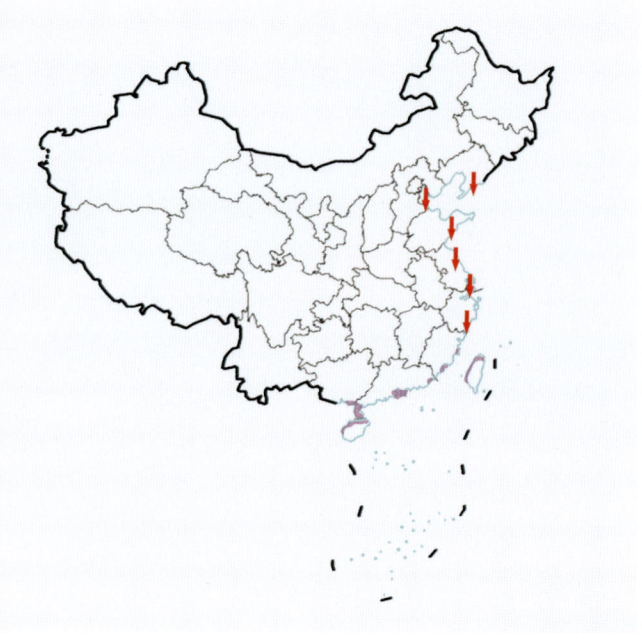

国内分布图
Map of Domestic Distribution

种群 Population

种群数量 Population Size	估计种群数量不超过 1,000 只 / The population size is estimated to be less than 1,000 individuals
种群趋势 Population Trend	目前缺乏种群数量的监测数据,但在最近十年间快速下降 / There is no exact monitoring on the species, while it is undergoing a rapid decline

生境与生态系统 Habitat(s) and Ecosystem(s)

生　　境 Habitat(s)	繁殖于西伯利亚东北部的滨海苔原地带。营巢于沼泽、湖泊、水塘边和海岸苔原与草地上。迁徙停歇期和越冬期在滨海及河口滩涂湿地活动。繁殖期主要以昆虫等无脊椎动物为食,迁徙停歇期和越冬期主要以滨海湿地的底栖动物为食,但具体食物组成尚不了解 / *C. pygmaea* breeds in coastal tundra and grassland along the marshes, lakes and lagoons in northeast Siberia. It uses coastal areas and estuary mudflats during winters and migration. It feeds on insects when breeding and benthic invertebrate during winters and migration, while the exact diets are not clear
生态系统 Ecosystem(s)	湿地生态系统 / Wetland Ecosystem

威胁 Threat(s)

主要威胁 Major Threat(s)	早期由于缺乏资料,难以估计勺嘴鹬的准确数量。根据近年调查,全球勺嘴鹬总数量不超过 1,000 只,且数量仍在快速下降。江苏如东在 2011 年秋季曾一次记录到勺嘴鹬 103 只,2013 年 10 月调查记录到的数量达 143 只,说明该区域对迁徙勺嘴鹬的重要性。滩涂围垦、开发以及外来植物互花米草入侵所造成的栖息地丧失和退化是其面临的主要威胁。另外,非法捕猎对其种群也有很大影响。由于种群数量极少,生存状况非常危险 / It is difficult to estimate the population size in history without sufficient data. The global population is estimated to be less than 1,000 individuals based on recent surveys. 103 and 143 individuals were recorded in 2011 and 2013 in Rudong in Jiangsu, which revealed the importance of the site. The major threat is habitat loss owing to development and reclamation of mudflat, and invasion of *Spartina alterniflora*. Furthermore, illegal hunting also exerts negative effect. The population size is extremely small and the survival of *C. pygmaea* is at stake

保护级别与保护行动 Protection Category and Conservation Action(s)

IUCN 红色名录 (2016) IUCN Red List (2016)	极危 / CR A2abcd+3bcd+4abcd; C2a (i)
保护行动 Conservation Action(s)	列入国家 I 级重点保护野生动物、《国家保护的有益的或者有重要经济、科学研究价值的陆生野生动物名录》和迁徙物种保护公约 (CMS) 附录 I。勺嘴鹬在我国的主要迁徙停歇地和越冬地闽江口已建立了自然保护区。但在我国记录数量最多的江苏如东,目前仍未建立自然保护区 / It has been listed as the first class in National Key Protected Wild Animal List, *National Protected List of Terrestrial Wild Animals with Good Benefits or Important Economic and Scientific Values*, and the Appendix I of the Convention on Migratory Species. There is Minjiangkou Nature Reserve in its main stopover and wintering sites in China, while no nature reserve has been set up in Rudong, with the largest recorded population

勺嘴鹬 *Calidris pygmeus*　　　　　　　　　　　　　　袁晓 摄　By Xiao Yuan

中华凤头燕鸥
Thalasseus bernsteini

极危　CR C2a(i)+b

| 数据缺乏 DD | 无危 LC | 近危 NT | 易危 VU | 濒危 EN | **极危 CR** | 区域灭绝 RE | 野外灭绝 EW | 灭绝 EX |

分类地位 Taxonomic Status

动物界 Animalia	脊索动物门 Chordata	鸟纲 Aves	鸻形目 Charadriiformes	鸥科 Laridae
学名 Scientific Name		*Thalasseus bernsteini*		
命名人 Species Authority		Schlegel, 1864		
英文名 English Name(s)		Chinese Crested Tern		
同物异名 Synonym(s)		*Sterna bernsteini*		
种下单元评估 Infra-specific Taxa Assessed		无 / None		

评估信息 Assessment Information

评估年份 Year Assessed	2016
评定人 Assessor(s)	陈水华 / Shuihua Chen
其他贡献人 Other Contributor(s)	无 / None

理由 Justification: 目前中华凤头燕鸥仅有3个繁殖群体，总个体数量约100只，且存在一定的波动。单个群体的成熟个体数低于50只。主要威胁因素依然普遍存在，未得到消除。因此，列为极危等级 / There are about 100 *Thalasseus bernsteini* individuals in only three breeding populations, and the mature individuals are less than 50. The major threats have not been eliminated. Thus, it is listed as Critically Endangered

评估历史 Assessment History:《国家重点保护野生动物名录》：Ⅱ (1989)，Ⅰ (2021)；《中国濒危动物红皮书·鸟类》(1998)：易危；《中国物种红色名录：第二卷 脊椎动物 下册》(2009)：极危 / *National Key Protected Wild Animal List*: Ⅱ (1989), Ⅰ (2021); *China Red Data Book of Endangered Animals*: Aves (1998): V; *China Species Red List*: Vol. Ⅱ Vertebrates Part 2 (2009): CR

地理分布 Geographical Distribution

国内分布 Domestic Distribution

主要繁殖于浙江和福建沿海，也见于天津、河北、山东、江苏、上海、广东、台湾沿海以及海南西沙 / It breeds mainly in the coastal of Zhejiang and Fujian, and also occurs in coastal of Tianjin, Hebei, Shandong, Jiangsu, Shanghai, Guangdong, Taiwan and Xisha in Hainan

分布标注 Distribution Note

非特有种 / Non-endemic

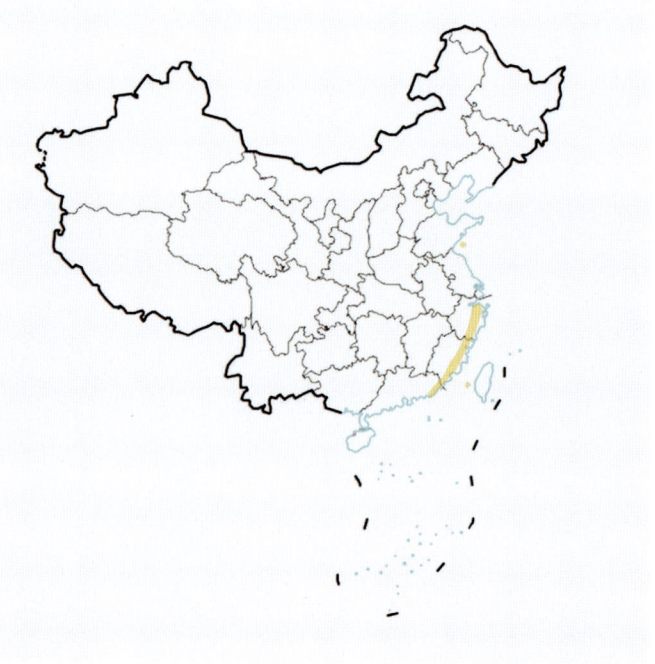

国内分布图
Map of Domestic Distribution

种群 Population

种群数量 Population Size	种群数量约 100 只 / The population size is about 100 individuals
种群趋势 Population Trend	下降 / Decreasing

生境与生态系统 Habitat (s) and Ecosystem (s)

生　　境 Habitat(s)	在偏远的以低矮灌草为主要植被的无人小海岛上繁殖，在草上或岩礁地面产卵。在周边海域觅食 (Chen *et al.*, 2011)。有时在滨海滩涂湿地栖息。越冬生境还不明确 / *T. bernsteini* breeds in offshore islets with short grassland and shrubs. It nests on the grass or rocky ground and forages along the coastal area (Chen *et al.*, 2011). Sometimes it uses coastal mudflat and no clear information on its wintering habitat
生态系统 Ecosystem(s)	海洋生态系统 / Marine Ecosystem

威胁 Threat (s)

主要威胁 Major Threat(s)	渔民上岛捡蛋是中华凤头燕鸥繁殖成功的最主要威胁，这一现象在东部沿海非常普遍，尚未得到遏制 (Chen *et al.*, 2009)。东部沿海夏季盛行的台风也是繁殖燕鸥的主要威胁 (Chen *et al.*, 2015)。其他威胁还包括海洋污染、过度捕捞、人为干扰和鼠害等 (Chen *et al.*, 2009) / The major threat is eggs collection by fishermen, and the situation is widespread and has not been under control (Chen *et al.*, 2009). And many of the nest failures attributed to typhoons (Chen *et al.*, 2015). Besides, other threats include ocean pollution, over fishing, human disturbance and rodents (Chen *et al.*, 2009)

保护级别与保护行动 Protection Category and Conservation Action (s)

IUCN 红色名录 (2016) IUCN Red List (2016)	极危 / CR C2a (i,ii); D
保护行动 Conservation Action(s)	列入国家 I 级重点保护野生动物名录、迁徙物种保护公约 (CMS) 附录 I。在现有的三个繁殖地，均已建立了保护区，分别为马祖燕鸥保护区、浙江韭山列岛国家级自然保护区和浙江五峙山列岛鸟类省级自然保护区。对于在本区域内的繁殖个体，三个保护区均实施了长期的监测和严格的保护。同时，海峡两岸针对该珍稀物种的保育开展了深入的交流，保护宣传工作在三地也相继开展 (陈水华和范忠勇, 2013)。2013 年开始，浙江自然博物馆和美国俄勒冈州立大学合作，在韭山列岛的中华凤头燕鸥种群招引和恢复项目取得成功，为该珍稀物种的拯救和保护带来了希望 (陈水华等, 2015) / It has been listed as the first class in National Key Protected Wild Animal List, and in the Appendix I of the Convention on Migratory Species. There have been set up three nature reserves in its main breeding sites, specifically, Matsu Islands Nature Reserve, Jiushan Islands National Nature Reserve, and Wuzhishan Islands Provincial Nature Reserve. Long term monitoring and strict conservation have been taken for the breeding individuals in these three nature reserves. Communications and exchanges of conservation on rare species across the Taiwan Straits have kept increasing, and the public education has been promoted in the three sites (Chen and Fan, 2013). The great achievement of the introduction and restoration project on *T. bernsteini* in Jiushan Islands by Zhejiang Museum of Natural History and Oregon State University in 2013 has brought hope to the conservation of the species (Chen *et al.*, 2015)

中华凤头燕鸥 *Thalasseus bernsteini*　　　　陈水华 摄　By Shuihua Chen

中国生物多样性红色名录

黑头白鹮
Threskiornis melanocephalus

极危　CR A1acd+2bcd; C1+2(i); D

| 数据缺乏 DD | 无危 LC | 近危 NT | 易危 VU | 濒危 EN | 极危 CR | 区域灭绝 RE | 野外灭绝 EW | 灭绝 EX |

分类地位 Taxonomic Status

动物界 Animalia	脊索动物门 Chordata	鸟纲 Aves	鹈形目 Pelecaniformes	鹮科 Threskiornithidae
学名 Scientific Name		*Threskiornis melanocephalus*		
命名人 Species Authority		Latham, 1790		
英文名 English Name(s)		Black-headed Ibis		
同物异名 Synonym(s)		*Tantalus melanocephalus*		
种下单元评估 Infra-specific Taxa Assessed		东亚种群不足 100 只 / The east Asia population is less than 100 individuals		

评估信息 Assessment Information

评估年份 Year Assessed	2016
评定人 Assessor(s)	丁长青 / Changqing Ding
其他贡献人 Other Contributor(s)	刘垚 / Yao Liu

理由 Justification: 黑头白鹮东亚种群数量不足 100 只，受生境丧失、污染、捕猎和捡蛋等威胁，数量下降明显。至今国内已罕见。因此，列为极危等级 / The east Asia population of *Threskiornis melanocephalus* is less than 100 individuals. The population is undergoing a rapid decline owing to the habitat loss, pollution, hunting, and egg collection. It is already rarely seen in China. Thus, it is listed as Critically Endangered

评估历史 Assessment History: 《国家重点保护野生动物名录》：Ⅱ (1989)，Ⅰ (2021);《中国濒危动物红皮书·鸟类》(1998)：稀有;《中国物种红色名录：第二卷 脊椎动物 下册》(2009)：濒危 / National Key Protected Wild Animal List: Ⅱ (1989), Ⅰ (2021); China Red Data Book of Endangered Animals: Aves (1998): R; China Species Red List: Vol. Ⅱ Vertebrates Part 2 (2009): EN

地理分布 Geographical Distribution

国内分布 Domestic Distribution

繁殖于黑龙江。偶见于我国中、东部和西南地区等地。东亚种群数量不足 100 只 (Wetlands International, 2014)，近年在中国已罕见 / It breeds in Heilongjiang, and occurs in central, east and southwest of China. The east Asia population is less than 100 individuals (Wetlands International, 2014). It is already rarely seen in China

分布标注 Distribution Note

非特有种 / Non-endemic

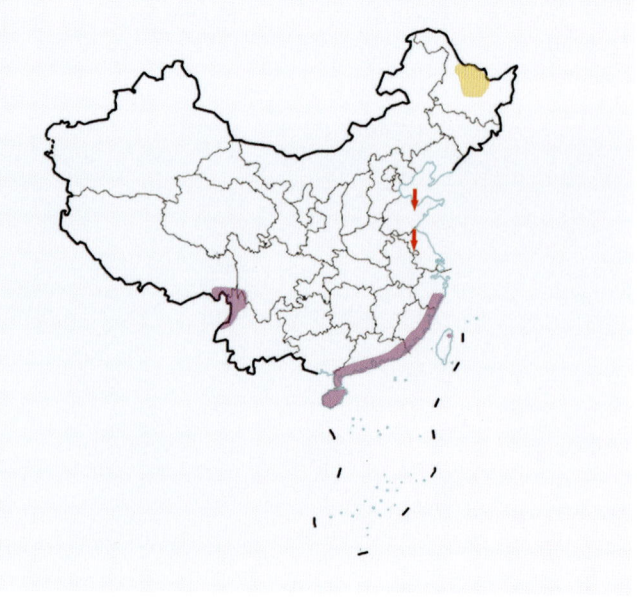

国内分布图
Map of Domestic Distribution

种群 Population

种群数量 Population Size	全球种群数量 19,000～21,800 只，其中成熟个体 13,000～15,000 只 (IUCN, 2012)。目前在中国偶见 / The global population size is 19,000～21,800 individuals, with 13,000～15,000 mature individuals. It is rare in China
种群趋势 Population Trend	下降 / Decreasing

生境与生态系统 Habitat(s) and Ecosystem(s)

生境 Habitat(s)	淡水沼泽、湖泊、河流、洪泛平原、潮间带、泥地滩涂和潟湖等 / Freshwater marshes, lakes, rivers, floodplains, intertidal flats, muddy flats and lagoons
生态系统 Ecosystem(s)	湿地生态系统 / Wetland Ecosystem

威胁 Threat(s)

主要威胁 Major Threat(s)	过度捕捞、割芦苇、污染和捡蛋，对其食物资源、筑巢、产卵和育雏产生影响 / Major threats include overfishing, reed harvest, pollution, food resource, and egg collection, which might influence food resource, nesting, and breeding

保护级别与保护行动 Protection Category and Conservation Action(s)

IUCN 红色名录 (2016) IUCN Red List (2016)	近危 / NT
保护行动 Conservation Action(s)	列入国家 I 级重点保护野生动物 / It has been listed into the first class in National Key Protected Wild Animal List

黑头白鹮 *Threskiornis melanocephalus* 　　　　　　　　　　　　　郑灼 摄　By Zhuo Zheng

黑兀鹫
Sarcogyps calvus

极危 CR A2abce+3bce+4abce; D

| 数据缺乏 DD | 无危 LC | 近危 NT | 易危 VU | 濒危 EN | 极危 CR | 区域灭绝 RE | 野外灭绝 EW | 灭绝 EX |

分类地位 Taxonomic Status

动物界 Animalia	脊索动物门 Chordata	鸟纲 Aves	鹰形目 Accipitriformes	鹰科 Accipitridae
学名 Scientific Name		*Sarcogyps calvus*		
命名人 Species Authority		Scopoli, 1786		
英文名 English Name(s)		Red-headed Vulture		
同物异名 Synonym(s)		*Vultur calvus*		
种下单元评估 Infra-specific Taxa Assessed		无 / None		

评估信息 Assessment Information

评估年份 Year Assessed	2016
评定人 Assessor(s)	马强 / Qiang Ma
其他贡献人 Other Contributor(s)	无 / None

理由 Justification: 黑兀鹫仅见于云南省，最后的记录是在20世纪60年代末，数量十分稀少，其国内外种群下降趋势显著。因此，列为极危等级 / *Sarcogyps calvus* occurs only in Yunnan. The last record was in 1960s, the population size was small. The population is undergoing a rapid decline. Thus, it is listed as Critically Endangered

评估历史 Assessment History: 《国家重点保护野生动物名录》：Ⅱ (1989)，Ⅰ (2021)；《中国濒危动物红皮书·鸟类》(1998)：濒危；《中国物种红色名录：第二卷 脊椎动物 下册》(2009)：濒危 / *National Key Protected Wild Animal List*: Ⅱ (1989), Ⅰ (2021); *China Red Data Book of Endangered Animals*: Aves (1998): E; *China Species Red List*: Vol. Ⅱ Vertebrates Part 2 (2009): EN

地理分布 Geographical Distribution

国内分布 Domestic Distribution	
仅云南西部有分布记录 / There are only records in west Yunnan	
分布标注 Distribution Note	
非特有种 / Non-endemic	

国内分布图
Map of Domestic Distribution

种群 Population

种群数量 Population Size	全球种群数量 3,500 ～ 15,000 只 / The global population size is estimated to be 3,500 ～ 15,000 individuals
种群趋势 Population Trend	下降 / Decreasing

生境与生态系统 Habitat(s) and Ecosystem(s)

生　　　境 Habitat(s)	植被良好的山地（海拔 2,500 m 以下）、开阔的低山丘陵生境 / *S. calvus* appeared well-wooded mountains (less than 2,500 m), and open small hills
生态系统 Ecosystem(s)	陆地生态系统 / Terrestrial Ecosystem

威胁 Threat(s)

主要威胁 Major Threat(s)	森林采伐和农业开发导致其栖息地破碎化，栖息地质量下降；大范围灭鼠导致的二次中毒；残留在死亡家畜体内的药物导致其慢性中毒死亡；农药、除草剂等污染物通过生物富集作用带来严重威胁；密集的高压输电设施增加了撞击受伤风险；对鸟巢的破坏和对其巢区的干扰会降低繁殖成功率，种群恢复困难 / Habitat fragmentation and degradation owing to deforestation and development of agriculture; adversely affected by the drug of rats extermination and dead domestic animals; bioamplification of toxin by pesticide and herbicide; dense powerlines increase the risk of attack and hurt; decrease of breeding success owing to destroy and disturbance of nests

保护级别与保护行动 Protection Category and Conservation Action(s)

IUCN 红色名录 (2016) IUCN Red List (2016)	极危 / CR A2abce+3bce+4abce
保护行动 Conservation Action(s)	列入国家 I 级重点保护野生动物和 CITES 公约附录 II / It has been listed as the first class in National Key Protected Wild Animal List and the Appendix II of CITES

黑兀鹫 *Sarcogyps calvus*　　　　　　　　　高宏颖 摄　By Hongying Gao

中国生物多样性 红色名录
China's Red List of Biodiversity

毛腿雕鸮
Bubo blakistoni

极危　CR A1ac; B1b(ii, iii); C1+2a(i)

| 数据缺乏 DD | 无危 LC | 近危 NT | 易危 VU | 濒危 EN | **极危 CR** | 区域灭绝 RE | 野外灭绝 EW | 灭绝 EX |

分类地位 Taxonomic Status

动物界 Animalia	脊索动物门 Chordata	鸟纲 Aves	鸮形目 Strigiformes	鸱鸮科 Strigidae
学　名 Scientific Name		*Bubo blakistoni*		
命名人 Species Authority		Seebohm, 1884		
英文名 English Name(s)		Blakiston's Eagle Owl		
同物异名 Synonym(s)		*Ketupa blakistoni*		
种下单元评估 Infra-specific Taxa Assessed		无 / None		

评估信息 Assessment Information

评估年份 Year Assessed	2016
评定人 Assessor(s)	朱磊 / Lei Zhu
其他贡献人 Other Contributor(s)	诸川汇、杨小农 / Chuanhui Zhu, Xiaonong Yang

理由 Justification: 毛腿雕鸮为远东地区特有大型鸮类。全球种群数量1,500～4,000只，国内种群数量少于50对。主要生活在高纬度河流两岸的成熟原生林中，冬季在不结冰的开放水域捕食，繁殖季需要较大的树洞以营巢，对生境质量要求很高。砍伐森林、沿河流进行的开发及水利设施修建等活动，导致该种的适宜栖息地片断化或丧失，国内已有相当长时间未见该种的报道。但2014年4月发现在吉林有非法交易该种雏鸟的记录，表明该种在国内仍有分布，并且正面临偷猎的威胁。因此，列为极危等级 / Large endemic owl in far east. The global population of *Bubo blakistoni* is estimated to number 1,500～4,000 individuals, and the population of China is less than 50 pairs. It uses the mature and primitive forests along large rivers at high latitude. *B. blakistoni* forages at open and non-frozen water during winters and nests in hollows of large trees, with high quality of habitat. Deforestation, development along the rivers and river regulation projects might deduce the habitat loss and fragmentation. No individual of the species was seen for a long time while there was a record of illegal trading of its subadult in April, 2014, which indicated the distribution of the species and the threat of illegal hunting. Thus, it is listed as Critically Endangered

评估历史 Assessment History: 《国家重点保护野生动物名录》：Ⅱ (1989), Ⅰ (2021);《中国物种红色名录：第二卷 脊椎动物 下册》(2009): 濒危 / *National Key Protected Wild Animal List*: Ⅱ (1989), Ⅰ (2021); *China Species Red List*: Vol. Ⅱ Vertebrates Part 2 (2009): EN

地理分布 Geographical Distribution

国内分布 Domestic Distribution
分布于黑龙江(小兴安岭)、吉林(延边), 内蒙古东北部 / It occurs in Heilongjiang (Xiao Hinggan Ling), Jilin (Yanbian) and northeast Inner Mongolia (Nei Mongol)
分布标注 Distribution Note
非特有种 / Non-endemic

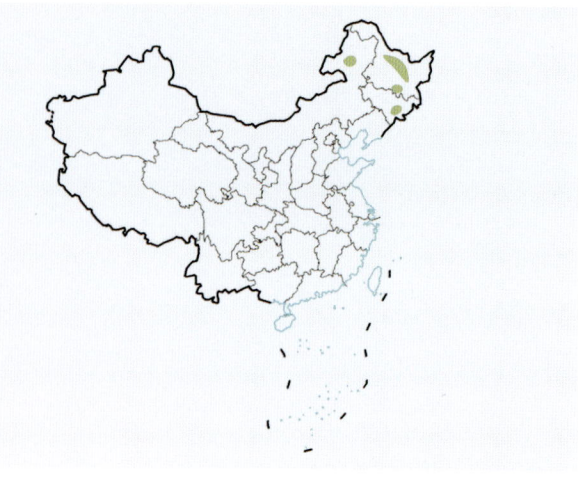

国内分布图
Map of Domestic Distribution

种群 Population

种群数量 Population Size	种群数量不足 100 只 / The population size is less than 100 individuals
种群趋势 Population Trend	下降 / Decreasing

生境与生态系统 Habitat (s) and Ecosystem (s)

生　　境 Habitat(s)	低山阔叶林、针阔混交林中的河流、河谷等生境，偏好在水流速度较快、水生动物食物资源丰富且隐蔽条件好的林区河流活动。在俄罗斯和日本的分布区内也有在多岩的海滨活动的报道 / *B. blakistoni* inhabits in riverside and valleys in broad-leaved forests, mixed coniferous broad-leaved forests. It prefers high speed rivers, with abundant food resources and dense forests. It is also seen in rocky coastal areas in Russia and Japan
生态系统 Ecosystem(s)	陆地生态系统 / Terrestrial Ecosystem

威胁 Threat (s)

主要威胁 Major Threat(s)	伐木及农业开发导致的栖息地丧失与斑块化是影响种群数量的主要原因；兴建水坝导致洄游性鱼类减少，渔业滥捕也加重了其食物资源的减少；偷猎及其他人类干扰 / Habitat loss and fragmentation owing to deforestation and agricultural development along the rivers probably reduce the population size. Other threats include shortage of food resources owing to dam construction and overfishing, illegal hunting and human disturbance

保护级别与保护行动 Protection Category and Conservation Action (s)

IUCN 红色名录 (2016) IUCN Red List (2016)	濒危 / EN C2a(i)
保护行动 Conservation Action(s)	列入国家 I 级重点保护野生动物和 CITES 公约附录 II，其分布区内已建立多个自然保护区。应继续加强栖息地保护，恢复和扩大栖息地面积。加强公众教育，对分布区种群数量和分布状况进行动态监测，控制其分布区内水坝等设施的修建，对捕鱼行为进行规范管理 / It has been listed as the first class in National Key Protected Wild Animal List and the Appendix II of the CITES. Several nature reserves have been set up in its distribution areas. Promotion of habitat conservation (restoration to enlarge habitats) are suggested. Other recommendations include improvement of public education, monitoring on its population size and distribution, control of dam construction and management on fishing

毛腿雕鸮 *Bubo blakistoni*　　　　　　　　　　　　　　　　　　　　　　龚本亮 摄　By Benliang Gong

冠斑犀鸟
Anthracoceros albirostris

极危　CR B1ab(iii); C1+2(i)

| 数据缺乏 DD | 无危 LC | 近危 NT | 易危 VU | 濒危 EN | 极危 CR | 区域灭绝 RE | 野外灭绝 EW | 灭绝 EX |

分类地位 Taxonomic Status

动物界 Animalia	脊索动物门 Chordata	鸟　纲 Aves	犀鸟目 Bucerotiformes	犀鸟科 Bucerotidae
学　　名 Scientific Name		*Anthracoceros albirostris*		
命　名　人 Species Authority		Shaw, 1808		
英　文　名 English Name(s)		Oriental Pied Hornbill		
同物异名 Synonym(s)		*Buceros albirostris*		
种下单元评估 Infra-specific Taxa Assessed		无 / None		

评估信息 Assessment Information

评估年份 Year Assessed	2016
评　定　人 Assessor(s)	周放 / Fan Zhou
其他贡献人 Other Contributor(s)	蒋爱伍 / Aiwu Jiang

理由 Justification：30 年前冠斑犀鸟还常见于广西西南部和云南的各大林区。由于偷猎和森林砍伐导致营巢树减少，尤其是自 20 世纪 80 年代以来，云南原始森林遭到大面积的砍伐而种植橡胶林，该物种的种群数量急剧下降，面临较大的灭绝风险。因此，列为极危等级 / *Anthracoceros albirostris* was common in forests in southwest Guangxi and Yunnan 30 years ago, while it is undergoing a rapid decline especially since 1980s, as a result of illegal trapping, deforestation of primary forests and planation of rubber forests. The probability of extinction is high for the species. Thus, it is listed as Critically Endangered

评估历史 Assessment History：《国家重点保护野生动物名录》：Ⅱ (1989)，Ⅰ (2021)；《中国濒危动物红皮书·鸟类》(1998)：易危；《中国物种红色名录：第二卷 脊椎动物 下册》(2009)：近危 / *National Key Protected Wild Animal List*: Ⅱ (1989), Ⅰ (2021); *China Red Data Book of Endangered Animals*: *Aves* (1998): V; *China Species Red List*: *Vol. II Vertebrates Part 2* (2009): NT

地理分布 Geographical Distribution

国内分布 Domestic Distribution
仅分布于广西西南部、云南的西部及南部和西藏东南部 / It only occurs in southwest Guangxi, west and south Yunnan, and southeast Tibet (Xizang)
分布标注 Distribution Note
非特有种 / Non-endemic

国内分布图
Map of Domestic Distribution

种群 Population

种群数量 Population Size	种群数量可能少于 500 只。1996～2001 年的全国陆生野生动物资源调查结果显示，冠斑犀鸟的种群数量约为 250 只，其中广西 150 只，云南 100 只，西藏未调查。广西境内的冠斑犀鸟数量以西大明山自然保护区种群最为集中，2009 年，西大明山自然保护区调查到的冠斑犀鸟的绝对数量为 43 只，估计其种群数量为 50～60 只 (罗益奎等，2013)，这也是目前已知的我国现存的最大种群。2014～2015 年，在云南和西藏开展的专项调查表明，云南记录到的冠斑犀鸟最大群有 20 多只 (勐腊县)，西藏墨脱县曾观察到的最大群为 10 只 / It might be less than 500 individuals. 250 individuals (150 in Guangxi, 100 in Yunnan, and no data in Tibet) were recorded in Terrestrial Wildlife Resources Survey of China during 1996 ～ 2001. In Guangxi, most of *A. albirostris* occur in Xidamingshan Nature Reserve. 43 individuals were recorded in 2009, and the population was estimated to be 50 ～ 60 individuals (Luo *et al.*, 2013), which is the largest known population in China. 20 individuals in Mengla of Yunnan and the largest group of 10 individuals in Motuo of Tibet were recorded in a specific survey during 2014 ～ 2015
种群趋势 Population Trend	下降 / Decreasing

生境与生态系统 Habitat (s) and Ecosystem (s)

生境 Habitat(s)	热带雨林和南亚热带常绿阔叶林，对原始森林的栖息环境具有严格的依赖性 / *A. albirostris* inhabits in rainforests and tropical evergreen broad-leaved forests. It is highly dependent on primeval forests
生态系统 Ecosystem(s)	陆地生态系统 / Terrestrial Ecosystem

威胁 Threat (s)

主要威胁 Major Threat(s)	由于偷猎和森林砍伐导致营巢树减少。自 20 世纪 80 年代以来，云南原始森林遭到大面积的砍伐而种植橡胶林，该物种的种群数量急剧下降。栖息地面积减少和质量下降，尤其是缺乏可用来营巢的大树，是导致冠斑犀鸟种群数量下降的主要原因 (罗益奎等，2013)。偷猎是导致冠斑犀鸟种群数量减少的直接原因 (陈天波等，2007)。此外，栖息地片断化已经影响到冠斑犀鸟不同种群的交流 / Major threats include illegal hunting and reduce of nesting tree owing to deforestation. Since 1980s, the population has decreased with the deforestation of primary forests and planation of rubber forests. Habitat loss and deterioration, especially shortage of large nesting trees, are the major factors of population decreasing (Luo *et al.*, 2013). Illegal hunting is the direct factor (Chen *et al.*, 2007). Besides, habitat fragmentation reduces the flows of different populations

保护级别与保护行动 Protection Category and Conservation Action (s)

IUCN 红色名录 (2016) IUCN Red List (2016)	无危 / LC
保护行动 Conservation Action(s)	列入国家 I 级重点保护野生动物和 CITES 公约附录 II。冠斑犀鸟的主要栖息地已经位于自然保护区中，建议对其栖息地进行严格保护，并在不同保护区之间建立生态廊道。广西西大明山保护区保存有我国唯一已知最大的冠斑犀鸟种群，建议将其晋升为国家级自然保护区 / It has been listed as the first class in National Key Protected Wild Animal List and Appendix II of the CITES. Nature reserves have covered its main habitats. Strict protection and construction ecological corridors are recommended. The largest known population of China was recorded in Xidamingshan Nature Reserve, which is suggested to upgrade to be national nature reserve

冠斑犀鸟 *Anthracoceros albirostris*　　　　　　　　何启海 摄　By Qihai He

双角犀鸟 Buceros bicornis

极危 CR B1ab(ii, iii); D

| 数据缺乏 DD | 无危 LC | 近危 NT | 易危 VU | 濒危 EN | 极危 CR | 区域灭绝 RE | 野外灭绝 EW | 灭绝 EX |

分类地位 Taxonomic Status

动物界 Animalia	脊索动物门 Chordata	鸟纲 Aves	犀鸟目 Bucerotiformes	犀鸟科 Bucerotidae
学名 Scientific Name		*Buceros bicornis*		
命名人 Species Authority		Linnaeus, 1758		
英文名 English Name(s)		Great Hornbill		
同物异名 Synonym(s)		无 / None		
种下单元评估 Infra-specific Taxa Assessed		无 / None		

评估信息 Assessment Information

评估年份 Year Assessed	2016
评定人 Assessor(s)	杨晓君 / Xiaojun Yang
其他贡献人 Other Contributor(s)	吴飞、常云艳等 / Fei Wu, Yunyan Chang *et al.*

理由 Justification: 近年来的记录显示双角犀鸟分布区仅见于相隔甚远的云南铜壁关省级自然保护区、南滚河国家级自然保护区和西双版纳国家级自然保护区。其种群数量可能少于100只，分布区也有明显向边境地区退缩的趋势。因此，列为极危等级 / *Buceros bicornis* was only seen in Tongbiguan Provincial Nature Reserve, Nangunhe National Nature Reserve, and Xishuangbanna National Nature Reserve in recent years. The population size might be less than 100 individuals, and the distribution area has been reduced to the boundary. Thus, it is listed as Vulnerable

评估历史 Assessment History:《国家重点保护野生动物名录》：Ⅱ (1989)，Ⅰ (2021)；《中国濒危动物红皮书·鸟类》(1998)：濒危；《中国物种红色名录：第二卷 脊椎动物 下册》(2009)：易危 / *National Key Protected Wild Animal List*: Ⅱ (1989), Ⅰ (2021); *China Red Data Book of Endangered Animals*: *Aves* (1998): E; *China Species Red List*: *Vol. II Vertebrates Part 2* (2009): VU

地理分布 Geographical Distribution

国内分布 Domestic Distribution
分布于云南德宏、临沧和西双版纳 / It occurs in Dehong, Lincang and Xishuangbanna of Yuannan
分布标注 Distribution Note
非特有种 / Non-endemic

国内分布图
Map of Domestic Distribution

种群 Population

种群数量 Population Size	在中国境内准确的种群数量没有做过具体调查，但其种群数量在分布区的绝大部分区域都显著下降，部分区域局部灭绝 / No survey has been conducted on the population size, while it is decreasing in most distribution areas, and locally extinct in some sites
种群趋势 Population Trend	下降 / Decreasing

生境与生态系统 Habitat(s) and Ecosystem(s)

生　　境 Habitat(s)	分布于海拔 600～1,500 m 的原始热带雨林及季雨林 / *B. bicornis* inhabits in primeval tropical rainforests and seasonal rainforests at 600～1,500m
生态系统 Ecosystem(s)	热带雨林生态系统 / Tropical Rainforest Ecosystem

威胁 Threat(s)

主要威胁 Major Threat(s)	双角犀鸟通常栖息于原始林，其密度与其栖息区域大树的数量有一定的相关性 (James and Kannan，2009)，故森林砍伐造成的生境破坏、退化和片断化是其主要的威胁因子。由于双角犀鸟体型大，飞行和鸣叫声音巨大，容易被猎杀，其巨大的喙也经常作为装饰品销售，捕猎和贸易是双角犀鸟的重要威胁因子 / *B. bicornis* inhabits in primeval forests, and population density is relative with number of large trees in its habitat (James and Kannan, 2009). Therefore, major threats include habitat loss, and deterioration and fragmentation due to deforestation. Its large body and loud calling songs make itself easily caught and killed. The casque is targeted as adornments. Illegal hunting and trade are key threats

保护级别与保护行动 Protection Category and Conservation Action(s)

IUCN 红色名录 (2016) IUCN Red List (2016)	近危 / NT
保护行动 Conservation Action(s)	列入国家 I 级重点保护野生动物和 CITES 公约附录 II，分布区主要在自然保护区内，从而得到保护 / It has been listed as the first class in National Key Protected Wild Animal List and Appendix II of the CITES. It is under protection, and nature reserves have covered most of its distribution area

双角犀鸟 *Buceros bicornis*　　　　龚本亮 摄　By Benliang Gong

棕颈犀鸟
Aceros nipalensis

中国生物多样性红色名录 China's Red List of Biodiversity

极危 CR B1 ab(I, ii, iii); D

| 数据缺乏 DD | 无危 LC | 近危 NT | 易危 VU | 濒危 EN | **极危 CR** | 区域灭绝 RE | 野外灭绝 EW | 灭绝 EX |

分类地位 Taxonomic Status

动物界 Animalia	脊索动物门 Chordata	鸟纲 Aves	犀鸟目 Bucerotiformes	犀鸟科 Bucerotidae

学名 Scientific Name	*Aceros nipalensis*
命名人 Species Authority	Hodgson, 1829
英文名 English Name(s)	Rufous-necked Hornbill
同物异名 Synonym(s)	*Buceros nipalensis*
种下单元评估 Infra-specific Taxa Assessed	无 / None

评估信息 Assessment Information

评估年份 Year Assessed	2016
评定人 Assessor(s)	杨晓君 / Xiaojun Yang
其他贡献人 Other Contributor(s)	吴飞、常云艳等 / Fei Wu, Yunyan Chang *et al.*

理由 Justification: 棕颈犀鸟在中国的分布区极为狭窄，近10年来，关于该物种的记录极少，仅有的3次记录为云南的西双版纳国家级自然保护区、铜壁关保护区和西藏东南部的墨脱。因此，列为极危等级 / The distribution area of *Aceros nipalensis* is extremely small in China. There are only two records in recent ten years, in Xishuangbanna National Nature Reserve and Tongbiguan Nature Reserve of Yunnan, and Motuo of southeast Tibet (Xizang) specifically. Thus, it is listed as Critically Endangered

评估历史 Assessment History: 《国家重点保护野生动物名录》：II (1989)，I (2021)；《中国濒危动物红皮书·鸟类》(1998)：稀有；《中国物种红色名录：第二卷 脊椎动物 下册》(2009)：易危 / *National Key Protected Wild Animal List*: II (1989), I (2021); *China Red Data Book of Endangered Animals: Aves* (1998): R; *China Species Red List: Vol. II Vertebrates Part 2* (2009): VU

地理分布 Geographical Distribution

国内分布 Domestic Distribution

分布于云南省德宏、西双版纳和红河，西藏东南部 / It occurs mainly in Dehong, Xishuangbanna and Honghe in Yunnan, and southeast Tibet (Xizang)

分布标注 Distribution Note

非特有种 / Non-endemic

国内分布图
Map of Domestic Distribution

种群 Population

种群数量 Population Size	在中国境内准确的种群数量没有做过具体调查，但其种群数量下降明显，近10年来棕颈犀鸟在中国境内记录极少 / No survey has been conducted on the population size, while it is undergoing a significant decline. And the records are extremely small in recent ten years
种群趋势 Population Trend	下降 / Decreasing

生境与生态系统 Habitat(s) and Ecosystem(s)

生境 Habitat(s)	分布于海拔 600～1,800 m 的热带雨林 / *A. nipalensis* inhabits in tropical rainforests at 600～1,800 m
生态系统 Ecosystem(s)	陆地生态系统 / Terrestrial Ecosystem

威胁 Threat(s)

主要威胁 Major Threat(s)	适宜棕颈犀鸟的区域同时也适宜种植香蕉、橡胶和甘蔗等热带作物，由于这些热带作物的大面积种植，大量适宜棕颈犀鸟栖息的森林被砍伐造成其生境破坏和片断化。由于棕颈犀鸟体型大，叫声响亮，捕猎和贸易也是其主要的威胁因子 / Major threats include habitat loss and fragmentation owing to deforestation and planation of bananas, sugar canes and rubber trees. Its large body and loud calling songs make itself easily caught and killed. Illegal hunting and trade are key threats

保护级别与保护行动 Protection Category and Conservation Action(s)

IUCN 红色名录 (2016) IUCN Red List (2016)	易危 / VU A2cd+3cd+4cd
保护行动 Conservation Action(s)	列入国家 I 级重点保护野生动物和 CITES 公约附录 I，其主要分布区在自然保护区内，从而得到保护 / It has been listed as the first class in National Key Protected Wild Animal List and Appendix I of the CITES. It is under protection, and nature reserves have covered most of its distribution area

棕颈犀鸟 *Aceros nipalensis*　　　　郭益民 摄　By Yimin Guo

蓝冠噪鹛 *Garrulax courtoisi*

极危 CR B2ab(i-v); C2a(ii)

| 数据缺乏 DD | 无危 LC | 近危 NT | 易危 VU | 濒危 EN | 极危 CR | 区域灭绝 RE | 野外灭绝 EW | 灭绝 EX |

分类地位 Taxonomic Status

动物界 Animalia	脊索动物门 Chordata	鸟纲 Aves	雀形目 Passeriformes	噪鹛科 Leiothrichidae
学名 Scientific Name		*Garrulax courtoisi*		
命名人 Species Authority		Menegaux, 1923		
英文名 English Name(s)		Blue-crowned Laughingthrush		
同物异名 Synonym(s)		无 / None		
种下单元评估 Infra-specific Taxa Assessed		无 / None		

评估信息 Assessment Information

评估年份 Year Assessed	2016
评定人 Assessor(s)	卢欣 / Xin Lu
其他贡献人 Other Contributor(s)	无 / None

理由 Justification：蓝冠噪鹛分布区域狭窄，且种群极小，估计种群成熟个体数少于250，并有可能继续下降，极小的种群可能丢失基因的多样性。估计占有面积少于100 km² 且已知只有一个分布地点。因此，列为极危等级 / *Garrulax courtoisi* only occurs in one site, and the distribution area is small and estimated to be less than 100 km². The population size is estimated to be less than 250 mature individuals, and might be undergoing a decline. The gene diversity probably be lost in an extremely small population. Thus, it is listed as Critically Endangered

评估历史 Assessment History：《国家重点保护野生动物名录》：Ⅰ (2021)；《中国物种红色名录：第二卷 脊椎动物 下册》(2009)：极危 / *National Key Protected Wild Animal List*: I (2021); *China Species Red List: Vol. II Vertebrates Part 2* (2009): CR

地理分布 Geographical Distribution

国内分布 Domestic Distribution
只分布于中国江西婺源县 / It only occurs in Wuyuan, Jiangxi
分布标注 Distribution Note
特有种 / Endemic

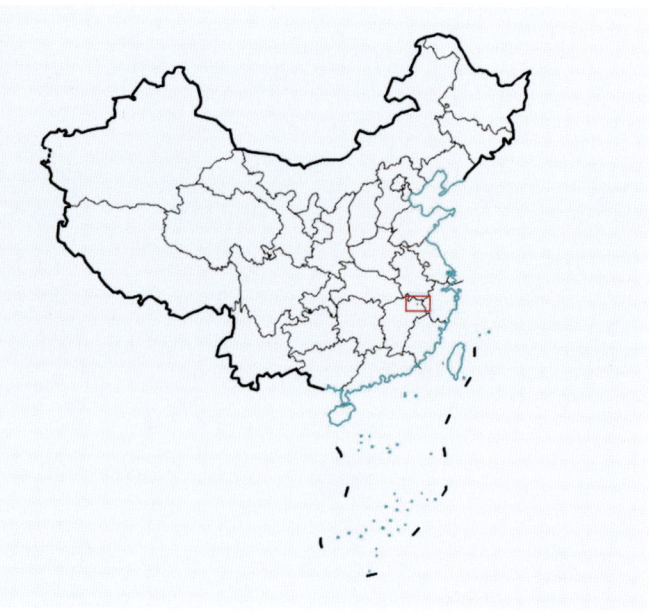

国内分布图
Map of Domestic Distribution

种群 Population

种群数量 Population Size	成熟个体的种群数量少于 250 只 / The population size is estimated to be less than 250 matured individuals
种群趋势 Population Trend	下降 / Decreasing

生境与生态系统 Habitat(s) and Ecosystem(s)

生境 Habitat(s)	主要分布于江西婺源月亮湾紧邻村庄的常绿树林里，树种以枫杨、香樟等为主 / The evergreen forest comprised of *Pterocarya stenoptera* and *Cinnamomum camphora*, adjacent to village in Wuyuan County, Jiangxi Province
生态系统 Ecosystem(s)	陆地生态系统 / Terrestrial Ecosystem

威胁 Threat(s)

主要威胁 Major Threat(s)	捕获和贸易是其主要的威胁之一，道路建设等工程破坏了繁殖地点的巢和栖息地。更多零散繁殖地并未受到保护 / Illegal trapping and trading are the major threats. Besides, the destroy of nests and loss of habitats due to the infrastructure construction projects such as highway is another threat. Most breeding sites are not under protection

保护级别与保护行动 Protection Category and Conservation Action(s)

IUCN 红色名录 (2016) IUCN Red List (2016)	极危 / CR C2a (i, ii)
保护行动 Conservation Action(s)	列入国家 I 级重点保护野生动物。已建立保护区，但对本物种的宣传和保护工作有待加强 / It has been listed as the first class in National Key Protected Wild Animal List. There is nature reserve aimed to protect the species, while the activities and effects of publicity and conservation are to be improved

蓝冠噪鹛 *Garrulax courtoisi* — 周佳俊 摄 By Jiajun Zhou

黑冠薮鹛
Liocichla bugunorum
极危 CR B1ab(i)

| 数据缺乏 DD | 无危 LC | 近危 NT | 易危 VU | 濒危 EN | 极危 CR | 区域灭绝 RE | 野外灭绝 EW | 灭绝 EX |

分类地位 Taxonomic Status

动物界 Animalia	脊索动物门 Chordata	鸟纲 Aves	雀形目 Passeriformes	噪鹛科 Leiothrichidae
学名 Scientific Name		*Liocichla bugunorum*		
命名人 Species Authority		Athreya, 2006		
英文名 English Name(s)		Bugun Liocichla		
同物异名 Synonym(s)		无 / None		
种下单元评估 Infra-specific Taxa Assessed		无 / None		

评估信息 Assessment Information

评估年份 Year Assessed	2016
评定人 Assessor(s)	董路 / Lu Dong
其他贡献人 Other Contributor(s)	无 / None

理由 Justification: 黑冠薮鹛的分布区非常狭窄且高度破碎化，已知分布点不超过5个，成熟种群数量不足250只，适宜栖息地内人类干扰较大。因此，列为极危等级 / The distribution area of *Liocichla bugunorum* is small and highly fragmented. The known sites are less than five and the mature individuals are less than 250. The suitable habitats are under heavily human disturbance. Thus, it is listed as Critically Endangered

评估历史 Assessment History: 《国家重点保护野生动物名录》：Ⅰ (2021) / *National Key Protected Wild Animal List*: Ⅰ (2021)

地理分布 Geographical Distribution

国内分布 Domestic Distribution
仅分布于西藏东南部 / It only occurs in southeast Tibet (Xizang)
分布标注 Distribution Note
非特有种 / Non-endemic

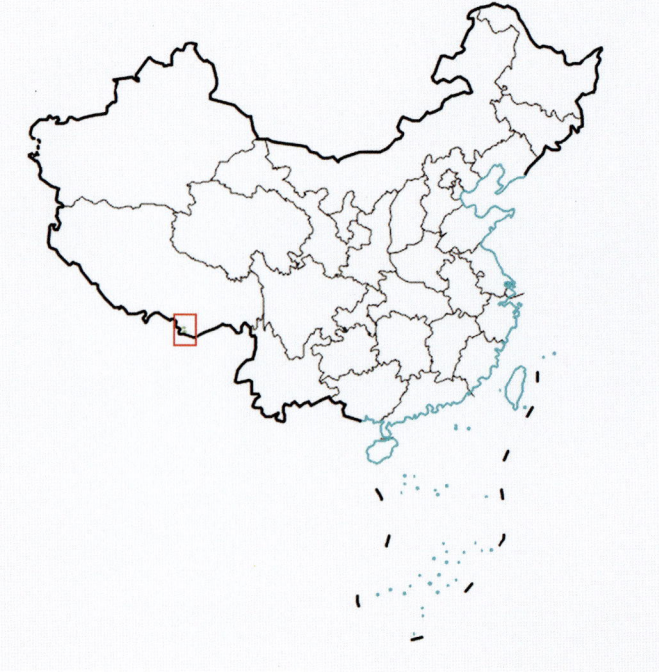

国内分布图
Map of Domestic Distribution

种群 Population

种群数量 Population Size	根据适宜栖息地面积及 3 个地点的种群密度估算,种群数量不足 400 只,成熟个体数不足 250 只,我国种群数量不详 / The population size is estimated to be less than 400 individuals, and the mature individuals are less than 250, based on the suitable habitat area and the population density in three sites. No survey in China
种群趋势 Population Trend	下降 / Decreasing

生境与生态系统 Habitat(s) and Ecosystem(s)

生境 Habitat(s)	主要分布于海拔 2,000 m 以上的茂密灌丛和竹林,多见于连续森林中的林窗 / *L. bugunorum* uses the dense shrubs and bamboos above 2,000 m, and mainly occurs in the gaps between forests
生态系统 Ecosystem(s)	陆地生态系统 / Terrestrial Ecosystem

威胁 Threat(s)

主要威胁 Major Threat(s)	伐木和人为干扰可能是本物种的主要威胁因子。尚需进一步调查 / The logging and human disturbance might be the major threats, which is need to be confirmed

保护级别与保护行动 Protection Category and Conservation Action(s)

IUCN 红色名录 (2016) IUCN Red List (2016)	极危 / CR C2a (i)
保护行动 Conservation Action(s)	列入国家 I 级重点保护野生动物。需加强调查和监测,以了解种群数量的变动趋势 / It has been listed as the first class in National Key Protected Wild Animal List. Survey and monitoring need to be improved to get knowledge on the population dynamic

黑冠薮鹛 *Liocichla bugunorum*   Sasidhar Akkiraju 摄 By Sasidhar Akkiraju

中国生物多样性红色名录 China's Red List of Biodiversity

黄胸鹀
Emberiza aureola
极危 CR A1acd

| 数据缺乏 DD | 无危 LC | 近危 NT | 易危 VU | 濒危 EN | **极危 CR** | 区域灭绝 RE | 野外灭绝 EW | 灭绝 EX |

分类地位 Taxonomic Status

动物界 Animalia	脊索动物门 Chordata	鸟 纲 Aves	雀形目 Passeriformes	鹀 科 Emberizidae
学 名 Scientific Name		*Emberiza aureola*		
命 名 人 Species Authority		Pallas, 1773		
英 文 名 English Name(s)		Yellow-breasted Bunting		
同物异名 Synonym(s)		*Schoeniclus aureolus*		
种下单元评估 Infra-specific Taxa Assessed		无 / None		

评估信息 Assessment Information

评估年份 Year Assessed	2016
评定人 Assessor(s)	张正旺 / Zhengwang Zhang
其他贡献人 Other Contributor(s)	无 / None

理由 Justification: 过度捕猎导致黄胸鹀在国内的许多分布地点的数量都出现了大幅度的下降。因此，列为极危等级 / The populations of *Emberiza aureola* in most sites in China are undergoing a rapid decline owing to the over trapping. Thus, it is listed as Critically Endangered

评估历史 Assessment History: 《国家重点保护野生动物名录》：Ⅰ (2021)；《中国物种红色名录：第二卷 脊椎动物 下册》(2009)：近危(近乎易危) / *National Key Protected Wild Animal List*: I (2021); *China Species Red List: Vol. II Vertebrates Part 2* (2009): NT (nearly VU)

地理分布 Geographical Distribution

国内分布 Domestic Distribution

繁殖于内蒙古东部、黑龙江、吉林、辽宁、河北、新疆等地，迁徙期间经过中东部大部分地区，一直往南到东南沿海和华南地区，在华南和西南地区越冬 / It breeds in east Inner Mongolia (Nei Mongol), Heilongjiang, Jilin, Liaoning, Hebei, Xinjiang, and occurs in majority province of central and east China during Migration. It winters in South China and Southwest China

分布标注 Distribution Note

非特有种 / Non-endemic

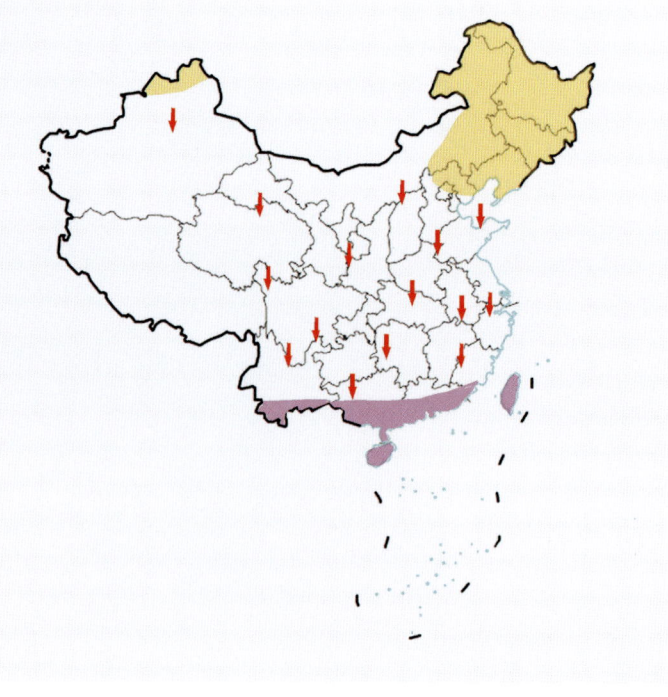

国内分布图
Map of Domestic Distribution

种群 Population

种群数量 Population Size	全球种群数量 10,000 ～ 13,900 只 (BirdLife International, 2004), 我国缺乏系统数量调查, 但种群数量下降严重 / The global population size is estimated to be 10,000 ～ 13,900 individuals (BirdLife International, 2004). No systematic survey in China, while its population is undergoing a decline in a serious way
种群趋势 Population Trend	下降 / Decreasing

生境与生态系统 Habitat(s) and Ecosystem(s)

生境 Habitat(s)	栖息于低山丘陵和开阔平原地带的灌丛、草甸、草地和林缘地带, 尤其喜欢溪流、湖泊和沼泽附近的灌丛、草地, 也栖息于有稀疏柳树、桦树、杨树的灌丛草地和田间, 是典型的河谷草甸灌丛草地鸟类 / E. aureola uses shrubs, meadows, grasslands and edge of forests, especially the shrubs and grassland along the streams, lakes, and marshes. It also uses farmland and shrubs with willow, birch and poplar. It is a typical valley shrub and grassland bird
生态系统 Ecosystem(s)	陆地生态系统 / Terrestrial Ecosystem

威胁 Threat(s)

主要威胁 Major Threat(s)	由于过度捕猎导致该物种在许多分布地点的数量都出现了大幅度的下降, 尤其是在迁徙路线上和越冬地更为显著。中国广东地区秋冬季有食用"禾花雀"的习俗, 每年有大量的黄胸鹀被捕捉贩卖, 如广东佛山一个交易市场一天就卖出了约 1 万只 (Chan, 2004) / The populations of E. aureola in most sites in China are undergoing a rapid decline owing to the over trapping, especially in the migratory route and wintering sites. A huge number of birds are trapped and sold each year due to the custom of cooking and eating Emberiza aureola in winter, e.g. around 10,000 individuals were sold one day in a single market in Foshan, Guangdong (Chan, 2004)

保护级别与保护行动 Protection Category and Conservation Action(s)

IUCN 红色名录 (2016) IUCN Red List (2016)	濒危 / EN A2acd+3cd+4acd
保护行动 Conservation Action(s)	列入国家 I 级重点保护野生动物、《国家保护的有益的或者有重要经济、科学研究价值的陆生野生动物名录》和迁徙物种保护公约 (CMS) 附录 I, 广东省将之列为省级重点保护野生动物 / It has been listed as the first class in National Key Protected Wild Animal List, *National Protected List of Terrestrial Wild Animals with Good Benefits or Important Economic and Scientific Values*, and Appendix I of the Convention on Migratory Species, and in Guangdong Key Protected Wild Animal List

黄胸鹀 *Emberiza aureola* 陈保利 摄 By Baoli Chen

四川山鹧鸪
Arborophila rufipectus

濒危　EN B2ab(iii); C2a(i)

| 数据缺乏 DD | 无危 LC | 近危 NT | 易危 VU | 濒危 EN | 极危 CR | 区域灭绝 RE | 野外灭绝 EW | 灭绝 EX |

分类地位 Taxonomic Status

动物界 Animalia	脊索动物门 Chordata	鸟纲 Aves	鸡形目 Galliformes	雉科 Phasianidae
学名 Scientific Name		*Arborophila rufipectus*		
命名人 Species Authority		Boulton, 1932		
英文名 English Name(s)		Sichuan Partridge		
同物异名 Synonym(s)		无 / None		
种下单元评估 Infra-specific Taxa Assessed		无 / None		

评估信息 Assessment Information

评估年份 Year Assessed	2016
评定人 Assessor(s)	冉江洪 / Jianghong Ran
其他贡献人 Other Contributor(s)	窦亮、戴波等 / Liang Dou, Bo Dai *et al.*

理由 Justification: 四川山鹧鸪的分布区狭窄，种群数量小，且栖息地片断化严重。受偷猎和适宜生境损失的影响预计种群还会持续下降。因此，列为濒危等级 / The distribution area and population size of *Arborophila rufipectus* is small and habitat is in high fragmentation. It's assumed that the population is undergoing a decline owing to the illegal trapping and habitat loss. Thus, it is listed as Endangered

评估历史 Assessment History:《国家重点保护野生动物名录》：Ⅰ (1989，2021);《中国濒危动物红皮书·鸟类》(1998)：濒危;《中国物种红色名录：第二卷 脊椎动物 下册》(2009)：濒危 / *National Key Protected Wild Animal List*: Ⅰ (1989, 2021); *China Red Data Book of Endangered Animals*: *Aves* (1998): E; *China Species Red List*: *Vol. II Vertebrates Part 2* (2009): EN

地理分布 Geographical Distribution

国内分布 Domestic Distribution
仅分布于四川屏山、甘洛、马边、峨边、美姑、沐川、雷波、金口河及云南绥江、永善 / It only occurs in Pingshan, Ganluo, Mabian, Ebian, Meigu, Muchuan, Leibo, and Jinkou River in Sichuan, and Suijiang, Yongshan in Yunnan
分布标注 Distribution Note
特有种 / Endemic

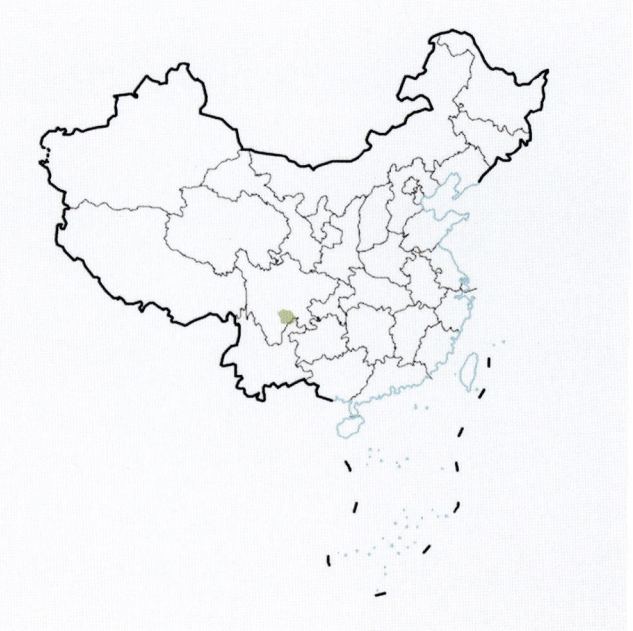

国内分布图
Map of Domestic Distribution

种群 Population

种群数量 Population Size	估计种群数量是 1,500 ～ 4,000 只 / The population size is estimated to be 1,500 ～ 4,000 individuals
种群趋势 Population Trend	下降 / Decreasing

生境与生态系统 Habitat (s) and Ecosystem (s)

生　　　境 Habitat(s)	主要栖息于海拔 1,000 ～ 2,200 m 的栎、杜鹃、栗、油茶等构成的常绿阔叶林内的浓密竹丛和灌丛中。每年 4 ～ 6 月进行繁殖，营地面巢 / *A. rufipectus* occurs in bamboo and shrubs in evergreen broad-leaved forests with *Quercus* spp., *Rhododendron* spp., *Castanea mollissima, Camellia oleifera* at 1,000 ～ 2,200 m. It breeds from April to June, and nests on the ground
生态系统 Ecosystem(s)	陆地、亚热带森林生态系统 / Terrestrial, Subtropical Forest Ecosystem

威胁 Threat (s)

主要威胁 Major Threat(s)	主要生活在低海拔的常绿阔叶林区，本区域也是当地社区居民主要的放牧、采集和薪材收集区域，这些活动对常绿阔叶林及林下植被破坏较大，致使其栖息地面积减少或质量下降。居民点扩展、道路建设及河流（金沙江）的影响，使其栖息地被分隔为多个斑块，栖息地破碎化严重。本物种也常是当地社区居民偷猎对象 / *A. rufipectus* occurs in lowland evergreen broad-leaved forests, and overlapped with living area of local community, *e.g.* grazing and wood collection, which might lead to forests destroy and habitat loss. The habitats are fragmented by the increase of residential area and roads construction, as well as natural patterns of rivers and mountains. And it is also threatened by illegal trapping of local people

保护级别与保护行动 Protection Category and Conservation Action (s)

IUCN 红色名录 (2016) IUCN Red List (2016)	濒危 / EN C2a (i)
保护行动 Conservation Action(s)	列入国家 I 级重点保护野生动物。目前，在四川分布区域的栖息地大多位于自然保护区内。屏山县的四川老君山国家级自然保护区和沐川县的四川芹菜坪自然保护区，均是以四川山鹧鸪为主要保护对象的保护区。云南省的栖息地还没有纳入自然保护区 / It has been listed as the first class in National Key Protected Wild Animal List. The main habitats in Sichuan are mostly located in nature reserves. The conservation target of Sichuan Laojunshan National Nature Reserve in Pingshan County and Sichuan Qincaiping Nature Reserve is *A. rufipectus*. While the habitats in Yunnan have not been designated as protected areas

四川山鹧鸪 *Arborophila rufipectus*　　　戴波 摄　By Bo Dai

海南山鹧鸪
Arborophila ardens

濒危 EN B1ab(iii)

| 数据缺乏 DD | 无危 LC | 近危 NT | 易危 VU | **濒危 EN** | 极危 CR | 区域灭绝 RE | 野外灭绝 EW | 灭绝 EX |

分类地位 Taxonomic Status

动物界 Animalia	脊索动物门 Chordata	鸟纲 Aves	鸡形目 Galliformes	雉科 Phasianidae
学名 Scientific Name		*Arborophila ardens*		
命名人 Species Authority		Styan, 1982		
英文名 English Name(s)		Hainan Partridge		
同物异名 Synonym(s)		*Arboricola ardens*		
种下单元评估 Infra-specific Taxa Assessed		无 / None		

评估信息 Assessment Information

评估年份 Year Assessed	2016
评定人 Assessor(s)	梁伟 / Wei Liang
其他贡献人 Other Contributor(s)	杨灿朝 / Canchao Yang

理由 Justification: 分布区小且严重分割，仅分布于海南天然雨林中。部分保护区存在不同程度的旅游和开发等，造成栖息地质量持续衰退。因此，列为濒危等级 / *Arborophila ardens* only occurs in rainforests in Hainan. The distribution area is small and fragmented. The habitat quality is declining owing to the tourism and development in nature reserves. Thus, it is listed as Endangered

评估历史 Assessment History: 《国家重点保护野生动物名录》: Ⅰ (1989, 2021); 《中国濒危动物红皮书·鸟类》(1998): 濒危; 《中国物种红色名录: 第二卷 脊椎动物 下册》(2009): 易危 / *National Key Protected Wild Animal List*: Ⅰ (1989, 2021); *China Red Data Book of Endangered Animals*: *Aves* (1998): E; *China Species Red List*: *Vol. II Vertebrates Part 2* (2009): VU

地理分布 Geographical Distribution

国内分布 Domestic Distribution
分布于海南中部和南部山区 / It occus in the mountain areas in middle and south Hainan
分布标注 Distribution Note
特有种 / Endemic

国内分布图
Map of Domestic Distribution

种群 Population

种群数量 Population Size	估计种群数量 3,900 ～ 5,200 只 (BirdLife International, 2013) / The population size is estimated to be 3,900 ～ 5,200 individuals (BirdLife International, 2013)
种群趋势 Population Trend	下降 / Decreasing

生境与生态系统 Habitat (s) and Ecosystem (s)

生　　境 Habitat(s)	热带山地雨林 / The tropical rainforests in mountain areas
生态系统 Ecosystem(s)	陆地、热带雨林生态系统 / Terrestrial, Tropical Rainforest Ecosystem

威胁 Threat (s)

主要威胁 Major Threat(s)	海南天然林面积的持续下降、栖息地质量的降低和非法捕猎 (Liang *et al.*, 2013) / The major threats are decreasing of natural forests in Hainan, reduction of habitat quality and illegal trapping (Liang *et al.*, 2013)

保护级别与保护行动 Protection Category and Conservation Action (s)

IUCN 红色名录 (2016) IUCN Red List (2016)	易危 / VUA2cd; B1ab (ii,iii,iv,v); C2a (i)
保护行动 Conservation Action(s)	列入国家 I 级重点保护野生动物。建议加大对天然林的保护和对非法捕猎的执法力度 / It has been listed as the first class in National Key Protected Wild Animal List. The protection of natural forests and the enforcement of laws on illegal trapping are suggested

海南山鹧鸪 *Arborophila ardens* 　　卢刚 摄　By Gang Lu

中国生物多样性红色名录

松鸡
Tetrao urogallus

濒危 EN C2a(i)

| DD 数据缺乏 | LC 无危 | NT 近危 | VU 易危 | **EN 濒危** | CR 极危 | RE 区域灭绝 | EW 野外灭绝 | EX 灭绝 |

分类地位 Taxonomic Status

动物界 Animalia	脊索动物门 Chordata	鸟纲 Aves	鸡形目 Galliformes	雉科 Phasianidae
学名 Scientific Name		*Tetrao urogallus*		
命名人 Species Authority		Linnaeus, 1758		
英文名 English Name(s)		Western Capercaillie		
同物异名 Synonym(s)		无 / None		
种下单元评估 Infra-specific Taxa Assessed		无 / None		

评估信息 Assessment Information

评估年份 Year Assessed	2016
评定人 Assessor(s)	孙悦华 / Yuehua Sun
其他贡献人 Other Contributor(s)	无 / None

理由 Justification: 尽管全球种群未受威胁，但在中国分布区狭窄，栖息地破碎化严重并面临捕猎的威胁。因此，列为濒危等级 / The global population of *Tetrao urogallus* is not threatened, while the distribution areas in China are small and habitats are fragmentated. And it is also threatened by the illegal trapping. Thus, it is listed as Endangered

评估历史 Assessment History: 《国家重点保护野生动物名录》：Ⅱ (2021); 《中国物种红色名录：第二卷 脊椎动物 下册》(2009)：无危 / *National Key Protected Wild Animal List*: Ⅱ (2021); *China Species Red List: Vol. Ⅱ Vertebrates Part 2* (2009): LC

地理分布 Geographical Distribution

国内分布 Domestic Distribution
仅见于新疆阿勒泰地区 / It only occurs in Altai area of Xinjiang in China
分布标注 Distribution Note
非特有种 / Non-endemic

国内分布图
Map of Domestic Distribution

种群 Population

种群数量 Population Size	不详 / No data
种群趋势 Population Trend	下降 / Decreasing

生境与生态系统 Habitat (s) and Ecosystem (s)

生境 Habitat(s)	典型的针叶林鸟类，多在林中空地、林缘和河谷地带，不喜茂密森林和阔叶林，育雏时可活动于草地或灌丛。冬季在雪穴中过夜，以松树、杨树、

桦树嫩枝、芽、花及植物浆果为食，觅食多在树上。求偶在固定的求偶场进行 / *T. urogallus* is typical species in coniferous forests, and occurs in the edge and open lands in forests, and valleys. It roosted in snow cave in winter, and feeds on twigs, shoots and buds of pines, poplars and birches, as well as berries. Courtship happens in fixed areas

生态系统 Ecosystem(s)	陆地生态系统 / Terrestrial Ecosystem

威胁 Threat (s)

主要威胁 Major Threat(s)	栖息地的丧失和破碎化及人为过度猎捕是影响其数量的主要因素 / The major threats are habitat loss and fragmentation, and over hunting

保护级别与保护行动 Protection Category and Conservation Action (s)

IUCN 红色名录 (2016) IUCN Red List (2016)	无危 / LC
保护行动 Conservation Action(s)	列入国家 II 级重点保护野生动物、《国家保护的有益的或者有重要经济、科学研究价值的陆生野生动物名录》。部分分布区位于保护区内，应加强

种群数量调查、栖息地保护及监管 / It was listed as the second class in National Key Protected Wild Animal List, *National Protected List of Terrestrial Wild Animals with Good Benefits or Important Economic and Scientific Values*. Some of its distribution area locates in nature reserves. Surveys on population size and habitat protection are suggested

松鸡 *Tetrao urogallus* 许传辉 摄 By Chuanhui Xu

黑嘴松鸡
Tetrao urogalloides

濒危 EN A2c+3c+4c

| 数据缺乏 DD | 无危 LC | 近危 NT | 易危 VU | 濒危 EN | 极危 CR | 区域灭绝 RE | 野外灭绝 EW | 灭绝 EX |

分类地位 Taxonomic Status

| 动物界 Animalia | 脊索动物门 Chordata | 鸟 纲 Aves | 鸡形目 Galliformes | 雉 科 Phasianidae |

学 名 Scientific Name	*Tetrao urogalloides*
命 名 人 Species Authority	von Middendorff, 1853
英 文 名 English Name(s)	Black-billed Capercaillie
同物异名 Synonym(s)	*Tetrao parvirostris*
种下单元评估 Infra-specific Taxa Assessed	无 / None

评估信息 Assessment Information

评 估 年 份 Year Assessed	2016
评 定 人 Assessor(s)	王海涛 / Haitao Wang
其他贡献人 Other Contributor(s)	刘宇 / Yu Liu

理由 Justification: 黑嘴松鸡分布范围缩小，种群数量急剧下降，中国东北地区种群近10年下降已超过50%；河北省多次调查未见。因此，列为濒危等级 / The distribution area and population size of *Tetrao urogalloides* has decreased. The population in northeast China has decreased by 50% in recent decade and no records in Hebei Province in surveys. Thus, it is listed as Endangered

评估历史 Assessment History:《国家重点保护野生动物名录》: I (1989, 2021);《中国濒危动物红皮书·鸟类》(1998): 稀有;《中国物种红色名录：第二卷 脊椎动物 下册》(2009): 未评估 / *National Key Protected Wild Animal List*: I (1989, 2021); *China Red Data Book of Endangered Animals*: Aves (1998): R; *China Species Red List*: Vol. II Vertebrates Part 2 (2009): NA

地理分布 Geographical Distribution

国内分布 Domestic Distribution

分布于黑龙江和内蒙古的大小兴安岭地区，分布范围约 158,860 km² (赵作审和赵中琴, 1996; 贺福银等, 2004; 尹远新等, 2009)。在传统分布地河北省，多次调查未见 (李东明等, 2004) / It occurs in east and north in Da and Xiao Hinggan Ling in Heilongjiang and Inner Mongolia (Nei Mongol), and the distribution are is about 158,860 km² (Zhao and Zhao, 1996; He *et al.*, 2004; Yin *et al.*, 2009). No records in recent surveys in Hebei, the historical historic distribution site (Li *et al.*, 2004)

分布标注 Distribution Note

非特有种 / Non-endemic

国内分布图
Map of Domestic Distribution

种群 Population

种群数量 Population Size	1984～1985 年黑龙江省西北部调查，密度 0.43 只 /km²，估计总量 22,000 只（卢汰春，1991）；1996～1999 年大兴安岭地区调查，不足 1,200 只（贺福银等，2004）；2004～2006 年东北地区调查，密度 0.0611 只 /km²，种群数量 (5,811±808) 只（尹远新等，2009）/ The population size is estimated to be 22,000 individuals based on the survey in northwest Heilongjiang in 1984～1985, with a density of 0.43 individual per km² (Lu, 1991); The population in Da Hinggan Ling is less than 1,200 individuals in 1996～1999 (He et al., 2004); and was estimated to be 5,811±808 individuals in 2004～2006, with a density of 0.0611 individual per km² (Yin et al., 2009)
种群趋势 Population Trend	下降 / Decreasing

生境与生态系统 Habitat (s) and Ecosystem (s)

生　　境 Habitat(s)	以兴安落叶松为优势种的海拔 300～800 m 的寒温带针叶林 / *T. urogalloides* uses cold temperate coniferous forests at 300～800 m a.s.l., with *Larix gmelinii* as dominant species
生态系统 Ecosystem(s)	陆地生态系统 / Terrestrial Ecosystem

威胁 Threat (s)

主要威胁 Major Threat(s)	森林长期过度采伐、农业开发、过度放牧和草原荒漠化造成栖息地减少和质量下降，非法捕猎，紫貂 (*Martes zibellina*) 捕食（卢汰春，1991；徐利等，1996；赵作审和赵中琴，1996；尹远新等，2009）/ The habitats loss and deterioration due to deforestation, development of agriculture, over grazing and grasslands desertification; illegal trapping; predation of *Martes zibellina* (Lu, 1991; Xu et al., 1996; Zhao and Zhao, 1996; Yin et al., 2009)

保护级别与保护行动 Protection Category and Conservation Action (s)

IUCN 红色名录 (2016) IUCN Red List (2016)	无危 / LC
保护行动 Conservation Action(s)	列入国家 I 级重点保护野生动物。部分分布区域处于保护区内。缺乏繁殖生境选择、行为及生活史基础研究资料 / It has been listed as the first class in National Key Protected Wild Animal List. Some distribution areas are in nature reserves. There is little basic information of breeding habitat selection, behavior and life history

黑嘴松鸡 *Tetrao urogalloides* 　　　　　　　　　　商华 摄　By Hua Shang

黄腹角雉
Tragopan caboti

中国生物多样性 红色名录

濒危 EN B2ab(i); C2a(i)

| 数据缺乏 DD | 无危 LC | 近危 NT | 易危 VU | 濒危 EN | 极危 CR | 区域灭绝 RE | 野外灭绝 EW | 灭绝 EX |

分类地位 Taxonomic Status

动物界 Animalia	脊索动物门 Chordata	鸟纲 Aves	鸡形目 Galliformes	雉科 Phasianidae
学 名 Scientific Name		*Tragopan caboti*		
命 名 人 Species Authority		Gould, 1857		
英 文 名 English Name(s)		Cabot's Tragopan		
同物异名 Synonym(s)		无 / None		
种下单元评估 Infra-specific Taxa Assessed		浙江黄腹角雉种群生存力分析显示，种群未来20年的灭绝概率小于10% (Zhang and Zheng, 2007) / The Probability of extinction of *T. c. caboti* is estimated to be less than 10% based on the population viability analysis (Zhang and Zheng, 2007)		

评估信息 Assessment Information

评 估 年 份 Year Assessed	2016
评 定 人 Assessor(s)	张雁云 / Yanyun Zhang
其他贡献人 Other Contributor(s)	董路 / Lu Dong

理由 Justification: 黄腹角雉的种群数量总数 4,000～5,000 只，不存在种群数量超过 1,000 只的亚种群；占有面积少于 2,000 km²，且分割在 6 个区域。部分保护区存在不同程度的旅游和开发等，造成栖息地质量衰退。东西部的黄腹角雉种群基因流已经完全阻隔，形成 2 个独立的遗传单元 (Dong *et al.*, 2011)。因此，列为濒危等级 / The population size of *Tragopan caboti* is estimated to be 4,000～5,000 individuals and no sub-population over 1,000 individuals. The distribution area is less than 2,000 km², and fragmented into 6 areas. The habitat quality is undergoing a deterioration owing to tourism and development activities in some nature reserves. The gene flow of east and west population is totally separated and two genetic units have been formed (Dong *et al.*, 2011). Thus, it is listed as Endangered

评估历史 Assessment History: 《国家重点保护野生动物名录》: I (1989, 2021); 《中国濒危动物红皮书·鸟类》(1998): 易危; 《中国物种红色名录: 第二卷 脊椎动物 下册》(2009): 近危 / *National Key Protected Wild Animal List*: I (1989, 2021); *China Red Data Book of Endangered Animals*: *Aves* (1998): V; *China Species Red List: Vol. II Vertebrates Part 2* (2009): NT

地理分布 Geographical Distribution

国内分布 Domestic Distribution
分布于浙江南部和西南部，江西东北部、中部和南部，福建西部，湖南东南部，广东北部，广西东北部 / It occurs in South and southwest Zhejiang, northeast, middle and south Jiangxi, west Fujiang, southeast Hunan, north Guangdong, northeast Guangxi
分布标注 Distribution Note
特有种 / Endemic

国内分布图
Map of Domestic Distribution

种群 Population

种群数量 Population Size	种群数量 4,000～5,000 只 / The population size is 4,000～5,000 individuals
种群趋势 Population Trend	下降 / Decreasing

生境与生态系统 Habitat (s) and Ecosystem (s)

生　　境 Habitat(s)	我国东部亚热带山地森林内的海拔 500～1,800 m 的常绿阔叶林和常绿-落叶-针叶混交林内，建群树种有壳斗科 (Fagaceae)、樟科 (Lauraceae)、

山茶科 (Theaceae)、冬青科 (Aquiifoliaceae)、山矾科 (Symplocaceae)、蔷薇科 (Rosaceae) 和杜鹃花科 (Ericaceae) 等植物，黄山松 (*Pinus taiwannensis*) 为主要针叶树种 (郑光美等，1985，1986) / *T. caboti* uses evergreen broad-leaved forests and mixed deciduous-coniferous forests in subtropical forests at 500～1,800 m, dominated by plant species of the families Fagaceae, Lauraceae, Theaceae, Aquiifoliaceae, Symplocaceae, Rosaceae, and Ericaceae. The main coniferous species is *Pinus taiwannensis* (Zheng *et al.*, 1985, 1986)

生态系统 Ecosystem(s)	陆地生态系统 / Terrestrial Ecosystem

威胁 Threat (s)

主要威胁 Major Threat(s)	松鸦、青鼬、锦蛇等天敌在孵化和育雏期对卵和雏鸟的破坏比例很高 (达 40% 以上) (丁长青和郑光美，1997)，由于目前缺乏捕食者的控制，这些

天敌的数量呈上升趋势；栖息地的破坏与片断化；适宜巢址不足；保护区外至今仍存在的偷猎现象，以及保护区内旅游开发等增加，导致人类活动对其影响不断加大。保护区内尚未硬化的道路对其活动的影响不大 (Sun *et al.*, 2009a)，对于硬化后的道路，由于车辆行驶快、鸟类不适应等因素，很少发现鸟类穿越 / More than 40% of known nests have been destroyed by natural predators, in particular Eurasian Jay *Garrulus glandarius*, Yellow-throated Marten *Martes flavigula* and rat snakes (Ding and Zheng, 1997), the population size of which has been increased due to lack of predators. Other threats include: habitat loss and fragmentation; lack of suitable nests; illegal hunting outside nature reserves; human disturbance owing to the tourism in nature reserves. And the impact of roads may be limited to hard-surfaced (Sun *et al.*, 2009a); few birds would go across the road considering the high speed of vehicles and adaptiveness of birds

保护级别与保护行动 Protection Category and Conservation Action (s)

IUCN 红色名录 (2016) IUCN Red List (2016)	易危 / VU C2a (i)
保护行动 Conservation Action(s)	列入国家 I 级重点保护野生动物和 CITES 公约附录 I 物种。对栖息地、活动区、繁殖、行为、食性等进行了深入研究，获得了比较齐全的生活史

资料；成立了多个以黄腹角雉为旗舰物种的国家级和省级自然保护区，保护面积加大；在浙江种群的栖息地内悬挂黄腹角雉人工巢，巢数显著增加 (Deng *et al.*, 2005)，这一方式已经在武夷山等保护区得到推广；突破了限制人工繁殖的各个环节，在湖南和北京建立了稳定的人工种群，并相继在浙江和湖南成功地实施了向原产地再补充和再引入工作 (丁长青等，1997；Liu *et al.*, 2016) / It has been listed as the first class in National Key Protected Wild Animal List, and Appendix I of the CITES. Information on its life history is relatively complete, and habitats, home range, breeding, behavior, diet has been intensively studied. There are several national and provincial nature reserves set *T. caboti* as flagship species. Provision of artificial nesting platforms in Zhejiang significantly increased the nests (Deng *et al.*, 2005), and the method has been introduced to Wuyishan Nature Reserve, *etc*. The bottlenecks of breeding technology have been broken, and artificial populations are stable in Hunan and Beijing, which were reintroduced into Zhejiang and Hunan (Ding *et al.*, 1997; Liu *et al.*, 2016)

黄腹角雉
Tragopan caboti

包其敏 摄
By Qimin Bao

白尾梢虹雉
Lophophorus sclateri

濒危 EN C2a(i)

| 数据缺乏 DD | 无危 LC | 近危 NT | 易危 VU | 濒危 EN | 极危 CR | 区域灭绝 RE | 野外灭绝 EW | 灭绝 EX |

分类地位 Taxonomic Status

动物界 Animalia	脊索动物门 Chordata	鸟纲 Aves	鸡形目 Galliformes	雉科 Phasianidae
学 名 Scientific Name		*Lophophorus sclateri*		
命 名 人 Species Authority		Jerdon, 1870		
英 文 名 English Name(s)		Sclater's Monal		
同物异名 Synonym(s)		无 / None		
种下单元评估 Infra-specific Taxa Assessed		无 / None		

评估信息 Assessment Information

评估年份 Year Assessed	2016
评 定 人 Assessor(s)	张正旺 / Zhengwang Zhang
其他贡献人 Other Contributor(s)	无 / None

理由 Justification: 白尾梢虹雉分布区域狭小，以往调查十分有限。推测其种群规模小、栖息地破碎化以及处于显著的下降之中。该物种在中国的受威胁现状更为严重，分布区域小，种群数量估计在 2,000 只以下，目前面临栖息地的破坏和捕猎压力较大，推测这种趋势将会持续下去。因此，列为濒危等级 / The distribution area of *Lophophorus sclateri* is small, and the surveys were not fully covered. It's presumed that the population is small, habitats are fragmentated and declining. While it is more threatened in China, with population size estimated to be less than 2,000 individuals. The destroy of habitats and high pressure of predation probably continue. Thus, it is listed as Endangered

评估历史 Assessment History: 《国家重点保护野生动物名录》：Ⅰ (1989, 2021)；《中国濒危动物红皮书·鸟类》(1998)：稀有；《中国物种红色名录：第二卷 脊椎动物 下册》(2009)：易危 / National Key Protected Wild Animal List: I (1989, 2021); China Red Data Book of Endangered Animals: Aves (1998): R; China Species Red List: Vol. II Vertebrates Part 2 (2009): VU

地理分布 Geographical Distribution

国内分布 Domestic Distribution
分布于西藏东南部的墨脱、米林、察隅等地，以及云南西北部的腾冲、泸水、福贡和贡山一带 / It only occurs in Motuo, Milin, Zayü in southeast Tibet (Xizang), and Tengchong, Lushui, Fugong, Gongshan in northwest Yunnan
分布标注 Distribution Note
非特有种 / Non-endemic

国内分布图
Map of Domestic Distribution

种群 Population

种群数量 Population Size	基于对现有记录、分布范围极其丰富度的评估，全球种群数量估计为 3,500～15,000 只 (BirdLife, 2014)，我国种群数量估计在 2,000 只以下 / The global population size is estimated to be 3,500～15,000 mature individuals based on an assessment of known records, descriptions of abundance and range size (BirdLife, 2014). The population size is less than 2000 individuals in China
种群趋势 Population Trend	下降 / Decreasing

生境与生态系统 Habitat (s) and Ecosystem (s)

生境 Habitat(s)	主要栖息于海拔 2,500～4,000 m 的高山森林和林缘灌丛与草地，特别是亚高山针叶林、高山竹林灌丛、杜鹃灌丛等地带，有时也到高山草地和风化的裸岩地带活动。多单独活动，在云南高黎贡山的冬季曾见有 9 只生活在一起的群体。推断白尾梢虹雉雄鸟在海拔 3,300～3,600 m 存在垂直迁移现象。主要植物性食物包括贝母、多星韭、牛尾独活、高山羌活、紫花百合、七筋姑等 / *L. sclateri* inhabits alpine forests and shrubs and meadows at edges of forests from 2,500～4,000 m, especially subalpine coniferous forests, bamboo clumps and rhododendron shrubs. It also uses alpine meadows and rocky areas. It is solitary while a group of 9 individuals were recorded in winter in Gaoligong Shan. Male is presumed migrating vertically from 3,300 to 3,600 m. The main food includes *Frotillaria cirrhossa*, *Allium wallichii*, *Aralia apioides*, *Notopteryium forrestii*, *Lilium souliei*, and *Climtonia udensis*
生态系统 Ecosystem(s)	陆地生态系统 / Terrestrial Ecosystem

威胁 Threat (s)

主要威胁 Major Threat(s)	主要威胁包括栖息地破坏和捕猎，以及猛禽和小型食肉动物的捕食 / The major threats are habitat destroy, hunting, and predation of raptor and small carnivore

保护级别与保护行动 Protection Category and Conservation Action (s)

IUCN 红色名录 (2016) IUCN Red List (2016)	易危 / VU C2a (i)
保护行动 Conservation Action(s)	列入国家 I 级重点保护野生动物和 CITES 公约附录 I 物种。在其分布区已经建立了高黎贡山自然保护区，对野生种群的保护起到了重要作用。已经被国家林业和草原局列为极小种群濒危物种 / It has been listed as the first class in Chinese National Key Protected Wild Animal List, the Appendix I of the CITES, and listed as a minimally endangered species by the State Forestry Administration. Gaoligong Shan National Nature Reserve was set up, and has contributed to the wild population conservation

白尾梢虹雉 *Lophophorus sclateri*   许勇 摄 By Yong Xu

中国生物多样性红色名录

绿尾虹雉
Lophophorus lhuysii

濒危　EN B2ab(iii); C2a(i)

| 数据缺乏 DD | 无危 LC | 近危 NT | 易危 VU | 濒危 EN | 极危 CR | 区域灭绝 RE | 野外灭绝 EW | 灭绝 EX |

分类地位 Taxonomic Status

动物界 Animalia	脊索动物门 Chordata	鸟纲 Aves	鸡形目 Galliformes	雉科 Phasianidae
学名 Scientific Name		*Lophophorus lhuysii*		
命名人 Species Authority		Geoffroy Saint-Hilaire, 1866		
英文名 English Name(s)		Chinese Monal		
同物异名 Synonym(s)		无 / None		
种下单元评估 Infra-specific Taxa Assessed		无 / None		

评估信息 Assessment Information

评估年份 Year Assessed	2016
评定人 Assessor(s)	冉江洪 / Jianghong Ran
其他贡献人 Other Contributor(s)	窦亮等 / Liang Dou *et al.*

理由 Justification: 绿尾虹雉种群数量小、密度低，且栖息地片断化严重，受偷猎和生境损失影响，种群持续下降。在绿尾虹雉的模式标本产地宝兴县，高山区的放牧范围和强度一直处于扩张的态势，根据2008～2012年对宝兴县绿尾虹雉的野外调查发现，其种群数量较20世纪80年代有大幅下降。因此，列为濒危等级 / The population size and density of *Lophophorus lhuysii* is small and the habitats are fragmentated. The population is undergoing a decline owing to the habitat loss and illegal trapping. In Baoxing, where the type specimen was from, the extent and intensity of grazing is expanding. Based on the field survey at Baoxing in 2008～2012, the population size has declined compared to that in 1980s. Thus, it is listed as Endangered

评估历史 Assessment History:《国家重点保护野生动物名录》：Ⅰ(1989，2021)；《中国濒危动物红皮书·鸟类》(1998)：濒危；《中国物种红色名录：第二卷 脊椎动物 下册》(2009)：易危 / *National Key Protected Wild Animal List*: I (1989, 2021); *China Red Data Book of Endangered Animals*: *Aves* (1998): E; *China Species Red List*: *Vol. II Vertebrates Part 2* (2009): VU

地理分布 Geographical Distribution

国内分布 Domestic Distribution

分布于四川、甘肃南部、西藏东部和青海东南部 / It occurs in Sichuan, south Gansu, east Tibet (Xizang) and southeast Qinghai

分布标注 Distribution Note

特有种 / Endemic

国内分布图
Map of Domestic Distribution

种群 Population

种群数量 Population Size	估计野外种群数量为 10,000～25,000 只。四川宝兴密度为 1.0～1.33 只/km², 北川为 1.32 只/km² (卢汰春,1991); 四川松潘为 3.5 只/km² (隆廷伦和郭耕, 1998); 甘肃白水江自然保护区在高山灌丛草甸带最大为 3.25 只/km², 在针阔混交林里的密度为 0.5 只/km² (张涛, 1995) / The wild population size is estimated to be 10,000～25,000 individuals. In Sichuan, the density is 1.0～1.33 individuals per km² in Baoxing, 1.32 individuals per km² in Beichuan (Lu *et al.*, 1991), and 3.5 individuals per km² in Songpan (Long and Guo, 1998). In Gansu Baishuijiang Nature Reserve, the density is 3.25 individuals per km² in alpine meadows, 0.5 individual per km² in mixed coniferous broad-leaved forests (Zhang, 1995)
种群趋势 Population Trend	下降 / Decreasing

生境与生态系统 Habitat (s) and Ecosystem (s)

生境 Habitat(s)	栖息于高山草甸、灌丛和裸岩地带,尤其喜欢多陡崖和岩石的高山灌丛和灌丛草甸,冬季常下到海拔 2,500～3,000 m 的林缘灌丛等地带活动 / *L. lhuysii* inhabits alpine meadows, shrubs, and exposed cliffs and crags, especially the alpine shrubs and meadows with cliffs and rocks. It moves down to the shrubs at the edge of forests at 2,500～3,000 m
生态系统 Ecosystem(s)	陆地生态系统 / Terrestrial Ecosystem

威胁 Threat (s)

主要威胁 Major Threat(s)	绿尾虹雉的适宜栖息地是高海拔的草甸、灌丛和林缘,而这些地方也是当地居民放牧和采药的主要场所,随着畜牧业的发展,其栖息地受到的干扰越来越严重。由于高大山体和深切河谷的自然阻隔,以及道路等工程建设的干扰,其栖息地片断化严重,被分割成多个斑块。由于本物种体型较大,也常是偷猎的主要对象 / Its subalpine shrub and alpine meadow habitats have been disturbed by grazing and medicine collection. The habitats are fragmentated into patches owing to natural isolation and disturbance of projects (*e.g.* highway). Illegal hunting is also considered to be a threat

保护级别与保护行动 Protection Category and Conservation Action (s)

IUCN 红色名录 (2016) IUCN Red List (2016)	易危 / VU C2a (i)
保护行动 Conservation Action(s)	列入国家 I 级重点保护野生动物和 CITES 公约附录 I 物种。在主要分布的邛崃山系和岷山山系建立了许多自然保护区,绿尾虹雉大多数的栖息地被纳入了保护范围,但保护力度还有待加强。在四川宝兴县蜂桶寨国家级自然保护区内,建立有绿尾虹雉人工繁殖饲养场,进行绿尾虹雉人工繁育试验及对人工饲养种群的管理 / It has been listed as the first class in National Key Protected Wild Animal List, and listed into Appendix I of the CITES. Several nature reserves have been set up in its main distribution range in Qionglai Mountains and Minshan Mountains, and most of its habitats have been protected, while the conservation effectivity is to be improved. A breeding population has been established in Fengtongzhai National Nature Reserve in Baoxing, Sichuan province, and artificial breeding experiment and population management has been conducted

绿尾虹雉 *Lophophorus lhuysii*　　　　　　　　　　　　　　　　　　　　　　　董磊 摄　By Lei Dong

白冠长尾雉
Syrmaticus reevesii

中国生物多样性 红色名录

濒危　EN A2cd+3cd+4cd; C2a(i)

| 数据缺乏 DD | 无危 LC | 近危 NT | 易危 VU | 濒危 EN | 极危 CR | 区域灭绝 RE | 野外灭绝 EW | 灭绝 EX |

分类地位 Taxonomic Status

动物界 Animalia	脊索动物门 Chordata	鸟纲 Aves	鸡形目 Galliformes	雉科 Phasianidae
学名 Scientific Name		*Syrmaticus reevesii*		
命名人 Species Authority		Gray, 1829		
英文名 English Name(s)		Reeves's Pheasant		
同物异名 Synonym(s)		无 / None		
种下单元评估 Infra-specific Taxa Assessed		无 / None		

评估信息 Assessment Information

评估年份 Year Assessed	2016
评定人 Assessor(s)	张正旺 / Zhengwang Zhang
其他贡献人 Other Contributor(s)	徐基良、周春发 / Jiliang Xu, Chunfa Zhou

理由 Justification: 白冠长尾雉分布区退缩，种群数量下降显著。因此，列为濒危等级 / The population size and distribution areas of *Syrmaticus reevesii* have decreased significantly. Thus, it is listed as Endangered

评估历史 Assessment History: 《国家重点保护野生动物名录》：II (1989)，I (2021)；《中国濒危动物红皮书·鸟类》(1998)：易危；《中国物种红色名录：第二卷 脊椎动物 下册》(2009)：易危 / *National Key Protected Wild Animal List*: II (1989), I (2021); *China Red Data Book of Endangered Animals*: *Aves* (1998): V; *China Species Red List*: *Vol. II Vertebrates Part 2* (2009): VU

地理分布 Geographical Distribution

国内分布 Domestic Distribution

分布于河南、陕西、甘肃、四川、重庆、云南、贵州、湖北、湖南、安徽等地 / It occurs in Henan, Shaanxi, Gansu, Sichuan, Chongqing, Yunnan, Guizhou, Hubei, Hunan, Anhui, *etc.*

分布标注 Distribution Note

特有种 / Endemic

国内分布图
Map of Domestic Distribution

种群 Population

种群数量 Population Size	种群数量不足 10,000 只 / The population size is less than 10,000 individuals
种群趋势 Population Trend	下降 / Decreasing

生境与生态系统 Habitat (s) and Ecosystem (s)

生　　境 Habitat(s)	属于典型的森林鸟类，主要分布于海拔 200～2,600 m 的山地森林中，栖息在有高大乔木的常绿阔叶林、落叶阔叶林及针阔混交林内。这些林型的特点都是林上较为郁闭、林下较为空旷，有利于白冠长尾雉的活动和觅食。其栖息地的植被类型在不同地区之间略有差异。近年来的调查发现，白冠长尾雉也能在杉木林和马尾松林等人工林中栖息 (Xu et al., 2011)。一般在林缘和靠近林缘的农田觅食。全年以植物性食物为主，但随季节而有变化 / *S. reevesii* is typical forest species, and occurs in evergreen broad-leaved forests, deciduous broad-leaved forests and mixed coniferous broad-leaved forests at 200～2,600 m (Xu et al., 1991; Wu et al.,1994) with high canopy density while open in the undergrowth, which are suitable for foraging and other activities. The vegetation in habitats varies in different areas. It could roost in artificial forests (*e.g.* firs, pines) based on the recent surveys (Xu et al., 2011). It forages at the edge of forests or on the farmlands. It is herbivorous all year, while changing with the seasons
生态系统 Ecosystem(s)	陆地生态系统 / Terrestrial Ecosystem

威胁 Threat (s)

主要威胁 Major Threat(s)	森林面积的减少是导致白冠长尾雉数量下降的主要原因 (郑光美和王岐山，1998；张正旺等，2003)。森林采伐以及人工林的种植，致使一些地区白冠长尾雉的栖息地出现了片断化，对野生种群的生存产生负面影响。非法捕杀则是其数量迅速下降的直接原因，除了为了获取"雉鸡翎"而猎捕外，一些农民为了保护庄稼而在山区小块耕地周边投放毒饵，造成雉类集体中毒死亡的现象也十分严重 / The main cause of population decrease is loss of forests (Zheng and Wang, 1998; Zhang et al., 2003). The habitats fragmentation owing to deforestation and artificial forest plantation has exerted negative effect on the survival of wild population. The illegal hunting, mostly for its long tail feathers, which were used as a decoration in the Peking opera costumes, directly led to the rapid decline of the population. In addition, *S. reevesii* was poisoned by some local farmers in mountain areas for crops, and even a group of pheasants died occasionally

保护级别与保护行动 Protection Category and Conservation Action (s)

IUCN 红色名录 (2016) IUCN Red List (2016)	易危 / VU A2cd+3cd+4cd; C2a (i)
保护行动 Conservation Action(s)	列入国家 I 级重点保护野生动物。目前在白冠长尾雉分布地区已经建立了一批自然保护区，比较著名的有贵州梵净山、陕西佛坪、长青、太白、洋县，河南董寨、鸡公山、金刚台、太白顶、连康山，湖北神农架、五峰后河、大老岭，安徽天马(天堂寨)、鹞落坪等自然保护区，保护区总面积达到 3,000 km² 以上 / The species has been listed as the first class in National Key Protected Wild Animal List. Several nature reserves have been set up in its distributed areas, *i.e.* Fanjing Shan in Guizhou, Foping, Changqing, Taibai, Yangxian in Shaanxi; Dongzhai, Jigongshan, Jingangtai, Taibaiding, Liankangshan in Henan; Shennongjia, Wufenghouhe, Dalaoling in Hubei, Tianma, Yaoluoping in Anhui, with an area over 3,000 km²

白冠长尾雉
Syrmaticus reevesii

徐基良 摄
By Jiliang Xu

灰孔雀雉
Polyplectron bicalcaratum

濒危 EN C1+2a (i)

| 数据缺乏 DD | 无危 LC | 近危 NT | 易危 VU | 濒危 EN | 极危 CR | 区域灭绝 RE | 野外灭绝 EW | 灭绝 EX |

分类地位 Taxonomic Status

动物界 Animalia	脊索动物门 Chordata	鸟纲 Aves	鸡形目 Galliformes	雉科 Phasianidae
学名 Scientific Name		*Polyplectron bicalcaratum*		
命名人 Species Authority		Linnaeus, 1758		
英文名 English Name(s)		Grey Peacock Pheasant		
同物异名 Synonym(s)		无 / None		
种下单元评估 Infra-specific Taxa Assessed		无 / None		

评估信息 Assessment Information

评估年份 Year Assessed	2016
评定人 Assessor(s)	张正旺 / Zhengwang Zhang
其他贡献人 Other Contributor(s)	无 / None

理由 Justification: 由于栖息地的破坏和捕猎压力较大,灰孔雀雉的种群数量快速下降,栖息地片断化严重,推测这种趋势将会持续下去。云南的灰孔雀雉的数量已经比较稀少,估计不足1,000只。因此,列为濒危等级 / The population size of *Polyplectron bicalcaratum* is undergoing a rapid decline owing to the destroy of habitat and high pressure of hunting, and the trend is estimated to continue considering the fragmentated habitats. The population of *P. bicalcaratum* in Yunnan is less than 1,000 individuals. Thus, it is listed as Endangered

评估历史 Assessment History: 《国家重点保护野生动物名录》:Ⅰ(1989,2021);《中国濒危动物红皮书·鸟类》(1998):濒危;《中国物种红色名录:第二卷 脊椎动物 下册》(2009):未评估 / *National Key Protected Wild Animal List*: I (1989, 2021); *China Red Data Book of Endangered Animals*: Aves (1998): E; *China Species Red Lis*: *Vol. II Vertebrates Part 2 t* (2004): NA

地理分布 Geographical Distribution

国内分布 Domestic Distribution
仅分布于云南盈江、西双版纳、勐腊、景洪、思茅等地 / It only occurs in Yingjiang, Xishuangbanna, Mengla, Jinghong, Simao in Yunnan
分布标注 Distribution Note
非特有种 / Non-endemic

国内分布图
Map of Domestic Distribution

种群 Population

种群数量 Population Size	近几年在云南西双版纳自然保护区的调查，种群密度为 1.5～2 只 /km²。在保护区外种群密度更低，较之 20 世纪 50 年代种群数量明显减少 / The density is 1.5 ～ 2 individuals per km² in Xishuangbanna Nature Reserve in Yunnan. The density outside the nature reserves is much lower, compared to the population size in 1950s
种群趋势 Population Trend	下降 / Decreasing

生境与生态系统 Habitat (s) and Ecosystem (s)

生　　境 Habitat(s)	主要以昆虫、蠕虫以及植物茎、叶、果实、种子为食 / Carnivorous with insects and worms, while complementary with seedlings, leaves, fruits and seeds
生态系统 Ecosystem(s)	陆地生态系统 / Terrestrial Ecosystem

威胁 Threat (s)

主要威胁 Major Threat(s)	由于其赖以生存的季雨林和常绿季雨林生态环境遭到破坏，其分布区正在逐步减少和退缩；此外非法捕猎和农药中毒导致该物种在许多分布地点的数量都出现了下降，甚至出现局域性灭绝 / The ecosystem of seasonal rainforests and evergreen seasonal rainforests are being destroyed, and distribution areas are decreasing. In addition, its population decreased in some sites, or even local extinct, owing to illegal hunting and pesticide

保护级别与保护行动 Protection Category and Conservation Action (s)

IUCN 红色名录 (2016) IUCN Red List (2016)	无危 / LC
保护行动 Conservation Action(s)	列入国家 I 级重点保护野生动物和 CITES 公约附录 II / It has been listed as the first class National Key Protected Wild Animal List and Appendix II of the CITES

灰孔雀雉 Polyplectron bicalcaratum　　　　　　　　　　　　　　　　孙克信 摄　By Kexin Sun

棉凫
Nettapus coromandelianus

濒危 EN C2a(ii)

分类地位 Taxonomic Status

动物界 Animalia	脊索动物门 Chordata	鸟纲 Aves	雁形目 Anseriformes	鸭科 Anatidae
学名 Scientific Name		*Nettapus coromandelianus*		
命名人 Species Authority		Gmelin, 1789		
英文名 English Name(s)		Asian Pygmy-Goose		
同物异名 Synonym(s)		*Anas coromandelianus*		
种下单元评估 Infra-specific Taxa Assessed		无 / None		

评估信息 Assessment Information

评估年份 Year Assessed	2016
评定人 Assessor(s)	陈水华 / Shuihua Chen
其他贡献人 Other Contributor(s)	王思宇 / Siyu Wang

理由 Justification: 棉凫种群数量小，近年来未发现100只以上的大群，估计在中国境内繁殖的个体少于2,500只。因此，列为濒危等级 / The population size of *Nettapus coromandelianus* is small, and no flock over 100 individuals is recorded. The breeding population is estimated to be less than 2,500 individuals. Thus, it is listed as Endangered

评估历史 Assessment History: 《国家重点保护野生动物名录》：II (2021)；《中国濒危动物红皮书·鸟类》(1998)：稀有；《中国物种红色名录：第二卷 脊椎动物 下册》(2009)：濒危 / National Key Protected Wild Animal List: II (2021); China Red Data Book of Endangered Animals: Aves (1998): R; China Species Red List: Vol. II Vertebrates Part 2 (2009): EN

地理分布 Geographical Distribution

国内分布 Domestic Distribution

主要繁殖于四川中部至西南部、长江中下游以南地区，以及云南南部、海南岛以及广东和广西，偶见于华北。在广东、广西为留鸟，台湾为迷鸟 / It breeds in middle and southwest Sichuan, south areas of middle and lower Changjiang River, south Yunnan, Hainan, Guangdong, Guangxi, seldom in north China. It is resident in Guangdong and Guangxi, and vagrant visitor in Taiwan

分布标注 Distribution Note

非特有种 / Non-endemic

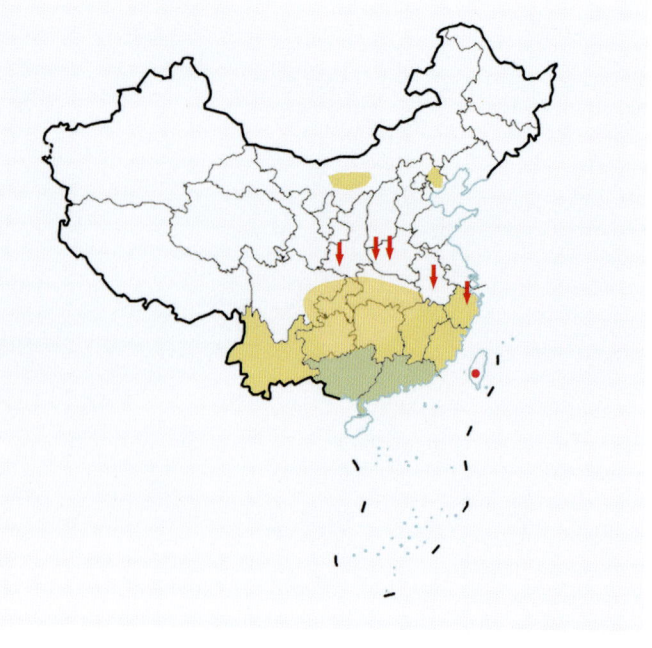

国内分布图
Map of Domestic Distribution

种群 Population

种群数量 Population Size	种群数量不足 2,500 只 / The population size is less than 2,500 individuals
种群趋势 Population Trend	下降 / Decreasing

生境与生态系统 Habitat(s) and Ecosystem(s)

生境 Habitat(s)	栖息于江河、湖泊、水塘和沼泽地带，特别是富有水生植物的开阔水域，有时也见于村庄附近的小水塘和水渠中 / *N. coromandelianus* inhabits in rivers, lakes, ponds and marshes, especially the open waters with abundant hydrophyte. It also occasionally occurs in small ponds or channels around villages
生态系统 Ecosystem(s)	湿地生态系统 / Wetland Ecosystem

威胁 Threat(s)

主要威胁 Major Threat(s)	适宜栖息地不断丧失，沿海地区可供棉凫迁徙途中停歇的湿地、湖泊不断被破坏，偷猎和捡蛋也是主要威胁之一 / The major threats are habitat loss owing to the destruction of wetlands and lakes along the migratory route, illegal hunting and eggs collection

保护级别与保护行动 Protection Category and Conservation Action(s)

IUCN 红色名录 (2016) IUCN Red List (2016)	无危 / LC
保护行动 Conservation Action(s)	列入中国国家 II 级重点保护野生动物、《国家保护的有益的或者有重要经济、科学研究价值的陆生野生动物名录》/ It has been listed as the second class in National Key Protected Wild Animal List, *National Protected List of Terrestrial Wild Animals with Good Benefits or Important Economic and Scientific Values*

棉凫 *Nettapus coromandelianus* 钱斌 摄 By Bin Qian

中华秋沙鸭
Mergus squamatus

濒危 EN C2a(i)

| 数据缺乏 DD | 无危 LC | 近危 NT | 易危 VU | **濒危 EN** | 极危 CR | 区域灭绝 RE | 野外灭绝 EW | 灭绝 EX |

分类地位 Taxonomic Status

动物界 Animalia	脊索动物门 Chordata	鸟纲 Aves	雁形目 Anseriformes	鸭科 Anatidae
学名 Scientific Name		*Mergus squamatus*		
命名人 Species Authority		Gould, 1864		
英文名 English Name(s)		Scaly-sided Merganser		
同物异名 Synonym(s)		无 / None		
种下单元评估 Infra-specific Taxa Assessed		无 / None		

评估信息 Assessment Information

评估年份 Year Assessed	2016
评定人 Assessor(s)	王海涛 / Haitao Wang
其他贡献人 Other Contributor(s)	刘宇、杨志杰、易国栋、左斌、王拓等 / Yu Liu, Zhijie Yang, Guodong Yi, Bin Zuo, Tuo Wang *et al.*

理由 Justification: 中华秋沙鸭种群数量小；适宜繁殖生境丧失；受非法狩猎和人为活动干扰，种群呈持续下降趋势。因此，列为濒危等级 / *Mergus squamatus* has a very small population which is suspected to be undergoing a continuing decline as a result of habitat loss, illegal hunting and human disturbance. Thus, it is listed as Endangered

评估历史 Assessment History: 《国家重点保护野生动物名录》：Ⅰ (1989, 2021)；《中国濒危动物红皮书·鸟类》(1998)：稀有；《中国物种红色名录：第二卷 脊椎动物 下册》(2009)：易危 / *National Key Protected Wild Animal List*: Ⅰ (1989, 2021); *China Red Data Book of Endangered Animals*: *Aves* (1998): R; *China Species Red List*: *Vol. II Vertebrates Part 2* (2009): VU

地理分布 Geographical Distribution

国内分布 Domestic Distribution

繁殖于吉林长白山和黑龙江小兴安岭，迁徙经过辽宁、北京、天津、河北、山东等地，主要在中国中部和南部以及长江中下游流域越冬，江西省为重要的越冬地 (刘宇等, 2008; 汪志如等, 2010; Barter *et al.*, 2014)，在台湾为旅鸟 / It breeds in Changbai Shan in Jilin and Xiao Hinggan Ling in Heilongjiang; migrates across Liaoning, Beijing, Tianjin, Hebei, Shandong *etc.*; and winters in south and central China, while mainly in middle and lower Changjiang River, with Jiangxi as the key wintering site (Liu *et al.*, 2008; Wang *et al.*, 2010; Barter *et al.*, 2014); and passage visitor in Taiwan

分布标注 Distribution Note

非特有种 / Non-endemic

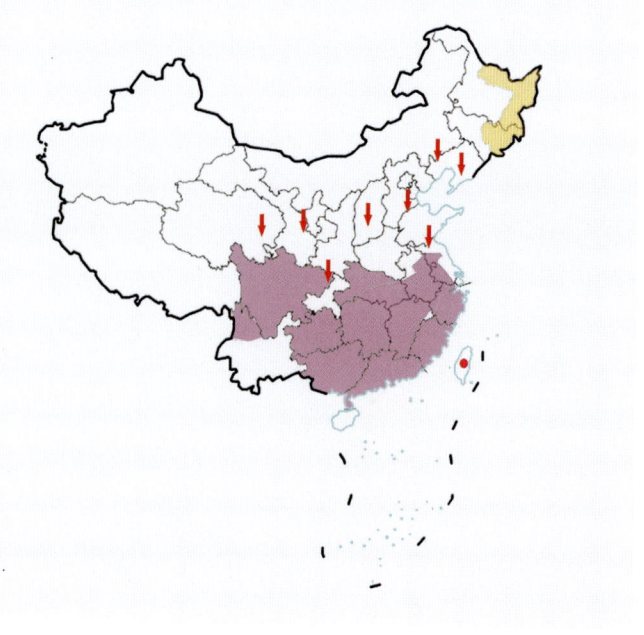

国内分布图
Map of Domestic Distribution

🦅 种群 Population

种群数量 Population Size 吉林长白山 2008～2009 年调查种群数量 170 繁殖对 (Liu *et al.*, 2010), 较东北繁殖地 200～250 对的数量下降 (赵正阶等, 1994); 江西省 2009 年调查 255 只 (汪志如等, 2010), 其他各地的报告多为迁徙和越冬的零散记录 (杨岚和杨晓君, 1997; 任巍等, 2008; 李小燕等, 2012)。2014～2015 年和 2015～2016 年朱雀会组织对全国越冬的中华秋沙鸭进行了调查, 分别发现 440 只和 634 只越冬个体 / The breeding population in northeast China was 200～250 pairs (Zhao *et al.*, 1994), and decreased into 170 breeding pairs recorded in Changbai Shan in Jilin in 2008～2009 (Liu *et al.*, 2010). In 2009, 255 individuals were recorded in Jiangxi (Wang *et al.*, 2010), and other wintering or migration records were in small population (Yang *et al.*, 1997; Ren *et al.*, 2008; Li *et al.*, 2012). 440 and 634 individuals were recorded in winter of 2014~2015 and 2015~2016, respectively, surveyed by China Birdwatching Association

种群趋势 Population Trend 下降 / Decreasing

🦅 生境与生态系统 Habitat (s) and Ecosystem (s)

生　　境 Habitat(s) 繁殖生境为温带海拔 1,000 m 以下低山阔叶林和针阔混交林带, 且临近水流急缓相间、水质清澈、河中常存在突兀砾石或滩岛的河流中游河段; 少数个体也在湖泊周边阔叶林和针阔混交林带繁殖。越冬地与繁殖地相似, 处于丘陵或山地河流的中游河段, 河道宽阔, 可保证足够的警戒距离或者两岸植被茂密, 人不易接近 / *M. squamatus* breeds below 1000 m in mountainous areas, mainly within the temperate broad-leaved forests or mixed coniferous and broad-leaved forests, along clear flowing rivers, with gravel bars or islands in middle reach. Several individuals were found breeding in the broad-leaved forests or mixed coniferous and broad-leaved forests around lakes. Similar to the breeding sites, the wintering sites are located in the middle reach of rivers in hilly or mountainous areas, and the rivers are wide enough for alerting or the vegetation on banks are dense and inaccessible

生态系统 Ecosystem(s) 湿地生态系统和陆地生态系统 / Wetland Ecosystem and Terrestrial Ecosystem

🦅 威胁 Threat (s)

主要威胁 Major Threat(s) 繁殖地: 大范围森林采伐, 尤其原生林砍采伐, 导致适宜繁殖的巢洞资源不足; 拦河筑坝; 大强度人类活动干扰 (赵正阶等, 1994; Liu *et al.*, 2010; Solovyeva *et al.*, 2014)。越冬地: 拦河筑坝、采砂、工业和生活污染、过度捕鱼和家禽饲养 (汪志如等, 2010; Barter *et al.*, 2014) / Breeding site: deforestation, especially the logging of primeval forests, which is likely to reduce suitable nest for breeding; dam construction; and human disturbance (Zhao *et al.*, 1994; Liu *et al.*, 2010; Solovyeva *et al.*, 2014). Wintering area: dam construction, sand extraction, industrial and domestic pollution, overfishing and poultry production (Wang *et al.*, 2010; Barter *et al.*, 2014)

🦅 保护级别与保护行动 Protection Category and Conservation Action (s)

IUCN 红色名录 (2016)
IUCN Red List (2016) 濒危 / EN C2a(ii)

保护行动 Conservation Action(s) 列入国家 I 级国家重点保护野生动物, 一些繁殖地和越冬地位于自然保护区中。越冬地分布与数量、越冬生态学研究在积极展开 / It has been listed as the first class in National Key Protected Wild Animal List, and some of the breeding and wintering sites are located in nature reserves. There have been positively undergoing studies on wintering distribution and population size, as well as wintering ecology

中华秋沙鸭 *Mergus squamatus*　　　　　　　　　　　　　　　　　　　　　　刘宇 摄　By Yu Liu

紫林鸽
Columba punicea

濒危 EN B1a+b(iii); D

| 数据缺乏 DD | 无危 LC | 近危 NT | 易危 VU | **濒危 EN** | 极危 CR | 区域灭绝 RE | 野外灭绝 EW | 灭绝 EX |

分类地位 Taxonomic Status

动物界 Animalia	脊索动物门 Chordata	鸟纲 Aves	鸽形目 Columbiformes	鸠鸽科 Columbidae
学名 Scientific Name		*Columba punicea*		
命名人 Species Authority		Blyth, 1842		
英文名 English Name(s)		Pale-capped Pigeon		
同物异名 Synonym(s)		*Alsocomus puniceus*		
种下单元评估 Infra-specific Taxa Assessed		无 / None		

评估信息 Assessment Information

评估年份 Year Assessed	2016
评定人 Assessor(s)	刘阳 / Yang Liu
其他贡献人 Other Contributor(s)	黄秦 / Qin Huang

理由 Justification: 紫林鸽在中国的分布区极为狭窄，且分布区不连续。种群数量随原始林砍伐或片断化而下降。因此，列为濒危等级 / The distribution area of *Columba punicea* in China is small and fragmentated. The population size is undergoing a decline with deforestation. Thus, it is listed as Endangered

评估历史 Assessment History: 《国家重点保护野生动物名录》：Ⅱ (2021)；《中国濒危动物红皮书·鸟类》(1998)：稀有；《中国物种红色名录：第二卷 脊椎动物 下册》(2009)：易危 / *National Key Protected Wild Animal List*: Ⅱ (2021); *China Red Data Book of Endangered Animals: Aves* (1998): R; *China Species Red List: Vol. II Vertebrates Part 2* (2009): VU

地理分布 Geographical Distribution

国内分布 Domestic Distribution
在西藏南部、云南西南部及海南岛为罕见留鸟 / It is rare resident in south Tibet (Xizang), southwest Yunnan and Hainan
分布标注 Distribution Note
非特有种 / Non-endemic

国内分布图
Map of Domestic Distribution

种群 Population

种群数量 Population Size	我国种群数量少 / The population size in China is small
种群趋势 Population Trend	下降 / Decreasing

生境与生态系统 Habitat (s) and Ecosystem (s)

生　　境 Habitat(s)	亚热带、热带山地森林内的常绿阔叶林、常绿-落叶混交林、竹林；也见于林缘地带、多植被的河畔甚至农耕地 / *C. punicea* uses tropical and subtropical evergreen broad-leaved forests, mixed deciduous broad-leaved forests, bamboo groves; also occurs at edge of forests, river banks with dense vegetation, or even farmlands
生态系统 Ecosystem(s)	陆地生态系统 / Terrestrial Ecosystem

威胁 Threat (s)

主要威胁 Major Threat(s)	栖息地丧失和破碎化以及非法捕猎 / Habitat loss and fragmentation, and illegal hunting

保护级别与保护行动 Protection Category and Conservation Action (s)

IUCN 红色名录 (2016) IUCN Red List (2016)	易危 / VU C2a (i)
保护行动 Conservation Action(s)	列入国家Ⅱ级重点保护野生动物、《国家保护的有益的或者有重要经济、科学研究价值的陆生野生动物名录》/ It has been listed as the second class in National Key Protected Wild Animal List, *National Protected List of Terrestrial Wild Animals with Good Benefits or Important Economic and Scientific Values*

紫林鸽 *Columba punicea*　　　　鸟网 提供　By birdnet.cn

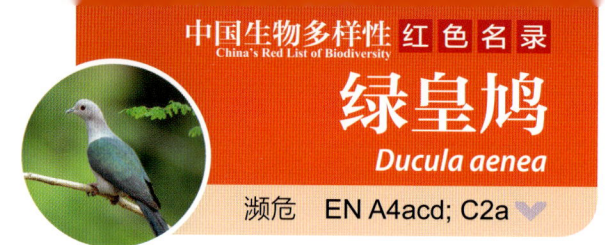

绿鸠 *Ducula aenea*

濒危 EN A4acd; C2a

| 数据缺乏 DD | 无危 LC | 近危 NT | 易危 VU | 濒危 EN | 极危 CR | 区域灭绝 RE | 野外灭绝 EW | 灭绝 EX |

分类地位 Taxonomic Status

动物界 Animalia	脊索动物门 Chordata	鸟纲 Aves	鸽形目 Columbiformes	鸠鸽科 Columbidae
学名 Scientific Name		*Ducula aenea*		
命名人 Species Authority		Linnaeus, 1766		
英文名 English Name(s)		Green Imperial Pigeon		
同物异名 Synonym(s)		*Mucadivora aenea, Columba aenea*		
种下单元评估 Infra-specific Taxa Assessed		无 / None		

评估信息 Assessment Information

评估年份 Year Assessed	2016
评定人 Assessor(s)	周放 / Fang Zhou
其他贡献人 Other Contributor(s)	陆舟 / Zhou Lu

理由 Justification: 绿鸠在中国的分布区呈零星分布,见于广东、海南等地,种群数量极其稀少。由于原生阔叶林被大量破坏,栖息地破碎化严重,适宜生境不断减少,导致其种群数量急剧下降。绿鸠过去常常被作为人类猎食的对象。因此,列为濒危等级 / The population size of *Ducula aenea* is extremely small and its distribution area is scattered. *e.g.* Guangdong and Hainan, and is undergoing a rapid decline as a result of destroy of primeval broad-leaved forests and habitat fragmentation. Furthermore, *D. aenea* is used to be a game bird. Thus, it is listed as Endangered

评估历史 Assessment History: 《国家重点保护野生动物名录》: II (1989, 2021); 《中国濒危动物红皮书·鸟类》(1998): 易危; 《中国物种红色名录: 第二卷 脊椎动物 下册》(2009): 未评估 / *National Key Protected Wild Animal List*: II (1989, 2021); *China Red Data Book of Endangered Animals: Aves* (1998): V; *China Species Red List: Vol. II Vertebrates Part 2*: NA

地理分布 Geographical Distribution

国内分布 Domestic Distribution
主要分布于广东、海南和云南南部 / It mainly occurs in Guangdong, Hainan and south Yunnan
分布标注 Distribution Note
非特有种 / Non-endemic

国内分布图
Map of Domestic Distribution

种群 Population

种群数量 Population Size	种群数量十分稀少 / The population size is extremely small
种群趋势 Population Trend	下降 / Decreasing

生境与生态系统 Habitat(s) and Ecosystem(s)

生　　境 Habitat(s)	热带雨林、亚热带常绿阔叶林和次生林 / Tropical rainforests, subtropical evergreen broad-leaved forests and secondary forests
生态系统 Ecosystem(s)	陆地生态系统 / Terrestrial Ecosystem

威胁 Threat(s)

主要威胁 Major Threat(s)	原生的常绿阔叶林被大量破坏、栖息地减少和非法捕猎可能是导致绿皇鸠受威胁的主要原因。在中国东南部地区的天然林和次生林逐渐改变成人工用材林，使该物种的栖息地受损严重，栖息地的持续丧失和破碎化范围不断扩大，使绿皇鸠的生存状况受到严重威胁 / The major threats of *D. aenea* might be destroy of primeval evergreen broad-leaved forests, habitat loss and illegal hunting. The natural forests and secondary forests in southeast China are changing into planted timber forests, which has induced the habitat loss and fragmentation of *D. aenea*, and threatened the survival of the species

保护级别与保护行动 Protection Category and Conservation Action(s)

IUCN 红色名录 (2016) IUCN Red List (2016)	无危 / LC
保护行动 Conservation Action(s)	列入国家Ⅱ级重点保护野生动物 / It has been listed as the second class in the National Key Protected Wild Animal List

绿皇鸠 *Ducula aenea*　　　　　　　　　　　　　　　　　　　　　　　　刘毅 摄　By Yi Liu

中国生物多样性红色名录 China's Red List of Biodiversity

大鸨
Otis tarda

濒危 EN B2b; C2b

| 数据缺乏 DD | 无危 LC | 近危 NT | 易危 VU | 濒危 EN | 极危 CR | 区域灭绝 RE | 野外灭绝 EW | 灭绝 EX |

分类地位 Taxonomic Status

动物界 Animalia	脊索动物门 Chordata	鸟 纲 Aves	鸨形目 Otidiformes	鸨 科 Otididae

学 名 Scientific Name	*Otis tarda*
命 名 人 Species Authority	Linnaeus, 1758
英 文 名 English Name(s)	Great Bustard
同物异名 Synonym(s)	无 / None
种下单元评估 Infra-specific Taxa Assessed	对内蒙古大鸨种群生存力分析显示，种群未来20年的灭绝概率小于10% / The analysis of population viability of *Otis tarda* in Inner Mongolia (Nei Mongol) indicated that the probability of extinction in 20 years is less than 10%

评估信息 Assessment Information

评 估 年 份 Year Assessed	2016
评 定 人 Assessor(s)	邢莲莲 / Lianlian Xing
其他贡献人 Other Contributor(s)	杨贵生、宋丽军、刘松涛、刘刚 / Guisheng Yang, Lijun Song, Songtao Liu, Gang Liu

理由 Justification: 20世纪90年代夏季在东北地区记录到约200只大鸨；新疆北部有2,500～3,000只（高行宜等，1994）。长江流域记录到冬候鸟700～800只，最大越冬种群307只，1998～2000年记录到的越冬种群为30只；2001年估计中国大鸨总数为3,000～4,000只（田秀华和王进军，2001），数量呈下降趋势。种群下降的主要原因是草原严重退化引起的巢区暴露，易受到天敌的攻击。因此，列为濒危等级 / 200 individuals of *Otis tarda* were recorded in northeast China in 1990s; 700～800 individuals were recorded wintering in Changjiang River with the largest flock of 307 individuals, and only 30 individuals were recorded wintering in 1998～2000; 2,500～3,000 individuals were recorded in north Xinjiang (Gao Xingyi *et al.*, 1994). The population size was estimated to be 3,000～4,000 individuals in 2001 (Tian and Wang, 2001), and was decreasing. The main cause might be the exposure of nests and vulnerability to predators as a result of grassland degeneration. Thus, it is listed as Endangered

评估历史 Assessment History: 《国家重点保护野生动物名录》：Ⅰ(1989，2021)；《中国濒危动物红皮书·鸟类》(1998)：易危；《中国物种红色名录：第二卷 脊椎动物 下册》(2009)：近危 / National Key Protected Wild Animal List: I (1989, 2021); *China Red Data Book of Endangered Animals*: Aves (1998):V; *China Species Red List*: *Vol. II Vertebrates Part 2*: NT

地理分布 Geographical Distribution

国内分布 Domestic Distribution	
繁殖于黑龙江西部、吉林西部、内蒙古东北部、新疆西北部，迁徙经过华北地区，越冬于山东、甘肃以及长江中下游地区 / It breeds in west Heilongjiang, west Jilin, northeast Inner Mongolia (Nei Mongol) and northwest Xinjiang; pass by North China when migration; and winters in Shandong, Gansu, middle and lower Changjiang River	
分布标注 Distribution Note	
非特有种 / Non-endemic	

国内分布图
Map of Domestic Distribution

种群 Population

种群数量 Population Size	全国种群数量 2,600～3,000 只，新疆 200～300 只 / The population size is estimated to be 2,600～3,000 individuals in China, and 200～300 individuals in Xinjiang
种群趋势 Population Trend	下降 / Decreasing

生境与生态系统 Habitat (s) and Ecosystem (s)

生　　境 Habitat(s)	我国各种类型草原，包括森林草原、典型草原及荒漠草原，最适宜近水草原地区 / Various grassland, including forest grassland, typical grassland and desert grassland, and grassland close to water are most suitable
生态系统 Ecosystem(s)	草原生态系统 / Grassland Ecosystem

威胁 Threat (s)

主要威胁 Major Threat(s)	草原退化以及狐狸、狼等对雏鸟和卵的掠食等 (邢莲莲，2006) / Grassland degeneration and predation of chicks and eggs by foxes and wolves (Xing, 2006)

保护级别与保护行动 Protection Category and Conservation Action (s)

IUCN 红色名录 (2016) IUCN Red List (2016)	易危 / VU A2cd+3cd+4cd
保护行动 Conservation Action(s)	列入国家 I 级重点保护野生动物、CITES 公约附录 II 和迁徙物种保护公约 (CMS) 附录 II。繁殖地已建立多个保护区，越冬地建议设立草原生态系统保护区和湿地自然保护区，严禁猎捕 / It has been listed into the first class of National Key Protected Wild Animal List, Appendix II of CITES and Appendix II of the Convention on Migratory Species (CMS). Grassland and wetland nature reserves should be set up in wintering sites and prohibition of hunting are suggested

大鸨 *Otis tarda* 　　　　　　　　　　　　　　　　　　　　　　　　　刘晶敏 摄　By Jingmin Liu

中国生物多样性红色名录

波斑鸨
Chlamydotis macqueenii

濒危 EN A2cd+3cd; C1

| 数据缺乏 DD | 无危 LC | 近危 NT | 易危 VU | 濒危 EN | 极危 CR | 区域灭绝 RE | 野外灭绝 EW | 灭绝 EX |

分类地位 Taxonomic Status

动物界 Animalia	脊索动物门 Chordata	鸟纲 Aves	鸨形目 Otidiformes	鸨科 Otididae

学名 Scientific Name	*Chlamydotis macqueenii*
命名人 Species Authority	Gray, 1832
英文名 English Name(s)	Macqueen's Bustard
同物异名 Synonym(s)	无 / None
种下单元评估 Infra-specific Taxa Assessed	无 / None

评估信息 Assessment Information

评估年份 Year Assessed	2016
评定人 Assessor(s)	丁长青 / Changqing Ding
其他贡献人 Other Contributor(s)	杨维康、刘垚 / Weikang Yang, Yao Liu

理由 Justification: 波斑鸨繁殖力低，幼鸟总存活率为10%（乔建芳等，2003）；由于畜牧业的发展和农业垦殖，导致栖息环境恶化和栖息地丧失，分布范围日益缩小，种群数量不断减少，估计过去20年内数量下降了至少30%，其中我国准噶尔盆地1998～2002年数量减少69%（Tourenq et al., 2005）。波斑鸨的种群数量在未来可能继续急剧下降（杨维康等，2005）。因此，列为濒危等级 / Reproduction of *Chlamydotis macqueenii* is low and only 10% chicks survival (Qiao *et al.*, 2003). With habitat loss and deterioration owing to development of grazing and reclamation, distribution range has continuingly declined and population size has decreased by 30% in 20 years. More seriously, population has decreased by 69% in Junggar Basin from 1998 to 2002 (Tourenq *et al.*, 2005). The population may continue to decline sharply in the future (Yang *et al.*, 2005). Thus, it is listed as Endangered

评估历史 Assessment History: 《国家重点保护野生动物名录》：Ⅰ（1989，2021）；《中国濒危动物红皮书·鸟类》（1998）：未定；《中国物种红色名录：第二卷 脊椎动物 下册》（2009）：近危 / *National Key Protected Wild Animal List*: Ⅰ (1989, 2021); *China Red Data Book of Endangered Animals*: *Aves* (1998): Ⅰ; *China Species Red List*: *Vol. II Vertebrates Part 2* (2009): NT

地理分布 Geographical Distribution

国内分布 Domestic Distribution
繁殖于内蒙古西部和中部、新疆北部，在河北为迷鸟 / It breeds in west and central Inner Mongolia (Nei Mongol), North Xinjiang; vagrant visitor in Hebei
分布标注 Distribution Note
非特有种 / Non-endemic

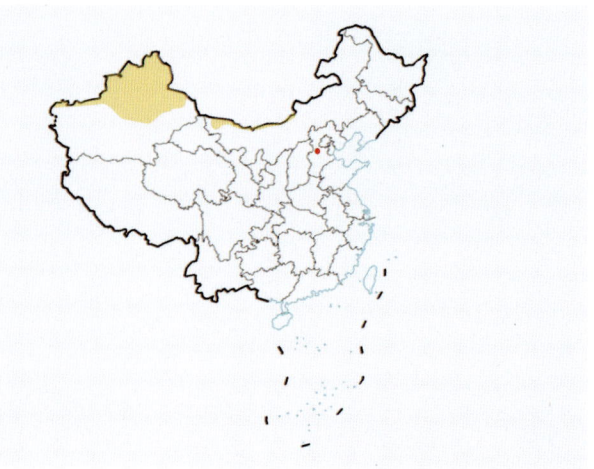

国内分布图
Map of Domestic Distribution

种群 Population

种群数量 Population Size 全球种群数量 106,000 ～ 110,000 只 (IUCN, 2012)，我国约 2,000 只 (杨维康等，2005；高行宜，2007) / The global population size is estimated to be 106,000 ～ 110,000 individuals (IUCN, 2012), and 2,000 individuals in China (Yang et al., 2005; Gao, 2007)

种群趋势 Population Trend 下降 / Decreasing

生境与生态系统 Habitat (s) and Ecosystem (s)

生　　境 Habitat(s) 栖息于乔木、灌木及草本都稀缺的荒漠和半干旱荒漠草原 (Snow and Perrins, 1998; Martí and del Moral, 2003) / *Chlamydotis macqueenii* inhabits deserts with few vegetation and semiarid desert grasslands (Snow and Perrins, 1998; Martí and del Moral, 2003)

生态系统 Ecosystem(s) 荒漠草原生态系统 / Desert Grassland Ecosystem

威胁 Threat (s)

主要威胁 Major Threat(s) 亚洲种群的主要受威胁因子为盗猎及阿拉伯传统捕猎活动 (Judas et al., 2009; Michler, 2009)。适宜栖息地减少也导致波斑鸨种群下降 (Goriup, 1997; Snow and Perrins, 1998; Combreau et al., 2001, 2002) / The principal threat is illegal hunting in Pakistan and traditional hunting in Arabia (Judas et al., 2009; Michler, 2009). The population also decreased with habitat loss (Goriup, 1997, Snow and Perrins, 1998, Combreau et al., 2001, 2002). The major threats in China are habitat loss and illegal hunting

保护级别与保护行动 Protection Category and Conservation Action (s)

IUCN 红色名录 (2016) IUCN Red List (2016) 易危 / VU A4acd

保护行动 Conservation Action(s) 列入国家 I 级重点保护野生动物、CITES 附录 II 和迁徙物种保护公约附录 II。对亚洲种群不同分布区的生态和迁徙进行了大量研究 (Combreau et al., 2001, 2002; Tourenq et al., 2004)；人工繁殖及再引入计划已经开始实施。建议基于迁徙物种保护公约 (CMS) 建立波斑鸨国际保护联盟，制订大范围的种群恢复计划，监测并控制盗猎捕猎行为；开展切实有效的持续性活动并建立反捕猎保护区，在波斑鸨主要繁殖区禁止一切有损环境的开发活动，逐步降低主要繁殖区的放牧强度，减少人为干扰和对栖息地的破坏 (Goriup, 1997; Combreau et al., 2001)；采取多种形式开展公众教育，落实各项保护措施 (高行宜，2007) / It has been listed as the first class in National Key Protected Wild Animal List, Appendix II of CITES and Appendix II of the Convention on Migratory Species (CMS). Studies have been conducted into the ecology and migration of the Asian population (Combreau et al., 2001, 2002; Tourenq et al., 2004); artificial breeding and reintroduction have taken place (Launay, 2004). Set up a global conservation network, based on agreement under the Convention on Migratory Species, and produce a range-wide action and recovery plan, to monitor and control the illegal hunting. Take active and consistent activities and create managed protected areas against illegal hunting, to stop detrimental activities at breeding area and reduce grazing and other human disturbance (Goriup, 1997; Combreau et al., 2001). Control of predators, *e.g.* trap and release elsewhere; improve public awareness and support protection activities (Gao, 2007)

波斑鸨
Chlamydotis macqueenii
陈光林 摄
By Guanglin Chen

中国生物多样性红色名录 China's Red List of Biodiversity

白枕鹤
Grus vipio

濒危 EN A2ace; C1

| 数据缺乏 DD | 无危 LC | 近危 NT | 易危 VU | **濒危 EN** | 极危 CR | 区域灭绝 RE | 野外灭绝 EW | 灭绝 EX |

分类地位 Taxonomic Status

动物界 Animalia	脊索动物门 Chordata	鸟纲 Aves	鹤形目 Gruiformes	鹤科 Gruidae
学名 Scientific Name		*Grus vipio*		
命名人 Species Authority		Pallas, 1811		
英文名 English Name(s)		White-naped Crane		
同物异名 Synonym(s)		*Antigone vipio*		
种下单元评估 Infra-specific Taxa Assessed		无 / None		

评估信息 Assessment Information

评估年份 Year Assessed	2016
评定人 Assessor(s)	丁长青 / Changqing Ding
其他贡献人 Other Contributor(s)	程雅畅、刘垚 / Yachang Cheng, Yao Liu

理由 Justification: 由于农业和经济发展造成的湿地退化，在过去 20 年国内白枕鹤种群下降了 60% 以上，并可能进一步下降。因此，列为濒危等级 / The population of *Grus vipio* has decreased 60% in the past 20 years owing to wetland degeneration by development of agriculture and economic, and the trend might be more serious in future. Thus, it is listed as Endangered

评估历史 Assessment History: 《国家重点保护野生动物名录》：Ⅱ (1989)，Ⅰ (2021)；《中国濒危动物红皮书·鸟类》(1998)：易危；《中国物种红色名录：第二卷 脊椎动物 下册》(2009)：易危 / *National Key Protected Wild Animal List*: Ⅱ (1989), Ⅰ (2021); *China Red Data Book of Endangered Animals: Aves* (1998): V; *China Species Red List: Vol. II Vertebrates Part 2* (2009): VU

地理分布 Geographical Distribution

国内分布 Domestic Distribution

繁殖于内蒙古、黑龙江等地，冬季主要在鄱阳湖越冬，在鄱阳湖越冬的种群迁徙时主要停歇于北京密云、黄河三角洲等地，在台湾为迷鸟 / It breeds in Inner Mongolia (Nei Mongol) and Heilongjiang, winters in Poyang Lake, and occurs in Miyun of Beijing and Yellow River Delta during migration. It is vagrant visitor in Taiwan

分布标注 Distribution Note

非特有种 / Non-endemic

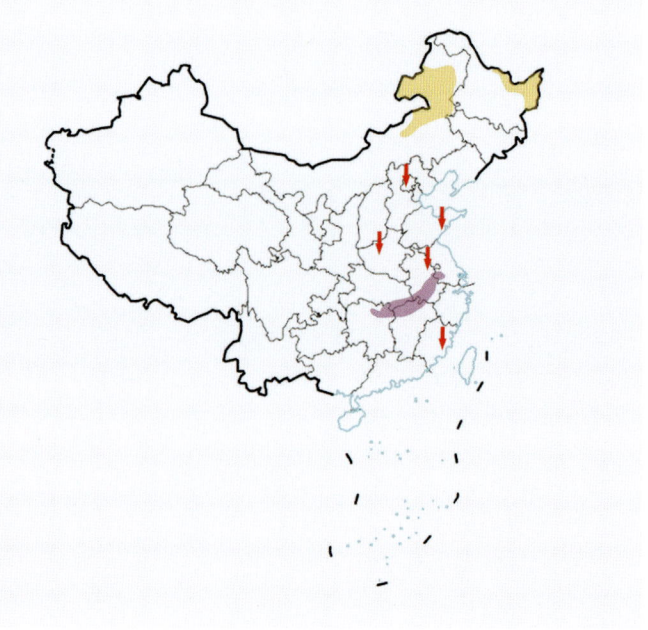

国内分布图 Map of Domestic Distribution

种群 Population

种群数量 Population Size	全球种群数量 5,500～6,500 只，其中东部种群数量呈上升趋势（主要是日本越冬种群上升明显），西部种群数量呈下降趋势 (Harris and Mirande, 2013)。中国 1988～1998 年调查数量为 3,000 只左右，2011 年调查数量为 1,000～1,500 只 (Wetlands International, 2014) / The global population size is estimated to be 5,500～6,500 individuals, and the east population size is increasing (Japanese population has been increasing obviously) and west population size is decreasing (Harris and Mirande, 2013). 3,000 individuals were recorded in the surveys in 1988～1998 and 1,000～1,500 were recorded in 2011 (Wetlands International, 2014)
种群趋势 Population Trend	下降 / Decreasing

生境与生态系统 Habitat (s) and Ecosystem (s)

生　　境 Habitat(s)	繁殖期在隐蔽的芦苇丛湿地中筑巢，在沼泽地、草地、河谷等处觅食 (Bradter et al., 2007)。越冬期在淡水湖泊、农田觅食，偶见于沿海 (IUCN, 2012) / *G. vipio* breeds in the wetlands with dense reeds where its nest can be concealed, and forages in marshes, grasslands, and valleys (Bradter et al., 2007). In winter, it frequents freshwater lakes, farmland and occasionally coastal flats (IUCN, 2012)
生态系统 Ecosystem(s)	湿地生态系统 / Wetland Ecosystem

威胁 Threat (s)

主要威胁 Major Threat(s)	在北方繁殖地的白枕鹤所受到的主要威胁是由于围垦及水资源过度开发造成的湿地沼泽面积缩小以及人类的干扰，如初冬把芦苇割光使早春迁徙而来的鹤无处营巢，过度捕鱼造成食物不足 (Bradter et al., 2005)；在南方越冬的鹤受到的主要威胁是湖水水位下降和人类的干扰，如行船、放牧、捕鱼、割草、捕猎和投毒等 / Breeding birds are threatened by decrease of marshes owing to reclamation and over use of water resource, as well as human disturbance, e.g. reeds collection which reduce the availability of suitable nesting habitat, and over fishing which reduce the food resources (Bradter et al., 2005). Wintering birds are threatened by decrease of water level and human disturbance, e.g. boating, grazing, fishing, mowing, hunting and poisoning

保护级别与保护行动 Protection Category and Conservation Action (s)

IUCN 红色名录 (2016) IUCN Red List (2016)	易危 / VU A2bcde+3bcde+4bcde
保护行动 Conservation Action(s)	列入国家 I 级重点保护野生动物和迁徙物种保护公约 (CMS) 附录 I。由于实施了大规模的生态补水工程，2001 年以来，白枕鹤的栖息繁殖空间得到一定的恢复，特别是经过 2005 年大规模补水后，白枕鹤的种群数量及繁殖巢数持续上升 (高忠燕等，2009)。建议加强繁殖地的调查，确定并保护新的繁殖地；加强越冬地的保护，开展湿地恢复工程；控制农药的使用，减少对湿地的污染 / It has been listed as the first class in *National Key Protected Wild Animal List* and Appendix I of the Convention on Migratory Species. Some habitats have been restored since 2001 as a result of large-scale ecological water regulation project, especially after water replenishing in 2005, the population size and breeding nests have been increasing (Gao et al., 2009). Surveys in breeding sites to confirm new breeding sites, protection and restoration in wintering sites, control of pesticide and pollution are suggested

白枕鹤 *Grus vipio* 赵明静 摄 By Mingjing Zhao

丹顶鹤
Grus japonensis

濒危 EN C1

| 数据缺乏 DD | 无危 LC | 近危 NT | 易危 VU | 濒危 EN | 极危 CR | 区域灭绝 RE | 野外灭绝 EW | 灭绝 EX |

分类地位 Taxonomic Status

动物界 Animalia	脊索动物门 Chordata	鸟纲 Aves	鹤形目 Gruiformes	鹤科 Gruidae
学名 Scientific Name		*Grus japonensis*		
命名人 Species Authority		Müller, 1776		
英文名 English Name(s)		Red-crowned Crane		
同物异名 Synonym(s)		无 / None		
种下单元评估 Infra-specific Taxa Assessed		目前丹顶鹤有 2 个种群，其中在日本的为留居种群，在我国的为迁徙种群 / Two populations: the population in Japan is resident and population in China is migratory		

评估信息 Assessment Information

评估年份 Year Assessed	2016
评定人 Assessor(s)	马志军 / Zhijun Ma
其他贡献人 Other Contributor(s)	苏立英、邹红菲 / Liying Su, Hongfei Zou

理由 Justification: 丹顶鹤种群数量少，尽管在日本的留居种群数量较稳定，但由于我国滨海湿地的围垦和开发，已导致丹顶鹤栖息地丧失与退化，丹顶鹤在我国的种群数量持续下降。在丹顶鹤的主要越冬地江苏盐城，其数量由 20 世纪末的近 1,200 只下降到目前的 600 只左右。因此，列为濒危等级 / *Grus japonensis* has a very small population, and although the population in Japan is stable, the population in China continues to decline owing to habitat loss and degradation by development and reclamation of coastal wetlands. The population has decreased from 1,200 in late 1990s to 600 recently in its main wintering sites, Yancheng in Jiangsu. Thus, it is listed as Endangered

评估历史 Assessment History:《国家重点保护野生动物名录》：Ⅰ (1989，2021)；《中国濒危动物红皮书·鸟类》(1998)：濒危；《中国物种红色名录：第二卷 脊椎动物 下册》(2009)：濒危 / *National Key Protected Wild Animal List*: Ⅰ (1989, 2021); *China Red Data Book of Endangered Animals*: *Aves* (1998): E; *China Species Red List*: *Vol. II Vertebrates Part 2* (2009): EN

地理分布 Geographical Distribution

国内分布 Domestic Distribution	分布标注 Distribution Note
繁殖于东北地区（黑龙江、吉林、辽宁）和内蒙古东部（达里诺尔湖）等地。主要越冬于江苏、山东等地的沿海滩涂，以及长江中游、下游地区，偶见于江西鄱阳湖和台湾。盐城自然保护区和黄河三角洲自然保护区是其主要越冬地。迁徙时经过东北南部、华北等地 / It breeds in Northeast China (Heilongjiang, Jilin, and Liaoning) and east Inner Mongolia (Nei Mongol) (DalaiNur Lake); and winters in coastal wetlands in Jiangsu, Shandong and middle and lower Changjiang River, and is occasionally seen in Poyang Lake in Jiangxi and Taiwan. Yancheng Nature Reserve and Yellow River Delta Nature Reserve are the main wintering sites. It is seen in south of Northeast China and North China when migrating	非特有种 / Non-endemic 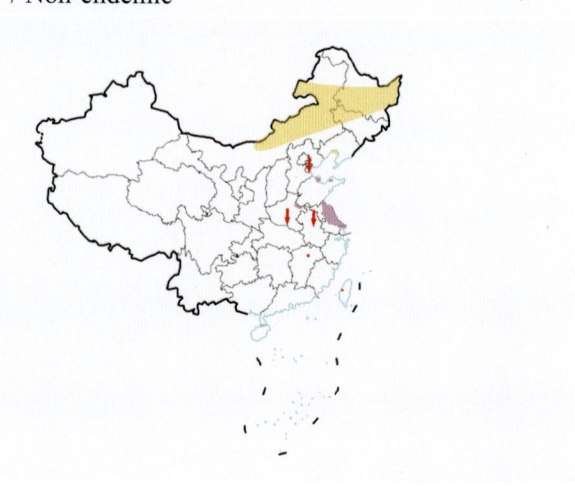

国内分布图
Map of Domestic Distribution

种群 Population

种群数量 Population Size	我国丹顶鹤的种群数量约 1,500 只 / The population size in China is estimated to be 1,500 individuals
种群趋势 Population Trend	下降 / Decreasing

生境与生态系统 Habitat(s) and Ecosystem(s)

生　　　境 Habitat(s)	繁殖于沼泽和草甸，特别是芦苇沼泽，越冬期主要在滩涂湿地以及水产养殖塘、盐田等人工湿地活动。冬季也会到滩涂附近的农田觅食农作物种子和幼苗。杂食性，主要以小鱼、底栖动物、昆虫等动物性食物为主，也吃莎草科、禾本科植物幼嫩的根、茎以及种子等 / *G. japonensis* breeds in marshes and meadow, especially reed marshes, and winters in coastal wetlands and artificial wetlands, *e.g.* aquaculture ponds, saltpans. It also forages in farmlands close to intertidal wetlands for seeds and seedlings. It is omnivorous and feeds on small fish, benthonic animals, and insects, as well as roots bulbs and seeds of sedges and grasses
生态系统 Ecosystem(s)	湿地生态系统 / Wetland Ecosystem

威胁 Threat(s)

主要威胁 Major Threat(s)	湿地丧失与退化导致的栖息地丧失和质量下降是其面临的主要威胁。繁殖地的沼泽湿地和越冬地的滩涂湿地被围垦、开发，导致丹顶鹤适宜的栖息地消失；在繁殖地，流域来水量减少使得大面积湿地干涸，收割芦苇和烧荒也使丹顶鹤适宜的营巢地减少。在迁徙停歇地与越冬地，农药和杀虫剂的使用导致在农田觅食的丹顶鹤中毒的现象也时有发生 / The key threat is the loss and degradation of wetlands. The suitable habitat continues to decrease owing to the reclamation and development of marshes in breeding sites and intertidal wetlands in wintering sites. At breeding sites, wetlands dry up as result of the water flow reduction, and reeds collection and spring fires destroy suitable nesting grounds. And the incidence of poisoning by pesticide when foraging in farmland has been increasing in wintering sites and stopovers

保护级别与保护行动 Protection Category and Conservation Action(s)

IUCN 红色名录 (2016) IUCN Red List (2016)	濒危 / EN C1
保护行动 Conservation Action(s)	列入国家 I 级重点保护野生动物、CITES 公约附录 I 和迁徙物种保护公约 (CMS) 附录 I。对其繁殖生态、越冬生态都有深入的研究，对其迁徙路线开展过卫星跟踪。在主要的繁殖地和越冬地已建立多个自然保护区。在多家动物园有人工饲养的个体或种群 / It has been listed as the first class in National Key Protected Wild Animal List, Appendix I of CITES and Appendix I of the Convention on Migratory Species. Breeding and wintering ecology has been deeply studied, and satellite tracking of its migration routes has been conducted. Several nature reserves have been set up in its breeding and wintering sites. There are artificial feeding individuals and populations in several zoos

丹顶鹤 *Grus japonensis* 　　　　袁晓 摄　By Xiao Yuan

白头鹤
Grus monacha

濒危 EN C1+2a(ii)

| 数据缺乏 DD | 无危 LC | 近危 NT | 易危 VU | 濒危 EN | 极危 CR | 区域灭绝 RE | 野外灭绝 EW | 灭绝 EX |

分类地位 Taxonomic Status

动物界 Animalia	脊索动物门 Chordata	鸟纲 Aves	鹤形目 Gruiformes	鹤科 Gruidae
学名 Scientific Name		*Grus monacha*		
命名人 Species Authority		Temmimck, 1835		
英文名 English Name(s)		Hooded Crane		
同物异名 Synonym(s)		无 / None		
种下单元评估 Infra-specific Taxa Assessed		无 / None		

评估信息 Assessment Information

评估年份 Year Assessed	2016
评定人 Assessor(s)	丁长青 / Changqing Ding
其他贡献人 Other Contributor(s)	郭玉民、刘垚 / Yumin Guo, Yao Liu

理由 Justification: 国内白头鹤种群数量小于 1,000 只，并且由于栖息地受到持续开发和威胁，种群数量在未来可能会持续下降。因此，列为濒危等级 / The population of *Grus monacha* in China is less than 1,000 individuals, and might continue to decrease in future owing to habitat loss. Thus, it is listed as Endangered

评估历史 Assessment History: 《国家重点保护野生动物名录》：Ⅰ (1989, 2021)；《中国濒危动物红皮书·鸟类》(1998)：易危；《中国物种红色名录：第二卷 脊椎动物 下册》(2009)：易危 / *National Key Protected Wild Animal List*: Ⅰ (1989, 2021); *China Red Data Book of Endangered Animals*: *Aves* (1998): V; *China Species Red List*: *Vol. II Vertebrates Part 2* (2009): VU

地理分布 Geographical Distribution

国内分布 Domestic Distribution
繁殖于黑龙江小兴安岭，冬季主要在长江中下游越冬 / It breeds in Xiao Hinggan Ling, Heilongjiang, and winters in middle and lower Changjiang River
分布标注 Distribution Note
非特有种 / Non-endemic

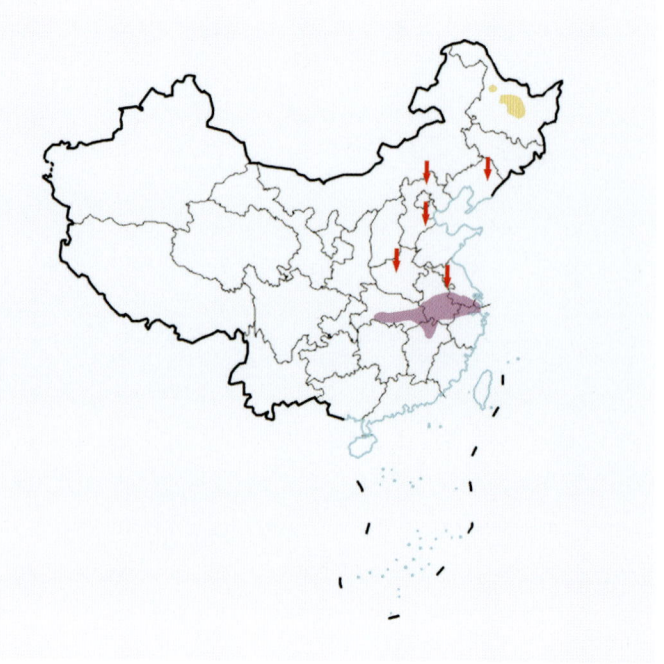

国内分布图
Map of Domestic Distribution

种群 Population

种群数量 Population Size	我国越冬种群 1,050～1,150 只，呈下降趋势，其中鄱阳湖 300～400 只，升金湖超过 600 只，崇明东滩超过 100 只 (Harris, 2012)；全球数量 11,550～11,650 只，呈上升趋势 (Harris and Mirande, 2013) / The population size in China is 1,050～1,150 individuals, and is undergoing a decline. There are 300～400 individuals in Poyang Lake, 600 individuals in Shengjin Lake, over 100 individuals in Chongming Dongtan (Harris, 2012). The global population is 11,550～11,650 individuals, and continues to increase (Harris and Mirande, 2013)
种群趋势 Population Trend	下降 / Decreasing

生境与生态系统 Habitat(s) and Ecosystem(s)

生境 Habitat(s)	营巢于偏远的山脚或河岸的林地，大多位于永冻地带；繁殖地和栖息地变化很大；活动时避开过于开阔或森林过于浓密的地带 (IUCN，2012) / *G. monacha* breeds in remote, wooded, upland bogs on gently sloping foothills and flat river terraces, mostly within the permafrost zone. The breeding habitats are various, while open or dense forests are avoided (IUCN, 2012)
生态系统 Ecosystem(s)	湿地生态系统 / Wetland Ecosystem

威胁 Threat(s)

主要威胁 Major Threat(s)	农药过度使用、人类干扰强度加大、水污染、互花米草入侵等使湿地受到破坏，越冬地主要食物海三棱藨草遭受毁灭性的破坏 / The key threat is wetland degeneration owing to abuse of pesticide, increase of human disturbance, water pollution, invasion of *Spartina alterniflora*, and devastatingly reduced food resource of *Scirpus mariqueter*

保护级别与保护行动 Protection Category and Conservation Action(s)

IUCN 红色名录 (2016) IUCN Red List (2016)	易危 / VU B2ab (i,ii,iii,iv)
保护行动 Conservation Action(s)	列入国家 I 级重点保护野生动物、CITES 附录 I 和迁徙物种保护公约 (CMS) 附录 I。建议加强湿地保护和恢复，控制水位，扩大崇明岛保护区面积；在越冬地建立鹤类保护组织进行长期监测；控制水污染及农药使用 / It has been listed as the first class in National Key Protected Wild Animal List, and listed into Appendix I in CITES and Appendix I in CMS. Control of water level, wetland protection and restoration and enlarge of Chongming Nature Reserve; organizations of crane conservation and longterm monitoring at wintering grounds; control of water pollution, pesticide and poisoning are all suggested

白头鹤 *Grus monacha* 楚云卢 摄 By Yunlu Chu

中国生物多样性 红色名录
China's Red List of Biodiversity

小青脚鹬
Tringa guttifer

濒危 EN C2a(i)

| 数据缺乏 DD | 无危 LC | 近危 NT | 易危 VU | 濒危 EN | 极危 CR | 区域灭绝 RE | 野外灭绝 EW | 灭绝 EX |

分类地位 Taxonomic Status

动物界 Animalia	脊索动物门 Chordata	鸟纲 Aves	鸻形目 Charadriiformes	鹬科 Scolopacidae

学名 Scientific Name	*Tringa guttifer*
命名人 Species Authority	Nordmann, 1835
英文名 English Name(s)	Nordmann's Greenshank
同物异名 Synonym(s)	无 / None
种下单元评估 Infra-specific Taxa Assessed	无 / None

评估信息 Assessment Information

评估年份 Year Assessed	2016
评定人 Assessor(s)	马志军 / Zhijun Ma
其他贡献人 Other Contributor(s)	中国沿海水鸟同步调查项目组 / Chinese Coastal Waterbirds Census Project

理由 Justification: 小青脚鹬种群数量少，由于滨海地区滩涂湿地围垦开发、环境污染等原因，栖息地丧失和退化，其种群数量可能正在下降。调查数据表明其繁殖地的种群数量正在快速下降。因此，列为濒危 / *Tringa guttifer* has a very small population which is declining as a result of habitat loss and degradation by development of coastal wetlands and pollution. The surveys indicate that the breeding population is undergoing a rapid decline. Thus, it is listed as Endangered

评估历史 Assessment History: 《国家重点保护野生动物名录》：Ⅱ (1989)，Ⅰ (2021)；《中国濒危动物红皮书·鸟类》(1998)：未定；《中国物种红色名录：第二卷 脊椎动物 下册》(2009)：濒危 / *National Key Protected Wild Animal List*: Ⅱ (1989), Ⅰ (2021); *China Red Data Book of Endangered Animals*: *Aves* (1998): Ⅰ; *China Species Red List*: *Vol. Ⅱ Vertebrates Part 2* (2009): EN

地理分布 Geographical Distribution

国内分布 Domestic Distribution

在我国主要为旅鸟，迁徙时经过我国东部地区，少数个体可能在华南沿海地区越冬 / It is passage migrant in China, and occurs in east of China during migration, with only a few individuals wintering in South China coast

分布标注 Distribution Note

非特有种 / Non-endemic

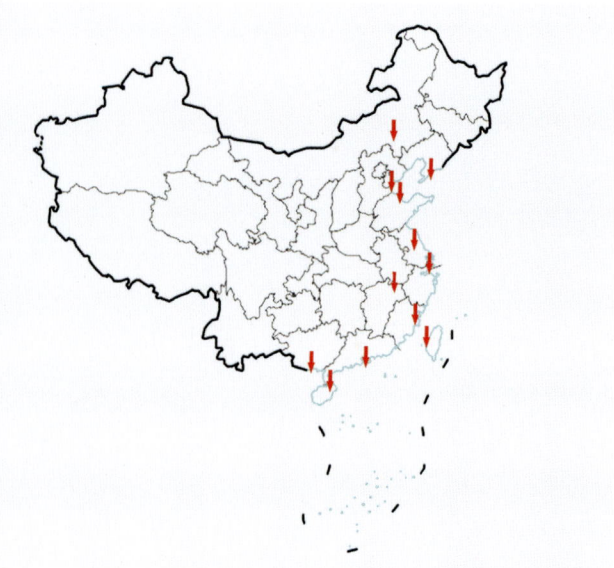

国内分布图
Map of Domestic Distribution

种群 Population

种群数量 Population Size	由于缺乏资料，难以确定小青脚鹬的数量。2012年曾估计全球种群数量不超过1,000只，但2013年秋季在江苏东台和如东附近一次记录到小青脚鹬1,117只。据此估计其种群数量约为1,500只 / It is difficult to estimate the population size of *T. guttifer*, without sufficient data. The global population size was estimated to be less than 1,000 individuals in 2012, while 1,117 individuals were recorded in Dongtai and Rudong in Jiangsu Province in autumn of 2013. Accordingly, the population size is presumed to be 1,500 individuals
种群趋势 Population Trend	缺乏种群数量的监测数据，估计呈下降趋势 / Whereas no sufficient monitoring data available, the population is estimated to be decreasing

生境与生态系统 Habitat (s) and Ecosystem (s)

生　　境 Habitat(s)	小青脚鹬在繁殖期主要栖息于苔原地区有稀疏植被的山地以及沼泽、水塘等湿地附近，迁徙停歇期和越冬期主要栖息于滨海及河口湿地的滩涂、沙洲等区域，有时也到水较浅或排水后的水产养殖塘活动 / *T. guttifer* breeds in mountainous areas, marshes and pools with sparse vegetation in tundra, and uses coastal and estuary intertidal flats and sand bars in winter or during migration, and sometimes occurs in shallow aquatic cultivation ponds
生态系统 Ecosystem(s)	湿地生态系统 / Wetland Ecosystem

威胁 Threat (s)

主要威胁 Major Threat(s)	小青脚鹬在非繁殖季节只在滨海地区的滩涂区域活动，滩涂围垦和开发、环境污染、外来植物互花米草入侵所造成的栖息地丧失和退化是其面临的主要威胁 / *T. guttifer* only occurs in intertidal flats in coastal area during non-breeding period. The major threats are habitat loss and degradation owing to reclamation and development of coastal wetlands, environmental pollution and invasion of *Spartina alterniflora*

保护级别与保护行动 Protection Category and Conservation Action (s)

IUCN 红色名录 (2016) IUCN Red List (2016)	濒危 / EN C2a (i)
保护行动 Conservation Action(s)	列入国家Ⅰ级重点保护野生动物名录、CITES公约附录Ⅰ和迁徙物种保护公约(CMS)附录Ⅰ。在我国沿海地区已建立了多个湿地类型的自然保护区。但由于缺乏数量及分布的详细调查，目前对其种群状况仍不清楚，难以采取针对性的保护行动 / It has been listed as the first class in National Key Protected Wild Animal List, the Appendix I of CITES, and the Appendix I of the Convention on Migratory Species. Several wetland nature reserves have been set up in coastal area. The situation of the population is not clear without detailed surveys on population and distribution

小青脚鹬 *Tringa guttifer* 　　　　　　　　　　　　　　　　　　　　　　　　　袁晓 摄　By Xiao Yuan

中国生物多样性红色名录

遗鸥
Ichthyaetus relictus

濒危　EN B1b(iii)+1c(iii)

| DD 数据缺乏 | LC 无危 | NT 近危 | VU 易危 | **EN 濒危** | CR 极危 | RE 区域灭绝 | EW 野外灭绝 | EX 灭绝 |

分类地位 Taxonomic Status

动物界 Animalia	脊索动物门 Chordata	鸟　纲 Aves	鸻形目 Charadriiformes	鸥　科 Laridae

学　　名 Scientific Name	*Ichthyaetus relictus*
命 名 人 Species Authority	Lönnberg, 1931
英 文 名 English Name(s)	Relict Gull
同物异名 Synonym(s)	*Larus relictus*
种下单元评估 Infra-specific Taxa Assessed	对种群生存力分析显示，种群未来 20 年的灭绝概率小于 10%（邢莲莲和杨贵生，2007）/ The analysis of population viability of *I. relictus* indicated that the probability of extinction in 20 years is less than 10% (Xing Lianlian and Yang Guisheng, 2007)

评估信息 Assessment Information

评 估 年 份 Year Assessed	2016
评 定 人 Assessor(s)	邢莲莲 / Lianlian Xing
其他贡献人 Other Contributor(s)	杨贵生等 / Guisheng Yang *et al.*

理由 Justification: 繁殖于鄂尔多斯高原的遗鸥种群数量总数约 17,000 只，20 世纪 90 年代内蒙古鄂尔多斯桃力庙 - 阿拉善湾海子湖心岛上最多时达 7,000 余只，但由于其巢建在湖心岛上，随着湖泊萎缩湖心岛消失，从 2001 年巢数开始减少，2005 年仅有 5 巢。鄂尔多斯高原陕西省境内的红碱淖 2010 年有 7,000 多巢，约 17,000 只，近几年来，红碱淖水位不断下降，加上旅游等人为干扰，繁殖地正在逐步丧失。因此，列为濒危等级 / The breeding population of *Ichthyaetus relictus* in Ordos Plateau is 17,000 individuals. There were more than 7,000 individuals in island of Taolimiao-Alx Lake in 1990s. The number of nests has decreased since 2001 to 5 in 2005 with the deterioration of lake and loss of island. There are 7,000 nests (c. 17,000 individuals) in Hongjiannao in Shaanxi in 2010, while the loss of breeding habitat in recent years with the decline of water level and increase of tourism has become a major threat. Thus, it is listed as Endangered

评估历史 Assessment History: 《国家重点保护野生动物名录》: Ⅰ (1989, 2021); 《中国濒危动物红皮书·鸟类》(1998): 易危; 《中国物种红色名录: 第二卷 脊椎动物 下册》(2009): 近危 / *National Key Protected Wild Animal List*: I (1989, 2021); *China Red Data Book of Endangered Animals*: *Aves* (1998): V; *China Species Red List*: *Vol. II Vertebrates Part 2* (2009): NT

地理分布 Geographical Distribution

国内分布 Domestic Distribution

繁殖于内蒙古中部、陕西北部和新疆西北部，迁徙经过新疆、陕西和东部各省 / It breeds in middle Inner Mongolia (Nei Mongol), north Shaanxi, northwest Xinjiang; occurs in Xinjiang, Shaanxi and eastern provinces when migrating

分布标注 Distribution Note

非特有种 / Non-endemic

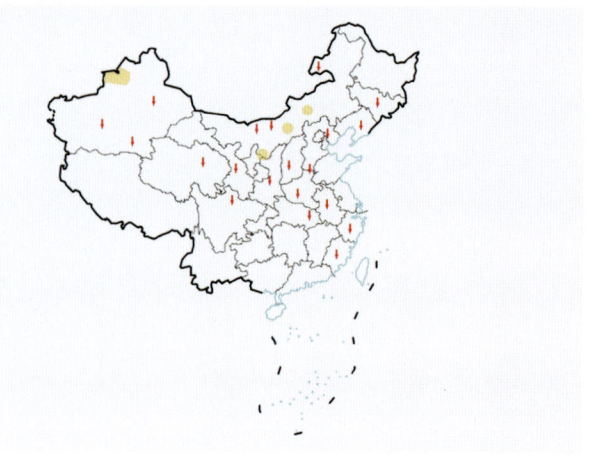

国内分布图
Map of Domestic Distribution

 种群 Population

种群数量 Population Size	种群数量约 17,000 只 / The population size is about 17,000 individuals
种群趋势 Population Trend	下降 / Decreasing

生境与生态系统 Habitat (s) and Ecosystem (s)

生　　境 Habitat(s)	栖息于河湖湿地 / *I. relictus* inhabits in wetlands, *e.g.* rivers, lakes
生态系统 Ecosystem(s)	湿地生态系统 / Wetland Ecosystem

威胁 Threat (s)

主要威胁 Major Threat(s)	湖泊萎缩和旅游等人为干扰 / The decrease of lakes and the human disturbance, *e.g.* tourism

保护级别与保护行动 Protection Category and Conservation Action (s)

IUCN 红色名录 (2016) IUCN Red List (2016)	易危 / VU D2
保护行动 Conservation Action(s)	列入国家 I 级重点保护野生动物、《国家保护的有益的或者有重要经济、科学研究价值的陆生野生动物名录》和 CITES 公约附录 I。主要繁殖地建立了自然保护区 / It has been listed as the first class in National Key Protected Wild Animal List, *National Protected List of Terrestrial Wild Animals with Good Benefits or Important Economic and Scientific Values* and Appendix I of CITES. Nature reserves have been set up in the majority breeding areas

遗鸥 *Ichthyaetus relictus* 　　　邢莲莲 摄　By Lianlian Xing

东方白鹳
Ciconia boyciana

中国生物多样性红色名录 / China's Red List of Biodiversity

濒危 EN C1+2a(ii)

| 数据缺乏 DD | 无危 LC | 近危 NT | 易危 VU | **濒危 EN** | 极危 CR | 区域灭绝 RE | 野外灭绝 EW | 灭绝 EX |

分类地位 Taxonomic Status

动物界 Animalia	脊索动物门 Chordata	鸟纲 Aves	鹳形目 Ciconiiformes	鹳科 Ciconiidae

学名 Scientific Name	*Ciconia boyciana*
命名人 Species Authority	Swinhoe, 1873
英文名 English Name(s)	Oriental White Stork
同物异名 Synonym(s)	无 / None
种下单元评估 Infra-specific Taxa Assessed	无 / None

评估信息 Assessment Information

评估年份 Year Assessed	2016
评定人 Assessor(s)	马志军 / Zhijun Ma
其他贡献人 Other Contributor(s)	周立志 / Lizhi Zhou

理由 Justification: 东方白鹳种群数量少，大部分个体集中在长江中下游地区越冬。缺乏数量调查但估计数量呈下降趋势。基于目前栖息地被大规模围垦和开发、食物资源过度收获、水体污染及其他人为干扰因素，未来其种群数量有可能进一步下降。因此，列为濒危等级 / The population size of *Ciconia boyciana* is small, and most individuals winter in middle and lower Changjiang River. There is no clear data while it is estimated to be decreasing. It might decrease in future, considering the habitat loss by large scale reclamation and development, over fishing, water pollution and other human disturbance. Thus, it is listed as Endangered

评估历史 Assessment History: 《国家重点保护野生动物名录》：Ⅰ (1989, 2021)；《中国濒危动物红皮书·鸟类》(1998)：濒危；《中国物种红色名录：第二卷 脊椎动物 下册》(2009)：濒危 / *National Key Protected Wild Animal List*: Ⅰ (1989, 2021); *China Red Data Book of Endangered Animals*: *Aves* (1998): E; *China Species Red List*: *Vol. II Vertebrates Part 2* (2009): EN

地理分布 Geographical Distribution

国内分布 Domestic Distribution

主要繁殖于黑龙江、吉林、河北、山东、江苏、江西、湖北、安徽。越冬于长江中下游地区，如江西鄱阳湖，湖南洞庭湖，湖北沉湖、洪湖、长湖，安徽升金湖及江苏沿海湿地等，在四川、贵州、西藏、福建、广东、香港和台湾也有少量的越冬记录；迁徙时经过辽宁、北京、天津、河北、山东等 / It breeds in Heilongjiang, Jilin, Hebei, Shandong, Jiangsu, Jiangxi, Hubei and Anhui. It winters in middle and lower Changjiang River, *e.g.* Poyang Lake in Jiangxi, Dongting Lake in Hunan, Chen Lake, Hong Lake, and Chang Lake in Hubei, Shengjin Lake in Anhui and coastal wetlands in Jiangsu, and there are several winter records in Sichuan, Guizhou, Tibet (Xizang), Fujian, Guangdong, Hong Kong and Taiwan. It occurs in Liaoning, Beijing, Tianjin, Hebei and Shandong when migrating

分布标注 Distribution Note

非特有种 / Non-endemic

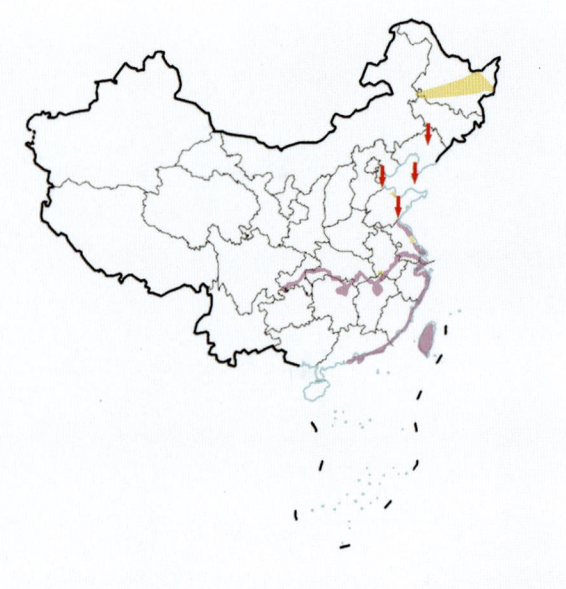

国内分布图 Map of Domestic Distribution

种群 Population

种群数量 Population Size	种群数量约 6,000 只 / The population size is about 6,000 individuals
种群趋势 Population Trend	目前缺乏种群数量的监测数据，但估计种群呈下降趋势 / There is no clear data while it is estimated to be decreasing

生境与生态系统 Habitat (s) and Ecosystem (s)

生 境 Habitat(s)	常在沼泽、湖泊以及水产养殖塘的浅水区域涉水觅食，主要以鱼类为食，也吃小型的两栖和爬行动物、底栖动物、昆虫等。繁殖期在树上营巢，近年来也在一些高大的人工建筑物 (如高压线铁塔) 上营巢 / *C. boyciana* forages in shallow water in marshes, lakes and aquaculture ponds, and mainly feeds on fishes, and small amphibian and reptile animals, benthic invertebrates, insects as well. It nests on trees, as well as some tall architecture, *e.g.* power tower
生态系统 Ecosystem(s)	湿地生态系统 / Wetland Ecosystem

威胁 Threat (s)

主要威胁 Major Threat(s)	由于繁殖地和越冬地的湿地丧失和退化，种群呈下降趋势。一些在人工建筑物上 (高压线铁塔) 的巢也常被捣毁。在迁徙停歇地和越冬地存在非法毒猎，其误食毒饵也会造成中毒。2012 年，天津北大港有 20 余只东方白鹳误食毒饵导致中毒死亡 / The population is decreasing owing to wetland loss of breeding and wintering sites. Some nests on architecture might be destroyed. *C. boyciana* might be killed by pesticide occasionally at stopovers or wintering sites. 20 individuals were killed by pesticide in 2012 at Beidagang of Tianjin

保护级别与保护行动 Protection Category and Conservation Action (s)

IUCN 红色名录 (2016) IUCN Red List (2016)	濒危 / EN C2a (ii)
保护行动 Conservation Action(s)	列入国家 I 级重点保护野生动物、CITES 公约附录 I 和列入迁徙物种保护公约 (CMS) 附录 I，对其繁殖生态、越冬生态都有过研究，对其迁徙路线曾开展过卫星跟踪。在主要的繁殖地和越冬地已建立多个自然保护区。在多个动物园有人工饲养的个体或种群 / It has been listed as the first class in National Key Protected Wild Animal List, and in the Appendix I of CITES and the Convention on Migratory Species, respectively. Studies on the breeding ecology, wintering ecology and satellite tracking of migration routes have been conducted. Several nature reserves have been set up, and there are captive individuals / populations in several zoos

东方白鹳 *Ciconia boyciana*　　　　　　　　　　　　　　　　　　　　　　　袁晓 摄　By Xiao Yuan

中国生物多样性 红色名录 China's Red List of Biodiversity

朱鹮
Nipponia nippon

濒危 EN B1ab(iii)

| 数据缺乏 DD | 无危 LC | 近危 NT | 易危 VU | **濒危 EN** | 极危 CR | 区域灭绝 RE | 野外灭绝 EW | 灭绝 EX |

分类地位 Taxonomic Status

动物界 Animalia	脊索动物门 Chordata	鸟纲 Aves	鹈形目 Pelecaniformes	鹮科 Threskiornithidae

学 名 Scientific Name	*Nipponia nippon*
命 名 人 Species Authority	Temminck, 1835
英 文 名 English Name(s)	Crested Ibis
同物异名 Synonym(s)	*Ibis nippon*
种下单元评估 Infra-specific Taxa Assessed	无 / None

评估信息 Assessment Information

评 估 年 份 Year Assessed	2016
评 定 人 Assessor(s)	丁长青 / Changqing Ding
其他贡献人 Other Contributor(s)	无 / None

理由 Justification: 朱鹮只有一个野生种群，分布区小于 5,000 km²，适宜栖息地范围有限，野生种群密度过高；高大营巢树遭砍伐；耕作制度改变导致冬水田面积逐年减少；使用农药化肥造成食物不足。野生种群的数量超过总数的 95%，各亚种群（再引入种群）数量远不及 250 只。2002 年开始出现农药中毒和暴发新城疫等事件，表明本种目前仍存在巨大的潜在威胁，状况不容乐观。因此，列为濒危等级 / There is only a wild population of *Nipponia nippon* with a distributed area less than 5,000 km². The suitable habitat is small and the density is high. Threats include the deforestation of tall nesting trees, the decrease of paddy field, the shortage of food resource by abuse of pesticides. The wild population takes 95% of the whole population, and the sub populations (reintroduced populations) are less than 250 individuals. Some birds were poisoned or killed by pesticide and newcastle disease since 2002, which indicated the great potential threat of *N. nippon*. Thus, it is listed as Endangered

评估历史 Assessment History: 《国家重点保护野生动物名录》：Ⅰ (1989, 2021)；《中国濒危动物红皮书·鸟类》(1998)：濒危；《中国物种红色名录：第二卷 脊椎动物 下册》(2009)：极危 / *National Key Protected Wild Animal List*: Ⅰ (1989, 2021); *China Red Data Book of Endangered Animals: Aves* (1998): E; *China Species Red List: Vol. II Vertebrates Part 2* (2009): CR

地理分布 Geographical Distribution

国内分布 Domestic Distribution

仅分布于陕西洋县及其附近地区。近年在河南信阳、浙江德清等地开展了野化放飞试验，已有再引入种群在野外成功繁殖的记录 / It only occurs in Yangxian of Shaanxi and adjacent area. There are successful breeding records of reintroduced individuals, after the rehabilitation project in Xinyang of Henan and Deqing of Zhejiang

分布标注 Distribution Note

非特有种 / Non-endemic

国内分布图
Map of Domestic Distribution

种群 Population

种群数量 Population Size	野生种群仅见于中国陕西南部，2012 年（同步调查结果）野生种群数量 1,100 ～ 1,200 只（王超等，2014）。在中国河南信阳、浙江德清、陕西铜川和宝鸡有少量野化放飞的再引入种群 / The wild population is only seen in southwest of Shaanxi. The population size is 1,100 ～ 1,200 individuals based on the synchronous survey in 2012 (Wang *et al.*, 2014). There are small reintroduced populations in Xinyang of Henan, Deqing of Zhejiang, Tongchuan and Baoji of Shaanxi
种群趋势 Population Trend	上升 / Increasing

生境与生态系统 Habitat (s) and Ecosystem (s)

生　　境 Habitat(s)	低山丘陵地带的针阔混交林、落叶阔叶林、河滩、溪流、水库、冬水田、稻田和旱地农田。冬水田、河滩和水库是其主要觅食地 / *N. nippon* habitats in mixed coniferous broad-leaved forests in low hills, deciduous broad-leaved forests, river shoals, streams, reservoirs, and paddy fields. It forages in the paddy fields, river shoals and reservoirs
生态系统 Ecosystem(s)	湿地生态系统 / Wetland Ecosystem

威胁 Threat (s)

主要威胁 Major Threat(s)	人为活动造成的适宜栖息地面积下降、高大营巢树遭砍伐、耕作制度改变导致冬季水田面积逐年减少，使用农药化肥造成食物不足等 / Major threats include the habitat loss by human activities, deforestation of tall nesting trees, the decrease of paddy field, the shortage of food resource by abuse of pesticides, *etc.*

保护级别与保护行动 Protection Category and Conservation Action (s)

IUCN 红色名录 (2016) IUCN Red List (2016)	濒危 / EN B1ab (iii)
保护行动 Conservation Action(s)	列入国家 I 级重点保护野生动物、列入 CITES 公约附录 I，进一步加强再引入强度和生境保护及恢复。在保护好现有野生种群及其栖息地的基础上，开展科学、严谨的再引入 / It has been listed as the first class in National Key Protected Wild Animal List, and in the Appendix I of the CITES. Based on the conservation of existing wild population and habitats, reintroduction and habitat protection and restoration should be improved, in a scientific and specific way

朱鹮 *Nipponia nippon*　　　　　　　　　　　　　　　　　　　　张雁云 摄　By Yanyun Zhang

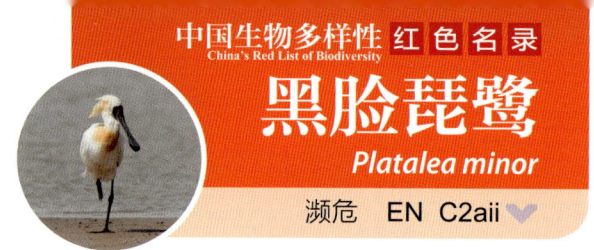

黑脸琵鹭 *Platalea minor*

濒危 EN C2aii

| 数据缺乏 DD | 无危 LC | 近危 NT | 易危 VU | 濒危 EN | 极危 CR | 区域灭绝 RE | 野外灭绝 EW | 灭绝 EX |

分类地位 Taxonomic Status

动物界 Animalia	脊索动物门 Chordata	鸟纲 Aves	鹈形目 Pelecaniformes	鹮科 Threskiornithidae
学名 Scientific Name		*Platalea minor*		
命名人 Species Authority		Temminck & Schlegel, 1849		
英文名 English Name(s)		Black-faced Spoonbill		
同物异名 Synonym(s)		无 / None		
种下单元评估 Infra-specific Taxa Assessed		无 / None		

评估信息 Assessment Information

评估年份 Year Assessed	2016
评定人 Assessor(s)	马志军 / Zhijun Ma
其他贡献人 Other Contributor(s)	余日东 / Ridong Yu

理由 Justification: 黑脸琵鹭种群单一且数量较少，在过去20年间种群数量明显增加，近年来数量保持稳定。考虑到当前和未来沿海地区的滩涂围垦开发、水体污染等造成的栖息地丧失和质量下降，以及种群集中分布所带来的疾病传播的威胁，其种群状况仍不容乐观。因此，列为濒危等级 / The population size of *Platalea minor* is small, while has significantly increased in the past 20 years, and population size remains stable in recent years. The situation is far from satisfactory considering the reclamation of coastal wetlands, habitat loss and deterioration by water pollution, and threats of diseases in high density population. Thus, it is listed as Endangered

评估历史 Assessment History: 《国家重点保护野生动物名录》：Ⅱ(1989)，Ⅰ(2021)；《中国濒危动物红皮书·鸟类》(1998)：濒危；《中国物种红色名录：第二卷 脊椎动物 下册》(2009)：濒危 / *National Key Protected Wild Animal List*: Ⅱ (1989), Ⅰ (2021); *China Red Data Book of Endangered Animals*: *Aves* (1998): E; *China Species Red List*: *Vol. II Vertebrates Part 2* (2009): EN

地理分布 Geographical Distribution

国内分布 Domestic Distribution

在辽宁庄河有少量繁殖个体。主要在华南地区越冬，台湾台江、香港米埔和深圳湾为其主要越冬地。江苏、上海、浙江、福建等地的沿海地区也有少量越冬种群 / It only occurs in coastal area of East Asia. Only a few breeds in Zhuanghe of Liaoning. It winters in South China, *e.g.* Taijiang of Taiwan, Mai Po of Hong Kong and Shenzhen Bay. There are small populations in coastal area of Jiangsu, Shanghai, Zhejiang, Fujian

分布标注 Distribution Note

非特有种 / Non-endemic

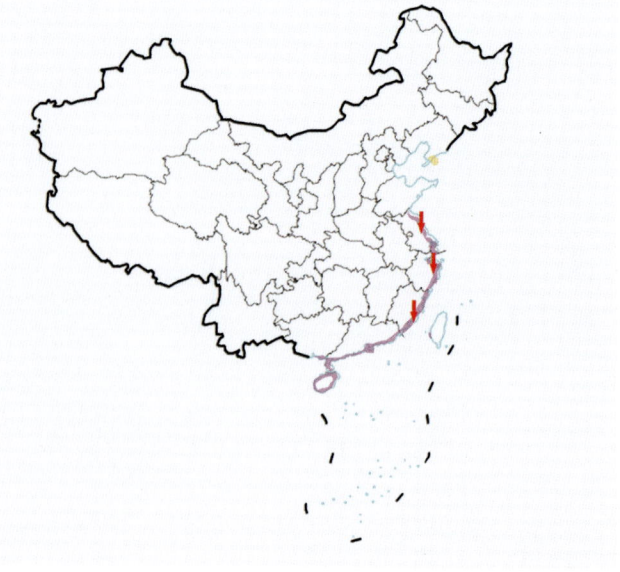

国内分布图
Map of Domestic Distribution

种群 Population

种群数量 Population Size	种群数量不足 3,000 只 (2014 年 1 月记录到 2,726 只) / The population size is less than 3,000 individuals. 2,726 individuals were recorded in January, 2014
种群趋势 Population Trend	稳定 / Stable

生境与生态系统 Habitat (s) and Ecosystem (s)

生　　境 Habitat(s)	繁殖于海岛，越冬期主要在沿海地区的浅水滩涂、红树林湿地以及水产养殖塘中涉水觅食。主要以小鱼、虾等水生生物为食 / P. minor breeds in islands and forages in shallow muddy flats in coastal area, mangrove wetlands and ponds during winters. Its diet include invertebrates, including fish, crayfish, etc
生态系统 Ecosystem(s)	湿地生态系统 / Wetland Ecosystem

威胁 Threat (s)

主要威胁 Major Threat(s)	黑脸琵鹭分布范围狭窄，仅在沿海地区分布。2012～2014 年全球种群数量调查的结果分别为 2,693 只、2,725 只、2,726 只。滨海湿地的围垦和开发导致的栖息地丧失和退化是其面临的主要威胁。20 世纪 90 年代记录的数量曾不足 500 只。过去 20 多年来通过开展保护活动，种群数量明显增加。近年来种群数量保持稳定。但由于种群数量较少，仍有可能出现波动。例如，2002 年冬季，在台湾越冬的黑脸琵鹭因受肉毒杆菌的感染，有 73 只中毒死亡 / The distribution area of P. minor is small, and it only occurs in coastal area. 2,693, 2,725, 2,726 individuals were recorded in global surveys in 2012～2014. The habitat loss owing to reclamation and development of coastal area is the major threat. The population size was less than 500 individuals in 1990s, while the population size has significantly increased with the conservation activities for the past 20 years, and keeps in a stable trend in recent years. It might be fluctuated considering the small global population, e.g. 73 individuals were killed after an infection of botulism in Taiwan during winter in 2002

保护级别与保护行动 Protection Category and Conservation Action (s)

IUCN 红色名录 (2016) IUCN Red List (2016)	濒危 / EN C2a (ii)
保护行动 Conservation Action(s)	列入国家 I 级重点保护野生动物和迁徙物种保护公约 (CMS) 附录 I。对其繁殖生态、越冬生态都有过研究，对其迁徙路线进行过卫星跟踪。在主要的繁殖地和越冬地已建立多个自然保护区。每年 1 月，香港观鸟会组织开展公众参与的黑脸琵鹭数量调查活动，以提高公众对黑脸琵鹭的关注度。在深圳、台湾等地也开展过多种形式的公众教育活动 / It has been listed as the first class in National Key Protected Wild Animal List, and in the Appendix I of the Convention on Migratory Species. Studies on breeding and wintering ecology and satellite tracking have been conducted. Several nature reserves have been set up in its main breeding and wintering sites. The survey in January organized by Hong Kong Birdwatching Society has exerted great effect on the promotion of the public education. There are also public education activities in Shenzhen and Taiwan

黑脸琵鹭 *Platalea minor* 　　　　　袁晓 摄　By Xiao Yuan

海南鳽
Gorsachius magnificus

濒危 EN C2a(i)

| 数据缺乏 DD | 无危 LC | 近危 NT | 易危 VU | 濒危 EN | 极危 CR | 区域灭绝 RE | 野外灭绝 EW | 灭绝 EX |

🦅 分类地位 Taxonomic Status

动物界 Animalia	脊索动物门 Chordata	鸟纲 Aves	鹈形目 Pelecaniformes	鹭科 Adeidae
学名 Scientific Name		*Gorsachius magnificus*		
命名人 Species Authority		Ogilvie Grant, 1899		
英文名 English Name(s)		White-eared Night Heron		
同物异名 Synonym(s)		*Nycticorax magnifica*		
种下单元评估 Infra-specific Taxa Assessed		无 / None		

🦅 评估信息 Assessment Information

评估年份 Year Assessed	2016
评定人 Assessor(s)	周放 / Fang Zhou
其他贡献人 Other Contributor(s)	余丽江 / Lijiang Yu

理由 Justification: 海南鳽栖息地片断化持续加重导致分布点零散，种群数量少，估计不足1,500只。森林砍伐和农业用地导致栖息地严重片断化，山区小水电开发严重影响山涧溪流生境，使得海南鳽的天然觅食地减少。现已知的分布记录点虽然比较多，但大多为零星个体的发现记录，不能充分说明种群数量有所提升。因此，列为濒危等级 / The distribution area of *Gorsachius magnificus* is seriously scattered by the habitat fragmentation owing to deforestation and agricultural land use. The population is less than 1,500 individuals. The stream habitats might be destroyed by small hydropower stations in mountain area, with the decrease of foraging areas. Most of the records of *G. magnificus* are in small numbers, which could not prove the increase of population size. Thus, it is listed as Endangered

评估历史 Assessment History:《国家重点保护野生动物名录》：Ⅱ (1989)，Ⅰ (2021)；《中国濒危动物红皮书·鸟类》(1998)：濒危；《中国物种红色名录：第二卷 脊椎动物 下册》(2009)：濒危 / *National Key Protected Wild Animal List*: Ⅱ (1989), Ⅰ (2021); *China Red Data Book of Endangered Animals*: *Aves* (1998): E; *China Species Red List*: *Vol. II Vertebrates Part 2* (2009): EN

🦅 地理分布 Geographical Distribution

国内分布 Domestic Distribution

广西中部和西南部、广东北部、江西南部、浙江有相对稳定的种群；零星散布于安徽、福建、湖北、湖南、贵州、四川、云南等地；海南自1962年之后一直无分布记录 / There is stable population in southwest and middle Guangxi, north Guangdong, south Jiangxi and Zhejiang. There are very few records in Anhui, Fujian, Hubei, Hunan, Guizhou, Sichuan, Yunnan, *etc.* and no record in Hainan since 1962

分布标注 Distribution Note

非特有种 / Non-endemic

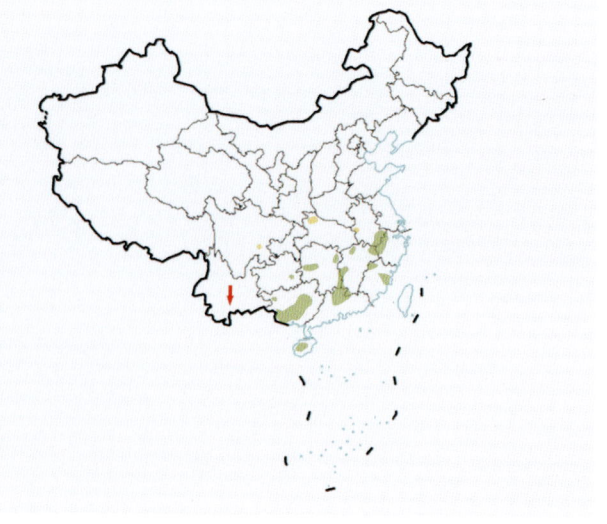

国内分布图
Map of Domestic Distribution

种群 Population

种群数量 Population Size	种群数量不足 1,500 只 / The population size is less than 1,500 individuals
种群趋势 Population Trend	下降 / Decreasing

生境与生态系统 Habitat (s) and Ecosystem (s)

生　　境 Habitat(s)	栖息于附近有山涧溪流、水库、山塘、水田等湿地的阔叶林、针阔混交林或针叶林。繁殖期偏好选择高大、植被盖度大的乔木营巢 (周放等，2005)。因森林砍伐导致的生境退化，海南鳽在次生林营巢和前往人工鱼塘觅食，增加了感染家禽疾病的风险。喜单独活动 / *G. magnificus* inhabits broad-leaved forests, mixed coniferous broad-leaved forests, coniferous forests adjacent to streams, reservoirs, ponds, paddy fields. It prefers tall trees with great cover for nesting (Zhou *et al.*, 2005). It was seen foraging at secondary forests and artificial ponds after the habitat deterioration by deforestation, which increases the risk of disease infection and predation. It usually occurs in single
生态系统 Ecosystem(s)	陆地生态系统 / Terrestrial Ecosystem

威胁 Threat (s)

主要威胁 Major Threat(s)	植被砍伐导致的栖息地严重片断化、质量下降是影响海南鳽种群的主要因素。尽管该物种也能在次生林或人工林繁殖，但存在因植被单一性引起的疾病风险 (Fellowes *et al.*，2001)。人类频繁的活动也容易导致繁殖失败 (Li *et al.*, 2007)。山区小水电开发严重影响山涧溪流生境，溪流电鱼的行为也会降低觅食地质量 / The main factor for the decrease of *G. magnificus* is habitat fragmentation and deterioration owing to deforestation. Although the species can nest in secondary or planted forests, vulnerability of monocultures to pest invasion is a risk (Fellowes *et al.*, 2001). Human disturbance may cause this secretive bird to abandon eggs while incubating (Li *et al.*, 2007). The stream habitats might be destroyed by small hydropower stations in mountain area, and overfishing could reduce the quality of foraging area

保护级别与保护行动 Protection Category and Conservation Action (s)

IUCN 红色名录 (2016) IUCN Red List (2016)	濒危 / EN C2a (i)
保护行动 Conservation Action(s)	列入国家 I 级重点保护野生动物。除了分布于保护区的种群在加强保护力度外，保护区外的分布区也采取了一些有效的保护措施。在种群数量最大和较稳定的广西，上思县政府已专门设立保护小区，并且还有志愿者们自发成立的民间保护协会，扶绥县的保护小区后来并入白头叶猴保护区一起管理。鉴于近 10 年来的零星分布记录有所提升，应开展专项调查以核实分布、数量现状和保护需求 / It has been listed as the first class in National Key Protected Wild Animal List. Besides the conservation of population in nature reserves, effective actions have been taken in distributed area outside the nature reserves. In its main distribution of Guangxi, specific sub protected areas have been set up by Shangsi County and local conservation society was set up by volunteers. The sub protected areas of *G. magnificus* in Fusui County has been jointly managed with nature reserves of *Trachypithecus poliocephalus*. Considering the increase of scattered records in recent ten years, specific surveys of distribution, population size and conservation need are suggested

海南鳽 *Gorsachius magnificus*　　　　　　　　　　　　　　　范忠勇 摄　By Zhongyong Fan

中国生物多样性红色名录 China's Red List of Biodiversity

白鹈鹕
Pelecanus onocrotalus

濒危 EN B1ab; D

| 数据缺乏 DD | 无危 LC | 近危 NT | 易危 VU | **濒危 EN** | 极危 CR | 区域灭绝 RE | 野外灭绝 EW | 灭绝 EX |

分类地位 Taxonomic Status

动物界 Animalia	脊索动物门 Chordata	鸟纲 Aves	鹈形目 Pelecaniformes	鹈鹕科 Pelecanidae
学名 Scientific Name		*Pelecanus onocrotalus*		
命名人 Species Authority		Linnaeus, 1758		
英文名 English Name(s)		Great White Pelican		
同物异名 Synonym(s)		无 / None		
种下单元评估 Infra-specific Taxa Assessed		无 / None		

评估信息 Assessment Information

评估年份 Year Assessed	2016
评定人 Assessor(s)	陈水华 / Shuihua Chen
其他贡献人 Other Contributor(s)	王思宇 / Siyu Wang

理由 Justification: 白鹈鹕在国外分布广泛，种群数量在 10,000 只以上，但国内数量很少，20 世纪 70 年代以前曾经在新疆、青海、黄河口等地有越冬或繁殖，现在已经十分罕见，且基本为单只或几只零星分布，没有稳定的分布区。在国内种群数量约 250 只。因此，列为濒危等级 / *Pelecanus onocrotalus* is widely distributed abroad, with a global population of 10,000 individuals, while the population size in China is small. It breeded or wintered in Xinjiang, Qinghai, Yellow River Delta before 1970s, while it has been rarely seen recently, with records of one or only several individuals. No stable distribution sites. The population in China is estimated to be 250 individuals. Thus, it is listed as Endangered

评估历史 Assessment History: 《国家重点保护野生动物名录》：Ⅱ (1989)，Ⅰ (2021)；中国濒危动物红皮书·鸟类 (1998)：未定；《中国物种红色名录：第二卷 脊椎动物 下册》(2009)：无危 / *National Key Protected Wild Animal List*: Ⅱ (1989)，Ⅰ (2021); *China Red Data Book of Endangered Animals*: *Aves* (1998): I; *China Species Red List*: *Vol. II Vertebrates Part 2* (2009): LC

地理分布 Geographical Distribution

国内分布 Domestic Distribution	分布标注 Distribution Note
在中国的分布范围十分狭窄并且数量稀少。近年见于北京、安徽、四川、甘肃、新疆北部、西藏、香港，2013 年 6 月新疆阿勒泰地区，有近年最大数量白鹈鹕种群的记录，有近百只个体在此停栖。除上述地区外，近 10 年来几乎没有在中国其他省市地区出现的确切记录 / The population size and distribution area are small in China. It occurred in Bejing, Anhui, Sichuan, Gansu, north Xinjiang, Tibet (Xizang), Hong Kong. The largest record of the species in China was nearly 100 individuals in Altai in Xinjiang in June, 2013. No exact record in other provinces in recent ten years	非特有种 / Non-endemic 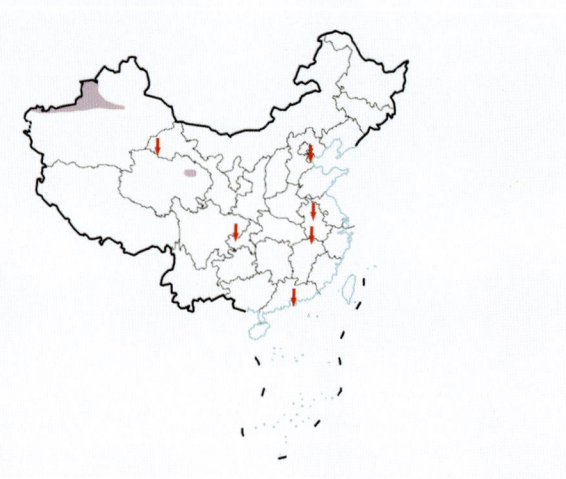 国内分布图 **Map of Domestic Distribution**

138

种群 Population

种群数量 Population Size	种群数量约 250 只 / The population size is about 250 individuals
种群趋势 Population Trend	下降 / Decreasing

生境与生态系统 Habitat (s) and Ecosystem (s)

生　　境 Habitat(s)	栖息于大型淡水或咸水湖泊、潟湖、沼泽地、三角洲地区、宽阔河道或海湾地区 / *P. onocrotalus* uses large freshwater or salty lakes, lagoons, marshes, delta, or large rivers and bays
生态系统 Ecosystem(s)	湿地生态系统 / Wetland Ecosystem

威胁 Threat (s)

主要威胁 Major Threat(s)	适宜栖息地自然条件不断恶化，草场退化，植被稀少，加上人为填湖、过度引水灌溉、污染、过度捕鱼、偷猎和捡蛋等干扰行为，导致该物种在 20 世纪数量急剧减少，大部分个体迁至其他地方越冬。在国内现已十分罕见 / The population of the species has been undergoing a rapid decline owing to the deterioration of suitable habitats, reduction of grasslands and vegetation, reclamation of lakes, over use of water in irrigation, water pollution, over fishing, illegal hunting, and egg collection. Most individuals have migrated to other sites. It was quite rare to be observed in China

保护级别与保护行动 Protection Category and Conservation Action (s)

IUCN 红色名录 (2016) IUCN Red List (2016)	无危 / LC
保护行动 Conservation Action(s)	列入国家 I 级重点保护野生动物和迁徙物种保护公约 (CMS) 附录 II。应进一步开展关于该物种生态习性等各方面的研究，同时通过野外调查掌握该物种在国内的种群数量、种群分布和种群动态。对其适宜栖息的环境加以保护，并加强管理和执法力度 / It has been listed as the first class in National Key Protected Wild Animal List, and in the Appendix II of the Convention on Migratory Species. Ecology studies on the species and surveys of distribution, population size and trend are suggested. Improve the habitat protection and the management and enforcement

白鹈鹕 *Pelecanus onocrotalus*　　　　　　　　　　　　　　　　　　　　　　　李维 摄　By Wei Li

斑嘴鹈鹕
Pelecanus philippensis

濒危 EN B1ab; D

| 数据缺乏 DD | 无危 LC | 近危 NT | 易危 VU | 濒危 EN | 极危 CR | 区域灭绝 RE | 野外灭绝 EW | 灭绝 EX |

分类地位 Taxonomic Status

动物界 Animalia	脊索动物门 Chordata	鸟纲 Aves	鹈形目 Pelecaniformes	鹈鹕科 Pelecanidae
学名 Scientific Name		*Pelecanus philippensis*		
命名人 Species Authority		Gmelin, 1789		
英文名 English Name(s)		Spot-billed Pelican		
同物异名 Synonym(s)		无 / None		
种下单元评估 Infra-specific Taxa Assessed		无 / None		

评估信息 Assessment Information

评估年份 Year Assessed	2016
评定人 Assessor(s)	陈水华、王思宇 / Shuihua Chen, Siyu Wang
其他贡献人 Other Contributor(s)	无 / None

理由 Justification: 斑嘴鹈鹕在中国极少见，必须加以重视和保护。因此，列为濒危等级 / *Pelecanus philippensis* is extremely rare in China, and focus and conservation are necessary. Thus, it is listed as Endangered

评估历史 Assessment History:《国家重点保护野生动物名录》：Ⅰ(2021)；《中国物种红色名录：第二卷 脊椎动物 下册》(2009)：易危 / *National Key Protected Wild Animal List*: Ⅰ (2021); *China Species Red List: Vol. II Vertebrates Part 2* (2009): VU

地理分布 Geographical Distribution

国内分布 Domestic Distribution

过去的一些记录可能是与卷羽鹈鹕混淆。2000年以后在江西永修发现4只（中国观鸟记录中心），2010年在新疆沙湾县安集海乡发现1只，这是目前比较确切的2笔记录。总的来说，该物种目前在中国没有稳定分布区，呈零星分布，记录到的种群很小 / Some of the records mistook *Pelecanus philippensis* with *P. crispus*. There are two exact records: 4 individuals in Yongxiu of Jiangxi after 2000 (China Birdwatching Record Center), 1 in Anjihai Village of Shawan County, Xinjiang in 2010. Generally, there are no stable sites in China, and only very small numbers were recorded

分布标注 Distribution Note

非特有种 / Non-endemic

国内分布图
Map of Domestic Distribution

种群 Population

种群数量 Population Size	种群数量不足 100 只 / The population size is less than 100 individuals
种群趋势 Population Trend	下降 / Decreasing

生境与生态系统 Habitat(s) and Ecosystem(s)

生 境 Habitat(s)	栖息于各种深水或浅水的湿地，也适应人工湿地或湖泊，在淡水和咸水中都可生活，栖息于开阔地或林地 / *P. philippensis* inhabits in shallow or deep marshes and artificial wetlands, and can use both freshwater and salty water. It roosts in open land or forestry area
生态系统 Ecosystem(s)	湿地生态系统 / Wetland Ecosystem

威胁 Threat(s)

主要威胁 Major Threat(s)	适宜栖息地丧失和人为干扰以及全球气温变化等原因，也会导致该物种迁徙行为的改变。主要的威胁来自森林砍伐导致的繁殖地和捕食地丧失，筑巢所需的高大乔木的减少，栖息地植被遭到外来物种入侵，人为捕猎、盗蛋、捕捉雏鸟等。此外还有对渔业资源的过度捕捞和禽流感等疾病的威胁 / Habitat loss, human disturbance and climate change might change its migration behavior. Major threats of the species are loss of breeding sites and foraging sites, reduction of tall trees for nesting, alien species invasion of local vegetation, hunting, egg collection, overfishing and diseases *e.g.* avian influenza

保护级别与保护行动 Protection Category and Conservation Action(s)

IUCN 红色名录 (2016) IUCN Red List (2016)	近危 / NT
保护行动 Conservation Action(s)	列入国家 I 级重点保护野生动物。对该物种的研究较少 / It has been listed as the first class in National Key Protected Wild Animal List. Little study has been conducted

斑嘴鹈鹕 *Pelecanus philippensis*　　　　　　　　　　　蔡长银 摄　By Changyin Cai

卷羽鹈鹕
Pelecanus crispus

濒危 EN A2ce+3ce+4ce; D

| 数据缺乏 DD | 无危 LC | 近危 NT | 易危 VU | 濒危 EN | 极危 CR | 区域灭绝 RE | 野外灭绝 EW | 灭绝 EX |

分类地位 Taxonomic Status

动物界 Animalia	脊索动物门 Chordata	鸟纲 Aves	鹈形目 Pelecaniformes	鹈鹕科 Pelecanidae

学 名 Scientific Name	*Pelecanus crispus*
命 名 人 Species Authority	Bruch, 1832
英 文 名 English Name(s)	Dalmatian Pelican
同物异名 Synonym(s)	无 / None
种下单元评估 Infra-specific Taxa Assessed	无 / None

评估信息 Assessment Information

评 估 年 份 Year Assessed	2016
评 定 人 Assessor(s)	张正旺 / Zhengwang Zhang
其他贡献人 Other Contributor(s)	无 / None

理由 Justification: 近年来欧洲的卷羽鹈鹕种群在逐渐增长，但其他地区的种群正在快速下降，中国沿海地区数量十分稀少。因此，列为濒危等级 / The European population of *Pelecanus crispus* has been increasing while other populations are decreasing. The population in coastal area of China is quite small. Thus, it is listed as Endangered

评估历史 Assessment History: 《国家重点保护野生动物名录》：Ⅰ (2021)；《中国物种红色名录：第二卷 脊椎动物 下册》(2009)：易危 / *National Key Protected Wild Animal List*: Ⅰ (2021); *China Species Red List: Vol. II Vertebrates Part 2* (2009): VU

地理分布 Geographical Distribution

国内分布 Domestic Distribution

主要繁殖于新疆北部，越冬于山东、江苏、浙江、福建、广东、香港、海南，迁徙时见于河北、山西、新疆等地，在台湾为迷鸟 / It breeds in north Xinjiang, and winters in Shandong, Jiangsu, Zhejiang, Fujian, Guangdong, Hong Kong, Hainan. It occurs in Hebei, Shanxi, Xinjiang *etc.* When migrating. It is vagrant visitor in Taiwan

分布标注 Distribution Note

非特有种 / Non-endemic

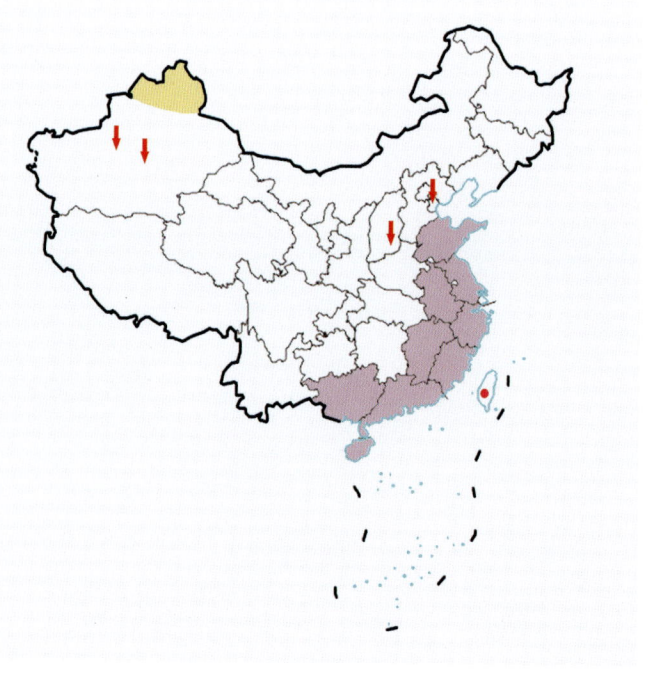

国内分布图
Map of Domestic Distribution

种群 Population

种群数量 Population Size	全球种群数量 10,000～13,900 只，包括 6,700～9,300 只成体。其中东亚 50 只（BirdLife，2014）/ The global population size is estimated to be 10,000～13,900 individuals, and 50 individuals of east Asia population (BirdLife, 2014)
种群趋势 Population Trend	尽管欧洲的种群相对稳定或有一定幅度的增长，但由于其他地区的鹈鹕种群受到各种人为活动的影响，总的种群数量仍处于下降之中 / The European population has been increasing while other populations are decreasing owing to human activities. The global population is estimated to be decreasing

生境与生态系统 Habitat(s) and Ecosystem(s)

生　　境 Habitat(s)	栖息于内陆湖泊、江河与沼泽以及沿海地带。主要以鱼类、甲壳类、软体动物、两栖动物等为食 / *P. crispus* inhabits in lakes, rivers, marshes, and coastal areas. Its main diets include fish, crustaceans, molluscs, and amphibians
生态系统 Ecosystem(s)	湿地生态系统 / Wetland Ecosystem

威胁 Threat(s)

主要威胁 Major Threat(s)	以前该物种的种群下降主要是由于湿地排水、狩猎和渔民的伤害所致 (Crivelli *et al.*, 1997; Mix and Bräunlich, 2000)。非法的猎杀现在仍时有

发生并且被认为是东亚种群的主要致危因素 (Shi *et al.*, 2008; Yu and Chen, 2008)。其他威胁还包括旅游和捕鱼活动的干扰、湿地的破坏或改变、水域污染、撞击高压线以及对渔业资源的过度利用 (Crivelli *et al.*, 1999; Hatzilacou, 1993; Mix and Bräunlich, 2000) / Former declines were primarily caused by wetland drainage, shooting and persecution by fishers (Crivelli *et al.*, 1997; Mix and Bräunlich, 2000). Hunting is considered one of the main threats for the east Asian population (Shi *et al.*, 2008; Yu and Chen, 2008). Other threats include disturbance from tourists and fishers, wetland alteration and destruction, water pollution, collision with overhead power-lines and over-exploitation of fish stocks (Crivelli *et al.*, 1999; Hatzilacou, 1993; Mix and Bräunlich, 2000)

保护级别与保护行动 Protection Category and Conservation Action(s)

IUCN 红色名录 (2016) IUCN Red List (2016)	易危 / VU A2ce+3ce+4ce
保护行动 Conservation Action(s)	列入国家 I 级重点保护野生动物、CITES 公约附录 I 和迁徙物种保护公约 (CMS) 附录 I。在欧洲已经针对其主要威胁开展了卓有成效的保护工作，

如对高压线进行标记或包裹、提供可作为繁殖巢址的平台、水位管理以及在关键地点开展公众教育等。均有助于减少卷羽鹈鹕的死亡率和提高繁殖成效。建议在关键地点加强对繁殖和越冬种群数量以及生态变化的长期监测，在东亚一些潜在的越冬地点进行实地调查，开展湿地的可持续管理，在主要繁殖地安排专人看守，对高压线进行标记或用更加醒目的电缆来替代。在其分布区依法保护卷羽鹈鹕及其栖息地，开展公众教育，制止偷猎和对鱼类资源的过度利用 / It has been listed as the first class in National Key Protected Wild Animal List, Appendix I of CITES and Appendix I of the Convention on Migratory Species. Conservation efforts have reduced the impact of the major threats in Europe, *e.g.* marking and dismantling of power-lines, provision of breeding platforms, water level management and education programmes at key sites, have reduced mortality and increased breeding success. Monitor breeding, wintering numbers and ecological changes at key sites. Survey potential wintering grounds in central and east Asia. Sustainably manage wetlands. Special guard at breeding colonies. Bury power-lines or replace with more visible cable. Legally protect the species and its habitat in range states. Conduct public awareness campaigns. Prevent poaching and overexploitation of fish

卷羽鹈鹕 *Pelecanus crispus*　　　　　　　　　　　　　　　　　　　　　　　许传辉 摄　By Chuanhui Xu

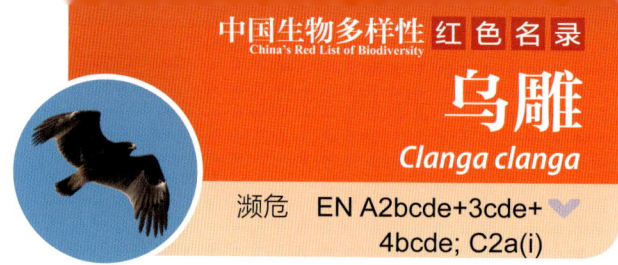

乌雕
Clanga clanga

濒危 EN A2bcde+3cde+4bcde; C2a(i)

| 数据缺乏 DD | 无危 LC | 近危 NT | 易危 VU | 濒危 EN | 极危 CR | 区域灭绝 RE | 野外灭绝 EW | 灭绝 EX |

分类地位 Taxonomic Status

动物界 Animalia	脊索动物门 Chordata	鸟纲 Aves	鹰形目 Accipitriformes	鹰科 Accipitridae
学名 Scientific Name		*Clanga clanga*		
命名人 Species Authority		Pallas, 1811		
英文名 English Name(s)		Greater Spotted Eagle		
同物异名 Synonym(s)		*Aguila clanga*		
种下单元评估 Infra-specific Taxa Assessed		无 / None		

评估信息 Assessment Information

评估年份 Year Assessed	2016
评定人 Assessor(s)	马强 / Qiang Ma
其他贡献人 Other Contributor(s)	无 / None

理由 Justification: 乌雕在我国数量稀少，种群数量有减少趋势。栖息地破碎化、质量下降。大量使用杀虫剂、盗猎行为直接导致种群数量减少。人为干扰的加剧导致其繁殖成功率下降。随着高压输电设施的增多，撞击死亡概率呈增长态势。因此，列为濒危等级 / The population size of *Clanga clanga* in China is small and it is undergoing a decline. The threats include habitat fragmentation and deterioration. Abuse of pesticide and illegal hunting have aggravated the decreasing trend of population directly. The breeding success is decreasing with the serious human disturbance and the fatality rate is increasing with the construction of high power cables and facilities. Thus, it is listed as Endangered

评估历史 Assessment History: 《国家重点保护野生动物名录》：Ⅱ (1989)，Ⅰ (2021)；《中国濒危动物红皮书·鸟类》(1998)：稀有；《中国物种红色名录：第二卷 脊椎动物 下册》(2009)：易危 / *National Key Protected Wild Animal List*: Ⅱ (1989), Ⅰ (2021); *China Red Data Book of Endangered Animals*: Aves (1998): R; *China Species Red List*: Vol. Ⅱ Vertebrates Part 2 (2009): VU

地理分布 Geographical Distribution

国内分布 Domestic Distribution
主要繁殖于新疆、内蒙古、黑龙江和吉林；越冬于我国东南、华南、西南地区 / It breeds in Xinjiang, Inner Mongolia (Nei Mongol), Heilongjiang and Jilin. It winters in Southeast, South, and Southwest of China
分布标注 Distribution Note
非特有种 / Non-endemic

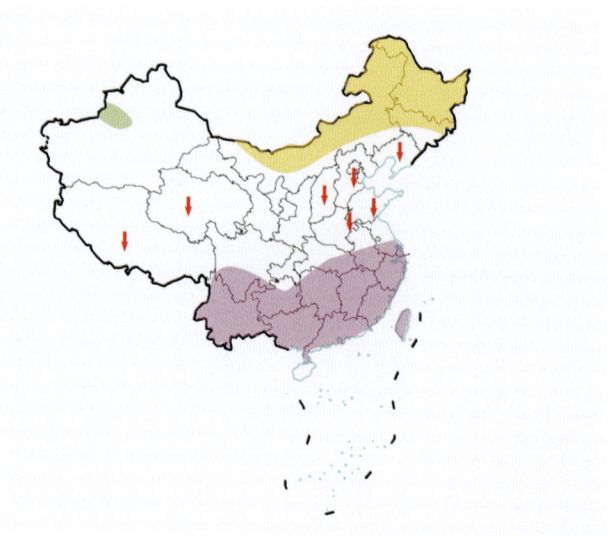

国内分布图
Map of Domestic Distribution

种群 Population

种群数量 Population Size	估计全球数量为 5,000～13,200 只。在我国，监测数据缺乏，迁徙期和越冬期的估计数量为 50～1,000 只 / The global population size is estimated to be 5,000～13,200 individuals. And the migratory and wintering population in China is estimated to be 50～1,000 individuals, while monitoring data is deficient
种群趋势 Population Trend	下降 / Decreasing

生境与生态系统 Habitat (s) and Ecosystem (s)

生　　境 Habitat(s)	喜栖息于开阔的湿地森林、湿地草甸、沼泽，也栖息于红树林地区。冬季偶尔到河流泥滩、河口地区活动；也到农田、垃圾场觅食。主要在海拔较低的地区繁殖，也在山地森林中筑巢繁殖，一般在海拔 1,000 m 以下。在繁殖季节，最高见于海拔 1,700 m / C. clanga inhabits open wetland forests, marshes, swamps, and mangroves. It seldomly occurs in mudflats and estuaries of rivers in winter, and sometimes could be found foraging in farmland or even garbage dump. It usually breeds at low elevation while also nests in forests in mountain area below 1,000 m. The highest record was at 1,700 m during breeding
生态系统 Ecosystem(s)	陆地和淡水生态系统 / Terrestrial and Freshwater Ecosystem

威胁 Threat (s)

主要威胁 Major Threat(s)	致危因素包括栖息地的破碎化、人为干扰与偷猎、环境污染以及由高压输电设备造成的电击死亡 / The major threats include habitat fragmentation, human disturbance, illegal hunting, environmental pollution and attack by high power facilities

保护级别与保护行动 Protection Category and Conservation Action (s)

IUCN 红色名录 (2016) IUCN Red List (2016)	易危 / VU C2a (ii)
保护行动 Conservation Action(s)	列入国家 I 级重点保护野生动物、CITES 公约附录 II 和迁徙物种保护公约 (CMS) 附录 I / It has been listed as the first class in National Key Protected Wild Animal List, in the Appendix II of the CITES and the Appendix I of the Convention on Migratory Species, respectively

乌雕 *Clanga clanga*　　　　　孙克信 摄　By Kexin Sun

白肩雕
Aquila heliaca

濒危 EN A2bcde+3cde+4bcde

| 数据缺乏 DD | 无危 LC | 近危 NT | 易危 VU | 濒危 EN | 极危 CR | 区域灭绝 RE | 野外灭绝 EW | 灭绝 EX |

分类地位 Taxonomic Status

动物界 Animalia	脊索动物门 Chordata	鸟纲 Aves	鹰形目 Accipitriformes	鹰科 Accipitridae
学名 Scientific Name		*Aquila heliaca*		
命名人 Species Authority		Savigny, 1809		
英文名 English Name(s)		Imperial Eagle		
同物异名 Synonym(s)		无 / None		
种下单元评估 Infra-specific Taxa Assessed		无 / None		

评估信息 Assessment Information

评估年份 Year Assessed	2016
评定人 Assessor(s)	马强 / Qiang Ma
其他贡献人 Other Contributor(s)	无 / None

理由 Justification: 白肩雕分布范围较大，但种群数量较少。许多分布区不连续，栖息地有明显片断化趋势，栖息地质量不断下降。白肩雕在我国主要繁殖于新疆西部。在我国东南、西南、华南等地越冬。因此，列为濒危等级 / The distribution area of *Aquila heliaca* is large, while population size is small. The habitats are fragmented and deteriorated. It breeds in west Xinjiang and winters in southeast, southwest, and south China. Thus, it is listed as Endangered

评估历史 Assessment History: 《国家重点保护野生动物名录》：Ⅰ (1989，2021)；《中国濒危动物红皮书·鸟类》(1998)：易危；《中国物种红色名录：第二卷 脊椎动物 下册》(2009)：易危 / National Key Protected Wild Animal List: Ⅰ (1989, 2021); China Red Data Book of Endangered Animals: Aves (1998): V; China Species Red List: Vol. II Vertebrates Part 2 (2009): VU

地理分布 Geographical Distribution

国内分布 Domestic Distribution

分布于吉林、辽宁、河北、北京、天津、河南、山东、陕西、内蒙古、甘肃、新疆、青海、云南西部、四川、重庆、贵州、湖北、江苏、上海、浙江、福建、广东、广西、香港、台湾等 / It occurs in Jilin, Liaoning, Hebei, Beijing, Tianjin, Henan, Shandong, Shaanxi, Inner Mongolia (Nei Mongol), Gansu, Xinjiang, Qinghai, west Yunnan, Sichuan, Chongqing, Guizhou, Hubei, Jiangsu, Shanghai, Zhejiang, Fujian, Guangdong, Guangxi, Hong Kong, Taiwan

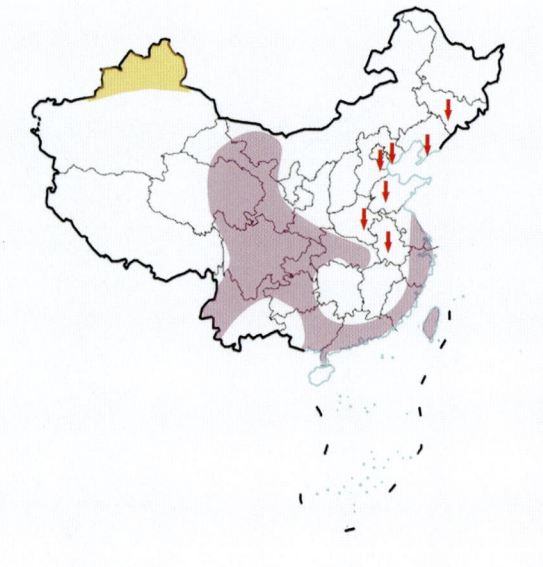

分布标注 Distribution Note

非特有种 / Non-endemic

国内分布图
Map of Domestic Distribution

种群 Population

种群数量 Population Size	全球种群数量 3,500～15,000 只 / The global population size is estimated to be 3,500～15,000 individuals
种群趋势 Population Trend	下降 / Decreasing

生境与生态系统 Habitat(s) and Ecosystem(s)

生　　境 Habitat(s)	白肩雕是栖息于较低海拔的大型猛禽。主要栖息于河岸森林、草原、半荒漠以及农耕区生境 / *A. heliaca* inhabits at low area: forest along river, steppe, semi-desert and agricultural land
生态系统 Ecosystem(s)	陆地生态系统 / Terrestrial Ecosystem

威胁 Threat(s)

主要威胁 Major Threat(s)	森林采伐和农业开发带来的影响十分严重，使其栖息地破碎化，栖息地质量下降。同时，也导致其猎物资源显著减少。农药、除草剂等污染物，通过生物富集作用，给白肩雕带来严重威胁；高压输电设施使白肩雕面临撞击受伤危险；对成鸟的盗猎降低了其种群规模；对鸟巢的破坏和对其巢区的干扰会降低繁殖成功率，进而影响种群的复壮 / It is highly vulnerable to the impacts of habitat fragmentation and deterioration owing to deforestation and agricultural development, which also decrease the predations. Contaminants such as pesticide, herbicide are also threats. And it could be attacked by the high-power facilities. The population has decreased by illegal hunting, and the disturbance and destroy of nests might decrease the breeding success

保护级别与保护行动 Protection Category and Conservation Action(s)

IUCN 红色名录 (2016) IUCN Red List (2016)	易危 / VU C2a (ii)
保护行动 Conservation Action(s)	列入国家 I 级重点保护野生动物、CITES 公约附录 II 和迁徙物种保护公约 (CMS) 附录 I / It has been listed as the first class in National Key Protected Wild Animal List, Appendix II of the CITES, Appendix I of the Convention on Migratory Species

白肩雕 *Aquila heliaca*　　　　邢锐 摄　By Rui Xing

玉带海雕
Haliaeetus leucoryphus

濒危　EN A2bcde+3cde+4bcde

| 数据缺乏 DD | 无危 LC | 近危 NT | 易危 VU | 濒危 EN | 极危 CR | 区域灭绝 RE | 野外灭绝 EW | 灭绝 EX |

分类地位 Taxonomic Status

动物界 Animalia	脊索动物门 Chordata	鸟纲 Aves	鹰形目 Accipitriformes	鹰科 Accipitridae
学名 Scientific Name		*Haliaeetus leucoryphus*		
命名人 Species Authority		Pallas, 1771		
英文名 English Name(s)		Pallas's Fish Eagle		
同物异名 Synonym(s)		*Aquila leucorypha*		
种下单元评估 Infra-specific Taxa Assessed		无 / None		

评估信息 Assessment Information

评估年份 Year Assessed	2016
评定人 Assessor(s)	马强 / Qiang Ma
其他贡献人 Other Contributor(s)	无 / None

理由 Justification: 栖息地丧失和环境污染等因素使玉带海雕面临更大生存压力，其种群数量在全球范围内呈下降趋势。在我国的分布范围较广，在新疆、西藏、青海、内蒙古和黑龙江等地繁殖，但数量稀少，亟待加强保护。因此，列为濒危等级 / *Haliaeetus leucoryphus* faces high pressure of habitat loss and environmental pollution. The global population size is undergoing a decline. The distribution area in China is large and breeding areas include Xinjiang, Tibet (Xizang), Qinghai, Inner Mongolia (Nei Mongol) and Heilongjiang, while the population size is small and conservation should be improved. Thus, it is listed as Endangered

评估历史 Assessment History: 《国家重点保护野生动物名录》：Ⅱ (1989)，Ⅰ (2021)；《中国濒危动物红皮书·鸟类》(1998)：易危；《中国物种红色名录：第二卷 脊椎动物 下册》(2009)：易危 / *National Key Protected Wild Animal List*: Ⅱ (1989), Ⅰ (2021); *China Red Data Book of Endangered Animals*: *Aves* (1998): V; *China Species Red List*: *Vol. II Vertebrates Part 2* (2009): VU

地理分布 Geographical Distribution

国内分布 Domestic Distribution

分布于黑龙江、吉林、辽宁、河北、天津、山东、河南、山西、陕西、内蒙古、宁夏、甘肃、新疆、西藏、青海、云南、四川西部、江苏、上海、浙江等 / It occurs in Heilongjiang, Jilin, Liaoning, Hebei, Tianjin, Shandong, Henan, Shanxi, Shaanxi, Inner Mongolia (Nei Mongol), Ningxia, Gansu, Xinjiang, Tibet (Xizang), Qinghai, Yunnan, west Sichuan, Jiangsu, Shanghai and Zhejiang etc.

分布标注 Distribution Note

非特有种 / Non-endemic

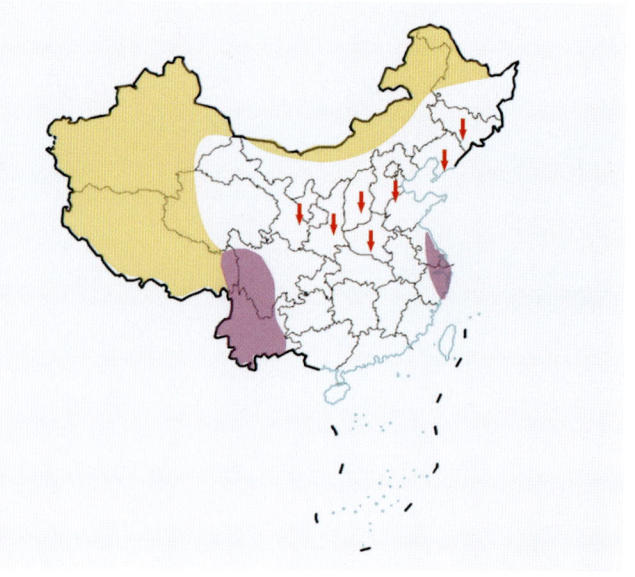

国内分布图
Map of Domestic Distribution

种群 Population

种群数量 Population Size	全球种群数量 3,500 ～ 15,000 只 (IUCN, 2004), 我国缺乏数量调查 / The global population size is estimated to number 3,500 ～ 15,000 individuals (IUCN, 2004), while no survey on population size in China
种群趋势 Population Trend	下降 / Decreasing

生境与生态系统 Habitat (s) and Ecosystem (s)

生　　境 Habitat(s)	栖息于河流、湖泊、淡水湿地、水塘等开阔水域环境，高原的湖泊地区亦可栖息，主要是在海拔 3,200 ～ 4,700 m 的地区繁殖，最高可分布到青藏高原海拔 5,200 m 的区域 / *H. leucoryphus* inhabits in open areas of rivers, lakes (lakes in plateau are also used), freshwater and ponds. It breeds at 3,200 ～ 4,700 m, and occurs at 5,200 m in Qinghai-Tibet (Xizang) Plateau
生态系统 Ecosystem(s)	陆地生态系统 / Terrestrial Ecosystem

威胁 Threat (s)

主要威胁 Major Threat(s)	栖息地减少、生境质量下降和人类干扰是最关键的威胁因子。湿地的开发使其丧失大量适宜生境，而高大树木的采伐减少了玉带海雕的适宜巢址和栖木。杀虫剂和其他工业污染物的累积降低了其本身的繁殖成功率，对玉带海雕的生存造成深远影响；渔业生产中普遍存在的过度捕捞也是一个重要威胁。部分地区的偷猎现象是不可忽视的威胁因素 / The major threats are habitat loss and deterioration and human disturbance. The development of wetlands has deduced the loss of the suitable habitats, and deforestation of tall tree reduced suitable sites for nesting. Contaminants such as pesticide might decrease the breeding success. Over fishing and illegal hunting are also threats

保护级别与保护行动 Protection Category and Conservation Action (s)

IUCN 红色名录 (2016) IUCN Red List (2016)	易危 / VU C2a (ii)
保护行动 Conservation Action(s)	列入国家 I 级重点保护野生动物、CITES 公约附录 II 和迁徙物种保护公约 (CMS) 附录 I / It has been listed as the first class in National Key Protected Wild Animal List, the Appendix II of the CITES and Appendix I of the Convention on Migratory Species

玉带海雕 *Haliaeetus leucoryphus*　　　　　王尧天 摄　By Yaotian Wang

中国生物多样性红色名录

虎头海雕
Haliaeetus pelagicus

濒危 EN A2bcde+3cde+4bcde; C2a(ii)

DD	LC	NT	VU	EN	CR	RE	EW	EX
数据缺乏	无危	近危	易危	濒危	极危	区域灭绝	野外灭绝	灭绝

分类地位 Taxonomic Status

动物界 Animalia	脊索动物门 Chordata	鸟纲 Aves	鹰形目 Accipitriformes	鹰科 Accipitridae

学 名 Scientific Name	*Haliaeetus pelagicus*
命 名 人 Species Authority	Pallas, 1811
英 文 名 English Name(s)	Steller's Sea Eagle
同物异名 Synonym(s)	*Aquila pelagica*
种下单元评估 Infra-specific Taxa Assessed	无 / None

评估信息 Assessment Information

评估年份 Year Assessed	2016
评 定 人 Assessor(s)	马强 / Qiang Ma
其他贡献人 Other Contributor(s)	无 / None

理由 Justification: 虎头海雕分布区相对较小。种群规模不大，且呈下降状态。森林采伐使其栖息地缩小、质量下降，环境污染危及成鸟的生存，严重制约了其种群的恢复。在我国，虎头海雕数量稀少，需进行严格保护。因此，列为濒危等级 / The distribution area *Haliaeetus pelagicus* is small. The population is small and in a decline. The major threats include habitat loss and deterioration owing to deforestation. And it is vulnerable to environmental pollution (Organic Chemicals, heavey metals, *etc.*). The population of *H. pelagicus* is small in China, and strict conservation is necessary. Thus, it is listed as Endangered

评估历史 Assessment History:《国家重点保护野生动物名录》：Ⅰ (1989，2021);《中国濒危动物红皮书·鸟类》(1998)：濒危;《中国物种红色名录：第二卷 脊椎动物 下册》(2009)：易危 / *National Key Protected Wild Animal List*: Ⅰ (1989, 2021); *China Red Data Book of Endangered Animals*: *Aves* (1998): E; *China Species Red List*: *Vol. II Vertebrates Part 2* (2009): VU

地理分布 Geographical Distribution

国内分布 Domestic Distribution

分布于黑龙江、吉林、辽宁、河北、北京、山西、内蒙古东北部、台湾，为不常见的冬候鸟 / It winters while rarely seen in Heilongjiang, Jilin, Liaoning, Hebei, Beijing, Shanxi, northeast Inner Mongolia (Nei Mongol) and Taiwan

分布标注 Distribution Note

非特有种 / Non-endemic

国内分布图
Map of Domestic Distribution

种群 Population

种群数量 Population Size	全球种群数量 4,600 ～ 5,100 只，其中包括 1,830 ～ 1,900 个繁殖对 / The global population size is 1,830 ～ 1,900 pairs, equivalent to 4,600 ～ 5,100 individuals
种群趋势 Population Trend	下降 / Decreasing

生境与生态系统 Habitat (s) and Ecosystem (s)

生　　境 Habitat(s)	栖息于海岸带、内陆湖泊、大河沿岸等生境 / *H. pelagicus* inhabits in coastal areas, inner lakes, banks of large rivers *etc.*
生态系统 Ecosystem(s)	陆地生态系统 / Terrestrial Ecosystem

威胁 Threat (s)

主要威胁 Major Threat(s)	环境污染对虎头海雕的生存影响较大，大规模的海岸带开发、水电工程建设、森林采伐，导致其栖息地的丧失。近海、内陆湖、大河等地区的过度捕捞，致使鱼类等食物资源减少，对其造成深远影响 / *H. pelagicus* is vulnerable to environmental pollution. Another threat is habitat loss owing to development of coastal areas, hydroelectricity project construction and deforestation. Overfishing in rivers, lakes and coastal waters have reduced its food resources

保护级别与保护行动 Protection Category and Conservation Action (s)

IUCN 红色名录 (2016) IUCN Red List (2016)	易危 / VU C2a (ii)
保护行动 Conservation Action(s)	列入国家 I 级重点保护野生动物、CITES 公约附录 II 和迁徙物种保护公约 (CMS) 附录 I / It has been listed as the first class in National Key Protected Wild Animal List, Appendix II of the CITES and Appendix I of the Convention on Migratory Species

虎头海雕 *Haliaeetus pelagicus*

韩笑 摄　By Xiao Han

褐渔鸮
Ketupa zeylonensis

濒危 EN A2bcd; C2a(i)

| 数据缺乏 DD | 无危 LC | 近危 NT | 易危 VU | 濒危 EN | 极危 CR | 区域灭绝 RE | 野外灭绝 EW | 灭绝 EX |

分类地位 Taxonomic Status

动物界 Animalia	脊索动物门 Chordata	鸟 纲 Aves	鸮形目 Strigidae	鸱鸮科 Strigidae
学 名 Scientific Name		*Ketupa zeylonensis*		
命 名 人 Species Authority		Gmelin, 1788		
英 文 名 English Name(s)		Brown Fish Owl		
同物异名 Synonym(s)		*Bubo zeylonensis*		
种下单元评估 Infra-specific Taxa Assessed		无 / None		

评估信息 Assessment Information

评估年份 Year Assessed	2016
评定人 Assessor(s)	朱磊 / Lei Zhu
其他贡献人 Other Contributor(s)	蒋爱伍、张强、诸川汇、杨小农 / Aiwu Jiang, Qiang Zhang, Chuanhui Zhu, Xiaonong Yang

理由 Justification: 褐渔鸮是仅分布于亚洲的大型鸮类，全球种群数量尚未进行过评估，在其分布区内不常见。见于我国南方，分布范围虽较广，但种群数量未知。1997～2004年于华南地区开展的调查中，仅在江西九连山自然保护区有1次记录。广东东莞和广西金钟山自然保护区有2次观鸟记录。2014年5月在广西有非法交易该种的记录，表明也受到了偷猎的威胁。因此，列为濒危等级 / *Ketupa zeylonensis* is massive owl endemic to Asia. There is no assessment on its global population size, and it is rarely seen. It was recorded in south China and the population size is unknown. There was only one record in Jiangxi Jiulian Shan Nature Reserve during the biodiversity rapid assessment of south China by Kadoorie Farm and Botanic Garden during 1997～2004. And only two records in the China Birdwatching Center: Dongguan of Guangdong and Jinzhong Shan of Guangxi. There was an illegal trading record in May 2014, which indicated the threat of illegal hunting. Thus, it is listed as Endangered

评估历史 Assessment History: 《国家重点保护野生动物名录》：Ⅱ(1989，2021)；《中国物种红色名录：第二卷 脊椎动物 下册》(2009)：无危 / *National Key Protected Wild Animal List*: Ⅱ (1989, 2021); *China Species Red List: Vol. Ⅱ Vertebrates Part 2* (2009): LC

地理分布 Geographical Distribution

国内分布 Domestic Distribution

分布于云南西部和南部、江西南部、湖北、广东、香港、澳门、广西及海南 / It occurs in west and south Yunnan, south Jiangxi, Hubei, Guangdong, Hong Kong, Macao, Guangxi and Hainan

分布标注 Distribution Note

非特有种 / Non-endemic

国内分布图
Map of Domestic Distribution

种群 Population

种群数量 Population Size	未知 / Unknown
种群趋势 Population Trend	下降 / Decreasing

生境与生态系统 Habitat(s) and Ecosystem(s)

生　　境 Habitat(s)	已知栖息于海拔 2,000 m 以下靠近水源的落叶、半落叶及常绿阔叶林中，也可见于人类居住区附近的稻田、鱼塘等生境。偏好在河边、水塘及河口等水体边稀疏但有较大胸径乔木的林地 / K. zeylonensis inhabits in deciduous, semi-deciduous, and evergreen forests close to waters below 2,000 m. It is also seen in rice fields, ponds close to residential areas. It prefers forests with sparse and large trees adjacent to waters, e.g. rivers, ponds and estuaries
生态系统 Ecosystem(s)	热带、亚热带森林生态系统 / Tropical, Subtropical Forest Ecosystem

威胁 Threat(s)

主要威胁 Major Threat(s)	栖息地丧失与斑块化是影响种群数量的主要原因；兴建水利设施和人类渔业滥捕可能导致其食物资源的减少；偷猎及其他人类干扰 / Habitat loss and fragmentation probably reduce the population size. Other threats include shortage of food resources owing to dam construction and overfishing, illegal hunting and human disturbance

保护级别与保护行动 Protection Category and Conservation Action(s)

IUCN 红色名录 (2016) IUCN Red List (2016)	无危 / LC
保护行动 Conservation Action(s)	列入国家 II 级重点保护野生动物和 CITES 公约附录 II，其分布区内已建立多个自然保护区。继续加强栖息地保护，恢复和扩大栖息地面积。加强公众教育，对分布区种群数量和分布状况进行动态监测，控制其分布区内水坝等设施的修建，对偷猎进行有效打击 / It has been listed as the second class in National Key Protected Wild Animal List and the Appendix II of the CITES. Several nature reserves have been set up in its distribution areas. Promotion of habitat conservation (restoration to enlarge habitats) are suggested. Other recommendations include improvement of public education, monitoring on its population size and distribution, control of dam construction and management on fishing

褐渔鸮 *Ketupa zeylonensis*　　巫嘉伟 摄　By Jiawei Wu

中国生物多样性红色名录
黄腿渔鸮
Ketupa flavipes

濒危 EN A2bcd+3bcd+4bcd; C2a(i)

| 数据缺乏 DD | 无危 LC | 近危 NT | 易危 VU | 濒危 EN | 极危 CR | 区域灭绝 RE | 野外灭绝 EW | 灭绝 EX |

分类地位 Taxonomic Status

动物界 Animalia	脊索动物门 Chordata	鸟纲 Aves	鸮形目 Strigiformes	鸱鸮科 Strigidae
学名 Scientific Name		*Ketupa flavipes*		
命名人 Species Authority		Hodgson, 1836		
英文名 English Name(s)		Tawny Fish Owl		
同物异名 Synonym(s)		*Bubo flavipes*		
种下单元评估 Infra-specific Taxa Assessed		无 / None		

评估信息 Assessment Information

评估年份 Year Assessed	2016
评定人 Assessor(s)	朱磊 / Lei Zhu
其他贡献人 Other Contributor(s)	诸川汇、杨小农 / Chuanhui Zhu, Xiaonong Yang

理由 Justification: 黄腿渔鸮是仅分布于亚洲的大型鸮类，全球种群数量尚未进行过评估，在其分布区内被认为非常稀少。是我国分布范围最广的渔鸮，但因其数量少，不易发现，以致对种群现状知之甚少（洪孝宇，2007a）。在台湾因捕食饲养鱼类而遭业主报复性捕杀（洪孝宇，2007b），在四川有捕食鱼类时溺亡的情况，浙江还有非法饲养该种作为宠物的记录。因此，列为濒危等级 / *Ketupa flavipes* is massive owl endemic to Asia. There is no assessment on its global population size, and it is rarely seen. It was recorded in south China with the largest distribution area compared with other fish owls. While the population size is small and it is difficult to observe, therefore the population status is far less than clear (Hong, 2007a). It was killed by the fishery farm owner during its predation of trout in Taiwan (Hong, 2007b). There were drowned record during its predation of trout in Sichuan, and record as illegal pets in Zhejiang. Thus, it is listed as Endangered

评估历史 Assessment History: 《国家重点保护野生动物名录》：II（1989，2021）；《中国濒危动物红皮书·鸟类》（1998）：稀有；《中国物种红色名录：第二卷 脊椎动物 下册》（2009）：无危 / *National Key Protected Wild Animal List*: II (1989, 2021); *China Red Data Book of Endangered Animals*: *Aves* (1998): R; *China Species Red List*: *Vol. II Vertebrates Part 2* (2009): LC

地理分布 Geographical Distribution

国内分布 Domestic Distribution

分布于云南、四川北部和东部、重庆、陕西南部、甘肃南部、贵州北部、湖北西北部、湖南、安徽、江西、江苏、浙江、上海、福建、广东北部、广西北部和台湾 / It occurs in Yunnan, east and north Sichuan, Chongqing, south Shaanxi, south Gansu, north Guizhou, northwest Hubei, Hunan, Anhui, Jiangxi, Jiangsu, Zhejiang, Shanghai, Fujian, north Guangdong, north Guangxi and Taiwan

分布标注 Distribution Note

非特有种 / Non-endemic

国内分布图
Map of Domestic Distribution

种群 Population

种群数量 Population Size	江苏和浙江的种群数量估计少于 100 对,台湾地区种群数量少于 1,000 只,其余分布区的种群数量不明 / The population size in Jiangsu and Zhejiang is estimated to be less than 100 pairs, and the population size in Taiwan is estimated to be less than 1,000 individuals. It is not clear in other areas
种群趋势 Population Trend	下降 / Decreasing

生境与生态系统 Habitat (s) and Ecosystem (s)

生 境 Habitat(s)	栖息于热带及亚热带溪流、河谷等水域附近的森林中,偏好中低海拔山区人迹罕至的天然林区。在四川唐家河保护区见于海拔 1,200 m 的溪流,在陕西秦岭南麓见于海拔 500 ~ 1,500 m 的混交林中,在台湾地区分布的海拔范围为 48 ~ 2,407 m,领域长约 6 km (洪孝宇,2007a,2007b;曾翌硕和林文隆,2010) / *K. flavipes* inhabits in streams and valleys close to tropical and subtropical forests. It prefers natural forests with little human disturbance at middle and low elevation. It is seen in streams at 1,200 m in Tangjiahe Nature Reserve, mixed forests at 500 ~ 1,500 m in Qinling, and 48 ~ 2,407 m in Taiwan. The range in Taiwan is 6 km along the stream (Hong, 2007a, 2007b; Zeng and Lin, 2010)
生态系统 Ecosystem(s)	陆地生态系统 / Terrestrial Ecosystem

威胁 Threat (s)

主要威胁 Major Threat(s)	栖息地丧失与斑块化是影响种群数量的主要原因;兴建水利设施和渔业滥捕可能会导致其食物资源的减少;因捕食养殖鱼类而遭报复性捕杀;偷猎及其他人类干扰 / Habitat loss and fragmentation are the major threats of decline of population size. Other threats include shortage of food resources owing to dam construction and overfishing, killing by fishery farm owner during its predation of fish, illegal hunting and human disturbance

保护级别与保护行动 Protection Category and Conservation Action (s)

IUCN 红色名录 (2016) IUCN Red List (2016)	无危 / LC
保护行动 Conservation Action(s)	列入国家 II 级重点保护野生动物和 CITES 公约附录 II,其分布区内已建立多个自然保护区。尚需继续加强栖息地保护,恢复和扩大栖息地面积。加强公众教育,对分布区种群数量和分布状况进行动态监测,控制其分布区内水坝等设施的修建,对偷猎进行有效打击 / It has been listed as the second class in National Key Protected Wild Animal List and Appendix II of the CITES. Several nature reserves have been set up in its distribution areas. Promotion of habitat conservation (restoration to enlarge habitats) are suggested. Other recommendations include improvement of public education, monitoring on its population size and distribution, control of dam construction and management on fishing

黄腿渔鸮 *Ketupa flavipes* 　　　　　　　　　　　　　　　　　　　董磊 摄　By Lei Dong

花冠皱盔犀鸟
Rhyticeros undulatus

濒危 EN B 1 ab(iii)+2 ab(iii); D

| 数据缺乏 DD | 无危 LC | 近危 NT | 易危 VU | 濒危 EN | 极危 CR | 区域灭绝 RE | 野外灭绝 EW | 灭绝 EX |

分类地位 Taxonomic Status

动物界 Animalia	脊索动物门 Chordata	鸟纲 Aves	犀鸟目 Bucerotiformes	犀鸟科 Bucerotidae
学名 Scientific Name		*Rhyticeros undulatus*		
命名人 Species Authority		Shaw, 1811		
英文名 English Name(s)		Wreathed Hornbill		
同物异名 Synonym(s)		*Buceros undulatus*		
种下单元评估 Infra-specific Taxa Assessed		无 / None		

评估信息 Assessment Information

评估年份 Year Assessed	2016
评定人 Assessor(s)	杨晓君 / Xiaojun Yang
其他贡献人 Other Contributor(s)	吴飞、常云艳等 / Fei Wu, Yunyan Chang *et al.*

理由 Justification: 花冠皱盔犀鸟在中国仅分布于云南德宏州中国和缅甸接壤的边境一带，分布区极为狭窄，近年来的记录显示种群数量稳定。因此，列为濒危等级 / *Rhyticeros undulatus* only occurs in Dehong, Yunnan, at the boundary of China and Myanmar. The distribution area is extremely small and the population is in a stable trend according to records in recent years. Thus, it is listed as Endangered

评估历史 Assessment History: 《国家重点保护野生动物名录》：Ⅱ (1989)，Ⅰ (2021)；《中国物种红色名录：第二卷 脊椎动物 下册》(2009)：无危 / *National Key Protected Wild Animal List*: Ⅱ (1989), Ⅰ (2021); *China Species Red List: Vol. II Vertebrates Part 2* (2009): LC

地理分布 Geographical Distribution

国内分布 Domestic Distribution

在国内仅记录于云南德宏州盈江县 (屈文政和杨岚, 1998) 和潞西市 / It was only recorded in Yingjiang (Qu and Yang, 1998) and Luxi of Dehong, Yunnan Province

分布标注 Distribution Note

非特有种 / Non-endemic

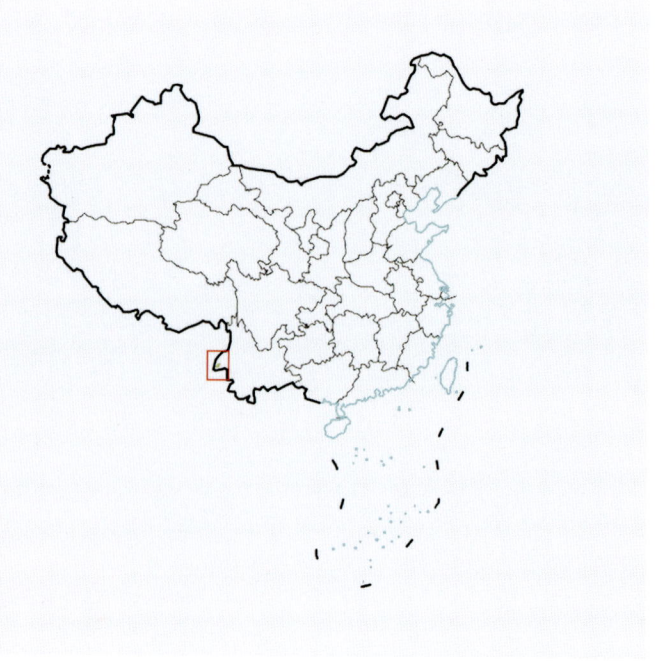

国内分布图
Map of Domestic Distribution

种群 Population

种群数量 Population Size	未知，但在极为狭窄的分布范围内容易见到 / Unknown, while easily seen in its extremely small distribution area
种群趋势 Population Trend	稳定 / Stable

生境与生态系统 Habitat (s) and Ecosystem (s)

生　　境 Habitat(s)	分布于海拔 400～2,000 m 的南亚热带湿性常绿阔叶林 / *R. undulatus* inhabits in tropical evergreen broad-leaved forests in south Asia at 400～2,000 m
生态系统 Ecosystem(s)	陆地生态系统 / Terrestrial Ecosystem

威胁 Threat (s)

主要威胁 Major Threat(s)	狩猎和栖息地破坏 / Illegal hunting and habitat loss

保护级别与保护行动 Protection Category and Conservation Action (s)

IUCN 红色名录 (2016) IUCN Red List (2016)	无危 / LC
保护行动 Conservation Action(s)	列入国家 I 级重点保护野生动物和 CITES 公约附录II，分布区主要在自然保护区内，从而得到保护 / It has been listed as the first class in National Key Protected Wild Animal List and the Appendix II of the CITES. It is under protection, and nature reserves have covered most of its distribution area

花冠皱盔犀鸟 *Rhyticeros undulatus*　　郑鹏 摄　By Peng Zheng

中国生物多样性红色名录

大黄冠啄木鸟
Chrysophlegma flavinucha

濒危 EN B1ab(iii)

| 数据缺乏 DD | 无危 LC | 近危 NT | 易危 VU | 濒危 EN | 极危 CR | 区域灭绝 RE | 野外灭绝 EW | 灭绝 EX |

分类地位 Taxonomic Status

动物界 Animalia	脊索动物门 Chordata	鸟纲 Aves	啄木鸟目 Piciformes	啄木鸟科 Picidae

学名 Scientific Name	*Chrysophlegma flavinucha*
命名人 Species Authority	Gould, 1834
英文名 English Name(s)	Greater Yellownape Woodpecker
同物异名 Synonym(s)	*Picus flavinucha*
种下单元评估 Infra-specific Taxa Assessed	无 / None

评估信息 Assessment Information

评估年份 Year Assessed	2016
评定人 Assessor(s)	周放 / Fang Zhou
其他贡献人 Other Contributor(s)	无 / None

理由 Justification: 由于生境丧失和片断化导致大黄冠啄木鸟种群数量锐减。近20年来，该种在原分布地已很难见到。1997～2004年在广西、广东、海南三省(自治区)设54个调查点，仅在海南霸王岭、尖峰岭和吊罗山3个森林生境较好的保护区发现大黄冠啄木鸟 (Lee et al., 2006)。周放等近10多年一直在广西西南部进行研究，发现由于适宜营巢的栖树减少、生境片断化等使得其种群数量显著减少，在原有分布地多数保护区内都很难觅其踪迹。江西东北部的保护区有零星分布记录，但观察到的数量十分少 (程松林和林剑声, 2011)。因此，列为濒危等级 / The population of *Chrysophlegma flavinucha* is undergoing a rapid decline, and it is rarely seen in its historical distributed sites. It is only recorded in three nature reserves of Hainan (Bawangling, Jianfengling and Diaoluoshan) in a survey of 54 sites in Guangxi, Guangdong and Hainan during 1997～2004 (Lee et al., 2006). The population size has significantly decreased with the habitat fragmentation and loss of suitable nesting trees based on a ten-year survey of Zhou Fang in southwest Guangxi. It was seldomly recorded in northeast Jiangxi, while the number was small (Cheng and Lin, 2011). Thus, it is listed as Endangered

评估历史 Assessment History: 《国家重点保护野生动物名录》：II (2021)；《中国物种红色名录：第二卷 脊椎动物 下册》(2009)：无危 / National Key Protected Wild Animal List: II (2021); China Species Red List: Vol. II Vertebrates Part 2 (2009): LC

地理分布 Geographical Distribution

国内分布 Domestic Distribution

在国内分布于西藏东部、云南、四川西南部、广西西南部、福建中部、海南。江西东北部有零星分布记录 / It occurs in east Tibet, Yunnan, southwest Sichuan, southwest Guangxi, middle Fujian and Hainan. It was seldom recorded in northeast Jiangxi

分布标注 Distribution Note

非特有种 / Non-endemic

国内分布图
Map of Domestic Distribution

种群 Population

种群数量 Population Size	种群数量不足 1,000 对 / The population size is less than 1,000 breeding pairs
种群趋势 Population Trend	下降 / Decreasing

生境与生态系统 Habitat (s) and Ecosystem (s)

生　　境 Habitat(s)	原生性季节雨林、常绿阔叶林、山地次生阔叶林 / Primeval seasonal rainforests, evergreen broad-leaved forests and secondary broad-leaved forests in mountain areas
生态系统 Ecosystem(s)	陆地生态系统 / Terrestrial Ecosystem

威胁 Threat (s)

主要威胁 Major Threat(s)	栖停、繁殖、育雏都依赖于大树。森林砍伐、开荒种地等人类活动导致栖树减少、适宜生境遭到破坏；现有的季节性雨林、常绿阔叶林片断化

严重，均是造成整个种群数量减少的原因 / The roosting, breeding and brooding are dependent on large trees. Major threats include habitat loss and reduce of suitable nesting trees owing to deforestation and reclamation. The population size has significantly decreased with the fragmentation of seasonal rainforests and evergreen broad-leaved forests

保护级别与保护行动 Protection Category and Conservation Action (s)

IUCN 红色名录 (2016) IUCN Red List (2016)	无危 / LC
保护行动 Conservation Action(s)	列入国家 II 级重点保护野生动物和《国家保护的有益的或者有重要经济、科学研究价值的陆生野生动物名录》。建议对该种的种群状况、栖息地等

进行系统的生态学研究，以便更深入地了解其数量减少的原因 / It has been listed as the second class in National Key Protected Wild Animal List and *National Protected List of Terrestrial Wild Animals with Good Benefits or Important Economic and Scientific Values*. Systematic studies on population status and habitat ecology are suggested, to get more information on the decrease of population

大黄冠啄木鸟 *Chrysophlegma flavinucha*　　陈浩 摄　By Hao Chen　　杨卫光 摄　By Weiguang Yang

猎隼 Falco cherrug

濒危 EN A2bcde

数据缺乏 DD	无危 LC	近危 NT	易危 VU	濒危 EN	极危 CR	区域灭绝 RE	野外灭绝 EW	灭绝 EX

分类地位 Taxonomic Status

动物界 Animalia	脊索动物门 Chordata	鸟纲 Aves	隼形目 Falconiformes	隼科 Falconidae
学名 Scientific Name		*Falco cherrug*		
命名人 Species Authority		Gray, 1834		
英文名 English Name(s)		Saker Falcon		
同物异名 Synonym(s)		无 / None		
种下单元评估 Infra-specific Taxa Assessed		无 / None		

评估信息 Assessment Information

评估年份 Year Assessed	2016
评定人 Assessor(s)	马鸣 / Ming Ma
其他贡献人 Other Contributor(s)	徐峰、吴逸群、Eugene Potapov、Andrew Dixon、Dimitar Ragyov、Nicholas C. Fox、陈莹 / Feng Xu, Yiqun Wu, Eugene Potapov, Andrew Dixon, Dimitar Ragyov, Nicholas C. Fox, Ying Chen

理由 Justification: 猎隼曾经广泛分布于中亚的荒漠和草原地区，1990年后的过度捕捉和走私贸易，导致种群数量持续下降，已不足10,000只。主要分布在中国西部地区新疆、甘肃、青海、西藏、内蒙古、四川等地。也出现在辽宁、河北、北京、天津、山东、山西、浙江等地。因此，列为濒危等级 / *Falco cherrug* was widely distributed in deserts and steppes of Middle Asia, while it is undergoing a continuing decline and the population size is less than 10,000 individuals owing to over hunting and illegal trading in 1990s. It occurs in Xinjiang, Gansu, Qinghai, Tibet (Xizang), Inner Mongolia (Nei Mongol), Sichuan, *etc*. It is also seen in Liaoning, Hebei, Beijing, Tianjin, Shandong, Shanxi, and Zhejiang. Thus, it is listed as Endangered

评估历史 Assessment History: 《国家重点保护野生动物名录》：Ⅱ (1989)，Ⅰ (2021)；《中国濒危动物红皮书·鸟类》(1998)：易危；《中国物种红色名录：第二卷 脊椎动物 下册》(2009)：无危 / *National Key Protected Wild Animal List*: II (1989), I (2021); *China Red Data Book of Endangered Animals*: *Aves* (1998): V; *China Species Red List*: *Vol. II Vertebrates Part 2* (2009): LC

地理分布 Geographical Distribution

国内分布 Domestic Distribution

繁殖于新疆、甘肃、青海、西藏、内蒙古、四川等地，也出现在辽宁、吉林、河北、北京、天津、山东、山西、陕西、宁夏、浙江等地 / It occurs in Xinjiang, Gansu, Qinghai, Tibet (Xizang), Inner Mongolia (Nei Mongol) and Sichuan *etc*. during breeding. It is also seen in Liaoning, Jilin, Hebei, Beijing, Tianjin, Shandong, Shanxi, Shaanxi, Ningxia and Zhejiang *etc*.

分布标注 Distribution Note

非特有种 / Non-endemic

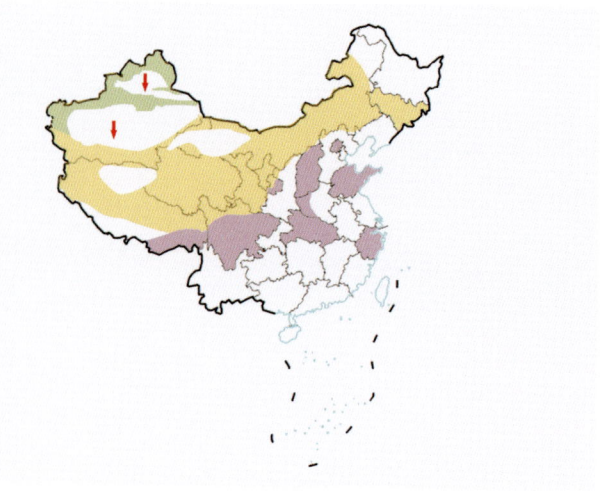

国内分布图 Map of Domestic Distribution

种群 Population

种群数量 Population Size	初步估计新疆有 1,700～2,100 只（繁殖或迁徙路过）；西藏有 1,300～1,900 只（繁殖与越冬）；青海有 800～1,100 只；内蒙古有 150～200 只；甘肃有 110～170 只（迁徙）；总计 4,300～5,300 只 / It is estimated that the population size is 1,700～2,100 individuals in Xinjiang (breeding and migrating); 1,300～1,900 individuals in Tibet (Xizang) (breeding and wintering); 800～1,100 individuals in Qinghai, 150～200 individuals in Inner Mongolia (Nei Mongol), and 110～170 individuals in Gansu (migrating), and 4,300～5,300 individuals in total
种群趋势 Population Trend	近 25 年猎隼在西部地区种群数量呈快速下降的趋势 (Ma, 1999；马鸣等，2005) / The west population is undergoing a rapid decline in recent 25 years (Ma, 1999; Ma *et al.*, 2005)

生境与生态系统 Habitat(s) and Ecosystem(s)

生境 Habitat(s)	栖息于开阔的草原、荒漠、丘陵地带。繁殖期为 3～7 月，多营巢于岩洞里或者悬崖凹陷之处。在中国西部，栖息地海拔 600～4,800 m。以鼠类、野兔、鼠兔、鸟类为主食，也捕食家鸽、信鸽等。在新疆、青海、西藏等地为繁殖鸟 / *F. cherrug* inhabits in open steppes, deserts, and hilly areas. It breeds in March to July, and nests in caves or on the cliffs. In the west, it occurs at 600～4,800 m. It feeds on rodent, hares, pikas and birds, and also pigeons. It breeds in Xinjiang, Qinghai, and Tibet (Xizang)
生态系统 Ecosystem(s)	陆地生态系统 / Terrestrial Ecosystem

威胁 Threat(s)

主要威胁 Major Threat(s)	持续几十年的草原灭鼠、西部大开发及采矿造成的栖息地和营巢地丧失，过度放牧、农业垦荒、高压线电击（梅宇等，2008）、非法捕捉、国际贸易以及工业发展带来的负面影响 / *F. cherrug* is vulnerable to the habitat loss owing to pest control, development of the west and mining, over foraging, reclamation, attack of high-power wires (Mei *et al.*, 2008), illegal hunting, international trading and development of industry

保护级别与保护行动 Protection Category and Conservation Action(s)

IUCN 红色名录 (2016) IUCN Red List (2016)	濒危 / EN A2bcde+3cde+4bcde
保护行动 Conservation Action(s)	列入国家 I 级重点保护野生动物、CITES 公约附录 II 和迁徙物种保护公约 (CMS) 附录 I。自 20 世纪 90 年代阿拉伯市场对猎隼的需求增加，利益驱动引发过度捕捉。中国政府采取措施，严厉打击走私，先后在北京、乌鲁木齐、喀什红其拉甫各大海关破获多起非法贸易案件，对震慑犯罪分子起到一定的效果。但一些地区仍然存在肆意滥捕滥猎，种群数量锐减，急需加强保护（马鸣等，2007）/ It has been listed as the first class in the National Key Protected Wild Animal List, Appendix II of the CITES and Appendix I of the Convention on Migratory Species. The increased demand of Arab market since 1990s has induced the over trapping of *F. cherrug*. The Chinese government has taken measures to crack down on smuggling. Customs of Beijing, Urumqi, Kashi, Hongqi Lafu have uncovered several illegal trading to shock and awe the criminals. While the over trapping in some local sites has not been eliminated and the population is undergoing a rapid decline, and the conservation is to be improved (Ma *et al.*, 2007)

猎隼 *Falco cherrug* 马鸣 摄 By Ming Ma

蓝背八色鸫
Pitta soror

濒危 EN B1b(ii); C1

| 数据缺乏 DD | 无危 LC | 近危 NT | 易危 VU | **濒危 EN** | 极危 CR | 区域灭绝 RE | 野外灭绝 EW | 灭绝 EX |

分类地位 Taxonomic Status

| 动物界 Animalia | 脊索动物门 Chordata | 鸟纲 Aves | 雀形目 Passeriformes | 八色鸫科 Pittidae |

学名 Scientific Name	*Pitta soror*
命名人 Species Authority	Wardlaw Ramsay, 1881
英文名 English Name(s)	Blue-rumped Pitta
同物异名 Synonym(s)	*Hydrornis soror*
种下单元评估 Infra-specific Taxa Assessed	无 / None

评估信息 Assessment Information

评估年份 Year Assessed	2016
评定人 Assessor(s)	周放 / Fang Zhou
其他贡献人 Other Contributor(s)	蒋爱伍 / Aiwu Jiang

理由 Justification: 蓝背八色鸫在我国仅见于云南东南部、海南中南部及广西中南部保存完好的常绿阔叶林中。国内两个亚种的分布区均较为狭窄，其中 *Pitta soror tonkinensis* 的分布区面积估计约仅有 3,000 km²；*Pitta soror douglasi* 是我国特有亚种，其分布区面积估计小于 1,500 km²。近几年来仅有少量有关野外蓝背八色鸫的观察记录，种群数已经极为稀少。因此，列为濒危等级 / *Pitta soror* only occurs in evergreen broad-leaved forest in southeast Yunnan, middle and south Hainan, and middle and south Guangxi. The distribution of the two subspecies are small, with 3,000 km² for *P. s. tonkinensis* and 1,500 km² for *P. s. douglasi*, the endemic subspecies in China. There are only a few records of *Pitta soror* in the wild in recent years, and the population is estimated to be small. Thus, it is listed as Endangered

评估历史 Assessment History: 《国家重点保护野生动物名录》：II (1989，2021)；《中国物种红色名录：第二卷 脊椎动物 下册》(2009)：近危 / *National Key Protected Wild Animal List*: II (1989, 2021); *China Species Red List*: Vol. II Vertebrates Part 2 (2009): NT

地理分布 Geographical Distribution

国内分布 Domestic Distribution
分布于海南、广西、云南东南部 / It occurs in Hainan, Guangxi and southeast Yunnan
分布标注 Distribution Note
非特有种 / Non-endemic

国内分布图
Map of Domestic Distribution

种群 Population

种群数量 Population Size	种群数量估计不足 2,000 对 / The population size is less than 2,000 pairs
种群趋势 Population Trend	下降 / Decreasing

生境与生态系统 Habitat(s) and Ecosystem(s)

生　　境 Habitat(s)	热带雨林和亚热带常绿阔叶林 / Tropical rainforest and subtropical evergreen broad-leaved forest
生态系统 Ecosystem(s)	陆地生态系统 / Terrestrial Ecosystem

威胁 Threat(s)

主要威胁 Major Threat(s)	原生的常绿阔叶林被大量破坏导致栖息地减少和质量下降，可能是导致蓝背八色鸫受威胁的主要原因 / The habitat loss and quality reduction owing to the destroy of natural evergreen broad-leaved forest might be the major threats of *P. soror*

保护级别与保护行动 Protection Category and Conservation Action(s)

IUCN 红色名录 (2016) IUCN Red List (2016)	无危 / LC
保护行动 Conservation Action(s)	列入国家 II 级重点保护野生动物。应加大对蓝背八色鸫的研究和调查力度，以准确获得其种群数量和栖息地需求资料 / It has been listed as the second class in National Key Protected Wild Animal List. The survey and research should be improved to get more knowledge on its population and habitat needs

蓝背八色鸫 *Pitta soror*　　　　屠彦博 摄　By Yanbo Tu

鹊鹂
Oriolus mellianus

濒危 EN C2a(ii)

| 数据缺乏 DD | 无危 LC | 近危 NT | 易危 VU | 濒危 EN | 极危 CR | 区域灭绝 RE | 野外灭绝 EW | 灭绝 EX |

分类地位 Taxonomic Status

动物界 Animalia	脊索动物门 Chordata	鸟纲 Aves	雀形目 Passeriformes	黄鹂科 Oriolidae
学名 Scientific Name		*Oriolus mellianus*		
命名人 Species Authority		Stresemann, 1922		
英文名 English Name(s)		Silver Oriole		
同物异名 Synonym(s)		无 / None		
种下单元评估 Infra-specific Taxa Assessed		无 / None		

评估信息 Assessment Information

评估年份 Year Assessed	2016
评定人 Assessor(s)	丁平 / Ping Ding
其他贡献人 Other Contributor(s)	无 / None

理由 Justification: 由于栖息地的减少和高度片断化，鹊鹂数量下降。种群数量总数约2,500只成熟个体。但四川等地保护力度的增加以及历史调查的局限，该物种数量也可能多于预期。因此，列为濒危等级 / The population of *Oriolus mellianus* is undergoing a decline due to the habitat loss and fragmentation. The population size is estimated to be 2,500 mature individuals, which might be underestimated due to the improvement of conservation effort in Sichuan and the limit of survey. Thus, it is listed as Endangered

评估历史 Assessment History: 《国家重点保护野生动物名录》：II (2021)；《中国濒危动物红皮书·鸟类》(1998)：易危；《中国物种红色名录：第二卷 脊椎动物 下册》(2009)：易危 / *National Key Protected Wild Animal List*: II (2021); *China Red Data Book of Endangered Animals: Aves* (1998): V; *China Species Red List: Vol. II Vertebrates Part 2* (2009): VU

地理分布 Geographical Distribution

国内分布 Domestic Distribution
分布于湖南、云南、四川中南部、贵州南部、广西中部和广东北部 / It occurs in Hunan, Yunnan, middle and south Sichuan, south Guizhou, middle Guangxi and north Guangdong
分布标注 Distribution Note
非特有种 / Non-endemic

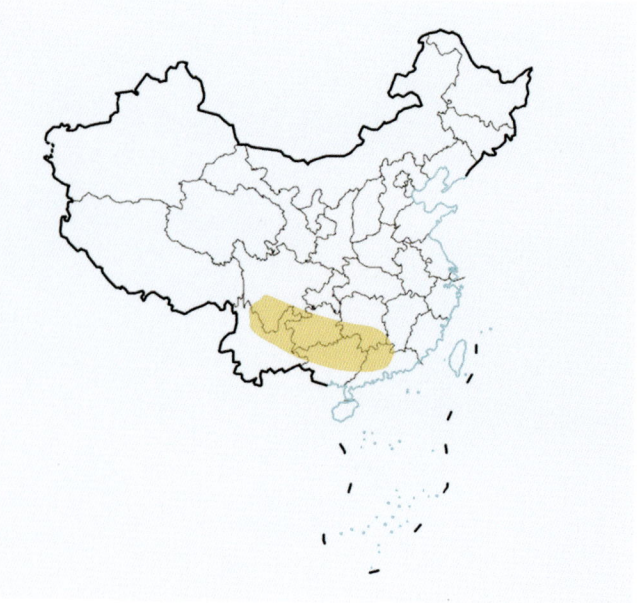

国内分布图
Map of Domestic Distribution

种群 Population

种群数量 Population Size	种群数量可能是 2,500 ～ 9,999 只成熟个体 (BirdLife International，2013)。但由于该物种在四川等地有栖息地进行了良好的保护，中国种群数量可能已多于 2,500 只成熟个体 / The population size was previously estimated to be 2,500 ～ 9,999 mature individuals (BirdLife International, 2013). The population in China might more than 2,500 mature individuals, due to the improvement of conservation effort in Sichuan
种群趋势 Population Trend	下降 / Decreasing

生境与生态系统 Habitat (s) and Ecosystem (s)

生　　境 Habitat(s)	栖息于海拔 600 ～ 1,700 m 的亚热带和热带湿润的低地或山地森林 / The lower mountain or mountain forest in tropical and subtropical, with the elevation of 600 ～ 1,700 m
生态系统 Ecosystem(s)	陆地生态系统 / Terrestrial Ecosystem

威胁 Threat (s)

主要威胁 Major Threat(s)	繁殖地和越冬地生境的丧失和片断化，林木的砍伐以及农田的侵蚀和山火都威胁着该物种的生存 / The breeding and wintering habitat loss and fragmentation, disafforestation, cropland expanding and mountain fires are all threats to the species

保护级别与保护行动 Protection Category and Conservation Action (s)

IUCN 红色名录 (2016) IUCN Red List (2016)	濒危 / EN C2a (ii)
保护行动 Conservation Action(s)	列入国家 II 级重点保护野生动物。中国境内至少 7 个保护区有过关于鹊鹂的记录。可通过一些林业工程，减少其栖息地的片断化程度 / It has been listed as the second class in National Key Protected Wild Animal List. There are records in at least 7 nature reserves. Some forestry projects are suggested, such as set up corridors between primitive forests to improve the situation of habitat fragmentation

鹊鹂 *Oriolus mellianus*　　　　　　　　　　　　　　黄秦 摄　By Qin Huang

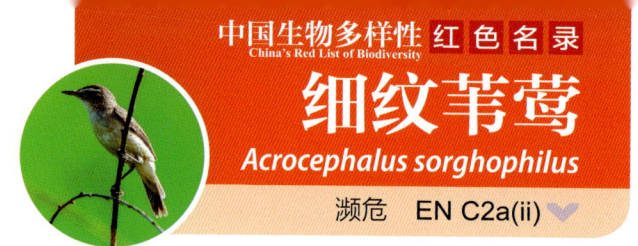

细纹苇莺
Acrocephalus sorghophilus

濒危 EN C2a(ii)

| 数据缺乏 DD | 无危 LC | 近危 NT | 易危 VU | 濒危 EN | 极危 CR | 区域灭绝 RE | 野外灭绝 EW | 灭绝 EX |

分类地位 Taxonomic Status

动物界 Animalia	脊索动物门 Chordata	鸟纲 Aves	雀形目 Passeriformes	苇莺科 Acrocephalidae
学名 Scientific Name		*Acrocephalus sorghophilus*		
命名人 Species Authority		Swinhoe, 1863		
英文名 English Name(s)		Streaked Reed Warbler		
同物异名 Synonym(s)		*Calamodyta sorghophila*		
种下单元评估 Infra-specific Taxa Assessed		无 / None		

评估信息 Assessment Information

评估年份 Year Assessed	2016
评定人 Assessor(s)	刘阳 / Yang Liu
其他贡献人 Other Contributor(s)	危骞、黄秦 / Qian Wei, Qin Huang

理由 Justification: 细纹苇莺的繁殖地点可能位于中国东北部地区，确切范围和生境尚未知。中国东部沿海是其主要迁徙时的生境，这些地区正面临着生境退化及非法捕猎的风险，对本种种群构成直接威胁。因此，列为濒危等级 / The breeding sites of *Acrocephalus sorghophilus* might be located in Northeast China, while the specific range and habitat are unknown. The coastal areas in East China might be the main habitat during migration, which are in the risk of habitat deterioration and illegal hunting. Thus, it is listed as Endangered

评估历史 Assessment History: 《国家重点保护野生动物名录》：II (2021)；《中国物种红色名录：第二卷 脊椎动物 下册》(2009)：易危 / National Key Protected Wild Animal List: II (2021); China Species Red List: Vol. II Vertebrates Part 2 (2009): VU

地理分布 Geographical Distribution

国内分布 Domestic Distribution

繁殖于我国东北，迁徙时经过华北、华中、福建、台湾 / It breeds in Northeast China. It migrates across North and Central China, Fujian and Taiwan

分布标注 Distribution Note

非特有种 / Non-endemic

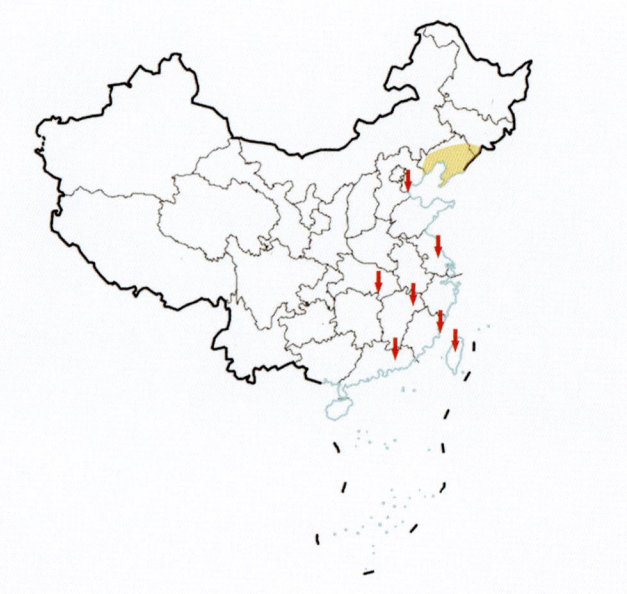

国内分布图
Map of Domestic Distribution

种群 Population

种群数量 Population Size	种群数量 100～400 只 / The population size is 100～400 individuals
种群趋势 Population Trend	下降 / Decreasing

生境与生态系统 Habitat (s) and Ecosystem (s)

生　　境 Habitat(s)	栖息于水域附近的芦苇丛、草丛、稻田中 / *A. sorghophilus* uses the reeds, grasslands and the croplands
生态系统 Ecosystem(s)	湿地生态系统 / Wetland Ecosystem

威胁 Threat (s)

主要威胁 Major Threat(s)	迁徙季节经过的中国东部沿海正面临湿地生境的丧失及退化，以及非法猎捕鸟类等威胁 / *A. sorghophilus* migrates across coastal areas in East China, which are in the risk of wetland deterioration and loss, as well as illegal hunting

保护级别与保护行动 Protection Category and Conservation Action (s)

IUCN 红色名录 (2016) IUCN Red List (2016)	濒危 / EN C2a (ii)
保护行动 Conservation Action(s)	列入国家 II 级重点保护野生动物、《国家保护的有益的或者有重要经济、科学研究价值的陆生野生动物名录》和迁徙物种保护公约 (CMS) 附录 I / It has been listed as the second class in National Key Protected Wild Animal List, *National Protected List of Terrestrial Wild Animals with Good Benefits or Important Economic and Scientific Values*, and Appendix I of The Convention on Migratory Species

细纹苇莺 *Acrocephalus sorghophilus*　　东北老张 摄　By Dongbeilaozhang

灰冠鸦雀
Sinosuthora przewalskii
濒危 EN C2a(i)

| 数据缺乏 DD | 无危 LC | 近危 NT | 易危 VU | 濒危 EN | 极危 CR | 区域灭绝 RE | 野外灭绝 EW | 灭绝 EX |

分类地位 Taxonomic Status

动物界 Animalia	脊索动物门 Chordata	鸟纲 Aves	雀形目 Passeriformes	莺鹛科 Sylviidae
学名 Scientific Name		*Sinosuthora przewalskii*		
命名人 Species Authority		Berezowski & Bianchi, 1891		
英文名 English Name(s)		Rusty-throated Parrotbill		
同物异名 Synonym(s)		*Paradoxornis przewalskii*		
种下单元评估 Infra-specific Taxa Assessed		无 / None		

评估信息 Assessment Information

评估年份 Year Assessed	2016
评定人 Assessor(s)	孙悦华 / Yuehua Sun
其他贡献人 Other Contributor(s)	无 / None

理由 Justification: 灰冠鸦雀分布范围狭窄，自然状态下种群密度低，适宜栖息地呈现破碎化，已知分布点少于5个，但由于调查的缺乏可能有所遗漏。因此，列为易危等级 / The distribution area of *Sinosuthora przewalskii* is small and the population density is low in natural condition. The suitable habitats are fragmented and *Sinosuthora przewalskii* only occurs in 5 sites, which might also be related to the survey intensity. Thus, it is listed as Vulnerable

评估历史 Assessment History: 《国家重点保护野生动物名录》：Ⅰ（2021）；《中国濒危动物红皮书·鸟类》（1998）：稀有；《中国物种红色名录：第二卷 脊椎动物 下册》（2009）：易危 / *National Key Protected Wild Animal List*: I (2021); *China Red Data Book of Endangered Animals: Aves* (1998): R; *China Species Red List: Vol. II Vertebrates Part 2* (2009): VU

地理分布 Geographical Distribution

国内分布 Domestic Distribution
分布于甘肃南部和东南部的舟曲、卓尼、文县、武都及四川北部九寨沟、松潘、青川、平武等地 / It occurs in Zhouqu, Zhuoni, Wenxian, Wudu in south and southeast Gansu; Jiuzhaigou, Songpan, Qingchuan, Pingwu *etc.* in north Sichuan
分布标注 Distribution Note
特有种 / Endemic

国内分布图
Map of Domestic Distribution

种群 Population

种群数量 Population Size	种群数量估计为 2,500 ~ 9,999 只 / The population size is estimated to be 2,500 ~ 9,999 individuals
种群趋势 Population Trend	下降 / Decreasing

生境与生态系统 Habitat (s) and Ecosystem (s)

生　　境 Habitat(s)	主要栖息于海拔 2,000 ~ 3,500 m 的针叶林和针阔叶混交林的林下竹丛中，也栖息于林缘疏林灌丛和草丛中 (董磊和孙悦华，2007；赵正阶，2001；张琼，2012)。近年在四川唐家河发现巢址，繁殖于竹丛中 / *S. przewalskii* uses the bamboos under the coniferous forests and mixed coniferous broad-leaved forests with elevation of 2,000 ~ 3,500 m, as well as the shrubs and grasses at the edge of forests (Dong and Sun, 2007; Zhao, 2001; Zhang, 2012). It breeds in the bamboos, and nests were found in Tangjiahe in recent years
生态系统 Ecosystem(s)	陆地生态系统 / Terrestrial Ecosystem

威胁 Threat (s)

主要威胁 Major Threat(s)	栖息地的丧失和破碎化及人为因素的干扰，所依赖的竹丛大规模开花死亡也可能是影响因素之一 / The major threats are habitat loss and fragmentation, and the human disturbance. The blooming and death of bamboo might also be a threat

保护级别与保护行动 Protection Category and Conservation Action (s)

IUCN 红色名录 (2016) IUCN Red List (2016)	易危 / VU C2a (i)
保护行动 Conservation Action(s)	列入国家 I 级重点保护野生动物、《国家保护的有益的或者有重要经济、科学研究价值的陆生野生动物名录》。部分分布区位于保护区内。需要估计其种群数量及分布情况，开展对其生活史的研究 / It has been listed as the first class in National Key Protected Wild Animal List, *National Protected List of Terrestrial Wild Animals with Good Benefits or Important Economic and Scientific Values*. Some occurs in the nature reserve, and the estimates of population size and distribution and the studies on life history are needed

灰冠鸦雀 *Sinosuthora przewalskii*　　　　　浣花一痴 摄　By Huanhuayichi

中国生物多样性 红色名录

弄岗穗鹛
Stachyris nonggangensis

濒危 EN B1a; C2a(ii)

| 数据缺乏 DD | 无危 LC | 近危 NT | 易危 VU | 濒危 EN | 极危 CR | 区域灭绝 RE | 野外灭绝 EW | 灭绝 EX |

🕊 分类地位 Taxonomic Status

动物界 Animalia	脊索动物门 Chordata	鸟纲 Aves	雀形目 Passeriformes	林鹛科 Timaliidae
学 名 Scientific Name		*Stachyris nonggangensis*		
命 名 人 Species Authority		Zhou and Jiang, 2008		
英 文 名 English Name(s)		Nonggang Babbler		
同物异名 Synonym(s)		无 / None		
种下单元评估 Infra-specific Taxa Assessed		无 / None		

🕊 评估信息 Assessment Information

评估年份 Year Assessed	2016
评 定 人 Assessor(s)	周放 / Fang Zhou
其他贡献人 Other Contributor(s)	蒋爱武 / Aiwu Jiang

理由 Justification: 弄岗穗鹛种群数量少，分布区极狭窄。仅见于喀斯特地区，对喀斯特森林生境的依赖性极高，目前发现仅分布在广西的弄岗保护区（弄岗和陇呼）、青龙山保护区（春秀）和邦亮保护区等4个区域 (Li and Zhou, 2013)，如不加强保护，种群数量会持续下降。因此，列为濒危等级 / The population size and distribution area of *Stachyris nonggangensis* are small. It is rare and highly dependent on karst forest, and only occurs in 4 sites: Nonggang and Longhu in Nonggang Nature Reserve, Chunxiu in Qinglong Shan Nature Reserve, and Bangliang Nature Reserve (Li and Zhou, 2013). It might be upgrade as critically endangered without effective protection. Thus, it is listed as Endangered

评估历史 Assessment History: 《国家重点保护野生动物名录》：Ⅱ (2021) / *National Key Protected Wild Animal List*: Ⅱ (2021)

🕊 地理分布 Geographical Distribution

国内分布 Domestic Distribution

模式产地为广西龙州县弄岗自然保护区。分布区狭窄，主要分布在邻近中越边境的桂西南喀斯特地区 (Li and Zhou, 2013)，为中国喀斯特地区特有种 / The type specimens were collected in Nonggang Nature Reserve in Guangxi, China. *S. nonggangensis* is the endemic species of China, and appears to occupy a limited distribution in the Sino-Vietnamese border region in southwest Guangxi, China, and adjacent Viet Nam (Li and Zhou, 2013)

分布标注 Distribution Note

特有种 / Endemic

国内分布图
Map of Domestic Distribution

种群 Population

种群数量 Population Size	种群数量不足 1000 对 / The population size is less than 1000 pairs
种群趋势 Population Trend	下降 / Decreasing

生境与生态系统 Habitat (s) and Ecosystem (s)

生　　　境 Habitat(s)	典型喀斯特鸟类，只分布于喀斯特山区，生活于季节雨林中的下层，对喀斯特森林高度依赖 / *S. nonggangensis* is a typical karst bird and only occurs in karst mountain area. It uses the bottom of the seasonal rainforest, and is highly dependent on karst forest
生态系统 Ecosystem(s)	陆地生态系统 / Terrestrial Ecosystem

威胁 Threat (s)

主要威胁 Major Threat(s)	只栖息于喀斯特季雨林，并且只营巢于喀斯特林区的峭壁石缝中，对喀斯特森林生境的依赖性极高。喀斯特森林是一种十分脆弱的生态系统，一旦遭到破坏就难以恢复。喀斯特森林的破坏和片断化是该种鸟的主要威胁 / *S. nonggangensis* only occurs in seasonal rainforest in karst area. It nests in cavities in a limestone cliff and is highly dependent on karst forest, which is extremely vulnerable and therefore difficult to restore if destroyed. The habitat loss and fragmentation of karst forest are the major threats

保护级别与保护行动 Protection Category and Conservation Action (s)

IUCN 红色名录 (2016) IUCN Red List (2016)	易危 / VU C2a (ii)
保护行动 Conservation Action(s)	列入国家II级重点保护野生动物。加强分布区和种群现状调查并开展生态生物学研究。有关管理部门应制定有效保护措施，减少砍伐、盗猎和保护区内的人为活动干扰 / It has been listed as the second class in National Key Protected Wild Animal List. The ecological and biological studies on distribution and population is needed. And effective protection is needed to control the logging, illegal hunting and human disturbance

弄岗穗鹛 *Stachyris nonggangensis*　　周敏 摄　By Min Zhou

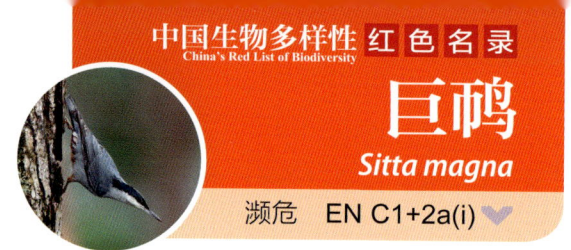

巨䴓
Sitta magna

濒危 EN C1+2a(i)

| 数据缺乏 DD | 无危 LC | 近危 NT | 易危 VU | 濒危 EN | 极危 CR | 区域灭绝 RE | 野外灭绝 EW | 灭绝 EX |

分类地位 Taxonomic Status

动物界 Animalia	脊索动物门 Chordata	鸟纲 Aves	雀形目 Passeriformes	䴓科 Sittidae
学名 Scientific Name		*Sitta magna*		
命名人 Species Authority		Wardlaw Ramsay, 1876		
英文名 English Name(s)		Giant Nuthatch		
同物异名 Synonym(s)		无 / None		
种下单元评估 Infra-specific Taxa Assessed		无 / None		

评估信息 Assessment Information

评估年份 Year Assessed	2016
评定人 Assessor(s)	杨晓君 / Xiaojun Yang
其他贡献人 Other Contributor(s)	吴飞、常云艳等 /Fei Wu, Yunyan Chang *et al.*

理由 Justification: 巨䴓主要栖息于成熟云南松林中，在胸径大于 15 cm 的云南松上觅食 (韩联宪等，2012)，而云南松林多位于林场，由于采伐等原因，成熟云南松林面积较小。近年来巨䴓仅在云南极少的几个地点有记录，估计种群数量在 800～2,000 只 (BirdLife International, 2014)。其种群数量面临进一步下降的威胁。因此，列为濒危等级 / Recent studies show that *Sitta magna* uses the forests of *Pinus yunnanensis*, and forages in the pine tree with DBH above 15 cm (Han *et al.*, 2012). The area of mature *Pinus yunnanensis* forest is small due to the logging. There are only a few records of *Sitta magna* at several sites in Yunnan, and the population size is estimated to be 800～2,000 individuals (BirdLife International, 2014), which is undergoing a decline. Thus, it is listed as Endangered

评估历史 Assessment History:《国家重点保护野生动物名录》: II (2021); 《中国物种红色名录: 第二卷 脊椎动物 下册》(2009): 易危 / *National Key Protected Wild Animal List*: II (2021); *China Species Red List*: *Vol. II Vertebrates Part 2* (2009): VU

地理分布 Geographical Distribution

国内分布 Domestic Distribution
分布于云南、四川西南部、贵州西南部 (杨岚和杨晓君，2004) / It occurs in Yunnan, southwest Sichuan and southwest Guizhou
分布标注 Distribution Note
非特有种 / Non-endemic

国内分布图
Map of Domestic Distribution

种群 Population

种群数量 Population Size	种群整体数量未知，但主要分布区云南的种群数量估计为 800～2,000 只（BirdLife International，2014）/ The total population size is unknown. The population in Yunnan, its main distribution area, is estimated to be 800～2,000 individuals (BirdLife International, 2014)
种群趋势 Population Trend	下降 / Decreasing

生境与生态系统 Habitat (s) and Ecosystem (s)

生境 Habitat(s)	栖息于以针叶林为主的针阔混交林 / *S. magna* uses mixed coniferous broad-leaved forests with most coniferous woods
生态系统 Ecosystem(s)	陆地生态系统 / Terrestrial Ecosystem

威胁 Threat (s)

主要威胁 Major Threat(s)	因为采伐和林地更新造成的云南松成熟林面积不断减小（邓章文等，2012）/ The major threat is the decline of the area of *Pinus yunnanensis* forests owing to the logging and forest regeneration (Deng, 2012)

保护级别与保护行动 Protection Category and Conservation Action (s)

IUCN 红色名录 (2016) IUCN Red List (2016)	濒危 / EN C2a (i)
保护行动 Conservation Action(s)	列入国家 II 级重点保护野生动物、《国家保护的有益的或者有重要经济、科学研究价值的陆生野生动物名录》/ It has been listed as the second class in National Key Protected Wild Animal List, *National Protected List of Terrestrial Wild Animals with Good Benefits or Important Economic and Scientific Values*

巨䴓 *Sitta magna*　　　　　　　　　　　　　　　许哥 摄　By Ge Xu

丽䴓
Sitta formosa

濒危 EN B1ab(iii)+2ab(iii); D

| 数据缺乏 DD | 无危 LC | 近危 NT | 易危 VU | 濒危 EN | 极危 CR | 区域灭绝 RE | 野外灭绝 EW | 灭绝 EX |

分类地位 Taxonomic Status

| 动物界 Animalia | 脊索动物门 Chordata | 鸟纲 Aves | 雀形目 Passeriformes | 䴓科 Sittidae |

学 名 Scientific Name	*Sitta formosa*
命 名 人 Species Authority	Blyth, 1843
英 文 名 English Name(s)	Beautiful Nuthatch
同物异名 Synonym(s)	无 / None
种下单元评估 Infra-specific Taxa Assessed	无 / None

评估信息 Assessment Information

评 估 年 份 Year Assessed	2016
评 定 人 Assessor(s)	杨晓君 / Xiaojun Yang
其他贡献人 Other Contributor(s)	吴飞、常云艳等 / Fei Wu, Yunyan Chang *et al.*

理由 Justification: 丽䴓分布区极小，估计种群数量极小，由于生境破坏且有限的分布区相对独立，其种群数量有下降趋势。因此，列为濒危等级 / The distribution area of *Sitta formosa* is quite small, and the population is estimated to be extremely small. Its population is undergoing a decline owing to the habitat loss and fragmentation of distribution area. Thus, it is listed as Endangered

评估历史 Assessment History: 《国家重点保护野生动物名录》：II (2021)；《中国物种红色名录：第二卷 脊椎动物 下册》(2009)：易危 / *National Key Protected Wild Animal List*: II (2021); *China Species Red List: Vol. II Vertebrates Part 2* (2009): VU

地理分布 Geographical Distribution

国内分布 Domestic Distribution

分布于云南屏边、绿春、泸水、贡山 / It occurs in Pingbian, Lüchun, Lushui, Gongshan in Yunnan

分布标注 Distribution Note

非特有种 / Non-endemic

国内分布图
Map of Domestic Distribution

种群 Population

种群数量 Population Size 种群数量未知。由于分布区极为狭窄，推测种群数量极小，近年来仅在河口、泸水和贡山独龙江有少数几次记录 / The population size is unknown, while the distribution is quite small, the population size is estimated to be extremely small. There are only a few records in Hekou, Lushui, and Dulong River

种群趋势 Population Trend 下降 / Decreasing

生境与生态系统 Habitat(s) and Ecosystem(s)

生境 Habitat(s) 栖息于热带雨林、季雨林和山地常绿阔叶林中 / *S. formosa* uses the tropical rainforest, seasonal rainforest, and mountain evergreen broad-leaved forest

生态系统 Ecosystem(s) 陆地生态系统 / Terrestrial Ecosystem

威胁 Threat(s)

主要威胁 Major Threat(s) 栖息地破坏和斑块化 / Habitat loss and fragmentation

保护级别与保护行动 Protection Category and Conservation Action(s)

IUCN 红色名录 (2016) IUCN Red List (2016) 易危 / VU C2a (i)

保护行动 Conservation Action(s) 列入国家 II 级重点保护野生动物、《国家保护的有益的或者有重要经济、科学研究价值的陆生野生动物名录》。部分分布区位于自然保护区内而受到保护 / It has been listed as the second class in National Key Protected Wild Animal List, *National Protected List of Terrestrial Wild Animals with Good Benefits or Important Economic and Scientific Values*. Nature reserves cover some of its distribution area

丽䴓 *Sitta formosa*　　　　董文晓 摄　By Wenxiao Dong

中国生物多样性红色名录

黑喉歌鸲
Calliope obscura

濒危 EN C1+2a(i)

| 数据缺乏 DD | 无危 LC | 近危 NT | 易危 VU | 濒危 EN | 极危 CR | 区域灭绝 RE | 野外灭绝 EW | 灭绝 EX |

分类地位 Taxonomic Status

动物界 Animalia	脊索动物门 Chordata	鸟纲 Aves	雀形目 Passeriformes	鹟科 Muscicapidae

学名 Scientific Name	*Calliope obscura*
命名人 Species Authority	Berezowski and Bianchi, 1891
英文名 English Name(s)	Blackthroat
同物异名 Synonym(s)	*Luscinia obscura*, *Larvivora obscura*
种下单元评估 Infra-specific Taxa Assessed	无 / None

评估信息 Assessment Information

评估年份 Year Assessed	2016
评定人 Assessor(s)	丁长青 / Changqing Ding
其他贡献人 Other Contributor(s)	刘垚 / Yao Liu

理由 Justification: 黑喉歌鸲种群数量小；分布范围狭小且栖息地面积不断减少，将导致种群数量持续下降。因此，列为濒危等级 / The population size and distribution of *Calliope obscura* are small, and the habitat area and population size is undergoing a rapid decline. Thus, it is listed as Endangered

评估历史 Assessment History:《国家重点保护野生动物名录》：Ⅱ (2021)；《中国物种红色名录：第二卷 脊椎动物 下册》(2009)：易危 / *National Key Protected Wild Animal List*: Ⅱ (2021); *China Species Red List: Vol. II Vertebrates Part 2* (2009): VU

地理分布 Geographical Distribution

国内分布 Domestic Distribution

分布于甘肃岷县，陕西太白山、佛坪、长青，四川白河、九寨沟、卧龙，云南蒙自 / It occurs in Minxian in Gansu, Taibai Shan, Foping, Changqing in Shaanxi, Baihe, Jiuzhaigou, Wolong in Sichuan and Mengzi in Yunnan

分布标注 Distribution Note

特有种 / Endemic

国内分布图
Map of Domestic Distribution

种群 Population

种群数量 Population Size	种群数量不明，仅有零星调查发现。2011 年 Davies 在秦岭新发现两个繁殖区的 14 只个体，佛坪保护区样线调查也发现少数个体，但调查发现的个体总数不超过 1,000 只 / The population size is unknown due to lack of survey, with only a few records. 14 birds were recorded in two breeding sites in Qinling Mountain (Davies, 2011), and several birds were recorded in the Foping Nature Reserve. The total recorded birds were less than 1,000 individuals
种群趋势 Population Trend	未知 / Unknown

生境与生态系统 Habitat (s) and Ecosystem (s)

生 境 Habitat(s)	在甘肃分布于海拔 3,050 ～ 3,350 m 的竹林，在四川分布于温带森林，在陕西分布于海拔 2,400 ～ 2,500 m 的针叶林及针阔混交林内的浓密竹林间 (IUCN，2012) / According to the records, *C. obscura* uses the bamboos at 3,050 ～ 3,350 m in Gansu and temperate forests in Sichuan, bamboos in coniferous forests and mixed coniferous broad-leaved forests in Jiangxi at 2,400 ～ 2,500 m (IUCN, 2012)
生态系统 Ecosystem(s)	陆地生态系统 / Terrestrial Ecosystem

威胁 Threat (s)

主要威胁 Major Threat(s)	主要威胁可能是森林的丧失和破碎化。从 20 世纪 60 年代以来，四川等地的森林覆盖率迅速下降，可能使其生境丧失 (IUCN，2012) / The major threats might be forest loss and fragmentation. The forest coverage rate in Sichuan has been undergoing a rapid decline since 1960s, which might lead to the habitat loss (IUCN, 2012)

保护级别与保护行动 Protection Category and Conservation Action (s)

IUCN 红色名录 (2016) IUCN Red List (2016)	易危 / VU C2a (i)
保护行动 Conservation Action(s)	列入国家 II 级重点保护野生动物、《国家保护的有益的或者有重要经济、科学研究价值的陆生野生动物名录》。建议加强对该物种的数量调查、越冬地调查等研究，制定保护网络 / It has been listed as the second class in National Key Protected Wild Animal List, *National Protected List of Terrestrial Wild Animals with Good Benefits or Important Economic and Scientific Values*. Surveys on the population size and wintering habitat, improvement of conservation network

黑喉歌鸲 *Calliope obscura*　　　　　　　　　　　　　　　　　　　　　　　张永文 摄　By Yongwen Zhang

棕头歌鸲
Larvivora ruficeps

濒危 EN C2a(ii)

| 数据缺乏 DD | 无危 LC | 近危 NT | 易危 VU | 濒危 EN | 极危 CR | 区域灭绝 RE | 野外灭绝 EW | 灭绝 EX |

分类地位 Taxonomic Status

动物界 Animalia	脊索动物门 Chordata	鸟纲 Aves	雀形目 Passeriformes	鹟科 Muscicapidae
学名 Scientific Name		*Larvivora ruficeps*		
命名人 Species Authority		Hartert, 1907		
英文名 English Name(s)		Rufous-headed Robin		
同物异名 Synonym(s)		*Luscinia ruficeps*		
种下单元评估 Infra-specific Taxa Assessed		无 / None		

评估信息 Assessment Information

评估年份 Year Assessed	2016
评定人 Assessor(s)	董路 / Lu Dong
其他贡献人 Other Contributor(s)	无 / None

理由 Justification: 棕头歌鸲的繁殖分布区非常狭窄，种群数量可能不足 2,500 只，且大多存在于同一个亚种群中。因此，列为濒危等级 / The breeding area of *Larvivora ruficeps* is quite small. Its population might be less than 2,500, and most are one subspecies. Thus, it is listed as Endangered

评估历史 Assessment History:《国家重点保护野生动物名录》：Ⅰ (2021)；《中国物种红色名录：第二卷 脊椎动物 下册》(2009)：易危 / *National Key Protected Wild Animal List*: Ⅰ (2021); *China Species Red List: Vol. II Vertebrates Part 2* (2009): VU

地理分布 Geographical Distribution

国内分布 Domestic Distribution

仅繁殖于我国四川北部和陕西南部，偶见于云南中部 / It only breeds in north Sichuan and south Shaanxi in China, and occurs in central Yunnan occasionally

分布标注 Distribution Note

非特有种 / Non-endemic

国内分布图
Map of Domestic Distribution

种群 Population

种群数量 Population Size	繁殖个体数量在 1,000 ～ 2,499 只，且基本存在于同一个种群中 / The population is estimated to be 1,000 ～ 2,499 mature individuals, and most are one subspecies
种群趋势 Population Trend	下降 / Decreasing

生境与生态系统 Habitat(s) and Ecosystem(s)

生　　境 Habitat(s)	在繁殖地，主要栖息于海拔 2,400 ～ 2,800 m 的温带混交林及灌丛中，尤其偏好狭窄的河谷地带。越冬于海拔 2,000 m 左右的山地灌丛 / *L. ruficeps* breeds in the temperate mixed forest and shrubs at 2,400 ～ 2,800 m, especially in the valleys; and winters in the shrubs at 2000 m
生态系统 Ecosystem(s)	陆地生态系统 / Terrestrial Ecosystem

威胁 Threat(s)

主要威胁 Major Threat(s)	适宜栖息地面积的减少和片断化是本物种的主要威胁因子。此外，水电站大坝的建设对河谷生境的破坏也可能是威胁因子之一 / The major threats of this species are suitable habitat loss and fragmentation. The destroy of valley due to dam construction might be another threat

保护级别与保护行动 Protection Category and Conservation Action(s)

IUCN 红色名录 (2016) IUCN Red List (2016)	濒危 / EN C2a (ii)
保护行动 Conservation Action(s)	列入国家 I 级重点保护野生动物、《国家保护的有益的或者有重要经济、科学研究价值的陆生野生动物名录》。本物种的分布区内已建立了多个

国家级自然保护区，如四川省九寨沟自然保护区、王朗自然保护区和陕西太白山自然保护区，将有助于对本物种的保护 / It has been listed as the first class in National Key Protected Wild Animal List, *National Protected List of Terrestrial Wild Animals with Good Benefits or Important Economic and Scientific Values*. There are several nature reserves set up in its distribution area, *e.g.* Sichuan Jiuzhaigou Nature Reserve, Wanglang Nature Reserve, and Shannxi Taibai Shan Nature Reserve, which is beneficial for the species

棕头歌鸲 *Larvivora ruficeps*　　　　鸟网 提供　By birdnet.cn

贺兰山红尾鸲
Phoenicurus alaschanicus

濒危 EN B1b(ii,iii); C2a(i,ii) b

| 数据缺乏 DD | 无危 LC | 近危 NT | 易危 VU | **濒危 EN** | 极危 CR | 区域灭绝 RE | 野外灭绝 EW | 灭绝 EX |

分类地位 Taxonomic Status

动物界 Animalia	脊索动物门 Chordata	鸟纲 Aves	雀形目 Passeriformes	鹟科 Muscicapidae
学 名 Scientific Name		*Phoenicurus alaschanicus*		
命名人 Species Authority		Przewalski, 1876		
英文名 English Name(s)		Alashan Redstart		
同物异名 Synonym(s)		*Rutirilla alaschanica*		
种下单元评估 Infra-specific Taxa Assessed		无 / None		

评估信息 Assessment Information

评估年份 Year Assessed	2016
评定人 Assessor(s)	卢欣 / Xin Lu
其他贡献人 Other Contributor(s)	无 / None

理由 Justification: 由于贺兰山红尾鸲分布面积狭窄，加之植被丧失，栖息地质量下降，种群数量可能正在经历持续的衰退；可能不存在种群数超过 1,000 只个体的亚种群。因此，列为濒危等级 / The population size of *Phoenicurus alaschanicus* is undergoing a constant decline due to the small distribution area, vegetation loss and habitat deterioration. Each sub population might be less than 1,000 individuals. Thus, it is listed as Endangered

评估历史 Assessment History: 《国家重点保护野生动物名录》：II (2021)；《中国物种红色名录：第二卷 脊椎动物 下册》(2009)：易危 / *National Key Protected Wild Animal List*: II (2021); *China Species Red List: Vol. II Vertebrates Part 2* (2009): VU

地理分布 Geographical Distribution

国内分布 Domestic Distribution

中国中北部及西部特有种。分布于青海省西宁、天峻及柴达木盆地，宁夏贺兰山及甘肃东部。越冬于陕西南部、河北、山西，偶至北京 / It is endemic in west, middle and north China. It occurs in Xining, Tianjun, Qaidam Basin in Qinghai; Helan Shan in Ningxia and east Gansu. It winters in south Shaanxi, Hebei and Shanxi, and occasionally appeared in Beijing

分布标注 Distribution Note

特有种 / Endemic

国内分布图
Map of Domestic Distribution

种群 Population

种群数量 Population Size	分布区狭窄，种群数量估计有 3,000～6,000 对成熟个体 (IUCN，2012) / The distribution area is small and the population size is estimated to be 3,000～6,000 matured individuals (IUCN, 2012)
种群趋势 Population Trend	下降 / Decreasing

生境与生态系统 Habitat(s) and Ecosystem(s)

生　　境 Habitat(s)	海拔 3,300 m 至林线高山山地针叶林、山区稠密灌丛及多松散岩石的山坡、草地，以及以松柏科针叶林为主的河边平原地带，冬季至海拔 2,000 m 的地带 / *Phoenicurus alaschanicus* uses coniferous forest above 3,300 m, mountainside with dense shrubs and scattered rock, grassland, and flatland along the river with coniferous forest. It winters at 2,000 m
生态系统 Ecosystem(s)	陆地生态系统 / Terrestrial Ecosystem

威胁 Threat(s)

主要威胁 Major Threat(s)	推测主要威胁来源于栖息地丧失 / The major threat is habitat loss

保护级别与保护行动 Protection Category and Conservation Action(s)

IUCN 红色名录 (2016) IUCN Red List (2016)	近危 / NT
保护行动 Conservation Action(s)	列入国家 II 级重点保护野生动物、《国家保护的有益的或者有重要经济、科学研究价值的陆生野生动物名录》/ It has been listed as the second class in National Key Protected Wild Animal List, *National Protected List of Terrestrial Wild Animals with Good Benefits or Important Economic and Scientific Values*

贺兰山红尾鸲 *Phoenicurus alaschanicus*　　　　　　　　　　　王文娟 摄　By Wenjuan Wang

白喉石䳭
Saxicola insignis

濒危 EN C2a(ii)

| 数据缺乏 DD | 无危 LC | 近危 NT | 易危 VU | 濒危 EN | 极危 CR | 区域灭绝 RE | 野外灭绝 EW | 灭绝 EX |

分类地位 Taxonomic Status

动物界 Animalia	脊索动物门 Chordata	鸟纲 Aves	雀形目 Passeriformes	鹟科 Muscicapidae
学 名 Scientific Name		*Saxicola insignis*		
命 名 人 Species Authority		Gray and Gray, 1846		
英 文 名 English Name(s)		White-throated Bushchat		
同物异名 Synonym(s)		无 / None		
种下单元评估 Infra-specific Taxa Assessed		无 / None		

评估信息 Assessment Information

评估年份 Year Assessed	2016
评 定 人 Assessor(s)	丁平 / Ping Ding
其他贡献人 Other Contributor(s)	无 / None

理由 Justification: 估计全球种群数量为3,500～15,000只 (BirdLife International, 2012), 国内种群数量未见有评估。由于其栖息地不断被开垦、过度放牧，以及牧草收割等原因，其栖息地面积和种群数量正逐渐减小。因此，列为濒危等级 / The distribution area of *Saxicola insignis* is large and the world population size is estimated to be 3,500～15,000 (BirdLife International, 2012). The population size in China has not been assessed. The habitat and population is undergoing a constant decline owing to the reclaimation, over-grazing, and collection of forage grass. Thus, it is listed as Endangered

评估历史 Assessment History: 《国家重点保护野生动物名录》: II (2021); 《中国物种红色名录: 第二卷 脊椎动物 下册》(2009): 易危 / *National Key Protected Wild Animal List*: II (2021); *China Species Red List: Vol. II Vertebrates Part 2* (2009): VU

地理分布 Geographical Distribution

国内分布 Domestic Distribution

国内为陕西南部、内蒙古西部、宁夏、甘肃、新疆、青海中部和江西的旅鸟 / It occurs in south Shaanxi, west Inner Mongolia (Nei Mongol), Ningxia, Gansu, Xinjiang, middle Qinghai and Jiangxi when migrating

分布标注 Distribution Note

非特有种 / Non-endemic

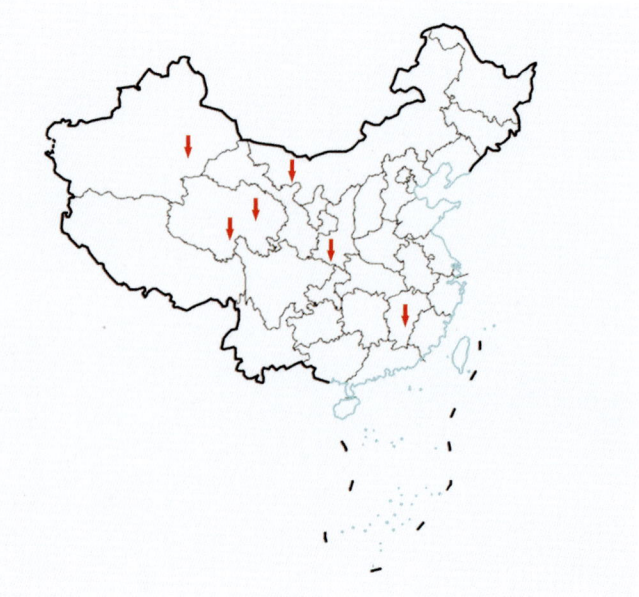

国内分布图
Map of Domestic Distribution

种群 Population

种群数量 Population Size	未知 / Unknown
种群趋势 Population Trend	下降 / Decreasing

生境与生态系统 Habitat (s) and Ecosystem (s)

生　　境 Habitat(s) 区域	在高山及亚高山上的草甸以及山林中的矮树上繁殖。冬季则在河床边的草地、芦苇以及柽柳等处觅食。在迁徙的过程中，也可到达海拔 4,500 m 的 / *S. insignis* breeds on short trees in grassland and forests in mountain and subalpine. It forages in grassland, reeds and tamarisk along the riverbed in winter. It could reach areas above 4,500 m while migration
生态系统 Ecosystem(s)	陆地生态系统 / Terrestrial Ecosystem

威胁 Threat (s)

主要威胁 Major Threat(s)	面临着栖息地草场的破坏、过度放牧、牧草收割，以及各种人为活动干扰等因素的胁迫 / The major threats of the species are loss of grassland, over grazing, collection of forage grass, and human disturbance

保护级别与保护行动 Protection Category and Conservation Action (s)

IUCN 红色名录 (2016) IUCN Red List (2016)	易危 / VU C2a (ii)
保护行动 Conservation Action(s)	列入国家 II 级重点保护野生动物、《国家保护的有益的或者有重要经济、科学研究价值的陆生野生动物名录》 / It has been listed as the second class in National Key Protected Wild Animal List, *National Protected List of Terrestrial Wild Animals with Good Benefits or Important Economic and Scientific Values*

白喉石䳭 *Saxicola insignis*　　　　张铭 摄　By Ming Zhang

中国生物多样性红色名录

栗斑腹鹀
Emberiza jankowskii

濒危 EN A2abc+3bc+4bc; B2ac(iii)

| 数据缺乏 DD | 无危 LC | 近危 NT | 易危 VU | **濒危 EN** | 极危 CR | 区域灭绝 RE | 野外灭绝 EW | 灭绝 EX |

 ### 分类地位 Taxonomic Status

动物界 Animalia	脊索动物门 Chordata	鸟纲 Aves	雀形目 Passeriformes	鹀科 Emberizidae

学 名 Scientific Name	*Emberiza jankowskii*
命 名 人 Species Authority	Taczanowski, 1888
英 文 名 English Name(s)	Jankowski's Bunting
同物异名 Synonym(s)	无 / None
种下单元评估 Infra-specific Taxa Assessed	无 / None

评估信息 Assessment Information

评估年份 Year Assessed	2016
评定人 Assessor(s)	王海涛 / Haitao Wang
其他贡献人 Other Contributor(s)	姜云垒、刘宇、张立世、秦博、李时、张博、陶慧娟、王琳等 / Yunlei Jiang, Yu Liu, Lishi Zhang, Bo Qin, Shi Li, Bo Zhang, Huijuan Tao, Lin Wang *et al.*

理由 Justification: 历史上吉林省东部的国内最大种群已经消失。其适宜栖息地因土地开垦和过度放牧等原因而快速丧失。2008～2014年在内蒙古陆续发现新分布区，现存分布区栖息地质量退化且严重破碎化，种群繁殖成功率低，导致种群急剧下降，目前尚未对该物种栖息地实施有效保护 (Jiang *et al.*，2008; Wang *et al.*，2010)。因此，列为濒危等级 / The east Jilin population of *Emberiza jankowskii*, the largest population in China, has been gone. The habitat lost rapidly owing to the conversion of its grassland habitat for agriculture and pasture. The new sites have been found in Inner Mongolia (Nei Mongol) in 2008～2014. All these habitats are in low quality and seriously fragmented. The population is undergoing a dramatical decline with the low reproductive success. There is no effective conservation on its habitat yet (Jiang *et al.*, 2008; Wang *et al.*, 2010). Thus, it is listed as Endangered

评估历史 Assessment History: 《国家重点保护野生动物名录》：Ⅰ (2021)；《中国濒危动物红皮书·鸟类》(1998)：稀有；《中国物种红色名录：第二卷 脊椎动物 下册》(2009)：易危 / *National Key Protected Wild Animal List*: Ⅰ (2021); *China Red Data Book of Endangered Animals*: *Aves* (1998): R; *China Species Red List*: *Vol. II Vertebrates Part 2* (2009): VU

 ### 地理分布 Geographical Distribution

国内分布 Domestic Distribution

分布于吉林西部的洮南、镇赉，内蒙古东部的扎赉特旗、科尔沁右翼中旗、科尔沁右翼前旗、扎鲁特旗，偶见于北京 / It occurs in Taonan and Zhenlai in western Jilin, Zhalaite, Keyouzhongqi, Keyouqianqi, Zhalute in east Inner Mongolia (Nei Mongol), and occasionally appeared in Beijing

分布标注 Distribution Note

非特有种 / Non-endemic

国内分布图
Map of Domestic Distribution

种群 Population

种群数量 Population Size	估计种群数量 7,000 ～ 15,000 只 / The population size is estimated to be 7,000 ～ 15,000 individuals
种群趋势 Population Trend	下降 / Decreasing

生境与生态系统 Habitat (s) and Ecosystem (s)

生　　境 Habitat(s)	现存种群主要在草本植物以贝加尔针茅、线叶菊和大油芒为主，优势灌木为西伯利亚山杏的生境中繁殖 (高玮，2002a)；冬季游荡于草甸、防护林和农田交错地带 / *E. jankowskii* breeds in shrubs of *Armeniaca sibirica* with a variety of herbaceous vegetation, e.g. *Stipa baicalensis, Filifolium sibiricum, Spodiopogon sibiricus* (Gao, 2002a), and uses the cross of meadow, shelter forest and croplands in winter
生态系统 Ecosystem(s)	陆地草原生态系统 / Terrestrial Grassland Ecosystem

威胁 Threat (s)

主要威胁 Major Threat(s)	适宜栖息地由于农田开发和过度放牧等影响而大面积丧失，且这种丧失趋势仍在继续；现存栖息地生境质量下降，植被覆盖度较低；另外，捕食、巢寄生和人为破坏等导致繁殖成功率较低 (Jiang *et al.*，2008；Wang *et al.*，2010) / The major threats are habitat loss owing to the development of agriculture and over grazing, habitat quality deterioration owing to the decline of vegetation coverage. Besides, low reproductive success owing to predation, nest parasitism and human disturbance is another threat

保护级别与保护行动 Protection Category and Conservation Action (s)

IUCN 红色名录 (2016) IUCN Red List (2016)	濒危 / EN A2bc+3bc+4bc
保护行动 Conservation Action(s)	列入国家 I 级重点保护野生动物、《国家保护的有益的或者有重要经济、科学研究价值的陆生野生动物名录》。已对栗斑腹鹀生境选择、巢址选择和繁殖行为等进行了研究，获得了较齐全的生活史等资料。新佳木、图牧吉较小面积的适宜生境位于保护区内。大部分栗斑腹鹀现存生境为公益林区，放牧等人为干扰仍较频繁，未制定针对性保护措施 / It has been listed as the first class in National Key Protected Wild Animal List, *National Protected List of Terrestrial Wild Animals with Good Benefits or Important Economic and Scientific Values*. The studies on habitat selection, nest site selection and breeding behavior have collected information on its life history. The suitable habitats in Xinjiamu, Tumuji are located in nature reserves. Most habitats of *E. jankowskii* are in public welfare forests, while are under high disturbance and without specific conservation action

栗斑腹鹀 *Emberiza jankowskii*　　　　　　　　　　　　　　　　　　　　董桂茗 摄　By Guijun Dong

白眉山鹧鸪
Arborophila gingica

易危 VU C2a(i)

| 数据缺乏 DD | 无危 LC | 近危 NT | 易危 VU | 濒危 EN | 极危 CR | 区域灭绝 RE | 野外灭绝 EW | 灭绝 EX |

分类地位 Taxonomic Status

动物界 Animalia	脊索动物门 Chordata	鸟纲 Aves	鸡形目 Galliformes	雉科 Phasianidae
学 名 Scientific Name		*Arborophila gingica*		
命名人 Species Authority		Gmelin, 1789		
英文名 English Name(s)		White-necklaced Partridge		
同物异名 Synonym(s)		*Tetrao gingicus*		
种下单元评估 Infra-specific Taxa Assessed		广西亚种 *Arborophila gingica guangxiensis* (濒危 EN)		

评估信息 Assessment Information

评估年份 Year Assessed	2016
评定人 Assessor(s)	周放 / Fang Zhou
其他贡献人 Other Contributor(s)	余丽江 / Lijiang Yu

理由 Justification: 生境片断化和丧失、传统狩猎习俗是导致白眉山鹧鸪种群数量下降的主要原因。虽然在福建的大部分地区可见，但基本都是零星散布的，而且其他分布区并不常见，种群数量少，片断化小种群现象明显。其中，仅分布于广西北部和中南部的广西亚种，栖息在相互隔离的区域内，数量不超过1,500只。相较于指名亚种，广西亚种种群数量下降速度更快。因此，白眉山鹧鸪列为易危等级，广西亚种应列为濒危等级 / The population size is decreasing owing to habitat loss and fragmentation, and traditional custom of hunting. It occurs in most areas in Fujian, while scattered in small number. It is not common in other distributed areas, and in small population. *Arborophila gingica guangxiensis*, is less than 1,500 individuals, and only occurs in north and south of middle in Guangxi, without corridor between habitats fragmented by human activities (residents, farmlands and roads). Compared to *A. g. gingica*, the population of *A. g. guangxiensis* is undergoing a more rapid decline. Thus, *A. g. gingica* is listed as Vulnerable, and *A. g. guangxiensis* as Endangered

评估历史 Assessment History: 《国家重点保护野生动物名录》：II (2021)；《中国濒危动物红皮书·鸟类》(1998)：稀有；《中国物种红色名录：第二卷 脊椎动物 下册》(2009)：易危 / National Key Protected Wild Animal List: II (2021); China Red Data Book of Endangered Animals: Aves (1998): R; China Species Red List: Vol. II Vertebrates Part 2 (2009): VU

地理分布 Geographical Distribution

国内分布 Domestic Distribution
中国南方山地和丘陵地区的特有种(雷富民和卢汰春，2006)。分布于浙江南部、江西南部、湖南东南部、福建、广东北部、广西北部和中部 / It is endemic to mountains and hills in south China (Lei and Lu, 2006). It occurs in south Zhejiang, south Jiangxi, southeast Hunan, Fujian, north Guangdong, north and central Guangxi
分布标注 Distribution Note
特有种 / Endemic

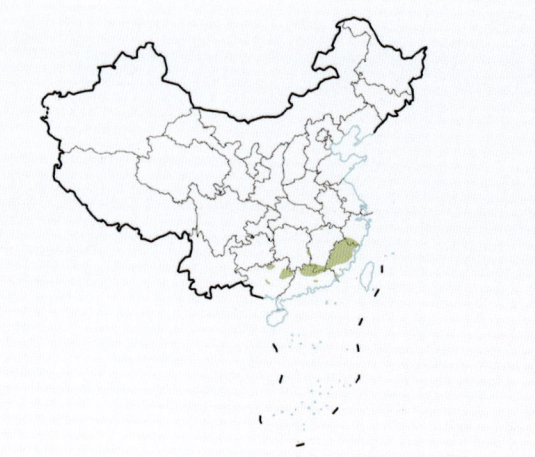

国内分布图
Map of Domestic Distribution

种群 Population

种群数量 Population Size

种群数量不足 10,000 只，其中广西亚种不超过 1,500 只。亚种 *A. g. gingica* 分布自东向东南呈狭长、斑块状，福建大部分地区可见，种群密度最高的地区在龙栖山中部海拔 900～1,200 m 的区域，密度可达 8～10 只/km² (何芬奇等，2007)；而浙江南部、江西南部、湖南东南部、广东北部、广西东部的数量较少，广东和浙江局部地区的种群密度是 0.31 只/km² (李小惠等，1990；郑光美和王岐山，1998) / The population size is estimated to be less than 10,000 individuals, while *A. g. guangxiensis* is estimated to be no more than 1,500. *A. g. gingica* occurs in most areas of Fujian with a narrow and patchy pattern from east to southeast. The populations in south Zhejiang, south Jiangxi, southeast Hunan, north Guangdong, and east Guangxi are small. Population densities vary in different parts of its range: from 0.31 individuals/km² in parts of Guangdong and Zhejiang (Li *et al.*, 1990; Zheng and Wang, 1998) to 8～10 individuals/km² at 900～1,200 m in the middle of Longqi Shan, Fujian (He *et al.*, 2007)

种群趋势 Population Trend

下降 / Decreasing

生境与生态系统 Habitat (s) and Ecosystem (s)

生境 Habitat(s)

栖息于山地阔叶林、针阔混交林、灌丛及竹林等，通常只见于山地的上部和中部。冬季偏好选择林下有空隙的林子作为觅食地 / *A. gingica* uses broad-leaved forests, mixed coniferous broad-leaved forests, shrubs and bamboo forests in mountain area, usually seen in top and middle of mountain. It forages in bamboo forests in winter

生态系统 Ecosystem(s)

陆地生态系统 / Terrestrial Ecosystem

威胁 Threat (s)

主要威胁 Major Threat(s)

农业耕作和伐木导致的栖息地丧失和片断化，是白眉山鹧鸪种群数量下降的主要原因，道路建设和采矿项目也会对种群造成影响。盗猎和贸易也是主要威胁因素。当地居民对林副产品的采摘，会干扰白眉山鹧鸪的正常觅食等行为。一些保护区对旅游业的发展及相应配套设施的建设也会增加干扰 / The major threats are habitat loss and fragmentation owing to farming and logging. The roads construction and mining also bring negative influence. The illegal trapping and trading as traditional game and bushmeat are another major threats. The collection of forest by-products, *e.g.* mushroom, by local residents may disturb the foraging of *A. gingica*. The development of tourism and infrastructure construction in nature reserve would bring disturbance

保护级别与保护行动 Protection Category and Conservation Action (s)

IUCN 红色名录 (2016) IUCN Red List (2016)

近危 / NT

保护行动 Conservation Action(s)

列入国家 II 级重点保护野生动物、《国家保护的有益的或者有重要经济、科学研究价值的陆生野生动物名录》。应协助管理部门制定有效保护措施，减少砍伐、盗猎等保护区内的人为活动干扰。分布于保护区之外的种群，可考虑建立保护小区。对于数量较少的广西亚种，应视为独立保护单元进行优先保护 / It has been listed as the second class in National Key Protected Wild Animal List, *National Protected List of Terrestrial Wild Animals with Good Benefits or Important Economic and Scientific Values*. Measures should be taken to decrease human disturbance, *e.g.* logging, illegal trapping, *etc. A. g. guangxiensis* should be given priority protection as independent group

白眉山鹧鸪 *Arborophila gingica*　　　　　　　　　　　　　　　　　　　　　　胡伟宁 摄　By Weining Hu

红胸山鹧鸪
Arborophila mandellii

易危 VU C2a(i)

| 数据缺乏 DD | 无危 LC | 近危 NT | 易危 VU | 濒危 EN | 极危 CR | 区域灭绝 RE | 野外灭绝 EW | 灭绝 EX |

分类地位 Taxonomic Status

动物界 Animalia	脊索动物门 Chordata	鸟纲 Aves	鸡形目 Galliformes	雉科 Phasianidae
学名 Scientific Name		*Arborophila mandellii*		
命名人 Species Authority		Hume, 1874		
英文名 English Name(s)		Chestnut-breasted Partridge		
同物异名 Synonym(s)		无 / None		
种下单元评估 Infra-specific Taxa Assessed		无 / None		

评估信息 Assessment Information

评估年份 Year Assessed	2016
评定人 Assessor(s)	董路 / Lu Dong
其他贡献人 Other Contributor(s)	无 / None

理由 Justification: 红胸山鹧鸪的种群数量较少，且呈现下降趋势。近年来其生境破碎化导致分布区片断化程度增加。因此，列为易危等级 / The population size of *Arborophila mandellii* is small and in decline. The habitat fragmentation has increased the fragmentation of its distribution area. Thus, it is listed as Vulnerable

评估历史 Assessment History: 《国家重点保护野生动物名录》：Ⅱ (2021)；《中国濒危动物红皮书·鸟类》(1998)：稀有；《中国物种红色名录：第二卷 脊椎动物 下册》(2009)：易危 / *National Key Protected Wild Animal List*: Ⅱ (2021); *China Red Data Book of Endangered Animals*: *Aves* (1998): R; *China Species Red List*: *Vol. Ⅱ Vertebrates Part 2* (2009): VU

地理分布 Geographical Distribution

国内分布 Domestic Distribution
主要分布于西藏东南部 / It mainly occurs in southeast Tibet (Xizang)
分布标注 Distribution Note
非特有种 / Non-endemic

国内分布图
Map of Domestic Distribution

种群 Population

种群数量 Population Size	种群数量估计为 3,999 ～ 9,999 只，我国缺乏数据调查 / The population size is estimated to be 3,999 ～ 9,999 individuals. No survey in China
种群趋势 Population Trend	下降 / Decreasing

生境与生态系统 Habitat (s) and Ecosystem (s)

生　　境 Habitat(s)	喜依水而居，主要栖息于常绿阔叶林的林下植被，如竹丛、灌木丛等。分布区域的海拔集中在 1,700 ～ 2,000 m / *A. mandellii* uses vegetation under the evergreen broad-leaved forest, *e.g.* bamboo, bushes, *etc.* The distribution area is at the elevation of 1,700 ～ 2,000 m (Spierenberg, 2005). Areas close to water are preferred
生态系统 Ecosystem(s)	陆地生态系统 / Terrestrial Ecosystem

威胁 Threat (s)

主要威胁 Major Threat(s)	林业开发、土地利用方式改变以及经济作物种植导致的森林退化和片断化是其主要威胁因子。偷猎盗猎行为也是主要威胁因子之一 / The major threats are fragmentation and deterioration of forests owing to development of forests, land use change and plantation of commercial crops. Illegal hunting and trapping is also the serious threats

保护级别与保护行动 Protection Category and Conservation Action (s)

IUCN 红色名录 (2016) IUCN Red List (2016)	易危 / VU C2a (i)
保护行动 Conservation Action(s)	列入国家 II 级重点保护野生动物、《国家保护的有益的或者有重要经济、科学研究价值的陆生野生动物名录》。在其分布区内已建立了一些以森林生态系统为保护对象的自然保护区，有助于对本物种的保护 / It has been listed as the second class in National Key Protected Wild Animal List, *National Protected List of Terrestrial Wild Animals with Good Benefits or Important Economic and Scientific Values*. There are some nature reserves targeted to conserve forest ecosystem in its distributed area, which would be helpful for the species protection

红胸山鹧鸪 *Arborophila mandellii*　　　　　　　　冯利民 摄　By Limin Feng

柳雷鸟
Lagopus lagopus

易危 VU C2a(i)

| 数据缺乏 DD | 无危 LC | 近危 NT | 易危 VU | 濒危 EN | 极危 CR | 区域灭绝 RE | 野外灭绝 EW | 灭绝 EX |

分类地位 Taxonomic Status

动物界 Animalia	脊索动物门 Chordata	鸟纲 Aves	鸡形目 Galliformes	雉科 Phasianidae
学 名 Scientific Name		*Lagopus lagopus*		
命 名 人 Species Authority		Linnaeus, 1758		
英 文 名 English Name(s)		Willow Grouse		
同物异名 Synonym(s)		*Tetrao lagopus*		
种下单元评估 Infra-specific Taxa Assessed		无 / None		

评估信息 Assessment Information

评 估 年 份 Year Assessed	2016
评 定 人 Assessor(s)	孙悦华 / Yuehua Sun
其他贡献人 Other Contributor(s)	无 / None

理由 Justification: 尽管全球范围内柳雷鸟数量并未受到威胁，但国内分布范围狭窄，种群数量较少，且受到猎捕的威胁，*L. lagopus okadai* 亚种多年未有记录，可能已绝迹。因此，列为易危等级 / The global population of *Lagopus lagopus* is not threatened, while the population size and distribution area in China are small, and threatened by hunting. No record of subspecies *L. lagopus okadai* for years, which probably has been extinct. Thus, it is listed as Vulnerable

评估历史 Assessment History: 《国家重点保护野生动物名录》：Ⅱ (1989，2021)；《中国濒危动物红皮书·鸟类》(1998)：未定；《中国物种红色名录：第二卷 脊椎动物 下册》(2009)：易危 / *National Key Protected Wild Animal List*: II (1989, 2021); *China Red Data Book of Endangered Animals*: *Aves* (1998): I; *China Species Red List*: *Vol. II Vertebrates Part 2* (2009): VU

地理分布 Geographical Distribution

国内分布 Domestic Distribution
分布于新疆阿勒泰地区和黑龙江北部 / It occurs in Altai in Xinjiang and north Heilongjiang
分布标注 Distribution Note
非特有种 / Non-endemic

国内分布图
Map of Domestic Distribution

种群 Population

种群数量 Population Size	不详 / No data
种群趋势 Population Trend	下降 / Decreasing

生境与生态系统 Habitat (s) and Ecosystem (s)

生　　境 Habitat(s)	栖于冻原地带及冻原灌丛森林，夏季多在生长有低矮灌丛，尤其是低矮桦树和柳属灌丛的潮湿冻原地带；冬季多在富有柳丛、灌丛和小块森林的沿河地区，有时甚至进入农田地带。以各种柳树、杨树和桦树的嫩枝、嫩叶、花苞和种子及浆果等植物性食物为食 / *L. lagopus* uses shrubs and forests in tundra. It occurs in shrubs with low birches and salix in summer, and occurs in riverside with salix, shrubs and forests or farmlands in winter. *L. lagopus* feeds on twigs, shoots, buds and seeds of salix, poplars, and birches, as well as berries
生态系统 Ecosystem(s)	陆地生态系统 / Terrestrial Ecosystem

威胁 Threat (s)

主要威胁 Major Threat(s)	人为猎捕及潜在的气候变化带来的栖息地丧失 / The major threats are hunting and habitat loss owing to climate change

保护级别与保护行动 Protection Category and Conservation Action (s)

IUCN 红色名录 (2016) IUCN Red List (2016)	无危 / LC
保护行动 Conservation Action(s)	列入国家 II 级重点保护野生动物。部分分布区位于保护区内。需要加强对猎捕的监管，调查种群生存现状，同时评估气候变化的影响 / It has been listed as the second class in National Key Protected Wild Animal List. Some distribution areas fall in nature reserves. The regulation of hunting should be improved, and surveys on population survival and assessment on effect of climate change are suggested

柳雷鸟 *Lagopus lagopus* 　　　　许传辉 摄　By Chuanhui Xu

红喉雉鹑
Tetraophasis obscurus

易危　VU B2ab (i); C2a(i)

| DD 数据缺乏 | LC 无危 | NT 近危 | **VU 易危** | EN 濒危 | CR 极危 | RE 区域灭绝 | EW 野外灭绝 | EX 灭绝 |

分类地位 Taxonomic Status

| 动物界 Animalia | 脊索动物门 Chordata | 鸟纲 Aves | 鸡形目 Galliformes | 雉科 Phasianidae |

学　　名 Scientific Name	*Tetraophasis obscurus*
命　名　人 Species Authority	Verreaux, 1869
英　文　名 English Name(s)	Chestnut-throated Partridge
同物异名 Synonym(s)	无 / None
种下单元评估 Infra-specific Taxa Assessed	无 / None

评估信息 Assessment Information

评 估 年 份 Year Assessed	2016
评　定　人 Assessor(s)	冉江洪 / Jianghong Ran
其他贡献人 Other Contributor(s)	窦亮等 / Liang Dou *et al.*

理由 Justification: 红喉雉鹑分布范围较广，种群数量较少，栖息地受人为干扰严重。因此，列为易危等级 / The distribution area of *Tetraophasis obscurus* is large while the population size is small, and the habitats are under high human disturbance. Thus, it is listed as Vulnerable

评估历史 Assessment History: 《国家重点保护野生动物名录》：Ⅰ (1989，2021)；《中国濒危动物红皮书·鸟类》(1998)：稀有；《中国物种红色名录：第二卷 脊椎动物 下册》(2009)：易危 / *National Key Protected Wild Animal List*: Ⅰ (1989, 2021); *China Red Data Book of Endangered Animals*: *Aves* (1998): R; *China Species Red List*: *Vol. II Vertebrates Part 2* (2009): VU

地理分布 Geographical Distribution

国内分布 Domestic Distribution

分布于四川平武、北川、宝兴、马尔康、小金、汶川、茂县、松潘、九寨沟、红原、若尔盖，甘肃天祝、肃南、临潭、卓尼、康乐、迭部、舟曲、文县，青海祁连、门源、同仁、泽库、班玛和玉树等县 / It occurs in Pingwu, Beichuan, Baoxing, Barkam, Xiaojin, Wenchuan, Maoxian, Songpan, Jiuzhaigou, Hongyuan, Zoigê in Sichuan; Tianzhu, Sunan, Lintan, Zhuoni, Kangle, Diebu, Zhouqu, Wenxian in Gansu; and Qilian, Menyuan, Tongren, Zeku, Banma and Yushu in Qinghai *etc.*

分布标注 Distribution Note

特有种 / Endemic

国内分布图
Map of Domestic Distribution

种群 Population

种群数量 Population Size	野外种群数量不清楚，种群数量预计在下降。在宝兴的调查密度为 7～8 只 /km²，茂县调查的密度为 3～11 只 /km²（文陇英等，2008）/ No clear information on field population size, and it is estimated to be decreasing. Based on the field surveys, the density in Baoxing was 7～8 individuals per km², and 3～11 individuals per km² in Maoxian (Wen *et al.*, 2008)
种群趋势 Population Trend	下降 / Decreasing

生境与生态系统 Habitat (s) and Ecosystem (s)

生境 Habitat(s)	生活在林线海拔 3,000～4,000 m 的多岩山地，常结小群活动于近林线的针叶林至高山杜鹃灌丛、高山草甸和碎石滩地带 / *T. obscurus* occurs in timberline at 3,000 m to rocky mountain areas at 4,000 m. It uses coniferous forests adjacent to the timberline, *Rhododendron* shrubs, alpine meadow, gravel bars in small flocks
生态系统 Ecosystem(s)	陆地生态系统 / Terrestrial Ecosystem

威胁 Threat (s)

主要威胁 Major Threat(s)	栖息地的海拔较高，由于地形、地貌的限制，使种群间不能连续分布。近年来，在亚高山地带人类的经济活动日益频繁，包括各种采集活动和放牧，使其生存环境受到越来越多的威胁和破坏，偷猎事件时有发生 / The habitats are at high altitude, and no consistent distribution of populations owing to the physical geography. Its habitats are destroyed or under threat by the increased intensity of human activities, including collection and foraging. Besides, illegal hunting is also a threat

保护级别与保护行动 Protection Category and Conservation Action (s)

IUCN 红色名录 (2016) IUCN Red List (2016)	无危 / LC
保护行动 Conservation Action(s)	列入国家 I 级重点保护野生动物。四川和甘肃南部的栖息地多已纳入保护区的保护管理中，但尚无以该物种为重点保护对象的保护区 / It has been listed as the first class in National Key Protected Wild Animal List. The habitats in Sichuan and south Gansu are under protection of nature reserves, while no nature reserve was targeted to protect this species

红喉雉鹑 *Tetraophasis obscurus*　　　　　　　　　　张铭 摄　By Ming Zhang

黄喉雉鹑
Tetraophasis szechenyii

易危　VU B2ab(i); C2a(i)

| 数据缺乏 DD | 无危 LC | 近危 NT | 易危 VU | 濒危 EN | 极危 CR | 区域灭绝 RE | 野外灭绝 EW | 灭绝 EX |

分类地位 Taxonomic Status

动物界 Animalia	脊索动物门 Chordata	鸟纲 Aves	鸡形目 Galliformes	雉科 Phasianidae
学　名 Scientific Name		*Tetraophasis szechenyii*		
命 名 人 Species Authority		von Madarász, 1885		
英 文 名 English Name(s)		Buff-throated Partridge		
同物异名 Synonym(s)		无 / None		
种下单元评估 Infra-specific Taxa Assessed		无 / None		

评估信息 Assessment Information

评 估 年 份 Year Assessed	2016
评 定 人 Assessor(s)	冉江洪 / Jianghong Ran
其他贡献人 Other Contributor(s)	窦亮、徐雨 / Liang Dou, Yu Xu

理由 Justification: 黄喉雉鹑分布范围较广，种群数量较少。因此，列为易危等级 / The distribution area of *Tetraophasis szechenyii* is large, while the population size is small. Thus, it is listed as Vulnerable

评估历史 Assessment History:《国家重点保护野生动物名录》：Ⅰ（1989，2021）；《中国濒危动物红皮书·鸟类》（1998）：易危；《中国物种红色名录：第二卷 脊椎动物 下册》（2009）：易危 / *National Key Protected Wild Animal List*: Ⅰ (1989, 2021); *China Red Data Book of Endangered Animals*: *Aves* (1998): V; *China Species Red List*: *Vol. II Vertebrates Part 2* (2009): VU

地理分布 Geographical Distribution

国内分布 Domestic Distribution

分布于四川西部、青海东南部、云南西北部和西藏东南部等 / It occurs in west Sichuan, southeast Qinghai, northwest Yunnan, and southeast Tibet (Xizang) etc.

分布标注 Distribution Note

特有种 / Endemic

国内分布图
Map of Domestic Distribution

种群 Population

种群数量 Population Size　野外种群数量不清楚。在四川雅江县帕姆岭估计密度是 (13.4±0.5) 只/km² (徐雨，2012)。西藏东部估计种群数量是 25,000 ～ 40,000 只 (刘少初和次仁，1993) / No clear information on its wild population size. The density is estimated to be 13.4±0.5 individuals per km² in Pamuling, Yajiang county in Sichuan (Xu, 2012). And the population size is estimated to be 25,000 ～ 40,000 in east Tibet (Xizang) (Liu and Ciren, 1993)

种群趋势 Population Trend　稳定 / Stable

生境与生态系统 Habitat (s) and Ecosystem (s)

生境 Habitat(s)　主要栖息于海拔 3,500 ～ 4,500 m 的针叶林、针阔混交林、高山栎林（灌丛）、杜鹃灌丛以及林线以上的高山草甸地带。可在树上和地面营巢 / *T. szechenyii* uses coniferous forests, mixed coniferous broad-leaved forests, oak forests, Rhododendron shrubs, and grasslands above the timberline. It could nest in the tree or on the ground

生态系统 Ecosystem(s)　陆地生态系统 / Terrestrial Ecosystem

威胁 Threat (s)

主要威胁 Major Threat(s)　主要面临的威胁是栖息地退化和片断化、偷猎和人类活动干扰（采集虫草、松茸等活动）/ The major threats are habitat fragmentation and deterioration, illegal hunting and human activities (Chinese caterpillar fungus and other fungus collection)

保护级别与保护行动 Protection Category and Conservation Action (s)

IUCN 红色名录 (2016) IUCN Red List (2016)　无危 / LC

保护行动 Conservation Action(s)　列入国家 I 级重点保护野生动物。黄喉雉鹑分布区域有许多藏区寺庙及神山，能得到一定的保护。在四川甘孜州，有许多自然保护区保护了它们的栖息地，如四川格西沟、贡嘎山、察青松多、海子山等国家级自然保护区 / It has been listed as the first class in National Key Protected Wild Animal List. There are Tibetan temples and holy mountains in its distribution area, and therefore *T. szechenyii* is under protection to some extent. In Ganzi in Sichuan, quite a few nature reserves (Gexigou, Gongga Shan, Chaqingsongduo, Haizi Shan) were set up for the habitat protection

黄喉雉鹑 *Tetraophasis szechenyii*　　董磊 摄　By Lei Dong

阿尔泰雪鸡
Tetraogallus altaicus

易危 VU B1b(ii,iii); C1

| 数据缺乏 DD | 无危 LC | 近危 NT | 易危 VU | 濒危 EN | 极危 CR | 区域灭绝 RE | 野外灭绝 EW | 灭绝 EX |

分类地位 Taxonomic Status

动物界 Animalia	脊索动物门 Chordata	鸟纲 Aves	鸡形目 Galliformes	雉科 Phasianidae
学名 Scientific Name		*Tetraogallus altaicus*		
命名人 Species Authority		Gebler, 1836		
英文名 English Name(s)		Altai Snowcock		
同物异名 Synonym(s)		无 / None		
种下单元评估 Infra-specific Taxa Assessed		无 / None		

评估信息 Assessment Information

评估年份 Year Assessed	2016
评定人 Assessor(s)	邓文洪 / Wenhong Deng
其他贡献人 Other Contributor(s)	无 / None

理由 Justification：阿尔泰雪鸡仅分布在新疆西北部的狭小区域内，野外种群数量稀少，分布范围小于 20,000 km²，野外种群数量正在下降。推断成熟个体数少于 10,000，并且推测近 10 年种群数量下降超过 10%。因此，列为易危等级 / The distribution area of *Tetraogallus altaicus* is in northwest Xinjiang, which is quite small, with area of 20,000 km². The wild population size is also small and estimated to be less than 10,000 mature individuals. The wild population is undergoing a decreasing. Thus, it is listed as Vulnerable

评估历史 Assessment History：《国家重点保护野生动物名录》：Ⅱ (1989，2021)；《中国物种红色名录：第二卷 脊椎动物 下册》(2009)：未评估 / *National Key Protected Wild Animal List*: Ⅱ (1989, 2021); *China Species Red List: Vol. Ⅱ Vertebrates Part 2* (2009): NA

地理分布 Geographical Distribution

国内分布 Domestic Distribution
仅分布于新疆西北部的北塔山 / It only occurs in Beita Shan in northwest Xinjiang
分布标注 Distribution Note
非特有种 / Non-endemic

国内分布图
Map of Domestic Distribution

种群 Population

种群数量 Population Size	估计我国种群数量不足 10,000 只 / The population size in China is estimated to be less than 10,000 individuals
种群趋势 Population Trend	近年来其种群可能呈下降趋势 / Probably decreasing

生境与生态系统 Habitat(s) and Ecosystem(s)

生　　境 Habitat(s)	常以家族为单位，小群活动于海拔 2,000 m 以上的苔原、草甸和裸岩地带。地方性留鸟 / *T. altaicus* uses tundra, meadow and bare rocky area above 2,000 m with small family flocks. It is local resident
生态系统 Ecosystem(s)	苔原生态系统 / Tundra Ecosystem

威胁 Threat(s)

主要威胁 Major Threat(s)	分布区狭窄，栖息地面积减小，乱捕滥猎严重 / The major threats are small distribution area with habitats decreasing, and the illegal trapping and over hunting

保护级别与保护行动 Protection Category and Conservation Action(s)

IUCN 红色名录 (2016) IUCN Red List (2016)	无危 / LC
保护行动 Conservation Action(s)	列入国家 II 级重点保护野生动物 / It has been listed as the second class in National Key Protected Wild Animal List

阿尔泰雪鸡 *Tetraogallus altaicus* 　　　　许传辉 摄　By Chuanhui Xu

红胸角雉
Tragopan satyra

易危 VU C2a(i)

| 数据缺乏 DD | 无危 LC | 近危 NT | 易危 VU | 濒危 EN | 极危 CR | 区域灭绝 RE | 野外灭绝 EW | 灭绝 EX |

分类地位 Taxonomic Status

动物界 Animalia	脊索动物门 Chordata	鸟纲 Aves	鸡形目 Galliformes	雉科 Phasianidae
学名 Scientific Name		*Tragopan satyra*		
命名人 Species Authority		Linnaeus, 1758		
英文名 English Name(s)		Satyr Tragopan		
同物异名 Synonym(s)		无 / None		
种下单元评估 Infra-specific Taxa Assessed		无 / None		

评估信息 Assessment Information

评估年份 Year Assessed	2016
评定人 Assessor(s)	董路 / Lu Dong
其他贡献人 Other Contributor(s)	张雁云 / Yanyun Zhang

理由 Justification: 红胸角雉的分布区狭窄，种群数量有限，分布区内伐木和开荒造成的栖息地丧失，导致种群数量快速下降。因此，列为易危等级 / The distribution area and population size of *Tragopan satyra* is small. The population is undergoing a rapid decrease owing to the logging and development of wasteland. Thus, it is listed as Vulnerable

评估历史 Assessment History:《国家重点保护野生动物名录》：Ⅰ (1989，2021)；《中国濒危动物红皮书·鸟类》(1998)：稀有；《中国物种红色名录：第二卷 脊椎动物 下册》(2009)：易危 / *National Key Protected Wild Animal List*: I (1989, 2021); *China Red Data Book of Endangered Animals*: Aves (1998): R; *China Species Red List*: Vol. II Vertebrates Part 2 (2009): VU

地理分布 Geographical Distribution

国内分布 Domestic Distribution
仅分布于西藏南部的亚热带森林中 / It only occurs in subtropical forests in south Tibet (Xizang)
分布标注 Distribution Note
非特有种 / Non-endemic

国内分布图
Map of Domestic Distribution

种群 Population

种群数量 Population Size 全球种群数量不超过 20,000 只，我国种群数量不超过 5,000 只 (Madge and McGowan, 2002)，近年在藏南林地有多次影像记录 / The global population size has been estimated as fewer than 20,000 individuals and not over 5,000 individuals in China (Madge and McGowan, 2002). There are several video records in forests in south Tibet (Xizang)

种群趋势 Population Trend 下降 / Decreasing

生境与生态系统 Habitat(s) and Ecosystem(s)

生境 Habitat(s) 主要生活于灌木层丰富的亚热带常绿阔叶林、杜鹃花灌丛及混交林中。栖息地海拔多在 2,200～4,300 m，冬季具垂直迁移，可下降至海拔 1,800 m 左右的区域越冬 / *T. satyra* uses subtropical evergreen broad-leaved forests undergrowth, *Rhododendron* forests and mixed forests, usually between 2,200 m and 4,250 m in the breeding season, sometimes moving down to 1,800 m in winter

生态系统 Ecosystem(s) 陆地生态系统 / Terrestrial Ecosystem

威胁 Threat(s)

主要威胁 Major Threat(s) 主要威胁因子为偷猎，当地人常捕捉本物种作为食物 (Choudhury, 2003)。此外由于林业开发、森林火灾、放牧和薪柴砍伐，导致栖息地衰退和破碎化加剧 (BirdLife International, 2001, Choudhury, 2003) / The major threats include excessive hunting - it is occasionally snared by local people for food (Choudhury, 2003), as well as habitat clearance and degradation due to timber harvesting, unplanned fires, fuelwood collection, and livestock grazing (BirdLife International, 2001; Choudhury, 2003)

保护级别与保护行动 Protection Category and Conservation Action(s)

IUCN 红色名录 (2016) IUCN Red List (2016) 近危 / NT

保护行动 Conservation Action(s) 列入国家 I 级重点保护野生动物。列入国家林业和草原局小种群调查项目，将对种群数量、分布和栖息地开展调查 / It has been listed as the first class in National Key Protected Wild Animal List. It has been listed into project on special surveys of species with small population by National Forestry and Grassland Administration, and its population size, distribution range and habitats will be systematically surveyed

红胸角雉 *Tragopan satyra* 陈林峰 摄 By Linfeng Chen

褐马鸡
Crossoptilon mantchuricum

易危　VU C2a(i)

| 数据缺乏 DD | 无危 LC | 近危 NT | **易危 VU** | 濒危 EN | 极危 CR | 区域灭绝 RE | 野外灭绝 EW | 灭绝 EX |

分类地位 Taxonomic Status

动 物 界 Animalia	脊索动物门 Chordata	鸟　纲 Aves	鸡 形 目 Galliformes	雉　科 Phasianidae
学　名 Scientific Name		*Crossoptilon mantchuricum*		
命 名 人 Species Authority		Swinhoe,1863		
英 文 名 English Name(s)		Brown Eared Pheasant		
同物异名 Synonym(s)		无 / None		
种下单元评估 Infra-specific Taxa Assessed		无 / None		

评估信息 Assessment Information

评 估 年 份 Year Assessed	2016
评 定 人 Assessor(s)	张正旺 / Zhengwang Zhang
其他贡献人 Other Contributor(s)	无 / None

理由 Justification: 褐马鸡分布区仅限于华北地区，而且被分割成3个区域，栖息地破碎化严重。尽管自然保护区的种群数量稳定或有显著增长，但保护区外的褐马鸡种群因为栖息地的破坏和人为干扰，推测整个种群处于下降之中。褐马鸡的遗传多样性很低，具有灭绝的潜在风险。因此，列为易危等级 / *Crossoptilon mantchuricum* is endemic to northern central China, and its habitats are fragmented into three areas. Although the population size in nature reserves are stable or have significantly increased, elsewhere remaining unprotected and isolated populations are declining through on-going habitat loss and human disturbance. The genetic diversity is low that indicates a potential of extinct. Thus, it is listed as Vulnerable

评估历史 Assessment History:《国家重点保护野生动物名录》：Ⅰ（1989，2021）；《中国濒危动物红皮书·鸟类》（1998）：易危；《中国物种红色名录：第二卷 脊椎动物 下册》（2009）：易危 / *National Key Protected Wild Animal List*: Ⅰ (1989, 2021); *China Red Data Book of Endangered Animals: Aves* (1998): V; *China Species Red List: Vol. II Vertebrates Part 2* (2009):VU

地理分布 Geographical Distribution

国内分布 Domestic Distribution
仅分布于河北、北京、山西和陕西中部 / It only occurs in Hebei, Beijing, Shanxi and central Shaanxi
分布标注 Distribution Note
特有种 / Endemic

国内分布图
Map of Domestic Distribution

种群 Population

种群数量 Population Size	种群数量 5,000 ～ 17,000 只 / The population size is 5,000 ～ 17,000 individuals
种群趋势 Population Trend	下降 / Decreasing

生境与生态系统 Habitat(s) and Ecosystem(s)

生　　　境 Habitat(s)	主要栖息在以华北落叶松、云杉次生林为主的林区和华北落叶松、云杉、杨树、桦树次生针阔混交森林中。白天多活动于灌草丛中，夜间栖宿在松树或桦树的大树枝杈上，冬季多活动于海拔 1,000 ～ 1,500 m 的高山地带，夏秋两季多在海拔 1,500 ～ 1,800 m 的山谷、山坡和有清泉的山坳里活动。除繁殖期外，常成群活动，特别是冬季，有时集群多达 30 ～ 50 只。主要在地面活动，尤其喜欢林间空地或林缘草地。活动场所和栖息地较为固定。主要以乔木、灌木和草本植物的叶、嫩茎、幼芽、花蕾、浆果、种子等植物性食物为食，也吃少量动物性食物 / *C. mantchuricum* inhabits mixed coniferous broad-leaved forests with pines, spruces, poplars and birches. It occurs in the shrubs and grassland during days and roosts in the branches of big trees (*e.g.* pines or birches) at night. It uses mountain areas at 1,000 ～ 1,500 m during winters, and uses valleys, and mountainsides, and basins with stream during summer. It is found in flocks, especially in winters, with as many as 30 ～ 50 individuals. It stays on the ground and open areas in forests and grasslands at the edge are preferred. It feeds on leaves, seedling, shoots, buds, berries and seeds of trees, shrubs and grasses, while carnivorous sometimes
生态系统 Ecosystem(s)	陆地生态系统 / Terrestrial Ecosystem

威胁 Threat(s)

主要威胁 Major Threat(s)	分布区狭小、栖息地破碎化以及遗传多样性偏低，是褐马鸡生存所面临的主要问题。调查发现，人类活动的干扰是褐马鸡致危的原因，主要体现在以下几个方面。①对森林资源的过度砍伐曾是导致其分布区面积减少的主要原因。由于人类对森林的长期利用，尤其是明、清以来对华北森林的几次大规模砍伐和破坏，造成森林面积的大幅度减少，从而导致褐马鸡分布区的面积急剧缩小 (何业恒和何文君，1990)。②非法捕猎曾是褐马鸡数量减少甚至在局部区域灭绝的重要原因。在过去许多地方都有一些农民以打猎为生，褐马鸡曾是他们的狩猎对象之一。人类对褐马鸡的捕杀一直持续到 20 世纪 70 年代末、80 年代初。即使是现在，少数地区尤其是在保护区的外围地带，仍有偷猎褐马鸡的现象。此外，用套子、农药狩猎环颈雉时误伤褐马鸡的情形也时有发生。③人类对褐马鸡繁殖的干扰是导致其繁殖成功率低的重要因素。一些农民春季上山挖羊肚菌和野菜期间，对褐马鸡的巢卵损害十分严重，已成为制约褐马鸡野生种群数量发展的重要因素之一 / The main problems are small distribution areas, fragmentated habitats and low genetic diversity. The major threat is human disturbance. ① the distribution areas have been declined due to the deforestation. The massive deforestation in north China eventuated rapid loss of forests and shrink of distribution areas of *C. mantchuricum* (He *et al.*, 1990). ② The population has decreased and even local extinct owing to illegal hunting. *C. mantchuricum* was one of the game and hunting was the main livelihood. The hunting was not stopped till early 1980s, while illegal hunting is still not eradicated among some minorities outside the nature reserves. And *C. mantchuricum* might be hurt by the traps, poisons used for *Phasianus colchicus*. ③ the breeding success is low owing to the human disturbance during breeding. The fungi (*e.g.* toadstool) and wild herbs collection in spring by farmers largely destroyed the nests and eggs, which leads to the population decreasing

保护级别与保护行动 Protection Category and Conservation Action (s)

IUCN 红色名录 (2016) **IUCN Red List (2016)**	易危 / VU C2a (i)
保护行动 **Conservation Action(s)**	列入国家 I 级重点保护野生动物和 CITES 公约附录 I 物种。目前在褐马鸡分布的地区已经建立了 8 个国家级自然保护区，即山西的庞泉沟、芦芽山、五鹿山、黑茶山，河北小五台山，北京百花山，陕西韩城、延安黄龙山国家级自然保护区。在山西五台山开展的褐马鸡再引入工作已经获得初步成功 / It has been listed as the first class in National Key Protected Wild Animal List, and the Appendix I of the CITES. Eight nature reserves have been set up in the three main distributed mountains, including Pangquangou, Luya Shan, Wulu Shan, Heicha Shan in Shanxi, Xiaowutai Shan in Hebei, Baihua Shan in Beijing; Hancheng and Huanglongshan National Nature Reserve in Shaanxi. The initial success of reintroduce in Wutai Shan in Shanxi would be a good start

褐马鸡 *Crossoptilon mantchuricum*　　邓文洪 摄　By Wenhong Deng

白颈长尾雉
Syrmaticus ellioti

易危　VU B2ab(iii); C2a(i)

| 数据缺乏 DD | 无危 LC | 近危 NT | **易危 VU** | 濒危 EN | 极危 CR | 区域灭绝 RE | 野外灭绝 EW | 灭绝 EX |

分类地位 Taxonomic Status

动物界 Animalia	脊索动物门 Chordata	鸟纲 Aves	鸡形目 Galliformes	雉科 Phasianidae
学名 Scientific Name		*Syrmaticus ellioti*		
命名人 Species Authority		Swinhoe, 1872		
英文名 English Name(s)		Elliot's Pheasant		
同物异名 Synonym(s)		无 / None		
种下单元评估 Infra-specific Taxa Assessed		无 / None		

评估信息 Assessment Information

评估年份 Year Assessed	2016
评定人 Assessor(s)	丁平 / Ping Ding
其他贡献人 Other Contributor(s)	无 / None

理由 Justification: 白颈长尾雉野生种群数量总数约为 28,000 只 (马福和张建龙, 2009), 估计少于 10,000 个繁殖对 (Brazil, 2009); 同时面临着栖息地丧失和片断化, 以及非法狩猎和各种人类活动干扰的影响, 使白颈长尾雉有效栖息地减少, 质量持续衰退, 种群数量有减少的趋势 (丁平等, 2000)。因此, 列为易危等级 / The population size of *Syrmaticus ellioti* is estimated to number 28,000 individuals (Ma and Zhang, 2009), and less than 10,000 pairs (Brazil, 2009); habitats loss and fragmentation, illegal hunting and human disturbance has led to the decline of population size (Ding *et al.*, 2007). Thus, it is listed as Vulnerable

评估历史 Assessment History: 《国家重点保护野生动物名录》: Ⅰ (1989, 2021); 中国濒危动物红皮书·鸟类》(1998): 易危; 《中国物种红色名录: 第二卷 脊椎动物 下册》(2009): 易危 / *National Key Protected Wild Animal List*: Ⅰ (1989, 2021); *China Red Data Book of Endangered Animals*: *Aves* (1998): V; *China Species Red List*: *Vol. II Vertebrates Part 2* (2009): VU

地理分布 Geographical Distribution

国内分布 Domestic Distribution

分布于重庆、贵州、湖北、湖南、安徽、江西、浙江、福建、广东和广西等地山区 / It occurs in the mountain areas in Chongqing, Guizhou, Hubei, Hunan, Anhui, Jiangxi, Zhejiang, Fujian, Guangdong and Guangxi *etc.*

分布标注 Distribution Note

特有种 / Endemic

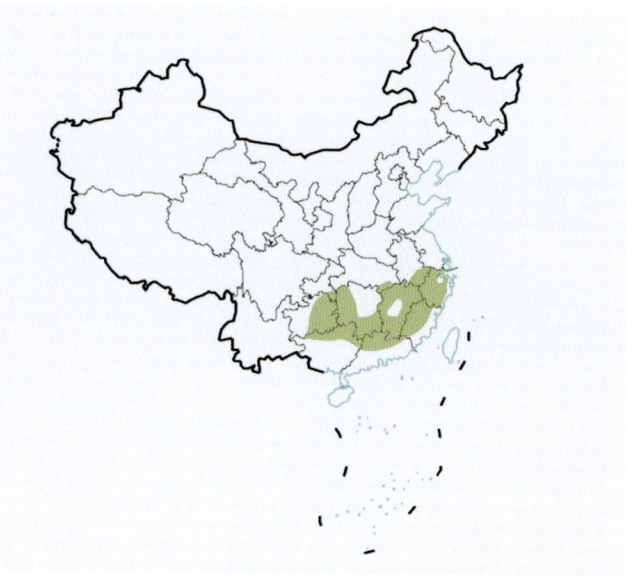

国内分布图
Map of Domestic Distribution

种群 Population

种群数量 Population Size	种群数量不足 10,000 对 / The population size is less than 10,000 pairs
种群趋势 Population Trend	下降 / Decreasing

生境与生态系统 Habitat(s) and Ecosystem(s)

生　　　境 Habitat(s)	多栖于海拔 300～1,500 m 的常绿阔叶林、常绿落叶阔叶混交林、落叶阔叶林、针阔混交林、针叶林、竹林和疏林灌丛等林内，其中阔叶林、混交林为其最适栖息地，针叶林为次适栖息地（丁平和诸葛阳，1989）。其适宜栖息地内乔木层盖度一般在 90% 左右，不低于 50%，且需具有良好的隐蔽条件和夜宿地（丁平等，2002）/ S. ellioti inhabits in evergreen broad-leaved forests, mixed evergreen deciduous broad-leaved forests, deciduous broad-leaved forests, mixed coniferous broad-leaved forests, coniferous forests, bamboo forests and shrubs at 300～1,500 m, among which, broad-leaved forests and mixed forests are the most suitable habitats, after with the coniferous forests (Ding and Zhuge, 1989). Its preferred habitat is forest with a tree cover of more than 90%, no less than 50%, and good for roosting and as a shelter (Ding et al., 2002)
生态系统 Ecosystem(s)	陆地生态系统 / Terrestrial Ecosystem

威胁 Threat(s)

主要威胁 Major Threat(s)	种群密度低，根据 1996～2000 年对浙江省内部分栖息地的调查，种群密度在 0.001～0.035 只 /hm² （丁平等，2002）。由于其主要活动于低海拔山地，因此栖息地破坏与片断化，以及非法狩猎和人类活动的干扰是其面临的最主要胁迫，一些地区的白颈长尾雉的栖息地已缩小（丁平等，2000）/ Based on the habitat survey in Zhejiang, the population density is as low as 0.001～0.035 individual per hm² (Ding et al., 2002). The major threats are habitat destroy and fragmentation, illegal hunting and human disturbance. The habitats have decreased in some areas (Ding et al., 2000)

保护级别与保护行动 Protection Category and Conservation Action(s)

IUCN 红色名录 (2016) IUCN Red List (2016)	近危 / NT
保护行动 Conservation Action(s)	列入国家 I 级重点保护野生动物和 CITES 附录 I 物种。据不完全统计，目前较为确定有白颈长尾雉分布的国家级自然保护区 33 个，其中浙江古田山、贵州雷公山和江西官山等，是以白颈长尾雉为主要保护对象的国家级自然保护区。白颈长尾雉人工饲养和繁殖工作已逐步开展（沈钧和余新华，1988；江惠敏，2005），并已建立了一些圈养种群 / It has been listed as the first class in National Key Protected Wild Animal List and Appendix I of the CITES. There are 33 nature reserves distributed with S. ellioti, among which, Zhejiang Gutian Shan National Nature reserve, Guizhou Leigong Shan National Nature Reserve, Jiangxi Guanshan National Nature Reserve are targeted to conserve the species. The breeding populations in captivity (Shen and Yu, 1988; Jiang, 2005) are ongoing, and some populations in captivity have been developed

白颈长尾雉 *Syrmaticus ellioti*　　龚本亮 摄　By Benliang Gong

黑颈长尾雉
Syrmaticus humiae

易危 VU C1+2a(i)

| 数据缺乏 DD | 无危 LC | 近危 NT | 易危 VU | 濒危 EN | 极危 CR | 区域灭绝 RE | 野外灭绝 EW | 灭绝 EX |

分类地位 Taxonomic Status

动物界 Animalia	脊索动物门 Chordata	鸟纲 Aves	鸡形目 Galliformes	雉科 Phasianidae
学名 Scientific Name		*Syrmaticus humiae*		
命名人 Species Authority		Hume, 1881		
英文名 English Name(s)		Hume's Pheasant		
同物异名 Synonym(s)		无 / None		
种下单元评估 Infra-specific Taxa Assessed		无 / None		

评估信息 Assessment Information

评估年份 Year Assessed	2016
评定人 Assessor(s)	张正旺 / Zhengwang Zhang
其他贡献人 Other Contributor(s)	无 / None

理由 Justification: 中国是黑颈长尾雉的主要分布区，目前在其分布的广西和云南都存在栖息地丧失和片断化问题，存在比较严重的盗猎现象，导致其种群数量下降显著。栖息地还在持续丧失，其种群处于下降之中，需要进行有效保护才能防止出现局域性灭绝。因此，列为易危等级 / China is the main distribution area of *Syrmaticus humiae*, and the population is undergoing a rapid decline owing to habitat loss and fragmentation in Guangxi and Yunnan, as well as the illegal hunting. Effective protection is needed to prevent local extinction. Thus, it is listed as Vulnerable

评估历史 Assessment History: 《国家重点保护野生动物名录》：Ⅰ (1989，2021)；《中国濒危动物红皮书·鸟类》(1998)：濒危；《中国物种红色名录：第二卷 脊椎动物 下册》(2009)：易危 / *National Key Protected Wild Animal List*: I (1989, 2021); *China Red Data Book of Endangered Animals*: *Aves* (1998): E; *China Species Red List*: Vol. II Vertebrates Part 2 (2009): VU

地理分布 Geographical Distribution

国内分布 Domestic Distribution
分布于云南西部、贵州西南部、广西西部 / It occurs in west Yunnan, southwest Guizhou and west Guangxi
分布标注 Distribution Note
非特有种 / Non-endemic

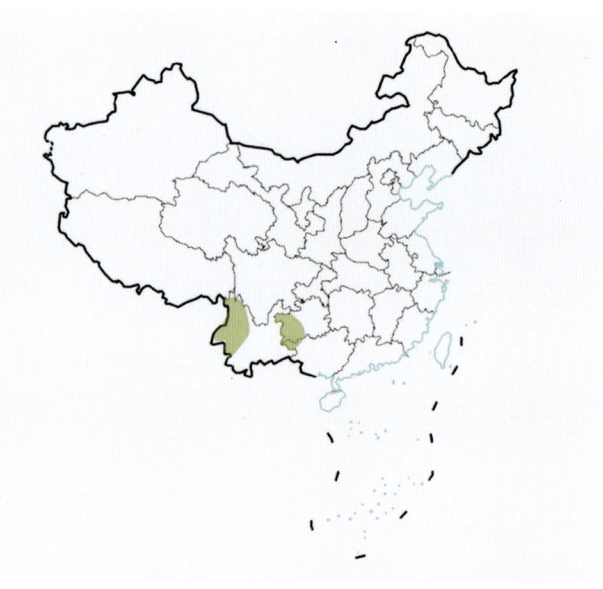

国内分布图
Map of Domestic Distribution

种群 Population

种群数量 Population Size	种群数量约 2,000 只 / The population size is about 2,000 individuals
种群趋势 Population Trend	下降 / Decreasing

生境与生态系统 Habitat(s) and Ecosystem(s)

生境 Habitat(s)	主要栖息于海拔 500～3,000 m 的阔叶林、针阔叶混交林以及疏林灌丛、草地和林缘地带，尤其喜欢在海拔 1,000～2,000 m 的林下蕨类、蒿草和灌丛植物发达而又多岩石的山坡混交疏林和林缘地带活动。常成对或小群游荡觅食。杂食性，主要以橡实、浆果、种子、根、嫩叶、幼芽等植物性食物为食，也吃昆虫等动物性食物，偶尔也到林缘耕地啄食农作物 / S. humiae inhabits the broad-leaved forests, mixed coniferous broad-leaved forests, and shrubs, meadows and areas at edge of forests at 500～3,000 m, especially mixed forests with ferns, wormwood, shrubs at rocky mountainside and forest edges at 1,000～2,000 m. It appears in pair or small flock. It is omnivorous, fed on acorns, berries, seeds, roots, leaves, buds, insects, or cultivated vegetation at the farmlands
生态系统 Ecosystem(s)	陆地生态系统 / Terrestrial Ecosystem

威胁 Threat(s)

主要威胁 Major Threat(s)	主要是盗猎和栖息地丧失对种群的影响较大 / Illegal hunting and habitat loss

保护级别与保护行动 Protection Category and Conservation Action(s)

IUCN 红色名录 (2016) IUCN Red List (2016)	近危 / NT
保护行动 Conservation Action(s)	列入国家 I 级重点保护野生动物和 CITES 公约附录 I 物种。在其分布区已经建立了多个自然保护区。今后应该进一步加强栖息地的保护和种群的监测 / It has been listed as the first class in National Key Protected Wild Animal List and the Appendix I of the CITES. There are several nature reserves in its distribution area. Habitat protection and monitoring on population are suggested to be improved

黑颈长尾雉 *Syrmaticus humiae* 　　廖国庆 摄　By Guoqing Liao

栗树鸭
Dendrocygna javanica
易危 VU D1

| 数据缺乏 DD | 无危 LC | 近危 NT | **易危 VU** | 濒危 EN | 极危 CR | 区域灭绝 RE | 野外灭绝 EW | 灭绝 EX |

分类地位 Taxonomic Status

动物界 Animalia	脊索动物门 Chordata	鸟纲 Aves	雁形目 Anseriformes	鸭科 Anatidae

学名 Scientific Name	*Dendrocygna javanica*
命名人 Species Authority	Horsfield, 1821
英文名 English Name(s)	Lesser Whistling Duck
同物异名 Synonym(s)	*Anas javanica*
种下单元评估 Infra-specific Taxa Assessed	无 / None

评估信息 Assessment Information

评估年份 Year Assessed	2016
评定人 Assessor(s)	梁伟 / Wei Liang
其他贡献人 Other Contributor(s)	杨灿朝 / Canchao Yang

理由 Justification: 由于栖息地的丧失和非法捕猎，栗树鸭在中国的种群数量持续下降。因此，列为易危等级 / The population of *Dendrocygna javanica* in China is undergoing a rapid decline owing to the habitat loss and illegal hunting. Thus, it is listed as Vulnerable

评估历史 Assessment History: 《国家重点保护野生动物名录》：Ⅱ (2021)；《中国濒危动物红皮书·鸟类》(1998)：易危；《中国物种红色名录：第二卷 脊椎动物 下册》(2009)：易危 / *National Key Protected Wild Animal List*: Ⅱ (2021); *China Red Data Book of Endangered Animals*: *Aves* (1998): V; *China Species Red List*: *Vol. II Vertebrates Part 2* (2009): VU

地理分布 Geographical Distribution

国内分布 Domestic Distribution

分布于云南西部和南部、江西、江苏、福建、广东、香港、广西、海南、台湾 / It occurs in west and south Yunnan, Jiangxi, Jiangsu, Fujian, Guangdong, Hong Kong, Guangxi, Hainan and Taiwan

分布标注 Distribution Note

非特有种 / Non-endemic

国内分布图
Map of Domestic Distribution

种群 Population

种群数量 Population Size	估计我国种群数量不足 100 繁殖对 (Brazil, 2009) / The population size in China is estimated to be less than 100 pairs (Brazil, 2009)
种群趋势 Population Trend	下降 / Decreasing

生境与生态系统 Habitat (s) and Ecosystem (s)

生境 Habitat(s)	湖泊、沼泽、红树林及稻田 / Lakes, swamps, mangroves, and paddy fields
生态系统 Ecosystem(s)	湿地生态系统 / Wetland Ecosystem

威胁 Threat (s)

主要威胁 Major Threat(s)	栖息地丧失、人为干扰和非法捕猎 / Habitat loss, human disturbance and illegal hunting

保护级别与保护行动 Protection Category and Conservation Action (s)

IUCN 红色名录 (2016) IUCN Red List (2016)	无危 / LC
保护行动 Conservation Action(s)	列入国家 II 级重点保护野生动物、《国家保护的有益的或者有重要经济、科学研究价值的陆生野生动物名录》。减少人为干扰，打击非法捕猎活动；加大对其适宜栖息地的保护 / It has been listed as the second class in National Key Protected Wild Animal List, *National Protected List of Terrestrial Wild Animals with Good Benefits or Important Economic and Scientific Values*. To minimize the human disturbance; strengthen the enforcement against illegal hunting; improve the habitat protection

栗树鸭 *Dendrocygna javanica*　　　　　　　　　　　　　　卢刚 摄　By Gang Lu

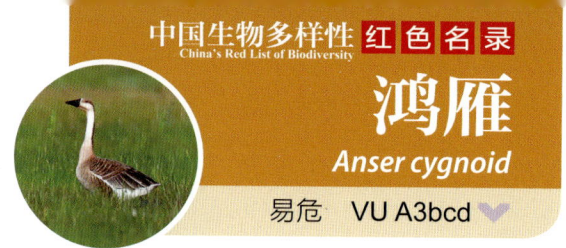

鸿雁
Anser cygnoid
易危 VU A3bcd

| 数据缺乏 DD | 无危 LC | 近危 NT | 易危 VU | 濒危 EN | 极危 CR | 区域灭绝 RE | 野外灭绝 EW | 灭绝 EX |

🕊 分类地位 Taxonomic Status

动物界 Animalia	脊索动物门 Chordata	鸟纲 Aves	雁形目 Anseriformes	鸭科 Anatidae

学 名 Scientific Name	*Anser cygnoid*
命 名 人 Species Authority	Linnaeus, 1758
英 文 名 English Name(s)	Swan Goose
同物异名 Synonym(s)	无 / None
种下单元评估 Infra-specific Taxa Assessed	无 / None

🕊 评估信息 Assessment Information

评 估 年 份 Year Assessed	2016
评 定 人 Assessor(s)	丁长青 / Changqing Ding
其他贡献人 Other Contributor(s)	刘垚 / Yao Liu

理由 Justification: 在过去10年间，由于栖息地破坏，食物资源减少，鸿雁主要越冬地之一的升金湖越冬种群下降超过90%，长江中下游其他越冬区同样面临着栖息地破坏的威胁，若不及时对湿地进行保护，其种群在未来可能会急剧下降。因此，列为易危等级 / The population size of *Anser cygnoid* at Shengjin Lake has decreased by 90% in the past ten years owing to the habitat loss and food shortage. Other overwintering areas in the middle and lower reaches of Changjiang River are also threatened with habitat destruction. If we do not protect wetlands in time, it may decline sharply in the future. Thus, it is listed as Vulnerable

评估历史 Assessment History: 《国家重点保护野生动物名录》：II (2021)；《中国物种红色名录：第二卷 脊椎动物 下册》(2009)：易危 / National Key Protected Wild Animal List: II (2021); China Species Red List: Vol. II Vertebrates Part 2 (2009): VU

🕊 地理分布 Geographical Distribution

国内分布 Domestic Distribution

主要繁殖于黑龙江、吉林和内蒙古，越冬在长江中下游和山东、江苏、福建、广东等沿海省份，偶见于台湾。迁徙时见于新疆北部阿尔泰山脉、西部天山、青海柴达木盆地、河北、河南等地 / It breeds in Heilongjiang, Jilin and Inner Mongolia (Nei Mongol), winters at the middle and lower Changjiang River and Shandong, Jiangsu, Fujian, Guangdong, seldom in Taiwan. It is seen in Altay Shan in north Xinjiang, Tian Shan in west Xinjiang, Qaidam Basin in Qinghai and Hebei, Henan *etc.* during migration

分布标注 Distribution Note

非特有种 / Non-endemic

国内分布图
Map of Domestic Distribution

种群 Population

种群数量 Population Size

全球种群数量约 90,000 只，大部分在中国越冬。2001 年在沙湖（鄱阳湖部分区域）发现 61,650 只个体，2004/2005 年冬季在长江中下游流域记录 61,000 只个体 (Cao et al., 2008)。2005～2008 年在鄱阳湖、升金湖等地记录 76,000～92,000 只个体 (Cao et al., 2010)。2011 年 1 月在长江流域记录 87,544 只个体 (IUCN, 2012)。安徽越冬种群下降趋势明显：2003～2006 年在升金湖有 10,000～20,000 只鸿雁，至 2008/2009 年冬季下降至 1,000 只。可能是因为水生植物的破坏导致食物资源减少，迁移到鄱阳湖等处越冬 (Zhang et al., 2011) / The global population size is estimated to be 90,000 individuals and most winters in China. 61,650 individuals were recorded in Shahu (sub-lake of Poyang Lake) in 2001 (Zhao, 2002); 61,000 individuals were recorded in middle and lower Changjiang River in winter of 2004 (Cao et al., 2008); 76,000～92,000 individuals were recorded in Poyang Lake, Shengjin Lake and Anhui Lakes cluster (Cao et al., 2010); and 87,544 individuals were recorded in Changjiang River in January in 2011 (IUCN, 2012). The decreasing trend in Anhui is significant: the population was 10,000～20,000 in winter of 2003～2006, while the number was 1,000 in winter of 2008. The reason might be the destroy of hydrophyte and food resource reduced, and most moved to other lakes, i.e. Poyang Lake (Zhang et al., 2011)

种群趋势 Population Trend

下降 / Decreasing

生境与生态系统 Habitat(s) and Ecosystem(s)

生境 Habitat(s)

繁殖于湿地草原和森林草原区，包括河流三角洲、河流山谷草地、半咸水和淡水湖泊的边缘，以及山区水流湍急的河流两侧。冬季在低地沼泽湖畔、稻田、河口和滩涂越冬 / *A. cygnoid* breeds in wetlands in the steppe and forest-steppe zones, including river deltas, river valleys with meadows, the margins of brackish and freshwater lakes, and in mountainous areas along narrow, fast-flowing rivers. In winter, it occurs in lowland lakeside marshes, rice-fields, estuaries and tidal flats

生态系统 Ecosystem(s)

湿地生态系统 / Wetland Ecosystem

威胁 Threat(s)

主要威胁 Major Threat(s)

繁殖地农业发展造成的湿地破坏与干扰加重，三江平原拾取鸟蛋以及栖息地破坏导致繁殖种群的下降 (IUCN, 2012)；非法捕猎 (Poyarkov, 2005)；越冬地食物资源破坏，三峡大坝影响长江中下游的水位，可能会加剧沉水植物的减少，使鸿雁食物资源减少 (Zhang, 2011)；高度密集的越冬群体增加了禽流感暴发的可能性 (Batbayar et al., 2011) / Habitat loss and increased disturbance owing to agricultural development, coupled with egg collection on Sanjiang plain, has probably resulted in a decline in the numbers of breeding population there (IUCN, 2012). Other threats include: illegal hunting (Poyarkov, 2005); the change of water level of middle and lower Changjiang River owing to the regulation of the Three Gorges Dam, which might induce reduced hydrophyte and food shortage in wintering sites (Zhang, 2011); The species was more concentrated at fewer key wintering localities, which increased the probability of influenza (Batbayar et al., 2011)

保护级别与保护行动 Protection Category and Conservation Action(s)

IUCN 红色名录 (2016) IUCN Red List (2016)

易危 / VU A2bcd+3bcd+4bcd

保护行动 Conservation Action(s)

列入国家 II 级重点保护野生动物、《国家保护的有益的或者有重要经济、科学研究价值的陆生野生动物名录》和迁徙物种保护公约 (CMS) 附录 I，停止对其活动区湿地的开发利用，禁止捕猎，加强对种群和栖息地环境的保护；开展对鸿雁分布的调查，了解其适宜生境及食物资源利用情况；加强沉水植物等食物资源的保护 / It has been listed as the second class in National Key Protected Wild Animal List, *National Protected List of Terrestrial Wild Animals with Good Benefits or Important Economic and Scientific Values*, and listed into Appendix I in the Convention on Migratory Species (CMS). Prevention of the development of wetlands at distribution areas, prohibition of hunting, conservation of the species and protection of habitats, surveys on its distribution areas to get knowledge on suitable habitats and food resources, protection of hydrophyte are suggested

鸿雁 *Anser cygnoid* 　　　　　　陈勇 摄　By Yong Chen

小白额雁
Anser erythropus

易危 VU A2abcd

| 数据缺乏 DD | 无危 LC | 近危 NT | 易危 VU | 濒危 EN | 极危 CR | 区域灭绝 RE | 野外灭绝 EW | 灭绝 EX |

分类地位 Taxonomic Status

动物界 Animalia	脊索动物门 Chordata	鸟纲 Aves	雁形目 Anseriformes	鸭科 Anatidae
学 名 Scientific Name		*Anser erythropus*		
命 名 人 Species Authority		Linnaeus, 1758		
英 文 名 English Name(s)		Lesser White-fronted Goose		
同物异名 Synonym(s)		*Anas erythropus*		
种下单元评估 Infra-specific Taxa Assessed		无 / None		

评估信息 Assessment Information

评 估 年 份 Year Assessed	2016
评 定 人 Assessor(s)	丁长青 / Changqing Ding
其他贡献人 Other Contributor(s)	刘垚 / Yao Liu

理由 Justification: 随着栖息地的破坏，小白额雁在中国的越冬地减少，在过去10年内种群数量约减少30%，目前集中于东洞庭湖。因此，列为易危等级 / The population size of *Anser erythropus* has decreased by 30% in 10 years owing to the habitat loss. It is concentrated at East Dongting Lake presently. Thus, it is listed as Vulnerable

评估历史 Assessment History: 《国家重点保护野生动物名录》：Ⅱ (2021)；《中国物种红色名录：第二卷 脊椎动物 下册》(2009)：易危 / *National Key Protected Wild Animal List*: Ⅱ (2021); *China Species Red List: Vol. Ⅱ Vertebrates Part 2* (2009): VU

地理分布 Geographical Distribution

国内分布 Domestic Distribution

越冬于中国长江中下游，以及广东、福建、台湾等东南沿海地区。迁徙时经过中国东北、新疆、内蒙古、河北、山东、河南等地 / It winters at middle and lower Changjiang River, and southeast coast of Guangdong, Fujiang, Taiwan *etc*. It is seen in northeast China, Xinjiang, Inner Mongolia (Nei Mongol), Hebei, Shandong, Henan *etc*. during migration

分布标注 Distribution Note

非特有种 / Non-endemic

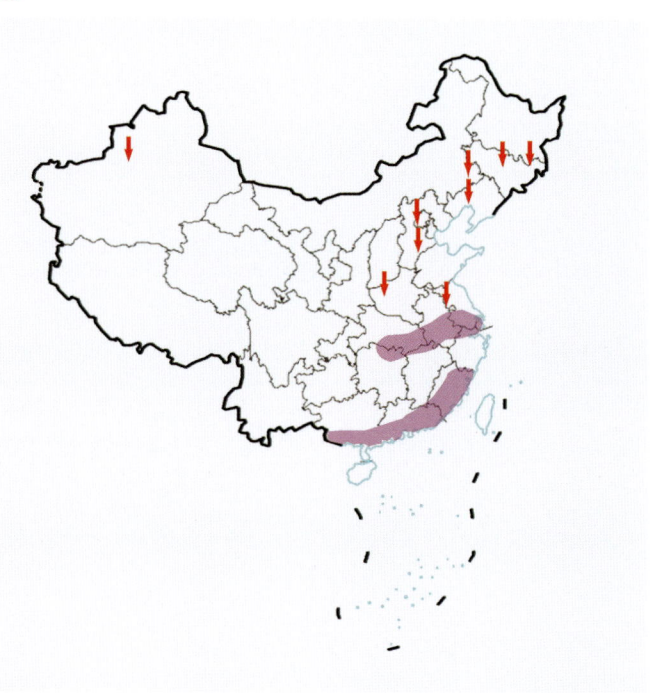

国内分布图
Map of Domestic Distribution

种群 Population

种群数量 Population Size 估计全球种群数量 35,000～49,000 只 (Wetlands International，2014)。我国 2002～2008 年记录长江中下游越冬小白额雁 15,000～19,000 只，2008～2010 年记录 12,600～13,000 只 (王鑫，2013) / The global population size is estimated to be 35,000～49,000 individuals (Wetlands International, 2014). 15,000～19,000 individuals were recorded in middle and lower Changjiang River in 2002～2008, and 12,600～13,000 individuals were recorded in 2008～2010 (Wang, 2013)

种群趋势 Population Trend 下降 / Decreasing

生境与生态系统 Habitat(s) and Ecosystem(s)

生境 Habitat(s) 东部种群在中国长江中下游越冬，活动于开阔少树地区的湿地、沼泽、半干旱盐田和草地，有时到农田觅食；夜宿于苇塘、灌丛或河湖的岸上 / The eastern population of *A. erythropus* winters in middle and lower Changjiang River, uses the open wetlands, marshes, semi-arid salt fields and grasslands with few trees, seldom forages in farmlands. It roosts in reed ponds, shrubs or banks of rivers or lakes

生态系统 Ecosystem(s) 湿地生态系统 / Wetland Ecosystem

威胁 Threat(s)

主要威胁 Major Threat(s) 水文变化、三峡大坝运行引起的下游泥沙沉积、非法捕猎以及日益严重的环境污染 / The change of hydrology, the sediment deduced by the Three Gorge Dam, illegal hunting and environmental pollution

保护级别与保护行动 Protection Category and Conservation Action(s)

IUCN 红色名录 (2016) IUCN Red List (2016) 易危 / VU A2bcd+3bcd+4bcd

保护行动 Conservation Action(s) 列入国家 II 级重点保护野生动物、《国家保护的有益的或者有重要经济、科学研究价值的陆生野生动物名录》和迁徙物种保护公约 (CMS) 附录 I。建议停止对湿地的开发利用，禁止捕猎，加强对种群和栖息地环境的保护。调查迁徙路线，了解中国种群的繁殖地，保护重要停歇地 / It has been listed as the second class in National Key Protected Wild Animal List, *National Protected List of Terrestrial Wild Animals with Good Benefits or Important Economic and Scientific Values* and Appendix I in the Convention on Migratory Species (CMS). The prevention of development of wetlands, prohibition of hunting, conservation of the population and protection of habitats are suggested. It is also recommended to survey the migration route, breeding sites, and the stopovers

小白额雁 *Anser erythropus* 包秀奇 摄 By Xiuqi Bao

红顶绿鸠
Treron formosae

易危 VU C1

| 数据缺乏 DD | 无危 LC | 近危 NT | 易危 VU | 濒危 EN | 极危 CR | 区域灭绝 RE | 野外灭绝 EW | 灭绝 EX |

分类地位 Taxonomic Status

动物界 Animalia	脊索动物门 Chordata	鸟纲 Aves	鸽形目 Columbiformes	鸠鸽科 Columbidae
学名 Scientific Name		*Treron formosae*		
命名人 Species Authority		Swinhoe, 1863		
英文名 English Name(s)		Whistling Green-pigeon		
同物异名 Synonym(s)		*Sphenurus formosae*		
种下单元评估 Infra-specific Taxa Assessed		无 / None		

评估信息 Assessment Information

评估年份 Year Assessed	2016
评定人 Assessor(s)	刘阳 / Yang Liu
其他贡献人 Other Contributor(s)	黄秦 / Qin Huang

理由 Justification: 红顶绿鸠分布区狭窄并由于栖息地原始林丧失或退化，种群受到威胁。因此，列为易危等级 / With small distribution range, the population of *Treron formosae* is threatened by habitat loss and degradation. Thus, it is listed as Vulnerable

评估历史 Assessment History:《国家重点保护野生动物名录》：II (1989, 2021);《中国濒危动物红皮书·鸟类》(1998): 未定;《中国物种红色名录: 第二卷 脊椎动物 下册》(2009): 无危 / *National Key Protected Wild Animal List*: II (1989, 2021); *China Red Data Book of Endangered Animals*: Aves (1998): I; *China Species Red List*: *Vol. II Vertebrates Part 2* (2009): LC

地理分布 Geographical Distribution

国内分布 Domestic Distribution

分布于台湾南部花莲、高雄、屏东、台东，以及兰屿和绿岛 / It occurs in Hualian, Gaoxiong, Pingdong, Taidong, Lanyu and Green Island in south Taiwan

分布标注 Distribution Note

非特有种 / Non-endemic

国内分布图
Map of Domestic Distribution

种群 Population

种群数量 Population Size	种群数量较少，近年在台湾屏东发现数量约 100 只的种群 / The population size is small. In recent years, a population of about 100 individuals has been found in Pingtung, Taiwan
种群趋势 Population Trend	下降 / Decreasing

生境与生态系统 Habitat(s) and Ecosystem(s)

生　　境 Habitat(s)	栖于热带及亚热带小型岛屿的浓密阴暗的原始阔叶林中 / *T. formosae* inhabits in dense and dark primeval broad-leaved forests in tropical and subtropical small islands
生态系统 Ecosystem(s)	陆地生态系统 / Terrestrial Ecosystem

威胁 Threat(s)

主要威胁 Major Threat(s)	栖息地的丧失和退化 / Habitat loss and degradation

保护级别与保护行动 Protection Category and Conservation Action(s)

IUCN 红色名录 (2016) IUCN Red List (2016)	近危 / NT
保护行动 Conservation Action(s)	列入国家 II 级重点保护野生动物和台湾地区第二类珍贵稀有的保育类动物，受到保护 / It has been listed as the second level in National Key Protected Wild Animal List, and II rare and precious conservation animal in Taiwan, and is under protection

红顶绿鸠 *Treron formosae*　　董文晓 摄　By Wenxiao Dong

花田鸡
Coturnicops exquisitus

中国生物多样性红色名录

易危 VU A2cd+3cd; C1

| 数据缺乏 DD | 无危 LC | 近危 NT | **易危 VU** | 濒危 EN | 极危 CR | 区域灭绝 RE | 野外灭绝 EW | 灭绝 EX |

分类地位 Taxonomic Status

动物界 Animalia	脊索动物门 Chordata	鸟纲 Aves	鹤形目 Gruiformes	秧鸡科 Rallidae
学名 Scientific Name		*Coturnicops exquisitus*		
命名人 Species Authority		Swinhoe, 1873		
英文名 English Name(s)		Swinhoe's Rail		
同物异名 Synonym(s)		无 / None		
种下单元评估 Infra-specific Taxa Assessed		无 / None		

评估信息 Assessment Information

评估年份 Year Assessed	2016
评定人 Assessor(s)	陈水华 / Shuihua Chen
其他贡献人 Other Contributor(s)	无 / None

理由 Justification: 花田鸡分布区狭窄，由于在繁殖区和越冬区栖息地的丧失和片断化，种群数量在持续下降中。据推测在过去10年，乃至未来10年种群数量下降的速度至少超过10%（Brazil，2009）。因此，列为易危等级 / The distribution range of *Coturnicops exquisitus* is small and population size has been undergoing a continuing decline owing to breeding and wintering habitat loss and fragmentation. It is estimated that the decrease rate could be over 10% in the past 10 years and even in the future 10 years (Brazil, 2009). Thus, it is listed as Vulnerable

评估历史 Assessment History: 《国家重点保护野生动物名录》：Ⅱ（1989，2021）；《中国物种红色名录：第二卷 脊椎动物 下册》(2009)：易危 / *National Key Protected Wild Animal List*: Ⅱ (1989, 2021); *China Species Red List*: *Vol. II Vertebrates Part 2* (2009): VU

地理分布 Geographical Distribution

国内分布 Domestic Distribution

在内蒙古东北部、黑龙江和吉林部分区域繁殖，在福建、广东等地越冬。迁徙经过辽宁、河北、山东、湖北、湖南、安徽、四川、云南、江苏、江西、上海等地 / It breeds in northeast Inner Mongolia (Nei Mongol), Heilongjiang, Jilin, and winters in Fujian, Guangdong, *etc*. It occurs in Liaoning, Hebei, Shandong, Hubei, Hunan, Anhui, Sichuan, Yunnan, Jiangsu, Jiangxi, and Shanghai *etc.* during migration

分布标注 Distribution Note

非特有种 / Non-endemic

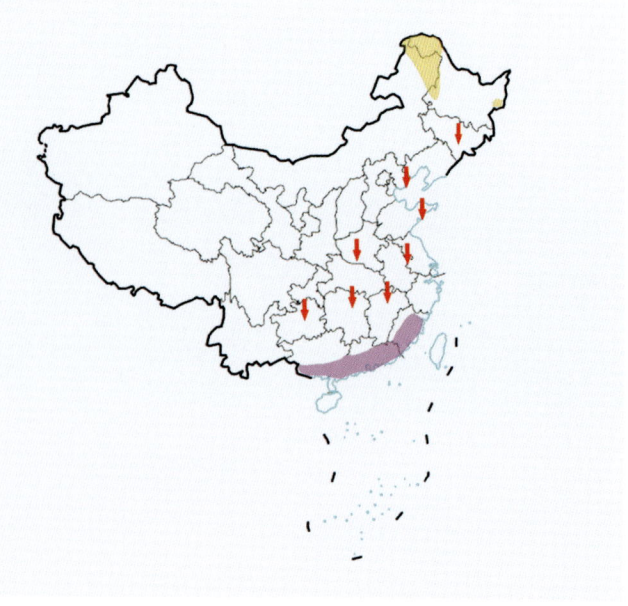

国内分布图
Map of Domestic Distribution

种群 Population

种群数量 Population Size	种群数量 2,500～9,999 只 / The population size is 2,500～9,999 individuals
种群趋势 Population Trend	下降 / Decreasing

生境与生态系统 Habitat (s) and Ecosystem (s)

生　　境 Habitat(s)	栖息于低山丘陵和林缘地带的水稻田、溪流、沼泽、草地、苇塘及其附近的草丛与灌丛中，有时也出现在林中草地和河流两岸的沼泽及草地上 / C. exquisitus inhabits in rice fields, streams, marshes, grasslands, reed ponds and close shrubs in hilly areas or edges of forests, sometimes occurs in grasslands in forests and marshes and grasslands along the rivers
生态系统 Ecosystem(s)	湿地生态系统 / Wetland Ecosystem

威胁 Threat (s)

主要威胁 Major Threat(s)	在繁殖区和越冬区，由于割草、耕作和开垦等农业生产导致的栖息地湿地破坏和减少，其他威胁包括筑坝、猎捕和杀虫剂的使用等 / The major threats are thought to be from the destruction and modification of wetlands in both its breeding and wintering ranges owing to agriculture activities, such as mowing, cultivation and reclamation. Besides, dams, hunting and pesticide also bring stress to the species

保护级别与保护行动 Protection Category and Conservation Action (s)

IUCN 红色名录 (2016) IUCN Red List (2016)	易危 / VU C2a (ii)
保护行动 Conservation Action(s)	列入国家 II 级重点保护野生动物。在主要繁殖区已建立了自然保护区，如黑龙江扎龙国家级自然保护区、兴凯湖国家级自然保护区等 / It has been listed as the second class in National Key Protected Wild Animal List. Some nature reserves have been set up in its main breeding sites, *e.g.* Heilongjiang Zhalong National Nature Reserve, Xingkai Lake National Nature Reserve, etc

花田鸡 *Coturnicops exquisitus*　　　　　　　　　　　　　　　　　王吉衣 摄　By Jiyi Wang

白喉斑秧鸡
Rallina eurizonoides

中国生物多样性红色名录

易危 VU A1cde

| 数据缺乏 DD | 无危 LC | 近危 NT | **易危 VU** | 濒危 EN | 极危 CR | 区域灭绝 RE | 野外灭绝 EW | 灭绝 EX |

分类地位 Taxonomic Status

动物界 Animalia	脊索动物门 Chordata	鸟纲 Aves	鹤形目 Gruiformes	秧鸡科 Rallidae

学名 Scientific Name	*Rallina eurizonoides*
命名人 Species Authority	Lafresnaye, 1845
英文名 English Name(s)	Slaty-legged Crake
同物异名 Synonym(s)	*Gallinula eurizonoïdes*
种下单元评估 Infra-specific Taxa Assessed	无 / None

评估信息 Assessment Information

评估年份 Year Assessed	2016
评定人 Assessor(s)	周放 / Fang Zhou
其他贡献人 Other Contributor(s)	无 / None

理由 Justification: 中国对白喉斑秧鸡一直缺乏关注，未见系统的种群数量和研究报道（王岐山等，2006）。1990年的中国水鸟统计时曾见到15只，近几年在野外的单次观察记录多为1～2只。台湾亚种*R. e. formosana*为中国特有亚种，亦缺乏相关的数量统计资料（刘小如等，2010）。近年来随着栖息地面积的快速缩减和水污染导致的湿地生境质量下降，在许多原有分布地已很难观察到其踪迹，据此推测该鸟种群数量随生境质量退化而下降。因此，列为易危等级 / Little attention has been paid to *Rallina eurizonoides* and there is no systematic study or report on its population size (Wang *et al.*, 2006). 15 individuals were recorded in the national waterbird census in 1990, and only one or two individuals were recorded in the wild in recent years. *R. e. formosana* is an endemic subspecies of China, while data of its population size is also deficient (Liu *et al.*, 2010). No individual has been found in its previous sites probably as result of habitat loss and wetland deterioration by water pollution. It is presumed that the population is undergoing a decline with the habitat quality. Thus, it is listed as Vulnerable

评估历史 Assessment History: 《中国濒危动物红皮书·鸟类》(1998)：未定；《中国物种红色名录：第二卷 脊椎动物 下册》(2009)：无危 / *China Red Data Book of Endangered Animals*: *Aves* (1998): I; *China Species Red List*: *Vol. II Vertebrates Part 2* (2009): LC

地理分布 Geographical Distribution

国内分布 Domestic Distribution

主要分布于南部的广西、海南、台湾、香港，偶见于湖南、河南，广东也有少量分布。江西、江苏、云南、贵州有零星分布记录 / It occurs in Guangxi, Hainan, Taiwan, Hong Kong in south China, occasionally in Hunan and Henan, and only a few in Guangdong. There are also scattered records in Jiangxi, Jiangsu, Yunnan, and Guizhou

分布标注 Distribution Note

非特有种 / Non-endemic

国内分布图
Map of Domestic Distribution

🐦 种群 Population

种群数量 Population Size	种群数量不足 10,000 只 / The population size is less than 10,000 individuals
种群趋势 Population Trend	下降 / Decreasing

🐦 生境与生态系统 Habitat (s) and Ecosystem (s)

生　　境 Habitat(s)	潮湿沼泽地、红树林、水稻田、灌丛、林缘地带等处 / Wet marshes, mangroves, rice fields, shrubs, areas at the edge of forest
生态系统 Ecosystem(s)	湿地生态系统 / Wetland Ecosystem

🐦 威胁 Threat (s)

主要威胁 Major Threat(s)	农药等化学制剂的滥用、沿海和内陆的湿地面积减少以及生境质量下降、水产养殖和农业种植等人类活动的加剧使得其适宜的生存环境不断遭受

干扰和破坏 / Major threats are habitat loss owing to the abuse of pesticide, decline and deterioration of inland and coastal wetlands, increased human disturbance of aquatic agriculture and plantation

🐦 保护级别与保护行动 Protection Category and Conservation Action (s)

IUCN 红色名录 (2016) IUCN Red List (2016)	无危 / LC
保护行动 Conservation Action(s)	列入《国家保护的有益的或者有重要经济、科学研究价值的陆生野生动物名录》。尚未被列为国家重点保护野生动物，建议增列为国家 II 级重点

保护野生动物。应对该种的分布地进行专项调查，同时保护好现有湿地及周边植被，特别是现有的大面积的适宜栖息地。加强对环境污染行为，尤其是对污染水体行为的执法力度 / It has been listed as *National Protected List of Terrestrial Wild Animals with Good Benefits or Important Economic and Scientific Values*. It has not been listed into *National Key Protected Wild Animal List*, and II level is suggested. Specific survey on its distribution and protection of large suitable habitat, wetlands and surrounding vegetation are recommended. Furthermore, the enforcement of regulation against environment and water pollution should be improved

白喉斑秧鸡 *Rallina eurizonoides*　　　　梁家登 摄　By Jiadeng Liang

长脚秧鸡
Crex crex

易危　VU A1cde

| 数据缺乏 DD | 无危 LC | 近危 NT | 易危 VU | 濒危 EN | 极危 CR | 区域灭绝 RE | 野外灭绝 EW | 灭绝 EX |

🦅 分类地位 Taxonomic Status

动物界 Animalia	脊索动物门 Chordata	鸟纲 Aves	鹤形目 Gruiformes	秧鸡科 Rallidae
学名 Scientific Name		*Crex crex*		
命名人 Species Authority		Linnaeus, 1758		
英文名 English Name(s)		Corncrake		
同物异名 Synonym(s)		无 / None		
种下单元评估 Infra-specific Taxa Assessed		无 / None		

🦅 评估信息 Assessment Information

评估年份 Year Assessed	2016
评定人 Assessor(s)	陈水华 / Shuihua Chen
其他贡献人 Other Contributor(s)	陆祎玮 / Yiwei Lu

理由 Justification: 长脚秧鸡在国外分布区域广阔。我国估计种群数量为 1,500～3,000 只。1998～2000 年的调查显示长脚秧鸡在新疆的适宜分布区域较为广阔（东西跨度 1,500 km，南北约 1,000 km），但由于生境遭受破坏，种群数量长期以来处于衰减状态。因此，列为易危等级 / The global distribution range of *Crex crex* is large. The population size in China is estimated to be 1,500～3,000 individuals. The suitable habitats in Xinjiang is large (1,500 km from east to west, and 1000 km from north to south), as indicated in a survey in 1998～2000. While the population is undergoing a decline for a long time as a result of habitat loss. Thus, it is listed as Vulnerable

评估历史 Assessment History:《国家重点保护野生动物名录》：Ⅱ (1989，2021)；《中国物种红色名录：第二卷 脊椎动物 下册》(2009)：易危 / *National Key Protected Wild Animal List*: Ⅱ (1989, 2021); *China Species Red List: Vol. Ⅱ Vertebrates Part 2* (2009): VU

🦅 地理分布 Geographical Distribution

国内分布 Domestic Distribution
繁殖于新疆西部和北部，偶见于云南和西藏 / It breeds in west and north Xinjiang, and occurs Yunnan and Tibet (Xizang) occasionally
分布标注 Distribution Note
非特有种 / Non-endemic

国内分布图
Map of Domestic Distribution

种群 Population

种群数量 Population Size	种群数量 1,500 ~ 3,000 只 / The population size is 1,500 ~ 3,000 individuals
种群趋势 Population Trend	下降 / Decreasing

生境与生态系统 Habitat (s) and Ecosystem (s)

生　　境 Habitat(s)	长脚秧鸡是草地生活型鸟类。在新疆栖息于海拔 600 ~ 3,700 m 的山前绿洲、河谷地带和亚高山草原（草甸）。常见于胡麻地、苜蓿、麦地和水域附近的杂草丛中 / *C. crex* inhabits grassland. It occurs in oasis, valleys and subalpine meadow at 600 ~ 3,700 m in Xinjiang. It is usually seen in fields with flax, clover and wheat, or the weeds close to the waters
生态系统 Ecosystem(s)	湿地生态系统 / Wetland Ecosystem

威胁 Threat (s)

主要威胁 Major Threat(s)	由于农业机械化的发展和大量的割草及放牧，致使生境遭到破坏，种群长期以来一直处于衰减状态 / The population is undergoing a decline for a long time as a result of habitat loss due to the development of agricultural, mowing and grazing

保护级别与保护行动 Protection Category and Conservation Action (s)

IUCN 红色名录 (2016) IUCN Red List (2016)	无危 / LC
保护行动 Conservation Action(s)	列入国家 II 级重点保护野生动物和迁徙物种保护公约 (CMS) 附录 II / It has been listed as the second class in National Key Protected Wild Animal List and Appendix II of the Convention on Migratory Species

长脚秧鸡 *Crex crex*　　　　快乐 摄　By Kuaile

斑胁田鸡
Zapornia paykullii
易危 VU A2cd

| 数据缺乏 DD | 无危 LC | 近危 NT | 易危 VU | 濒危 EN | 极危 CR | 区域灭绝 RE | 野外灭绝 EW | 灭绝 EX |

分类地位 Taxonomic Status

动物界 Animalia	脊索动物门 Chordata	鸟 纲 Aves	鹤形目 Gruiformes	秧鸡科 Rallidae
学 名 Scientific Name		*Zapornia paykullii*		
命 名 人 Species Authority		Ljungh, 1813		
英 文 名 English Name(s)		Band-bellied Crake		
同物异名 Synonym(s)		*Porzana paykullii*		
种下单元评估 Infra-specific Taxa Assessed		无 / None		

评估信息 Assessment Information

评估年份 Year Assessed	2016
评定人 Assessor(s)	陈水华 / Shuihua Chen
其他贡献人 Other Contributor(s)	王思宇 / Siyu Wang

理由 Justification: 斑胁田鸡在我国黑龙江、吉林、辽宁、内蒙古和河南北部为繁殖鸟，中部和南部区域为过境鸟。在我国的种群数量估计为100～10,000个繁殖对（IUCN，2014）。近10年记录很少，仅有2014年黑龙江牡丹江1只、2005年河北北戴河2只、2011年9月浙江台州1只等零星记录。因此，列为易危等级 / *Zapornia paykullii* breeds in Heilongjiang, Jilin, Liaoning, Inner Mongolia (Nei Mongol) and north Henan, and migrates across middle and south Henan. The population size in China is estimated to be 100～10,000 breeding pairs (IUCN, 2014). There are only a few records of *Z. paykullii* in recent ten years, one individual recorded in Mudanjiang of Heilongjiang in 2014, two individuals in Beidaihe in Hebei, and one in Taizhou of Zhejiang in September of 2011. Thus, it is listed as Vulnerable

评估历史 Assessment History:《国家重点保护野生动物名录》：II (2021);《中国物种红色名录：第二卷 脊椎动物 下册》(2009)：近危 / *National Key Protected Wild Animal List*: II (2021); *China Species Red List: Vol. II Vertebrates Part 2* (2009): NT

地理分布 Geographical Distribution

国内分布 Domestic Distribution

繁殖于黑龙江、吉林、辽宁、内蒙古、河北、北京、天津、河南等地，迁徙经过山东、四川、重庆、贵州、湖北、安徽、江西、江苏、上海、浙江、福建、广东、香港、澳门、广西和台湾等地 / It breeds in Heilongjiang, Jilin, Liaoning, Inner Mongolia (Nei Mongol), Hebei, Beijing, Tianjin and Henan etc., and migrates across Shandong, Sichuan, Chongqing, Guizhou, Hubei, Anhui, Jiangxi, Jiangsu, Shanghai, Zhejiang, Fujian, Guangdong, Hong Kong, Macao, Guangxi and Taiwan etc.

分布标注 Distribution Note

非特有种 / Non-endemic

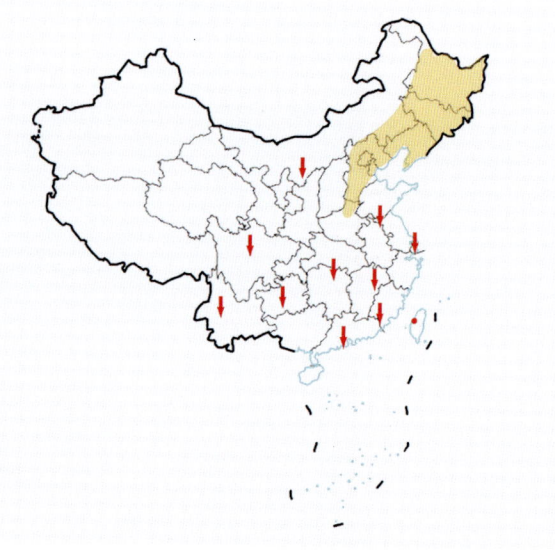

国内分布图 Map of Domestic Distribution

种群 Population

种群数量 Population Size	我国种群数量估计 100 ～ 10,000 个繁殖对。近 10 年记录很少，仅有 2014 年黑龙江牡丹江 1 只、2005 年河北北戴河 2 只、2011 年 9 月浙江台州 1 只等零星记录 / The population size in China is estimated to be 100 ～ 10,000 breeding pairs. There are only a few records in recent ten years, 1 individual recorded in Mudanjiang of Heilongjiang in 2014, 2 individuals in Beidaihe in Hebei, and 1 individual in Taizhou of Zhejiang in September of 2011
种群趋势 Population Trend	未知 / Unknown

生境与生态系统 Habitat (s) and Ecosystem (s)

生　　境 Habitat(s)	栖息于湿润的草甸和稻田 / *Z. paykullii* inhabits in wet meadows and rice fields
生态系统 Ecosystem(s)	湿地生态系统 / Wetland Ecosystem

威胁 Threat (s)

主要威胁 Major Threat(s)	农业生产、工业化发展造成的栖息地破坏 / Habitat loss owing to agricultural activities, development of industry

保护级别与保护行动 Protection Category and Conservation Action (s)

IUCN 红色名录 (2016) IUCN Red List (2016)	近危 / NT
保护行动 Conservation Action(s)	列入国家 II 级重点保护野生动物、《国家保护的有益的或者有重要经济、科学研究价值的陆生野生动物名录》/ It has been listed as the second class in National Key Protected Wild Animal List, *National Protected List of Terrestrial Wild Animals with Good Benefits or Important Economic and Scientific Values*

斑胁田鸡 *Zapornia paykullii*　　　　　　　　　　　　　　　　　　　　谷国强 摄　By Guoqiang Gu

中国生物多样性红色名录

紫水鸡
Porphyrio porphyrio

易危 VU A3cd

| DD 数据缺乏 | LC 无危 | NT 近危 | **VU 易危** | EN 濒危 | CR 极危 | RE 区域灭绝 | EW 野外灭绝 | EX 灭绝 |

分类地位 Taxonomic Status

动物界 Animalia	脊索动物门 Chordata	鸟纲 Aves	鹤形目 Gruiformes	秧鸡科 Rallidae

学名 Scientific Name	*Porphyrio porphyrio*
命名人 Species Authority	Linnaeus, 1758
英文名 English Name(s)	Purple Swamp Hen
同物异名 Synonym(s)	*Fulica porphyrio*
种下单元评估 Infra-specific Taxa Assessed	无 / None

评估信息 Assessment Information

评估年份 Year Assessed	2016
评定人 Assessor(s)	刘阳 / Yang Liu
其他贡献人 Other Contributor(s)	胡军华、黄秦 / Junhua Hu, Qin Huang

理由 Justification: 紫水鸡在中国的分布区彼此之间不连续，种群数量不多。一些小的种群由于栖息地的消失和退化面临消失的风险。因此，列为易危等级 / The distribution areas of *Porphyrio porphyrio* are fragmented and the population size is not large in China. Some small populations would vanish owing to the habitat loss and deterioration. Thus, it is listed as Vulnerable

评估历史 Assessment History: 《国家重点保护野生动物名录》: II (2021);《中国物种红色名录: 第二卷 脊椎动物 下册》(2009): 无危 / *National Key Protected Wild Animal List*: II (2021); *China Species Red List: Vol. II Vertebrates Part 2* (2009): LC

地理分布 Geographical Distribution

国内分布 Domestic Distribution

分布于云南、四川、贵州、西藏东南部、湖北、江西、上海、福建、广东、香港、广西、海南 / It occurs in Yunnan, Sichuan, Guizhou, southeast Tibet (Xizang), Hubei, Jiangxi, Shanghai, Fujian, Guangdong, Hong Kong, Guangxi, Hainan

分布标注 Distribution Note

非特有种 / Non-endemic

国内分布图
Map of Domestic Distribution

种群 Population

种群数量 Population Size	种群数量 500 ～ 1,000 只 / The population size is 500 ～ 1,000 individuals
种群趋势 Population Trend	下降 / Decreasing

生境与生态系统 Habitat (s) and Ecosystem (s)

生　　境 Habitat(s)	栖于多芦苇、水葱等植被的沼泽地及湖泊，在水上漂浮植物及芦苇地中行走。有时结小群到漫水的开阔草地、稻田或火烧过后的湿地上活动 / *P. porphyrio* inhabits marshes and lakes with vegetation *e.g.* reeds, *Scirpus validus*, *etc.* It walks on the floating plants and reeds, sometimes occurs in open meadows and rice fields inundated by water, or wetlands after burning
生态系统 Ecosystem(s)	湿地生态系统 / Wetland Ecosystem

威胁 Threat (s)

主要威胁 Major Threat(s)	栖息地丧失是种群数量下降的重要因素。栖息于村庄周围的湿地生境，频繁的人类生产活动导致栖息地萎缩和质量下降。虽然不断有新的分布点发现，但各个分布区并不连续，栖息地破碎化。此外非法狩猎亦有可能成为新的威胁因素 / The main threat of population decrease is habitat loss. Those who use wetlands around villages are facing habitat loss and degradation by heavy human activities. The habitats in distribution range are fragmentated, although new sites are found. Furthermore, illegal hunting might be another threat

保护级别与保护行动 Protection Category and Conservation Action (s)

IUCN 红色名录 (2016) IUCN Red List (2016)	无危 / LC
保护行动 Conservation Action(s)	列入国家 II 级重点保护野生动物、《国家保护的有益的或者有重要经济、科学研究价值的陆生野生动物名录》。目前部分栖息地，如云南丽江、广东海丰，已建立了自然保护区，对该种进行保护。厦门种群重新发现后，当地已经组织开展了专项的宣传和保育行动 / It has been listed as the second class in National Key Protected Wild Animal List, *National Protected List of Terrestrial Wild Animals with Good Benefits or Important Economic and Scientific Values*. Some habitats, *e.g.* Lijiang in Yunan, Haifeng in Guangdong, have set up nature reserves, and taken protection activities for this species. Specific activities on conservation and public education have been conducted after the re-discovery in Xiamen

紫水鸡 *Porphyrio porphyrio*　　　　　　　　　　　　　　　　　李文明 摄　By Wenming Li

黑颈鹤
Grus nigricollis

易危 VU C2a(ii)

| 数据缺乏 DD | 无危 LC | 近危 NT | **易危 VU** | 濒危 EN | 极危 CR | 区域灭绝 RE | 野外灭绝 EW | 灭绝 EX |

分类地位 Taxonomic Status

动物界 Animalia	脊索动物门 Chordata	鸟纲 Aves	鹤形目 Gruiformes	鹤科 Gruidae
学名 Scientific Name		*Grus nigricollis*		
命名人 Species Authority		Przevalski, 1876		
英文名 English Name(s)		Black-necked Crane		
同物异名 Synonym(s)		无 / None		
种下单元评估 Infra-specific Taxa Assessed		无 / None		

评估信息 Assessment Information

评估年份 Year Assessed	2016
评定人 Assessor(s)	马志军 / Zhijun Ma
其他贡献人 Other Contributor(s)	李凤山、杨晓君 / Fengshan Li, Xiaojun Yang

理由 Justification: 黑颈鹤种群数量较小，且由于繁殖地和越冬地的湿地丧失及退化、农业耕作方式改变等原因，种群数量可能呈下降趋势。虽然近年来的种群数量调查表明基本保持稳定，但种群数量的变化有待长期的监测来证实。因此，列为易危等级 / *Grus nigricollis* has a single small population that is in decline owing to the loss and degradation of wetlands, and changing agricultural practices in both its breeding and wintering grounds. The recent surveys indicate that the population of G. nigricollis is stable, while long term monitoring is needed to assess the trend. Thus, it is listed as Vulnerable

评估历史 Assessment History: 《国家重点保护野生动物名录》：Ⅰ (1989，2021)；《中国濒危动物红皮书·鸟类》(1998)：濒危；《中国物种红色名录：第二卷 脊椎动物 下册》(2009)：易危 / *National Key Protected Wild Animal List*: Ⅰ (1989, 2021); *China Red Data Book of Endangered Animals*: *Aves* (1998): E; *China Species Red List*: *Vol. II Vertebrates Part 2* (2009): VU

地理分布 Geographical Distribution

国内分布 Domestic Distribution

全球唯一在高原地区分布的鹤类。繁殖于青海、四川北部、甘肃南部、新疆东南部、西藏东南部。在西藏南部、云南北部、贵州越冬 / It is the only crane inhabits in plateau globally. It breeds in Qinghai, north Sichuan, south Gansu, southeast Xinjiang, and southeast Tibet (Xizang) Yunnan, Sichuan, Guizhou, and winters in south Xizang, north Yunnan, Guizhou

分布标注 Distribution Note

非特有种 / Non-endemic

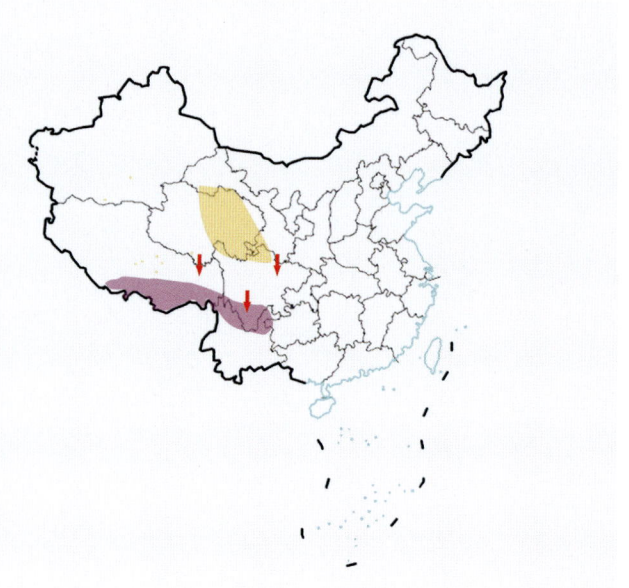

国内分布图
Map of Domestic Distribution

种群 Population

种群数量 Population Size	种群数量约 10,000 只 / The population size is estimated to be 10,000 individuals
种群趋势 Population Trend	目前缺乏种群数量的监测数据，但可能呈下降趋势 / It might be decreasing whereas no sufficient monitoring data

生境与生态系统 Habitat (s) and Ecosystem (s)

生 境 Habitat(s)	栖息于海拔 2,500～5,000 m 的高原地区，在沼泽、湖泊及草甸湿地繁殖，越冬期常到农田中取食农作物。主要以陆生和水生植物幼嫩的根、茎、叶、种子以及农作物的叶和种子为食，也取食昆虫等无脊椎动物 / *G. nigricollis* breeds in alpine bog meadows, lakes and riverine marshes in plateau at 2,500～5,000 m, and forages in farmlands during winter. It feeds on roots, tubers, leaves, seeds of vegetation and crops, as well as invertebrate, *e.g.* insects, *etc.*
生态系统 Ecosystem(s)	湿地生态系统、农田生态系统 / Wetland Ecosystem, Farmland Ecosystem

威胁 Threat (s)

主要威胁 Major Threat(s)	湿地丧失与退化导致的栖息地减少和质量下降是其面临的主要威胁。传统耕作模式的改变也使得黑颈鹤在越冬期的食物资源缺乏。黑颈鹤冬季在农田觅食导致与当地居民发生冲突 / The major threats are habitat loss and degeneration owing to degradation of wetlands. The change of traditional cultivation has reduced the availability of food resource in winters. The conflict between foraging *G. nigricollis* and local residents in winter is another threat

保护级别与保护行动 Protection Category and Conservation Action (s)

IUCN 红色名录 (2016) IUCN Red List (2016)	易危 / VU C2a (ii)
保护行动 Conservation Action(s)	列入国家 I 级重点保护野生动物、CITES 公约附录 I 和迁徙物种保护公约 (CMS) 附录 I。对其繁殖生态、越冬生态都有过研究，对其迁徙路线开展过卫星跟踪。在主要的繁殖地和越冬地已建立多个自然保护区。在多家动物园有人工饲养的个体或种群 / It has been listed as the first class in National Key Protected Wild Animal List, the Appendix I of CITES and the Appendix I of Convention on Migratory Species. There are studies on breeding and wintering ecology, and satellite tracking of its migration routes has been conducted. Several nature reserves have been set up in its breeding and wintering sites. There are artificial feeding individuals and populations in several zoos

黑颈鹤 *Grus nigricollis* 祝平 摄 By Ping Zhu

中国生物多样性红色名录

林沙锥
Gallinago nemoricola

易危 VU C2a(ii)

| 数据缺乏 DD | 无危 LC | 近危 NT | 易危 VU | 濒危 EN | 极危 CR | 区域灭绝 RE | 野外灭绝 EW | 灭绝 EX |

分类地位 Taxonomic Status

动物界 Animalia	脊索动物门 Chordata	鸟纲 Aves	鸻形目 Charadriiformes	鹬科 Scolopacidae
学　名 Scientific Name		*Gallinago nemoricola*		
命 名 人 Species Authority		Hodgson, 1836		
英 文 名 English Name(s)		Wood Snipe		
同物异名 Synonym(s)		无 / None		
种下单元评估 Infra-specific Taxa Assessed		无 / None		

评估信息 Assessment Information

评估年份 Year Assessed	2016
评 定 人 Assessor(s)	刘阳 / Yang Liu
其他贡献人 Other Contributor(s)	危骞 / Qian Wei

理由 Justification: 林沙锥全球种群数量2,500～10,000只，90%以上的分布记录集中在南亚和东南亚，中国未有针对其种群数量、分布面积和受威胁因素等内容的系统研究，推测种群数量不会高于2,500只(BirdLife International, 2014)，且分布生境极为脆弱，易受到环境变化和人为干扰的影响。因此，列为易危等级 / The global population of *Gallinago nemoricola* is estimated to be 2,500～10,000 individuals, and 90% of the records were in south Asia and southeast Asia. No systematic studies on population size, distribution range and threats. It is presumed the population size might be no more than 2,500 individuals. These habitats are fragile and suffer from environmental change and human disturbance. Thus, it is listed as Vulnerable

评估历史 Assessment History: 《国家重点保护野生动物名录》：Ⅱ (2021)；《中国濒危动物红皮书·鸟类》(1998)：未定；《中国物种红色名录：第二卷 脊椎动物 下册》(2009)：易危 / *National Key Protected Wild Animal List*: Ⅱ (2021); *China Red Data Book of Endangered Animals: Aves* (1998): Ⅰ; *China Species Red List: Vol. II Vertebrates Part 2* (2009): VU

地理分布 Geographical Distribution

国内分布 Domestic Distribution
分布于西藏南部和东部，云南西部和西北部，四川西部以及甘肃南部 / It occurs south and east Tibet (Xizang), west and northwest Yunnan, west Sichuan, and south Gansu
分布标注 Distribution Note
非特有种 / Non-endemic

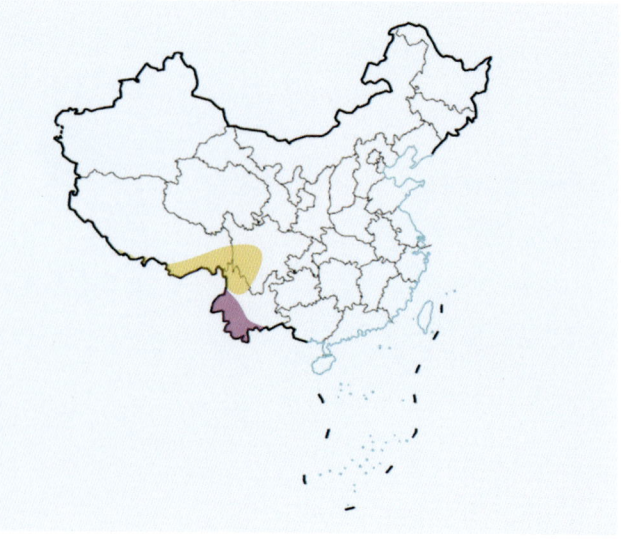

国内分布图
Map of Domestic Distribution

种群 Population

种群数量 Population Size	种群数量 1,000～2,500 只 / The population size is 1,000～2,500 individuals
种群趋势 Population Trend	下降 / Decreasing

生境与生态系统 Habitat (s) and Ecosystem (s)

生　　境 Habitat(s)	栖于海拔 1,000～5,000 m 的山地森林湿地、高草地和灌丛中的沼泽泥潭及池塘 / *G. nemoricola* inhabits mountainous forest wetlands at 1,000～5,000 m, alpine meadows, and marshes and pools in shrubs
生态系统 Ecosystem(s)	湿地和高山草甸生态系统 / Wetland and Alpine Meadow Ecosystem

威胁 Threat (s)

主要威胁 Major Threat(s)	栖息地的丧失及退化 / Habitat loss and degradation

保护级别与保护行动 Protection Category and Conservation Action (s)

IUCN 红色名录 (2016) IUCN Red List (2016)	易危 / VU C2a (ii)
保护行动 Conservation Action(s)	列入国家 II 级重点保护野生动物、《国家保护的有益的或者有重要经济、科学研究价值的陆生野生动物名录》。四川西部的分布区处于自然保护区中，栖息地得到一定程度的保护 / It has been listed as the second class in National Key Protected Wild Animal List, *National Protected List of Terrestrial Wild Animals with Good Benefits or Important Economic and Scientific Values*. Some of the distributed area in west Sichuan falls in the nature reserves and habitats are under protection to some extent

林沙锥 *Gallinago nemoricola*　　　　唐军 摄　By Jun Tang

中国生物多样性红色名录

大杓鹬
Numenius madagascariensis

易危　VU A2cd; C1

| 数据缺乏 DD | 无危 LC | 近危 NT | 易危 VU | 濒危 EN | 极危 CR | 区域灭绝 RE | 野外灭绝 EW | 灭绝 EX |

分类地位 Taxonomic Status

动物界 Animalia	脊索动物门 Chordata	鸟纲 Aves	鸻形目 Charadriiformes	鹬科 Scolopacidae

学　名 Scientific Name	*Numenius madagascariensis*
命名人 Species Authority	Linnaeus, 1766
英文名 English Name(s)	Eastern Curlew
同物异名 Synonym(s)	无 / None
种下单元评估 Infra-specific Taxa Assessed	无 / None

评估信息 Assessment Information

评估年份 Year Assessed	2016
评定人 Assessor(s)	马志军 / Zhijun Ma
其他贡献人 Other Contributor(s)	中国沿海水鸟同步调查项目组 / Chinese Coastal Waterbirds Census Project

理由 Justification: 目前缺乏对大杓鹬种群数量的监测数据，由于其所处的湿地生态系统的丧失和退化，其种群可能呈下降趋势，需加以关注。因此，列为易危等级 / No monitoring data on *Numenius madagascariensis*, the population might undergo a decrease due to the loss of wetland. Thus, it is listed as Vulnerable

评估历史 Assessment History: 《国家重点保护野生动物名录》：Ⅱ (2021)；《中国物种红色名录：第二卷 脊椎动物 下册》(2009)：近危 / *National Key Protected Wild Animal List*: Ⅱ (2021); *China Species Red List: Vol. Ⅱ Vertebrates Part 2* (2009): NT

地理分布 Geographical Distribution

国内分布 Domestic Distribution

繁殖于黑龙江、吉林、辽宁、河北和内蒙古东部；越冬于台湾。迁徙时经过中国东部沿海地区。在辽宁丹东、山东黄河三角洲、江苏如东均有大群迁徙的记录 / It breeds in Heilongjiang, Jilin, Liaoning, Hebei and east Inner Mongolia (Nei Mongol), and winters in Taiwan. It migrates across east coast of China with large population recorded in Dandong in Liaoning, Yellow River Delta in Shandong, and Rudong in Jiangsu

分布标注 Distribution Note

非特有种 / Non-endemic

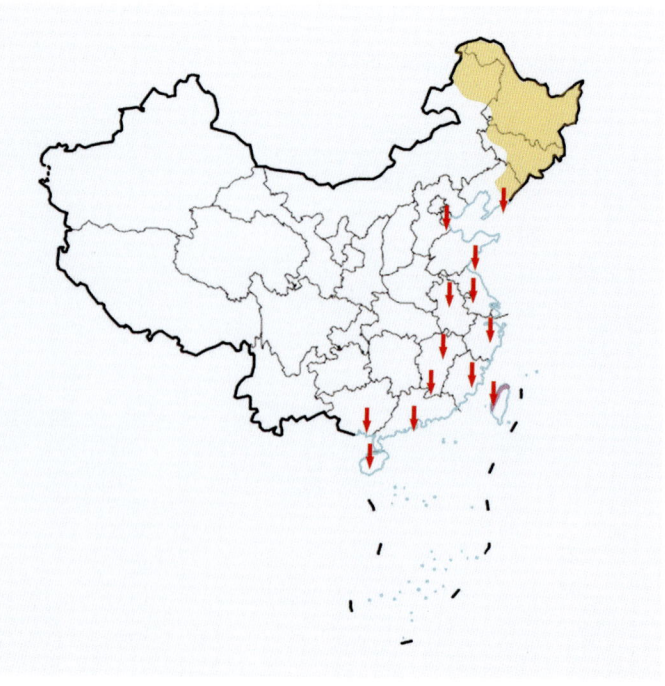

国内分布图
Map of Domestic Distribution

种群 Population

种群数量 Population Size	全球种群数量 38,000 只。我国的数量仍不清楚 / The global population size is estimated to be 38,000 individuals, and no clear information in China
种群趋势 Population Trend	目前缺乏种群数量的监测数据，但可能呈下降趋势 / Although no sufficient monitoring data available, the population size might continue to decline

生境与生态系统 Habitat (s) and Ecosystem (s)

生　　境 Habitat(s)	主要栖息于滨海和河口湿地、沼泽、湖滩以及附近的农田。主要食物包括蠕虫及其他节肢动物、甲壳类、腹足类、双壳类等底栖动物以及小鱼、小虾等 / *N. madagascariensis* inhabits in coastal and riverine wetlands, marshes, flats and close farmlands. It feeds on worms and arthropods, crustaceans, gastropods, bivalves and small fish and shrimps
生态系统 Ecosystem(s)	湿地生态系统 / Wetland Ecosystem

威胁 Threat (s)

主要威胁 Major Threat(s)	湿地的围垦和开发以及污染导致的栖息地丧失和质量下降是其面临的主要威胁 / The major threats are habitat loss and degradation owing to the reclamation, development and pollution of wetlands

保护级别与保护行动 Protection Category and Conservation Action (s)

IUCN 红色名录 (2016) IUCN Red List (2016)	濒危 / EN A2bc+3bc+4bc
保护行动 Conservation Action(s)	列入国家 II 级重点保护野生动物、《国家保护的有益的或者有重要经济、科学研究价值的陆生野生动物名录》和迁徙物种保护公约 (CMS) 附录 I，在迁徙经过的沿海地区已建立多个自然保护区，但缺乏针对性的保护措施 / It has been listed as the second class in National Key Protected Wild Animal List, *National Protected List of Terrestrial Wild Animals with Good Benefits or Important Economic and Scientific Values*, and the appendix I of the Convention on Migratory Species. Several nature reserves have been set up in coastal areas while no specific conservation actions have been taken

大杓鹬 *Numenius madagascariensis*　　　　　　　　　　　　　　　　　　　　袁晓 摄　By Xiao Yuan

中国生物多样性红色名录

大滨鹬
Calidris tenuirostris

易危 VU A4bcd

| 数据缺乏 DD | 无危 LC | 近危 NT | 易危 VU | 濒危 EN | 极危 CR | 区域灭绝 RE | 野外灭绝 EW | 灭绝 EX |

分类地位 Taxonomic Status

动物界 Animalia	脊索动物门 Chordata	鸟纲 Aves	鸻形目 Charadriiformes	鹬科 Scolopacidae
学名 Scientific Name		*Calidris tenuirostris*		
命名人 Species Authority		Horsfield, 1821		
英文名 English Name(s)		Great Knot		
同物异名 Synonym(s)		无 / None		
种下单元评估 Infra-specific Taxa Assessed		无 / None		

评估信息 Assessment Information

评估年份 Year Assessed	2016
评定人 Assessor(s)	马志军 / Zhijun Ma
其他贡献人 Other Contributor(s)	中国沿海水鸟同步调查项目组 / Chinese Coastal Waterbirds Survey Group

理由 Justification: 由于滨海地区滩涂湿地的围垦和开发导致的栖息地丧失，大滨鹬种群数量快速下降，且未来围垦计划可能导致其种群数量进一步下降。因此，列为易危等级 / The population of *Calidris tenuirostris* is undergoing a rapid decline owing to the habitat loss by reclamation and development of coastal mudflats. And the reclamation program in future might lead to further decrease. Thus, it is listed as Vulnerable

评估历史 Assessment History: 《国家重点保护野生动物名录》：Ⅱ (2021)；《中国物种红色名录：第二卷 脊椎动物 下册》(2009)：无危 / National Key Protected Wild Animal List: Ⅱ (2021); China Species Red List: Vol. Ⅱ Vertebrates Part 2 (2009): LC

地理分布 Geographical Distribution

国内分布 Domestic Distribution

在我国为旅鸟，迁徙期经过我国沿海地区，黄渤海的鸭绿江口、双台河口、渤海湾、黄河三角洲、崇明东滩等地区的滩涂湿地是其迁徙期的重要停歇地 / Passenger migrant in China. It migrates along the coastal area, and wetlands in Yalujiang Estuary, Shuangtai Estuary, Bohai Bay, Yellow River Delta, Chongming Dongtan are key stopovers

分布标注 Distribution Note

非特有种 / Non-endemic

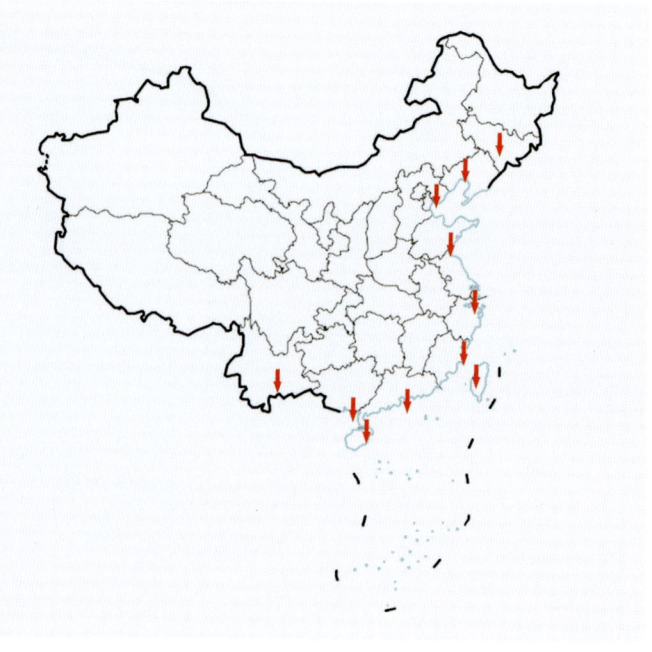

国内分布图
MAP of Domestic Distribution

种群 Population

种群数量 Population Size	估计春季迁徙经过我国的种群数量约 150,000 只 / The population size in China is 150,000 individuals during spring migration
种群趋势 Population Trend	目前缺乏种群数量的监测数据，但很可能呈下降趋势 / There has been few monitoring data on population size. It probably is in a decreasing trend

生境与生态系统 Habitat(s) and Ecosystem(s)

生境 Habitat(s)	繁殖地位于西伯利亚东北部的低山及丘陵地区，在覆盖地衣、石楠及草本植物的区域营巢。在迁徙期和越冬期，大滨鹬分布于沿海地区，极少到内陆地区活动。通常在沙质或泥质的河口和滨海滩涂湿地觅食，当潮水将滩涂淹没时，则到水产养殖塘等人类活动干扰较少的区域集群休息。繁殖期以昆虫等无脊椎动物为主要食物，也可食植物的浆果。迁徙停歇期和越冬期主要以双壳类、腹足类等底栖动物为主要食物 / *C. tenuirostris* breeds on hilly areas covered with lichen and patches of herbs, heather in northeast Siberia. It uses coastal areas during migration and in winters, and seldom occurs in inlands. It forages on the estuaries and coastal wetlands with mud and sandflats, and roosted in close pond ridges, bare lands, waste lands with low human disturbance during high tides. The adult breeding diet consists predominantly of invertebrate such as insects, and berries. During the winter and on passage the species takes invertebrate of bivalves and gastropods
生态系统 Ecosystem(s)	湿地生态系统 / Wetland Ecosystem

威胁 Threat(s)

主要威胁 Major Threat(s)	大滨鹬依赖于滨海和河口湿地，沿海地区大规模的围垦和开发造成的栖息地丧失是大滨鹬面临的主要威胁。韩国在朝鲜半岛西海岸进行的新万锦围垦项目，导致大滨鹬在黄渤海地区最大的迁徙停歇地消失，大滨鹬的种群数量减少了 90,000 只。环境污染、人类活动干扰等也给大滨鹬带来了不利影响 / Major threat of the species is habitat loss by large-scale reclamation and development of coastal mudflats. The largest stopover in Yellow Sea area has gone with the Saemangeum Reclamation Program in South Korea, and the population of *C. tenuirostris* has decreased by 90,000 individuals. The pollution and human disturbance also exert negative effect

保护级别与保护行动 Protection Category and Conservation Action(s)

IUCN 红色名录 (2016) IUCN Red List (2016)	濒危 / EN A2bc+3bc+4bc
保护行动 Conservation Action(s)	列入国家 II 级重点保护野生动物、《国家保护的有益的或者有重要经济、科学研究价值的陆生野生动物名录》和迁徙物种保护公约 (CMS) 附录 I。大滨鹬在我国的主要迁徙停歇地，如鸭绿江口、双台河口、崇明东滩等，已建立了自然保护区。近年来对其迁徙停歇生态开展了较深入的研究 / It has been listed as the second class in National Key Protected Wild Animal List, *National Protected List of Terrestrial Wild Animals with Good Benefits or Important Economic and Scientific Values*, and the Appendix I of the Convention on Migratory Species. Nature reserves have been set up in its main stopovers in Yalujiang Estuary, Shuangtai Estuary and Chongming Dongtan. There are studies on migration and stopover ecology

大滨鹬 *Calidris tenuirostris* 袁晓 摄 By Xiao Yuan

红腹滨鹬 *Calidris canutus*

易危 VU A4bcd

| 数据缺乏 DD | 无危 LC | 近危 NT | 易危 VU | 濒危 EN | 极危 CR | 区域灭绝 RE | 野外灭绝 EW | 灭绝 EX |

分类地位 Taxonomic Status

动物界 Animalia	脊索动物门 Chordata	鸟纲 Aves	鸻形目 Charadriiformes	鹬科 Scolopacidae

学名 Scientific Name	*Calidris canutus*
命名人 Species Authority	Linnaeus, 1758
英文名 English Name(s)	Red Knot
同物异名 Synonym(s)	*Tringa canutus*
种下单元评估 Infra-specific Taxa Assessed	无 / None

评估信息 Assessment Information

评估年份 Year Assessed	2016
评定人 Assessor(s)	马志军 / Zhijun Ma
其他贡献人 Other Contributor(s)	杨洪燕、张正旺 / Hongyan Yang, Zhengwang Zhang

理由 Justification: 红腹滨鹬在迁徙期曾广泛分布于渤海地区的滨海湿地，但由于滩涂围垦开发，目前40%以上的个体迁徙时在渤海湾北部的滩涂停歇，近年来数量呈下降趋势。鉴于滨海湿地的围垦和开发尚难以从根本扭转，红腹滨鹬仍面临着严重威胁，种群有可能继续下降。因此，列为易危等级 / *Calidris canutus* used to widely occur in Bohai Bay and 40% of these individuals use north Bohai Bay as stopover in recent years, while the population is undergoing a decline owing to mudflat reclamation. The situation is hardly to be changed considering the reclamation and development of coastal wetlands. The species is under significant threat with an on-going population decline. Thus, it is listed as Vulnerable

评估历史 Assessment History: 《中国物种红色名录：第二卷 脊椎动物 下册》(2009)：无危 / *China Species Red List*: Vol. II Vertebrates Part 2 (2009): LC

地理分布 Geographical Distribution

国内分布 Domestic Distribution

在我国为旅鸟，春季迁徙时经过黄渤海地区，其中河北唐山的沿海滩涂是红腹滨鹬最重要的迁徙停歇地。分布于吉林、辽宁、河北、山东、青海、湖北、江苏、上海、浙江、福建、广东、香港、澳门、广西、海南、台湾 / It migrates along the Yellow Sea area during spring and autumn, and the coastal mudflat in Tangshan of Hebei is the most important stopover. It occurs in Jilin, Liaoning, Hebei, Shandong, Qinghai, Hubei, Jiangsu, Shanghai, Zhejiang, Fujian, Guangdong, Hong Kong, Macao, Guangxi, Hainan, Taiwan

分布标注 Distribution Note

非特有种 / Non-endemic

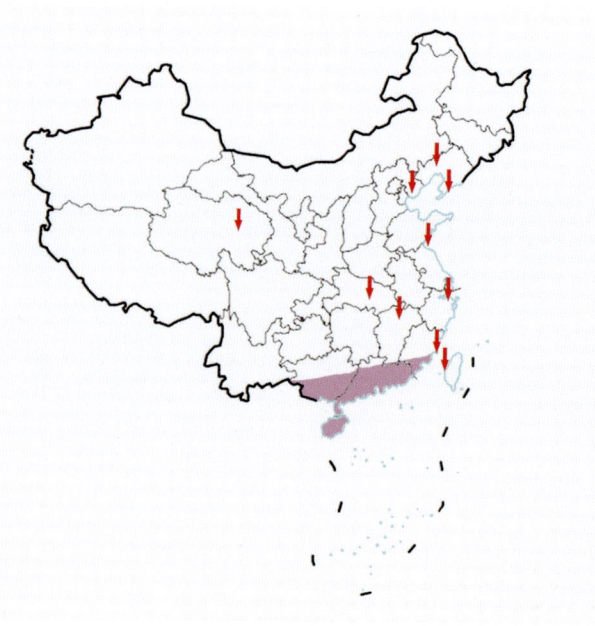

国内分布图
Map of Domestic Distribution

种群 Population

种群数量 Population Size	我国的种群数量约 50,000 只 / The population size in China is estimated to be 50,000 individuals
种群趋势 Population Trend	目前缺乏种群数量的监测数据，但可能呈下降趋势 / There is little monitoring data on the population size. It is probably undergoing a decline

生境与生态系统 Habitat(s) and Ecosystem(s)

生　　境 Habitat(s)	在高纬度苔原地区繁殖。迁徙停歇期和越冬期分布于沿海地区，极少到内陆地区活动。它们通常在沙质或泥质的河口和滨海滩涂湿地上觅食，当潮水将滩涂淹没时，则飞到觅食地附近的盐田或水产养殖塘等人类活动干扰较少的区域集群休息。繁殖期以昆虫等无脊椎动物为主要食物，也食植物的浆果。迁徙停歇期和越冬期主要以双壳类等底栖动物为主要食物 / *C. canutus* breeds on tundra in the high Arctic, and uses coastal areas during migration and in winters, and seldom occurs in inland. It forages on the estuaries and coastal wetlands with mud and sandflats, and roosted in close pond ridges, bare lands, waste lands with low human disturbance during high tides. The adult breeding diet consists predominantly of invertebrate such as insects, and berries. During the winter and on passage the species takes invertebrate of bivalves and gastropods
生态系统 Ecosystem(s)	湿地生态系统 / Wetland Ecosystem

威胁 Threat(s)

主要威胁 Major Threat(s)	由于红腹滨鹬依赖于滨海和河口湿地，沿海湿地大规模的围垦和开发给红腹滨鹬带来了极大威胁。渤海湾西部曾是红腹滨鹬的主要迁徙停歇地，该区域被围垦后，红腹滨鹬集中在渤海湾北部河北唐山的滨海滩涂湿地。但目前该区域还不是自然保护区，一旦该区域的滩涂湿地消失，将对红腹滨鹬的种群带来严重的不利影响。由于滨海湿地的过度围垦和开发还难以从根本扭转，红腹滨鹬未来的种群状况不容乐观 / Major threat of the species is habitat loss by large-scale reclamation and development of coastal mudflats. *C. canutus* turns to Tangshan in north Bohai Bay after the reclamation of west Bohai Bay, which used to be the main stopover. However, the north area has not been nature reserve, which might exert negative effect on the population if gone. The situation is hardly to be converted considering the reclamation and development of coastal wetlands. The population might be cheerless in future

保护级别与保护行动 Protection Category and Conservation Action(s)

IUCN 红色名录 (2016) IUCN Red List (2016)	近危 / NT
保护行动 Conservation Action(s)	列入《国家保护的有益的或者有重要经济、科学研究价值的陆生野生动物名录》和迁徙物种保护公约 (CMS) 附录 I。虽然红腹滨鹬在我国沿海地区的一些自然保护区都有分布，但其最重要的迁徙停歇地河北唐山的沿海滩涂未建立自然保护区。近年来对其迁徙停歇生态开展了较深入的研究 / It has been listed as *National Protected List of Terrestrial Wild Animals with Good Benefits or Important Economic and Scientific Values*, and the Appendix I of the Convention on Migratory Species. There are several nature reserves in its distribution area, while none occurs in its most important stopover, Tangshan. There are studies on migration and stopover ecology

红腹滨鹬 *Calidris canutus* 　　　　袁晓 摄　By Xiao Yuan

黑嘴鸥
Saundersilarus saundersi

易危　VU A2cde+B1b(iii)+C1

| 数据缺乏 DD | 无危 LC | 近危 NT | 易危 VU | 濒危 EN | 极危 CR | 区域灭绝 RE | 野外灭绝 EW | 灭绝 EX |

分类地位 Taxonomic Status

动物界 Animalia	脊索动物门 Chordata	鸟纲 Aves	鸻形目 Charadriiformes	鸥科 Laridae
学名 Scientific Name		*Saundersilarus saundersi*		
命名人 Species Authority		Swinhoe, 1871		
英文名 English Name(s)		Saunders's Gull		
同物异名 Synonym(s)		*Larus saundersi*		
种下单元评估 Infra-specific Taxa Assessed		无 / None		

评估信息 Assessment Information

评估年份 Year Assessed	2016
评定人 Assessor(s)	陈水华 / Shuihua Chen
其他贡献人 Other Contributor(s)	王思宇 / Siyu Wang

理由 Justification: 黑嘴鸥种群正在缩小，物种数下降，在3个世代内可能会有快速下降。因此，列为易危等级 / The population of *Saundersilarus saundersi* is undergoing a decline and may be subjected to rapidly decrease in three generations. Thus, it is listed as Vulnerable

评估历史 Assessment History: 《国家重点保护野生动物名录》：Ⅰ (2021)；《中国濒危动物红皮书·鸟类》(1998)：易危；《中国物种红色名录：第二卷 脊椎动物 下册》(2009)：易危 / *National Key Protected Wild Animal List*: Ⅰ (2021); *China Red Data Book of Endangered Animals*: *Aves* (1998): V; *China Species Red List*: *Vol. II Vertebrates Part 2* (2009): VU

地理分布 Geographical Distribution

国内分布 Domestic Distribution

在中国北部沿海地区繁殖，在南部沿海地区越冬，其中辽宁辽河口国家级自然保护区内栖息的种群数量达6,000余只，占全球种群分布数量8,000余只的70%以上，种群集中在9 km² 面积范围内 / It breeds in coast of northern China, and winters in coast of southern China. 6,000 individuals, more than 70% of its global population inhabit in Liaohe Estuary National Nature Reserve, and assemble in an area of 9 km²

分布标注 Distribution Note

非特有种 / Non-endemic

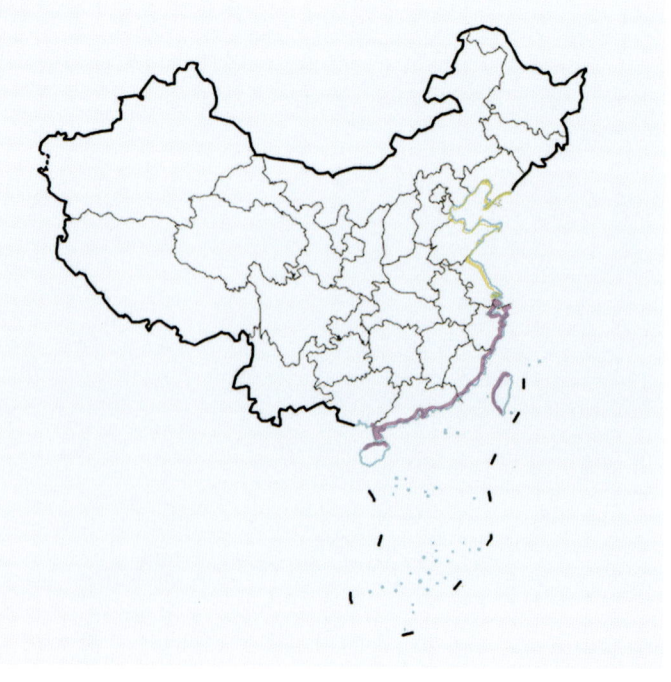

国内分布图
Map of Domestic Distribution

种群 Population

种群数量 Population Size	种群数量 6,000 ～ 10,000 只 / The population size is 6,000 ～ 10,000 individuals
种群趋势 Population Trend	下降 / Decreasing

生境与生态系统 Habitat (s) and Ecosystem (s)

生　　　境 Habitat(s)	栖息于河川、湖泊、沼泽地区，尤其是芦苇和碱蓬湿地 / *S. saundersi* inhabits in rivers, lakes and marshes, especially reed and seepweed wetlands
生态系统 Ecosystem(s)	淡水和海洋生态系统 / Freshwater and Marine Ecosystem

威胁 Threat (s)

主要威胁 Major Threat(s)	适宜栖息地减少 (Jiang *et al.*, 2010)、水质污染及捕猎和偷蛋是影响黑嘴鸥繁殖的主要因素。即使在已有的保护区内，也面临着外来物种入侵导致

植被退化、地下石油上溢导致水体污染（双台河口保护区）等威胁 / The major threats include habitat loss (Jiang *et al.*, 2010), water pollution, illegal hunting and egg collection. There are also threats, *e.g.* vegetation degradation owing to invasion of alien species, water pollution owing to oil spill in existing nature reserves

保护级别与保护行动 Protection Category and Conservation Action (s)

IUCN 红色名录 (2016) IUCN Red List (2016)	易危 / VU A3cde+4cde
保护行动 Conservation Action(s)	列入国家 I 级重点保护野生动物、《国家保护的有益的或者有重要经济、科学研究价值的陆生野生动物名录》和迁徙物种保护公约 (CMS) 附录 I。

在主要分布区建立了江苏盐城国家级自然保护区、山东东营黄河三角洲国家级自然保护区、辽宁辽河口国家级自然保护区、辽宁丹东鸭绿江口国家级自然保护区 / It has been listed as the first class in National Key Protected Wild Animal List, *National Protected List of Terrestrial Wild Animals with Good Benefits or Important Economic and Scientific Values*, and the Appendix I of the Convention on Migratory Species. National Nature reserves have been set up in the main distribution area in Jiangsu Yancheng, Shandong Dongying Yellow River Delta, Liaoning Liaohe Estuary and Liaoning Dandong Yalujiang Estuary

黑嘴鸥 *Saundersilarus saundersi*　　　　　　　　　　　　　　　　　　　　　　　　钱斌 摄　By Bin Qian

短尾信天翁
Phoebastria albatrus

中国生物多样性 红色名录

易危 VU D1

| 数据缺乏 DD | 无危 LC | 近危 NT | 易危 VU | 濒危 EN | 极危 CR | 区域灭绝 RE | 野外灭绝 EW | 灭绝 EX |

分类地位 Taxonomic Status

动物界 Animalia	脊索动物门 Chordata	鸟纲 Aves	鹱形目 Procellariiformes	信天翁科 Diomedeidae
学名 Scientific Name		*Phoebastria albatrus*		
命名人 Species Authority		Pallas, 1769		
英文名 English Name(s)		Short-tailed Albatross		
同物异名 Synonym(s)		*Diomedea albatrus*		
种下单元评估 Infra-specific Taxa Assessed		无 / None		

评估信息 Assessment Information

评估年份 Year Assessed	2016
评定人 Assessor(s)	刘阳 / Yang Liu
其他贡献人 Other Contributor(s)	黄秦 / Qin Huang

理由 Justification: 短尾信天翁在中国边缘分布，中国钓鱼岛是其重要的繁殖地之一，沿海的渔业作业可能会对本种造成威胁。因此，列为易危等级 / *Phoebastria albatrus* occurs at the edge of Chinese boundary, while Diaoyu Islands are its main breeding sites. The coastal fishery could pose a threat to this species. Thus, it is listed as Vulnerable

评估历史 Assessment History: 《国家重点保护野生动物名录》：Ⅰ (1989，2021)；《中国濒危动物红皮书·鸟类》(1998)：濒危；《中国物种红色名录：第二卷 脊椎动物 下册》(2009)：未评估 / *National Key Protected Wild Animal List*: I (1989, 2021); *China Red Data Book of Endangered Animals*: Aves (1998): E; *China Species Red List*: Vol. II Vertebrates Part 2 (2009): NA

地理分布 Geographical Distribution

国内分布 Domestic Distribution

分布于钓鱼岛海域、东海北部海域、山东、台湾附近海域 / It occurs in Diaoyu Dao, north of the East Sea, Shandong and Taiwan

分布标注 Distribution Note

非特有种 / Non-endemic

国内分布图
Map of Domestic Distribution

种群 Population

种群数量 Population Size	种群数量约 100 只 / The population size is about 100 individuals
种群趋势 Population Trend	下降 / Decreasing

生境与生态系统 Habitat (s) and Ecosystem (s)

生　　境 Habitat(s)	国内繁殖于钓鱼岛 (Brazil, 2009)，分布区扩展到整个北太平洋 (BirdLife International，2014) / *P. albatrus* breeds in Diaoyu Dao in China (Brazil, 2009). It could be seen across the north Pacific Ocean (BirdLife International, 2014)
生态系统 Ecosystem(s)	海洋生态系统 / Marine Ecosystem

威胁 Threat (s)

主要威胁 Major Threat(s)	繁殖地狭窄，早期可能受到人为干扰及捡拾鸟蛋的影响，台湾澎湖繁殖地的消失就是此原因导致的 (刘小如等，2010)。在繁殖地以外的活动区域，受到长线捕鱼的影响较大 (Piatt *et al.*，2006) / Its breeding sites is small and probably affected by human disturbance and egg collection, which has caused the loss of one breeding site, Penghu in Taiwan (Liu *et al.*, 2010). And in other areas, it might be caught and killed occasionally by longline fisheries (Piatt *et al.*, 2006)

保护级别与保护行动 Protection Category and Conservation Action (s)

IUCN 红色名录 (2016) IUCN Red List (2016)	易危 / VU D2
保护行动 Conservation Action(s)	列入国家 I 级重点保护野生动物、CITES 公约附录 I 和迁徙物种保护公约 (CMS) 附录 I / It has been listed as the first class in National Key Protected Wild Animal List, and in the Appendix I of CITES and the Convention on Migratory Species

短尾信天翁 *Phoebastria albatrus*　　　　　　　　　　　　　　王宁 摄　By Ning Wang

黑鹳
Ciconia nigra

易危 VU C2a(i)

| 数据缺乏 DD | 无危 LC | 近危 NT | 易危 VU | 濒危 EN | 极危 CR | 区域灭绝 RE | 野外灭绝 EW | 灭绝 EX |

分类地位 Taxonomic Status

动物界 Animalia	脊索动物门 Chordata	鸟纲 Aves	鹳形目 Ciconiiformes	鹳科 Ciconiidae

学 名 Scientific Name	*Ciconia nigra*
命 名 人 Species Authority	Linnaeus, 1758
英 文 名 English Name(s)	Black Stork
同物异名 Synonym(s)	无 / None
种下单元评估 Infra-specific Taxa Assessed	对西部种群生存力分析显示，种群未来 20 年的灭绝概率小于 10% / The analysis of population viability of west population indicated that the probability of extinction in 20 years is less than 10%

评估信息 Assessment Information

评 估 年 份 Year Assessed	2016
评 定 人 Assessor(s)	马鸣 / Ming Ma
其他贡献人 Other Contributor(s)	魏顺德、才代、苏化龙等 / Shunde Wei, Dai Cai, Hualong Su *et al.*

理由 Justification: 在过去的 30 多年里，国内黑鹳的种群数量维持在 1,800～2,600 只，虽然数量比较稀少，但分布广泛，记录次数较多，种群数量相对比较稳定。每年迁徙季节，在新疆的塔里木河流域、甘肃的黑河湿地等都有数百只的集群。近年，在云南的越冬数量也有增加。因此，列为易危等级 / In the past 30 years, the population of *Ciconia nigra* in China is 1,800～2,600 individuals, in a stable trend, with wide, nonconcentrated, and regular records. There are records of population of several hundreds in Tarim basin and Heihe Wetland during migration season. In addition, the wintering population has increased in Yunnan. Thus, it is listed as Vulnerable

评估历史 Assessment History:《国家重点保护野生动物名录》：Ⅰ (1989, 2021)；《中国濒危动物红皮书·鸟类》(1998)：濒危；《中国物种红色名录：第二卷 脊椎动物 下册》(2009)：无危 / *National Key Protected Wild Animal List*: I (1989, 2021); *China Red Data Book of Endangered Animals*: *Aves* (1998): E; *China Species Red List*: *Vol. II Vertebrates Part 2* (2009): LC

地理分布 Geographical Distribution

国内分布 Domestic Distribution
除西藏外，分布于全国其他各省（郑光美，2017) / It occurs all province of China except Tibet (Xizang) (Zheng, 2017)
分布标注 Distribution Note
非特有种 / Non-endemic

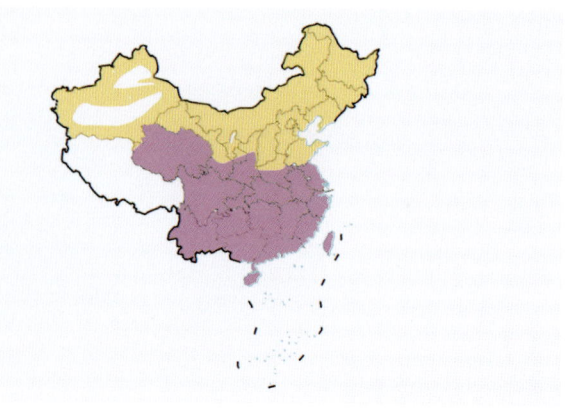

国内分布图
Map of Domestic Distribution

种群 Population

种群数量 Population Size	种群分布估计，新疆 800～1,100 只，河北 700～900 只，甘肃 450～750 只，山西 80～110 只，北京 25～50 只，内蒙古 50～100 只，黑龙江 25～30 只，总计 1,800～2,600 只 / It is estimated that there are 800～1,100 individuals in Xinjiang, 700～900 individuals in Hebei, 450～750 individuals in Gansu, 80～100 individuals in Shanxi 25～50 individuals in Beijing, 50～100 individuals in Inner Mongolia (Nei Mongol), 25～30 individuals in Heilongjiang, and totally 1,800～2,600 individuals
种群趋势 Population Trend	稳定 / Stable

生境与生态系统 Habitat (s) and Ecosystem (s)

生　　境 Habitat(s)	栖息于开阔水域（湿地）、草原、绿洲（胡杨林）、森林。营巢于大树或悬崖上。栖息地海拔可达 3,100 m。食物有鱼类、两栖类、昆虫等。在新疆等地为夏候鸟，冬季迁徙至南部各地，包括西藏、云南等，一部分黑鹳经过新疆迁徙到南亚（巴基斯坦、印度）越冬（马鸣等，2004）/ *C. nigra* inhabits in open waters (wetlands), grasslands, oasis (Tugai forests), and forests. It nests in big tree or cliffs at 0～3,100 m. The diet includes fish, amphibians and insects, *etc*. It breeds in Xinjiang, and migrates to the south, *e.g.* Tibet (Xizang), Yunnan, *etc*. Some winter in South Asia (Pakistan, India) based on the satellite tracking (Ma *et al.*, 2004)
生态系统 Ecosystem(s)	陆地生态系统 / Terrestrial Ecosystem

威胁 Threat (s)

主要威胁 Major Threat(s)	过度的森林砍伐、偷猎、栖息地改变、农业开垦以及高压线电击、风力发电机桨叶撞击、药物中毒、自然湖泊萎缩及改造成为水库（浅水区的消失）、过度捕捞导致食物资源匮乏等（苏化龙，1989；刘焕金和苏化龙，1990；魏顺德等，1990；马鸣等，1993，1997a；马鸣，2011）/ The major threats are over deforestation, illegal hunting, habitat loss, agricultural reclamation, attack by power-lines and fans of wind power generation, pesticides and other chemicals, loss of shallow water area by reduction of natural lakes and construction of dams, food resource shortage owing to over fishing (Su, 1989; Liu and Su, 1990; Wei *et al.*, 1990; Ma *et al.*, 1993, 1997; Ma, 2011)

保护级别与保护行动 Protection Category and Conservation Action (s)

IUCN 红色名录 (2016) IUCN Red List (2016)	无危 / LC
保护行动 Conservation Action(s)	列入国家 I 级重点保护野生动物、CITES 公约附录 II 和迁徙物种保护公约 (CMS) 附录 I。经过几十年的保护，种群数量相对稳定。过去只在少数省区有记录，现在几乎全国大部分省区都有记录（国家林业局，2009；郑光美，2017）。受到人们的关注和良好的保护 / It has been listed as the first class in National Key Protected Wild Animal List, in the Appendix II of CITES and Appendix I of the Convention on Migratory Species. The population is relatively stable under the conservation in the past tens of years. There are records in most of the provinces, while only a few in the past (SFA, 2009; Zheng Guangmei, 2017). The species receives high attention and is well conserved

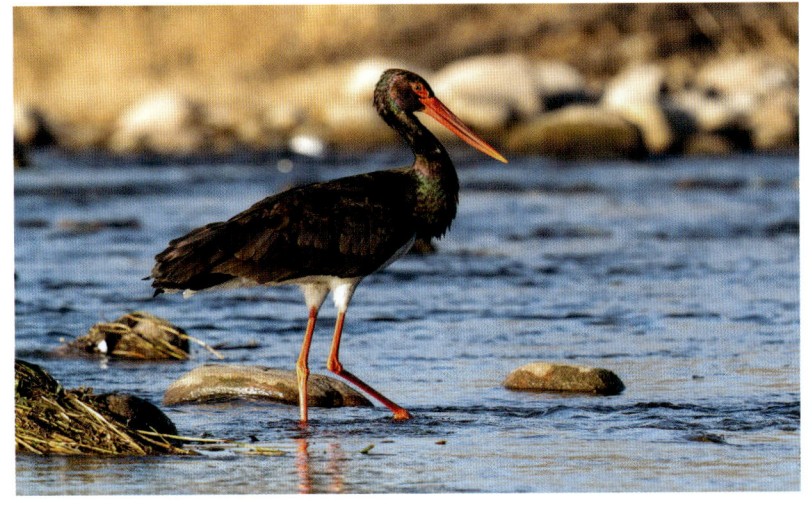

黑鹳 *Ciconia nigra*　　　　　　　　　　佳俊 摄　By Jiajun Zhou

黄嘴白鹭
Egretta eulophotes
易危 VU C2a(i)

| 数据缺乏 DD | 无危 LC | 近危 NT | **易危 VU** | 濒危 EN | 极危 CR | 区域灭绝 RE | 野外灭绝 EW | 灭绝 EX |

分类地位 Taxonomic Status

动物界 Animalia	脊索动物门 Chordata	鸟 纲 Aves	鹈形目 Pelecaniformes	鹭 科 Threskiornithidae
学 名 Scientific Name		*Egretta eulophotes*		
命名人 Species Authority		Swinhoe, 1860		
英文名 English Name(s)		Chinese Egret		
同物异名 Synonym(s)		*Herodias eulophotes*		
种下单元评估 Infra-specific Taxa Assessed		无 / None		

评估信息 Assessment Information

评估年份 Year Assessed	2016
评定人 Assessor(s)	马志军 / Zhijun Ma
其他贡献人 Other Contributor(s)	陈水华 / Shuihua Chen

理由 Justification: 黄嘴白鹭种群数量较少，且可能呈下降趋势。滨海湿地和海岛的围垦开发以及环境污染所导致的栖息地丧失和退化，人类活动干扰以及捡拾鸟卵等对其种群均带来不利影响。因此，列为易危等级 / The population size of *Egretta eulophotes* is small and might be undergoing a decline. The threats include habitat loss and deterioration owing to reclamation and development of coastal wetlands and islands, as well as environmental pollution, human disturbance, and egg collection. Thus, it is listed as Vulnerable

评估历史 Assessment History: 《国家重点保护野生动物名录》: II (1989, 2021); 《中国濒危动物红皮书·鸟类》(1998): 濒危; 《中国物种红色名录: 第二卷 脊椎动物 下册》(2009): 近危 / *National Key Protected Wild Animal List*: II (1989, 2021); *China Red Data Book of Endangered Animals*: *Aves* (1998): E; *China Species Red List*: Vol. II Vertebrates Part 2 (2009): NT

地理分布 Geographical Distribution

国内分布 Domestic Distribution

过去分布广泛，现已稀少。主要在沿海地区分布。在我国繁殖于辽东半岛、山东、江苏、浙江的沿海岛屿，在华东和华南地区越冬 / It was widely distributed while now is rare. It uses coastal areas, breeding at coastal islands in Liaodong Peninsula, Shandong, Jiangsu and Zhejiang, and wintering at East and South China

分布标注 Distribution Note

非特有种 / Non-endemic

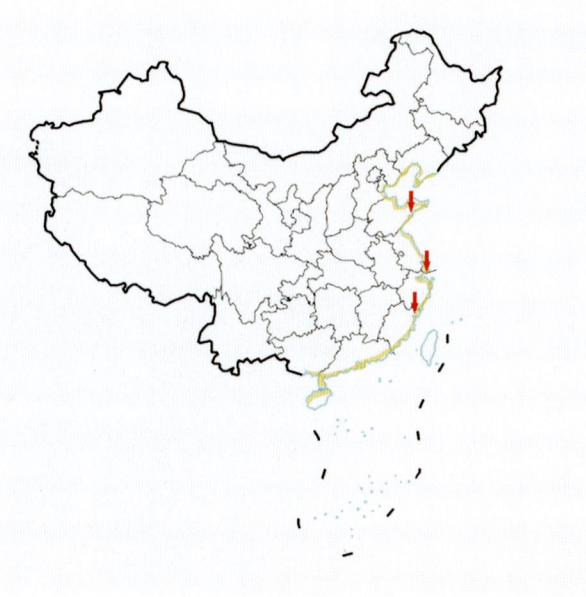

国内分布图
Map of Domestic Distribution

种群 Population

种群数量 Population Size	估计我国种群数量为 1,500～2,000 只，但缺乏具体的调查数据 / The population size is estimated to be 1,500～2,000 individuals, while no exact survey data
种群趋势 Population Trend	目前缺乏种群数量的监测数据，但可能呈下降趋势 / There is no monitoring on population size, while it is estimated to be decreasing

生境与生态系统 Habitat(s) and Ecosystem(s)

生　　境 Habitat(s)	栖息于滨海地区的潮间带、盐田以及附近的河岸、稻田，有集群营巢的习性，与其他鹭类集群繁殖。常在滩涂、河口、湖泊等浅水水域涉水觅食小鱼、虾以及水生无脊椎动物 / *E. eulophotes* inhabits in intertidal flats, salt fields at coastal area and adjacent river banks and fields. It nests in groups, and sometimes with other egrets or herons. It forages for fish, crayfish and invertebrates in shallow waters of muddy flats, estuaries, lakes
生态系统 Ecosystem(s)	湿地生态系统 / Wetland Ecosystem

威胁 Threat(s)

主要威胁 Major Threat(s)	历史上因采集黄嘴白鹭的羽毛作为饰品而导致其被大量捕杀。近年来，滨海地区滩涂湿地的围垦开发以及水体污染给黄嘴白鹭的栖息地带来严重威胁。在黄嘴白鹭繁殖季节，一些渔民捡拾鸟卵对其繁殖造成破坏性影响 / It has been listed as the second class in *National Key Protected Wild Animal List*. Large number of *E. eulophotes* were killed for its feather as ornaments in history. Recent years, the major threats are development and reclamation of coastal flats and water pollution. Besides, egg collection by fishmen would destroy the breeding of *E. eulophotes*

保护级别与保护行动 Protection Category and Conservation Action(s)

IUCN 红色名录 (2016) IUCN Red List (2016)	易危 / VU C2a (i)
保护行动 Conservation Action(s)	列入国家 I 级重点保护野生动物和迁徙物种保护公约 (CMS) 附录 II / It has been listed as the first class in National Key Protected Wild Animal List and in the Appendix II of the Convention on Migratory Species

黄嘴白鹭 *Egretta eulophotes*　　　　　　　　　　　　　　　　袁晓 摄　By Xiao Yuan

林雕
Ictinaetus malalensis
易危 VU A2cd

数据缺乏 DD	无危 LC	近危 NT	易危 VU	濒危 EN	极危 CR	区域灭绝 RE	野外灭绝 EW	灭绝 EX

分类地位 Taxonomic Status

动物界 Animalia	脊索动物门 Chordata	鸟纲 Aves	鹰形目 Accipitriformes	鹰科 Accipitridae
学 名 Scientific Name		*Ictinaetus malalensis*		
命名人 Species Authority		Temminck, 1822		
英文名 English Name(s)		Black Eagle		
同物异名 Synonym(s)		无 / None		
种下单元评估 Infra-specific Taxa Assessed		无 / None		

评估信息 Assessment Information

评估年份 Year Assessed	2016
评定人 Assessor(s)	董路 / Lu Dong
其他贡献人 Other Contributor(s)	无 / None

理由 Justification：林雕的分布区较广泛，但各种群的数量均不多，栖息地破碎化的情况较严重。中国的林雕种群隔离分布于台湾、华东山地和西南山地，种群数量约1万只，且存在快速下降的趋势。因此，列为易危等级 / The distribution area of *Ictinaetus malayensis* is large, while the size of each population is small and the habitats are fragmented. It occurs in Taiwan, mountain area in east and southeast China. The population in China is c. 10,000 individuals and is undergoing a rapid decline. Thus, it is listed as Vulnerable

评估历史 Assessment History：《国家重点保护野生动物名录》：II (1989, 2021)；《中国濒危动物红皮书·鸟类》(1998)：稀有；《中国物种红色名录：第二卷 脊椎动物 下册》(2009)：无危 / National Key Protected Wild Animal List: II (1989, 2021); China Red Data Book of Endangered Animals: Aves (1998): R; China Species Red List: Vol. II Vertebrates Part 2 (2009): LC

地理分布 Geographical Distribution

国内分布 Domestic Distribution

分布于陕西、西藏、青海、云南、四川、安徽、江西、江苏、浙江、福建、广东、海南、台湾 / It occurs in Shaanxi, Tibet (Xizang), Qinghai, Yunnan, Sichuan, Anhui, Jiangxi, Jiangsu, Zhejiang, Fujian, Guangdong, Hainan, Taiwan

分布标注 Distribution Note

非特有种 / Non-endemic

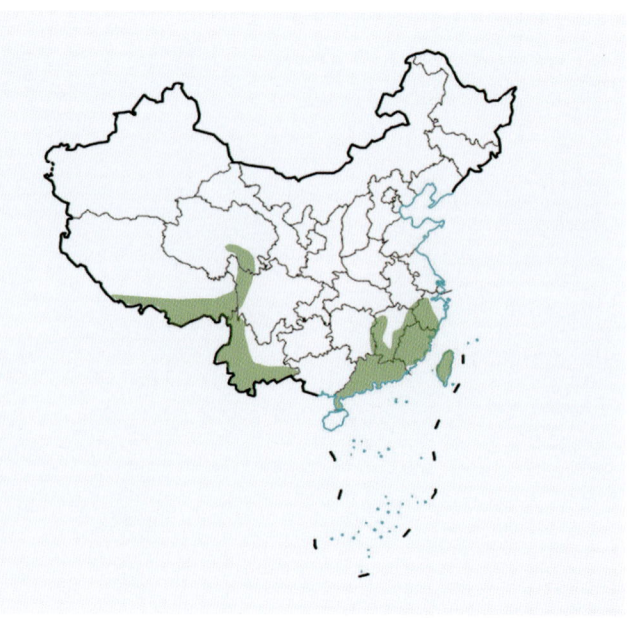

国内分布图
Map of Domestic Distribution

种群 Population

种群数量 Population Size	种群数量估计不足 10,000 只 (Brazil,2009) / The population size is estimated to be less than 10,000 individuals (Brazil, 2009)
种群趋势 Population Trend	下降 / Decreasing

生境与生态系统 Habitat (s) and Ecosystem (s)

生　　境 Habitat(s)	主要栖息于热带和亚热带的山地常绿阔叶林 / *I. malalensis* inhabits in tropical and subtropical evergreen broad-leaved forests
生态系统 Ecosystem(s)	陆地生态系统 / Terrestrial Ecosystem

威胁 Threat (s)

主要威胁 Major Threat(s)	适宜栖息地的快速丧失是导致生境破碎化和种群数量下降的主要原因 / The major threats are habitat loss and fragmentation

保护级别与保护行动 Protection Category and Conservation Action (s)

IUCN 红色名录 (2016) IUCN Red List (2016)	无危 / LC
保护行动 Conservation Action(s)	列入国家 II 级重点保护野生动物和 CITES 公约附录 II / It has been listed as the second class in National Key Protected Wild Animal List and the Appendix II of the CITES

林雕 *Ictinaetus malalensis*　　　巴赫 摄　By Bahe

靴隼雕
Hieraaetus pennatus

易危 VU A2cd; C1

数据缺乏 DD	无危 LC	近危 NT	**易危 VU**	濒危 EN	极危 CR	区域灭绝 RE	野外灭绝 EW	灭绝 EX

分类地位 Taxonomic Status

动物界 Animalia	脊索动物门 Chordata	鸟纲 Aves	鹰形目 Accipitriformes	鹰科 Accipitridae

学 名 Scientific Name	*Hieraaetus pennatus*
命 名 人 Species Authority	Gmelin, 1788
英 文 名 English Name(s)	Booted Eagle
同物异名 Synonym(s)	*Hieraaetus fasciatus*
种下单元评估 Infra-specific Taxa Assessed	无 / None

评估信息 Assessment Information

评 估 年 份 Year Assessed	2016
评 定 人 Assessor(s)	董路 / Lu Dong
其他贡献人 Other Contributor(s)	无 / None

理由 Justification: 靴隼雕的分布区非常广泛，国内分布区也较广，但数量较少，根据全球种群数量及国内适宜分布区所占比例估算，国内的种群数量不足 10,000 只，且其栖息地近几年来快速减少，在可预见的将来种群数量及成熟个体数可能有明显的下降。因此，列为易危等级 / The distribution of *Hieraaetus pennatus* is wide in China and globally, while population size is small. The population size is estimated to be 10,000 individuals in China based on the global population and suitable distribution area. Its habitat is undergoing a rapid decline in recent years, and the population size and mature individuals might be decreasing significantly in the foreseeable future. Thus, it is listed as Vulnerable

评估历史 Assessment History: 《国家重点保护野生动物名录》：Ⅱ (1989，2021)；《中国物种红色名录：第二卷 脊椎动物 下册》(2009)：无危 / *National Key Protected Wild Animal List*: II (1989, 2021); *China Species Red List*: *Vol. II Vertebrates Part 2* (2009): LC

地理分布 Geographical Distribution

国内分布 Domestic Distribution

分布于黑龙江、吉林、辽宁、北京、河南、内蒙古、甘肃、新疆北部、西藏南部、四川、江苏 / It occurs in Heilongjiang, Jilin, Liaoning, Beijing, Henan, Inner Mongolia (Nei Mongolia), Gansu, north Xinjiang, south Tibet (Xizang), Sichuan, Jiangsu.

分布标注 Distribution Note

非特有种 / Non-endemic

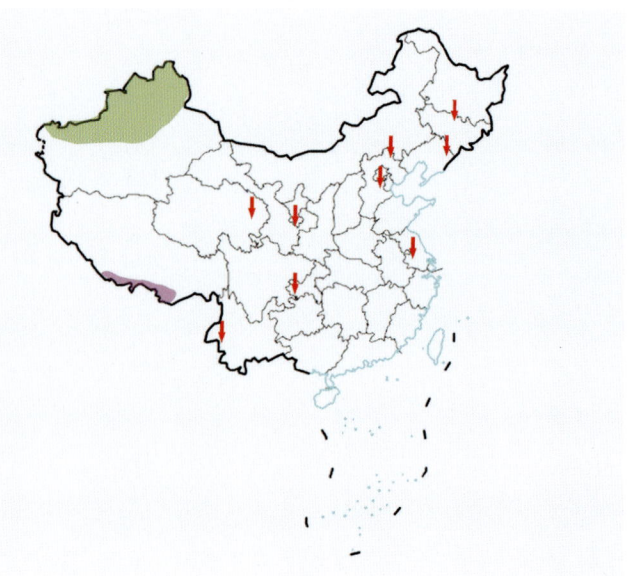

国内分布图
Map of Domestic Distribution

种群 Population

种群数量 Population Size	全球种群数量约 150,000 只，我国种群数量根据适宜栖息地面积估算，不足 10,000 只 / The global population size is estimated to be 150,000 individuals, and the population size in China is less than 10,000 individuals based on estimate of suitable distribution area
种群趋势 Population Trend	下降 / Decreasing

生境与生态系统 Habitat (s) and Ecosystem (s)

生　　境 Habitat(s)	主要栖息于开阔的森林，尤其是具有大面积林窗的区域，分布海拔可高至 3,000 m。筑巢于高大乔木上 (del Hoyo *et al.*，1994) / *H. pennatus* inhabits in open forests, especially with large forest window, and occurs as high as 3,000 m. It nests in tall woods (del Hoyo *et al.*, 1994)
生态系统 Ecosystem(s)	陆地生态系统 / Terrestrial Ecosystem

威胁 Threat (s)

主要威胁 Major Threat(s)	适宜森林栖息地的丧失、人类干扰是一些地理种群的主要威胁因子 (del Hoyo *et al.*,1994, Ferguson-Lees and Christie, 2001)。大型风力发电场也会对本物种造成威胁 / Major threats include the loss of suitable forest habitat and human disturbance (del Hoyo *et al.*,1994, Ferguson-Lees and Christie, 2001). It is also highly vulnerable to the impacts of potential wind energy developments

保护级别与保护行动 Protection Category and Conservation Action (s)

IUCN 红色名录 (2016) IUCN Red List (2016)	无危 / LC
保护行动 Conservation Action(s)	列入国家 II 级重点保护野生动物和 CITES 公约附录 II / It has been listed as the second class in National Key Protected Wild Animal List, and in the Appendix II of the CITES

靴隼雕 *Hieraaetus pennatus*　　　　　　　　　　　　　　　　　　张守玉 摄　By Shouyu Zhang

草原雕
Aquila nipalensis

易危 VU A2cd; C1+2b

| 数据缺乏 DD | 无危 LC | 近危 NT | **易危 VU** | 濒危 EN | 极危 CR | 区域灭绝 RE | 野外灭绝 EW | 灭绝 EX |

分类地位 Taxonomic Status

动物界 Animalia	脊索动物门 Chordata	鸟纲 Aves	鹰形目 Accipitriformes	鹰科 Accipitridae
学名 Scientific Name		*Aquila nipalensis*		
命名人 Species Authority		Hodgson, 1833		
英文名 English Name(s)		Steppe Eagle		
同物异名 Synonym(s)		无 / None		
种下单元评估 Infra-specific Taxa Assessed		无 / None		

评估信息 Assessment Information

评估年份 Year Assessed	2016
评定人 Assessor(s)	董路 / Lu Dong
其他贡献人 Other Contributor(s)	无 / None

理由 Justification: 国内分布区内近年来的观察记录表明，草原雕适宜栖息地的利用方式发生了显著改变，人为干扰等因素使其在可预见的未来种群数量及成熟个体数量将明显下降。因此，列为易危等级 / The use of suitable habitats has changed greatly according to the records of national distribution. The population size and mature individuals of *Aquila nipalensis* might be decreasing in the foreseeable future considering the human disturbance, *etc*. Thus, it is listed as Vulnerable

评估历史 Assessment History: 《国家重点保护野生动物名录》：Ⅱ (1989)，Ⅰ (2021)；《中国濒危动物红皮书·鸟类》(1998)：易危；《中国物种红色名录：第二卷 脊椎动物 下册》(2009)：无危 / *National Key Protected Wild Animal List*: Ⅱ (1989), Ⅰ (2021); *China Red Data Book of Endangered Animals: Aves* (1998): V; *China Species Red List: Vol. II Vertebrates Part 2* (2009): LC

地理分布 Geographical Distribution

国内分布 Domestic Distribution

繁殖于吉林、辽宁、北京、天津、河北北部、河南、山西、内蒙古、宁夏、甘肃、新疆、西藏、青海，越冬于长江以南地区，迁徙经过中部各省 / It breeds in Jilin, Liaoning, Beijing, Tianjin, north Hebei, Henan, Shanxi, Inner Mongolia (Nei Mongol), Ningxia, Gansu, Xinjiang, Tibet (Xizang), Qinghai, winters in south part of Changjiang River, and pass through the central provinces when migrating

分布标注 Distribution Note

非特有种 / Non-endemic

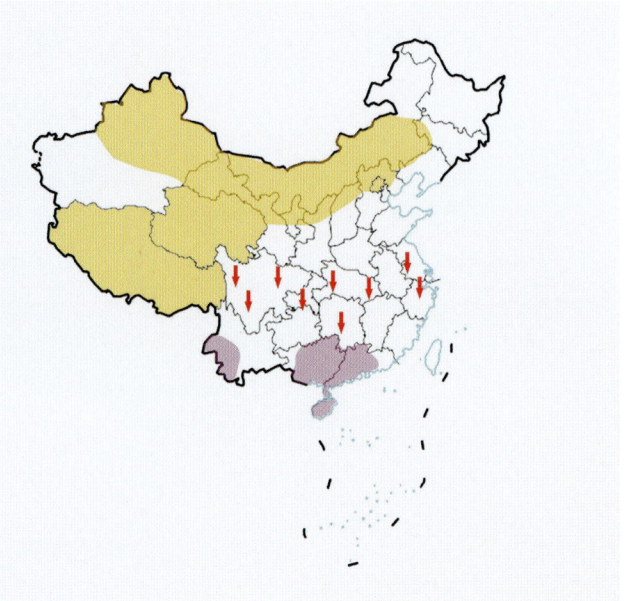

国内分布图
Map of Domestic Distribution

种群 Population

种群数量 Population Size	全球种群数量超过 100,000 只 (Ferguson-Lees and Christie, 2001)。我国种群数量尚无确切调查数据 / The global population size is more than 100,000 individuals (Ferguson-Lees and Christie, 2001). No specific survey data on population size in China yet
种群趋势 Population Trend	下降 / Decreasing

生境与生态系统 Habitat(s) and Ecosystem(s)

生　　境 Habitat(s)	主要栖息于草原及半干旱半湿润地区，海拔 200～2,300 m 的亚高山区域均有分布。营巢于地面上，近年来由于栖息地的破坏，巢址有向灌丛和乔木上转移的趋势 (del Hoyo et al., 1994) / *A. nipalensis* inhabits areas of steppe and semi-desert, and is recorded breeding at 200～2,300 m in mountainous regions. Nests have traditionally been built as large platforms on the ground, although recent habitat alterations seem to have caused a shift to building a few metres higher in bushes or trees (del Hoyo et al., 1994)
生态系统 Ecosystem(s)	陆地生态系统 / Terrestrial Ecosystem

威胁 Threat(s)

主要威胁 Major Threat(s)	草原利用方式的改变和农业开荒等人为活动的干扰，是草原雕的主要威胁因子。此外，新架设的电线和大型风力发电机组也对本物种的生存造成了一定的威胁 (del Hoyo et al., 1994；Strix，2012) / The major threats are change of use of steppe and human disturbance of agriculture development. Besides, it is also adversely affected by power lines and is very highly vulnerable to the impacts of potential wind energy developments (del Hoyo et al., 1994; Strix, 2012)

保护级别与保护行动 Protection Category and Conservation Action(s)

IUCN 红色名录 (2016) IUCN Red List (2016)	濒危 / EN A2abcd+3bcd+4abcd
保护行动 Conservation Action(s)	列入国家 I 级重点保护野生动物和 CITES 公约附录 II / It has been listed as the first class in National Key Protected Wild Animal List and in the Appendix II of the CITES

草原雕 *Aquila nipalensis*　　　　　　　　　　　　　　　　　　　　　李维 摄　By Wei Li

金雕
Aquila chrysaetos

易危 VU A2bcde+3bcde+4bcde; C2a(i)

| 数据缺乏 DD | 无危 LC | 近危 NT | 易危 VU | 濒危 EN | 极危 CR | 区域灭绝 RE | 野外灭绝 EW | 灭绝 EX |

分类地位 Taxonomic Status

动物界 Animalia	脊索动物门 Chordata	鸟纲 Aves	鹰形目 Accipitriformes	鹰科 Accipitridae
学名 Scientific Name		*Aquila chrysaetos*		
命名人 Species Authority		Linnaeus, 1758		
英文名 English Name(s)		Golden Eagle		
同物异名 Synonym(s)		无 / None		
种下单元评估 Infra-specific Taxa Assessed		青藏高原、帕米尔高原、天山、阿尔泰山等地金雕繁殖种群未来20年的灭绝概率小于10% (Ma *et al.*, 2010; Ding *et al.*, 2013) / The analysis of population viability of *A. chrysaetos* in Qinghai-Tibet (Xizang) Plateau, Pamir Plateau, Tianshan, and Altay Shan indicated that the probability of extinction in 20 years is less than 10% (Ma *et al.*, 2010; Ding *et al.*, 2013)		

评估信息 Assessment Information

评估年份 Year Assessed	2016
评定人 Assessor(s)	马鸣 / Ming Ma
其他贡献人 Other Contributor(s)	邢莲莲、丁鹏、张同、赵序茅、徐峰、陈莹、胡宝文、吴逸群、Andrew Dixon 等 / Lianlian Xing, Peng Ding, Tong Zhang, Xumao Zhao, Feng Xu, Ying Chen, Baowen Hu, Yiqun Wu, Andrew Dixon *et al.*

理由 Justification: 金雕曾经广泛分布于北半球的高山、森林、草原地区，由于人类的过度开发、环境破坏和滥捕滥猎，近35年来种群数量持续下降，国内的种群数量已不足10,000只 (Ma *et al.*, 2010)。因此，列为易危等级 / *Aquila chrysaetos* widely distributed in moutains, forests, grasslands in north hemisphere. The population has been decreasing for 35 years owing to the over development, environmental damage and illegal hunting (tradition of training eagles). The population in China is less than 10,000 individuals (Ma *et al.*, 2010). Thus, it is listed as Vulnerable

评估历史 Assessment History:《国家重点保护野生动物名录》：I (1989，2021)；《中国濒危动物红皮书·鸟类》(1998)：易危；《中国物种红色名录：第二卷 脊椎动物 下册》(2009)：近危 / *National Key Protected Wild Animal List*: I (1989, 2021); *China Red Data Book of Endangered Animals*: *Aves* (1998): V; *China Species Red List*: *Vol. II Vertebrates Part 2* (2009): NT

地理分布 Geographical Distribution

国内分布 Domestic Distribution
在吉林和辽宁为夏候鸟，是除广西、海南、台湾外各省的留鸟 / It's summer visitor in Jilin and Liaoning, and resident in other provinces except Guangxi, Hainan and Taiwan
分布标注 Distribution Note
非特有种 / Non-endemic

国内分布图
Map of Domestic Distribution

种群 Population

种群数量 Population Size

金雕的分布虽然广泛,但数量稀少。估计内蒙古呼伦贝尔草原分布 20 余只。自 2004 年以来,在新疆卡拉麦里山区金雕繁殖密度从 1.67 对(巢)/1,000 km² 减少到 2011 年的 0.37 对(巢)/1,000 km² (Ma, 2013)。分布密度比邻国要低一些 (Karyakin et al., 2010)。依此密度推算,在新疆的种群数量 4,000～5,000 只,乐观估计,我国的种群数量(主要分布在西部)大约为 10,000 只 (Ma et al., 2010)。这个数字远低于 2009 年公布的金雕种群数据(约 27,000 只)/ The distribution area of *A. chrysaetos* is large while the population is small. It is estimated that there are 20 individuals in Hunlunbuir Pasture of Inner Mongolia (Nei Mongol) (Liu, 2014). The density in Kalamaili Mountain area of Xinjiang has decreased from 1.67 pairs/1,000 km² in 2004 to 0.37 pairs/1,000 km² in 2011 (Ma, 2013). The density is less than the adjacent countries (Karyakin et al., 2010). It is estimated that the population size in Xinjiang is 4,000～5,000 individuals and the population in China is 10,000 individuals (Ma et al., 2010), which is far less than published data in 2009 (27,000 individuals)

种群趋势 Population Trend

近 30 年来,金雕种群数量呈快速下降的趋势,分布区域萎缩 / The population has been rapidly decreasing for 30 years and the distribution area has also decreased

生境与生态系统 Habitat (s) and Ecosystem (s)

生 境 Habitat(s)

生活于多山或丘陵地区,活动于开阔的草原、荒漠、丘陵地带。繁殖期为 3～7 月,多营巢于岩洞里或者山丘悬崖之处 (Zhao et al., 2013)。在中国西部,栖息地海拔 500～5,500 m。以蹄类(幼体)、野兔、旱獭、狐狸、鼠兔、鸟类为主食,也采食蛇类和老鼠 / *A. chrysaetos* inhabits in hilly or mountainous areas, especially the cliffs in valley. It occurs in open pastures, deserts, and hills in west China at 500～5,500 m. It breeds in March to July, and nests in cracks or the cliffs (Zhao et al., 2013). Its diet includes young ungulate, rabbit, marmot, fox, pika and birds, and sometimes snakes and rats

生态系统 Ecosystem(s)

陆地生态系统 / Terrestrial Ecosystem

威胁 Threat (s)

主要威胁 Major Threat(s)

近年来,人类对自然资源的过度开发利用,特别是过度放牧、农业垦荒、非法捕猎,对金雕种群造成强烈干扰。由于栖息地(营巢地)的迅速丧失,大规模的草原投毒灭鼠及兽药滥用,农药化肥、工业重金属中毒,疾病传染,动物迁徙路线上的高压电网电击及交通设施碰撞,及其他人为干扰因子等,不仅造成直接伤害,还会引发食物资源的持续匮乏,直接影响大型猛禽的繁衍 (Oaks et al., 2004;梅宇等, 2008;Ma et al., 2010;丁鹏, 2013) / Major threats include the over use of natural resources, especially over grazing, reclamation and over hunting. It is vulnerable to the rapid loss of breeding sties, abuse of pesticide and other agricultural chemicals, industrial heavy metal, infection of diseases, attack by high power and traffic facilities and other human disturbance, which might bring direct hurt as well as food shortage (Oaks et al., 2004; Li, 2004; Mei Yu et al., 2008; Ma et al., 2010; Ding, 2013)

保护级别与保护行动 Protection Category and Conservation Action (s)

IUCN 红色名录 (2016) / IUCN Red List (2016)

无危 / LC

保护行动 Conservation Action(s)

列入国家 I 级重点保护野生动物和 CITES 公约附录 II,西部地区的驯雕习俗与动物保护产生严重冲突。一些地区滥捕偷猎、草原灭鼠农药使其二次中毒等时有发生,需要重点关注 / It has been listed as the first class in National Key Protected Wild Animal List and the Appendix II of the CITES. With the resurgence of tradition of eagle training, there is serious conflict with the wildlife conservation. The over and illegal hunting, and poisoned by pesticide should also be paid highly attention

金雕
Aquila chrysaetos

李全民 摄
By Quanmin Li

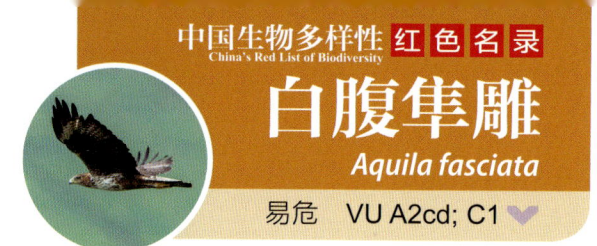

白腹隼雕
Aquila fasciata

易危 VU A2cd; C1

| 数据缺乏 DD | 无危 LC | 近危 NT | 易危 VU | 濒危 EN | 极危 CR | 区域灭绝 RE | 野外灭绝 EW | 灭绝 EX |

分类地位 Taxonomic Status

动物界 Animalia	脊索动物门 Chordata	鸟 纲 Aves	鹰形目 Accipitriformes	鹰 科 Accipitridae
学 名 Scientific Name		*Aquila fasciata*		
命 名 人 Species Authority		Vieillot, 1822		
英 文 名 English Name(s)		Bonelli's Eagle		
同物异名 Synonym(s)		无 / None		
种下单元评估 Infra-specific Taxa Assessed		无 / None		

评估信息 Assessment Information

评估年份 Year Assessed	2016
评 定 人 Assessor(s)	马强 / Qiang Ma
其他贡献人 Other Contributor(s)	无 / None

理由 Justification: 白腹隼雕在全球的分布区较广大；栖息地范围没有显著缩小或严重破碎化趋势；种群规模稍小，且有一定下降趋势，但尚未达到濒危的标准。在我国，其数量稀少，应加强保护。因此，列为易危等级 / The global distribution area of *A. fasciata* is wide, and area of habitat is not significantly decreasing or being fragmented. The population size is not large and undergoing a decline while not approaching the thresholds for Endangered or locally extinct. The conservation of the species should be improved considering the small population size of China. Thus, it is listed as Vulnerable

评估历史 Assessment History:《国家重点保护野生动物名录》：II (1989，2021);《中国濒危动物红皮书·鸟类》(1998)：稀有;《中国物种红色名录：第二卷 脊椎动物 下册》(2009)：无危 / *National Key Protected Wild Animal List*: II (1989, 2021); *China Red Data Book of Endangered Animals* (1998): R; *China Species Red List*: *Vol. II Vertebrates Part 2* (2009): LC

地理分布 Geographical Distribution

国内分布 Domestic Distribution

分布于北京、河北、河南、云南东部、四川、贵州、湖北、江西、江苏、上海、浙江、福建、广东、香港、澳门、广西 / It occurs Beijing, Hebei, Henan, east Yunnan, Sichuan, Guizhou, Hubei, Jiangxi, Jiangsu, Shanghai, Zhejiang, Fujian, Guangdong, Hong Kong, Macao, Guangxi

分布标注 Distribution Note

非特有种 / Non-endemic

国内分布图
Map of Domestic Distribution

种群 Population

种群数量 Population Size	未知 / Unknown
种群趋势 Population Trend	缺少长期监测数据，估计呈下降态势 / There is no long term monitoring data, and it is estimated to be decreasing

生境与生态系统 Habitat (s) and Ecosystem (s)

生　　境 Habitat(s)	栖息于开阔的生境中，从干旱到半湿润的山地均有分布，也栖息于森林生境。亚成体喜欢选择较开阔水域附近的栖息地 (Ferguson-Lees and Christie, 2001) / *A. fasciata* occupies mountainous, rocky, arid to semi-moist habitat. It generally occurs in open areas but also occupies woodland. Juveniles often occupy areas near large water bodies (Ferguson-Lees and Christie, 2001)
生态系统 Ecosystem(s)	陆地生态系统 / Terrestrial Ecosystem

威胁 Threat (s)

主要威胁 Major Threat(s)	森林采伐和农业开发使其栖息地破碎化、栖息地质量下降，同时也导致其猎物资源显著减少。农药、除草剂等污染物通过生物富集作用给其带来严重威胁；高压输电设施增大了其撞击受伤危险；对成鸟的盗猎减小了种群规模；而对鸟巢的破坏和对其巢区的干扰会降低繁殖成功率，进而影响种群的复壮 / The species is threatened due to habitat fragmentation and degradation caused by deforestation and development of agriculture, which might also decrease the food source. It is also vulnerable to chemicals (pesticide, herbicide, *etc.*), and attack by high power facilities. The population has decreased by illegal hunting, and the disturbance and destroy of nests might decrease the breeding success

保护级别与保护行动 Protection Category and Conservation Action (s)

IUCN 红色名录 (2016) IUCN Red List (2016)	无危 / LC
保护行动 Conservation Action(s)	列入国家 II 级重点保护野生动物和 CITES 公约附录 II / It has been listed as the second class in National Key Protected Wild Animal List and Appendix II of the CITES

白腹隼雕 *Aquila fasciata*　　　　　　　　　　　　　　　　陈光辉 摄　By Guanghui Chen

栗鸢
Haliastur indus
易危 VU C1

| 数据缺乏 DD | 无危 LC | 近危 NT | **易危 VU** | 濒危 EN | 极危 CR | 区域灭绝 RE | 野外灭绝 EW | 灭绝 EX |

分类地位 Taxonomic Status

动物界 Animalia	脊索动物门 Chordata	鸟纲 Aves	鹰形目 Accipitriformes	鹰科 Accipitridae
学 名 Scientific Name		*Haliastur indus*		
命 名 人 Species Authority		Boddaert, 1783		
英 文 名 English Name(s)		Brahminy Kite		
同物异名 Synonym(s)		无 / None		
种下单元评估 Infra-specific Taxa Assessed		无 / None		

评估信息 Assessment Information

评估年份 Year Assessed	2016
评 定 人 Assessor(s)	丁平 / Ping Ding
其他贡献人 Other Contributor(s)	无 / None

理由 Justification: 栗鸢分布广，种群趋势近于稳定，种群数量大，在世界范围内并未达到近危标准，但在中国繁殖对可能不足 100 对 (Brazil，2009)，种群数量推测为 200～5,000 只。因此，列为易危等级 / The distribution of *Haliastur indus* is wide, and global population size is large and in a stable trend, and does not approach the thresholds for Vulnerable, while the population in China has been estimated at less than 100 breeding pairs, equivalent to 200～5,000 individuals (Brazil 2009). Thus, it is listed as Vulnerable

评估历史 Assessment History: 《国家重点保护野生动物名录》：Ⅱ (1989，2021)；《中国物种红色名录：第二卷 脊椎动物 下册》(2009)：稀有 / *National Key Protected Wild Animal List*: Ⅱ (1989, 2021); *China Species Red List: Vol. II Vertebrates Part 2* (2009): R

地理分布 Geographical Distribution

国内分布 Domestic Distribution

分布于山东、西藏、云南、湖北、江西、江苏、浙江、福建、广东、香港、广西、台湾 / It occurs Shandong, Tibet (Xizang), Yunnan, Hubei, Jiangxi, Jiangsu, Zhejiang, Fujian, Guangdong, Hong Kong, Guangxi, Taiwan

分布标注 Distribution Note

非特有种 / Non-endemic

国内分布图
Map of Domestic Distribution

种群 Population

种群数量 Population Size	全球种群数量大于 100,000 只，我国种群数量可能在 200～5,000 只 / The global population size is estimated to be more than 100,000 individuals, while the population size in China has been estimated to be 200～5,000 individuals
种群趋势 Population Trend	下降 / Decreasing

生境与生态系统 Habitat (s) and Ecosystem (s)

生　　境 Habitat(s)	主要栖息于江河、湖泊、水塘、沼泽、沿海海岸和邻近的城镇与村庄。常单独在湖滨、海滨、河岸或水域与村庄上空长时间地盘旋 / *H. indus* inhabits in rivers, lakes, ponds, marshes, coastal areas and adjacent towns or villages. It might occur and soar solely in coastal areas, river banks or waters adjacent towns or villages
生态系统 Ecosystem(s)	陆地生态系统 / Terrestrial Ecosystem

威胁 Threat (s)

主要威胁 Major Threat(s)	未知 / Unknown

保护级别与保护行动 Protection Category and Conservation Action (s)

IUCN 红色名录 (2016) IUCN Red List (2016)	无危 / LC
保护行动 Conservation Action(s)	列入国家 II 级重点保护野生动物和 CITES 公约附录 II / It has been listed as the second class in National Key Protected Wild Animal List and Appendix II of the CITES

栗鸢 *Haliastur indus*　　　　　　　　　　　　　　　　　　蒲新强 摄　By Xinqiang Pu

白腹海雕
Haliaeetus leucogaster

易危 VU C2a(ii)

| 数据缺乏 DD | 无危 LC | 近危 NT | 易危 VU | 濒危 EN | 极危 CR | 区域灭绝 RE | 野外灭绝 EW | 灭绝 EX |

分类地位 Taxonomic Status

动 物 界 Animalia	脊索动物门 Chordata	鸟 纲 Aves	鹰形目 Accipitriformes	鹰 科 Accipitridae
学 名 Scientific Name		*Haliaeetus leucogaster*		
命 名 人 Species Authority		Gmelin, 1788		
英 文 名 English Name(s)		White-bellied Sea Eagle		
同物异名 Synonym(s)		无 / None		
种下单元评估 Infra-specific Taxa Assessed		无 / None		

评估信息 Assessment Information

评 估 年 份 Year Assessed	2016
评 定 人 Assessor(s)	马强 / Qiang Ma
其他贡献人 Other Contributor(s)	无 / None

理由 Justification: 在世界范围，白腹海雕分布区非常广大。栖息地缩小与破碎化程度等方面未达到易危的标准；但其种群数量较小，且呈下降趋势，在我国分布区很有限，数量稀少，需要加强保护。因此，列为易危等级 / *Haliaeetus leucogaster* has an extremely large range globally, and hence does not approach the thresholds for Vulnerable under the range criterion. The population size is small and the trend appears to be decreasing. The distribution area and population size in China is small and conservation should be improved. Thus, it is listed as Vulnerable

评估历史 Assessment History:《国家重点保护野生动物名录》：Ⅱ (1989)，Ⅰ (2021)；《中国濒危动物红皮书·鸟类》(1998)：未定；《中国物种红色名录：第二卷 脊椎动物 下册》(2009)：无危 / *National Key Protected Wild Animal List*: Ⅱ (1989), Ⅰ (2021); *China Red Data Book of Endangered Animals*: *Aves* (1998): Ⅰ; *China Species Red List*: *Vol. II Vertebrates Part 2* (2009): LC

地理分布 Geographical Distribution

国内分布 Domestic Distribution

在我国分布于内蒙古西部、福建、台湾、广东、香港、澳门、广西和海南 / It occurs in west Inner Mongolia (Nei Mongol), Fujian, Taiwan, Guangdong, Hong Kong, Macao, Guangxi and Hainan

分布标注 Distribution Note

非特有种 / Non-endemic

国内分布图
Map of Domestic Distribution

种群 Population

种群数量 Population Size	全球种群数量 10,000～100,000 只。在我国有 100～1,000 只，数量较少 / The global population size is estimated to be 10,000～100,000 individuals, while the population size in China is estimated to be 100～1,000 individuals
种群趋势 Population Trend	下降 / Decreasing

生境与生态系统 Habitat(s) and Ecosystem(s)

生境 Habitat(s)	是典型的海岸鸟类，栖息于大陆及岛屿沿海、港湾生境，有林地的内陆湿地，季雨林及红树林生境，也出现于离海岸不远的丘陵和水库上空。栖息于海拔 1,500 m 以下至海平面，一般栖息于海拔 900 m 以下。筑巢于森林或裸岩地带 / *H. leucogaster* is typical coastal species, and inhabits in coastal areas and bays, inner wetlands with forests, rainforests and mangroves. It is also seen in hills and reservoirs adjacent to the coast. It occurs at 0～1500 m, normally under 900 m. It nests in forests and rock areas
生态系统 Ecosystem(s)	陆地生态系统 / Terrestrial Ecosystem

威胁 Threat(s)

主要威胁 Major Threat(s)	环境污染对其生存影响很大，杀虫剂（如 DDT）的大量使用导致其卵壳变薄，降低孵化成功率；环境中的铅、汞和铜等重金属物质经食物链进入白腹海雕体内，对其生存构成威胁；栖息地破坏和对巢区的干扰也是重要的致危因素。白腹海雕选择悬崖、高大树木，甚至开放式的洞穴营巢繁殖，适宜巢址数量减少制约其繁殖密度，人为干扰的强度对其繁殖成功率有显著影响；渔业过度捕捞引起的食物资源减少将对其造成长远的不利影响 / It is high vulnerable to environmental pollution: abuse of pesticide (*e.g.* DDT) might attenuate eggshell and decrease the breeding success; the heavy metal (Pb, Hg, Cu, *etc.*) from food chain might threaten the survival. Besides, habitat loss and disturbance of nests are also high threats. *H. leucogaster* nests in cliffs, high trees, or even open caves. The nest density is related to suitable nesting sites and breeding success is significantly influenced by human disturbance. Over hunting and fishing might decrease food resource and exert negative effects

保护级别与保护行动 Protection Category and Conservation Action(s)

IUCN 红色名录 (2016) IUCN Red List (2016)	无危 / LC
保护行动 Conservation Action(s)	列入国家 I 级重点保护野生动物和 CITES 公约附录 II / It has been listed as the second class in National Key Protected Wild Animal List and the Appendix II of the CITES

白腹海雕 *Haliaeetus leucogaster*　　　　　沈越 摄　By Yue Shen

白尾海雕
Haliaeetus albicilla

易危　VU C1

| 数据缺乏 DD | 无危 LC | 近危 NT | **易危 VU** | 濒危 EN | 极危 CR | 区域灭绝 RE | 野外灭绝 EW | 灭绝 EX |

分类地位 Taxonomic Status

动物界 Animalia	脊索动物门 Chordata	鸟纲 Aves	鹰形目 Accipitriformes	鹰科 Accipitridae

学名 Scientific Name	*Haliaeetus albicilla*
命名人 Species Authority	Linnaeus, 1758
英文名 English Name(s)	White-tailed Sea Eagle
同物异名 Synonym(s)	*Falco albicilla*
种下单元评估 Infra-specific Taxa Assessed	无 / None

评估信息 Assessment Information

评估年份 Year Assessed	2016
评定人 Assessor(s)	王海涛 / Haitao Wang
其他贡献人 Other Contributor(s)	刘宇 / Yu Liu

理由 Justification: 白尾海雕分布范围广，但种群数量小，受人为活动干扰和非法狩猎威胁较大。因此，列为易危等级 / The distribution area of *Haliaeetus albicilla* is large while the population size is small, and it is high threatened by human disturbance and illegal hunting. Thus, it is listed as Vulnerable

评估历史 Assessment History:《国家重点保护野生动物名录》：Ⅰ (1989, 2021);《中国濒危动物红皮书·鸟类》(1998)：未定;《中国物种红色名录：第二卷 脊椎动物 下册》(2009)：近危 / *National Key Protected Wild Animal List*: Ⅰ (1989, 2021); *China Red Data Book of Endangered Animals: Aves* (1998): Ⅰ; *China Species Red List: Vol. Ⅱ Vertebrates Part 2* (2009): NT

地理分布 Geographical Distribution

国内分布 Domestic Distribution

除海南外，遍布全国。在黑龙江、内蒙古和新疆繁殖，其他地区为旅鸟或冬候鸟 / It occurs throughout of China except Hainan. It breeds in Heilongjiang, Inner Mongolia (Nei Mongol) and Xinjiang, and passenger or winter birds in other provinces

分布标注 Distribution Note

非特有种 / Non-endemic

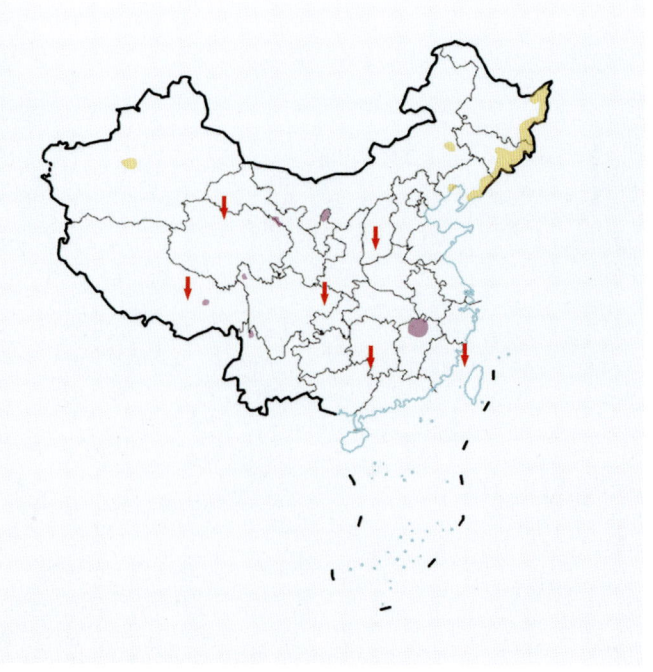

国内分布图
Map of Domestic Distribution

种群 Population

种群数量 Population Size	种群数量不明。吉林图们江下游 1988～1998 年调查为常见种 (杨兴家和金玄善，2000)；黑龙江兴凯湖 1989 年记录到 15 只 (李文发等，1994)；辽宁大连和锦州 1996～1998 年调查时记录到 1～4 只 (万冬梅等，2002)；云南纳帕海 2006～2007 年冬季为常见种，记录到 31 只 (冯理等，2008)；宁夏贺兰山地区 2009～2012 年调查估计 9～14 只 (梁军等，2013) / The population size is not clear. It was a common species in lower Tumen River, Jilin (Yang and Jin, 2000), and 15 individuals were recorded in Xingkai Lake in 1989 (Li *et al.*, 1994), and 1～4 individuals were recorded in Dalian and Jinzhou in 1996～1998 (Wan *et al.*, 2002). In winter of 2006～2007, it was reported that it was a common species in Napahai, Yunnan, and the total number was 31 (Feng *et al.*, 2008). There was about 9～14 individuals in Helan Shan in 2009～2012 (Liang *et al.*, 2013)
种群趋势 Population Trend	未知 / Unknown

生境与生态系统 Habitat (s) and Ecosystem (s)

生　　境 Habitat(s)	栖息于海岸、宽阔开放的河口、大型内陆湖泊和江河等近水地带，也见于草原和山地森林 / *H. albicilla* inhabits near the coast, wide estuaries, large lakes and rivers, and also are observed in grasslands and mountain forests
生态系统 Ecosystem(s)	陆地生态系统 / Terrestrial Ecosystem

威胁 Threat (s)

主要威胁 Major Threat(s)	人类活动对卵和雏鸟的破坏，导致繁殖失败 (李文发等，1994；傅承钊，1986)，非法狩猎和保护区旅游开发带来的环境压力 (冯理等，2008) / Eggs and chicks are used to be destroyed by human activity (Li *et al.*, 1994; Fu, 1986), and illegal hunting and tourism development in protected areas cause more environmental problems to the species (Feng *et al.*, 2008)

保护级别与保护行动 Protection Category and Conservation Action (s)

IUCN 红色名录 (2016) IUCN Red List (2016)	无危 / LC
保护行动 Conservation Action(s)	列入国家 I 级重点保护野生动物、CITES 公约附录 II 和迁徙物种保护公约 (CMS) 附录 I。缺少繁殖、生境和种群动态等基础研究资料，部分分布区处于自然保护区内 / It has been listed as the first class in National Key Protected Wild Animal List, the Appendix II of the CITES and Appendix I of the Convention on Migratory Species. Foundamental research data, such as breeding, habitat and population dynamics, are lacking, and part of the distribution areas are not covered by natural reserve

白尾海雕 *Haliaeetus albicilla*　　　　　　　　　　刘宇 摄　By Yu Liu

大鵟
Buteo hemilasius

易危 VU A2ac

| 数据缺乏 DD | 无危 LC | 近危 NT | 易危 VU | 濒危 EN | 极危 CR | 区域灭绝 RE | 野外灭绝 EW | 灭绝 EX |

分类地位 Taxonomic Status

动物界 Animalia	脊索动物门 Chordata	鸟纲 Aves	鹰形目 Accipitriformes	鹰科 Accipitridae
学名 Scientific Name		*Buteo hemilasius*		
命名人 Species Authority		Temminck and Schlegel, 1844		
英文名 English Name(s)		Upland Buzzard		
同物异名 Synonym(s)		*Aquila hemilasius*		
种下单元评估 Infra-specific Taxa Assessed		无 / None		

评估信息 Assessment Information

评估年份 Year Assessed	2016
评定人 Assessor(s)	董路 / Lu Dong
其他贡献人 Other Contributor(s)	无 / None

理由 Justification: 大鵟的分布区广泛，全球种群数量存在波动。在国内分布区，由于适宜栖息地（尤其是越冬栖息地）的丧失和杀虫剂、鼠药等有毒制剂的使用，根据观察记录推测，过去十余年内种群数量快速下降，且可预见的将来仍存在快速下降的趋势。因此，列为易危等级 / The distribution area of *Buteo hemilasius* is large, and the global population is fluctuating. The population in China is undergoing a rapid decline as a result of the suitable habitat loss (especially wintering sites) and over use of pesticide and other chemicals. The decreasing trend might continue in the foreseeable future. Thus, it is listed as Vulnerable

评估历史 Assessment History: 《国家重点保护野生动物名录》：II (1989, 2021)；《中国物种红色名录：第二卷 脊椎动物 下册》(2009)：无危 / *National Key Protected Wild Animal List*: II (1989, 2021); *China Species Red List: Vol. II Vertebrates Part 2* (2009): LC

地理分布 Geographical Distribution

国内分布 Domestic Distribution

国内东北地区为留鸟，青藏高原为夏候鸟，华北、华东、台湾和西南地区为冬候鸟 / It is resident in Northeast China, occurs in Qinghai-Tibet (Xizang) Plateau as summer visitor, and winters in Taiwan, North, East and Southwest China

分布标注 Distribution Note

非特有种 / Non-endemic

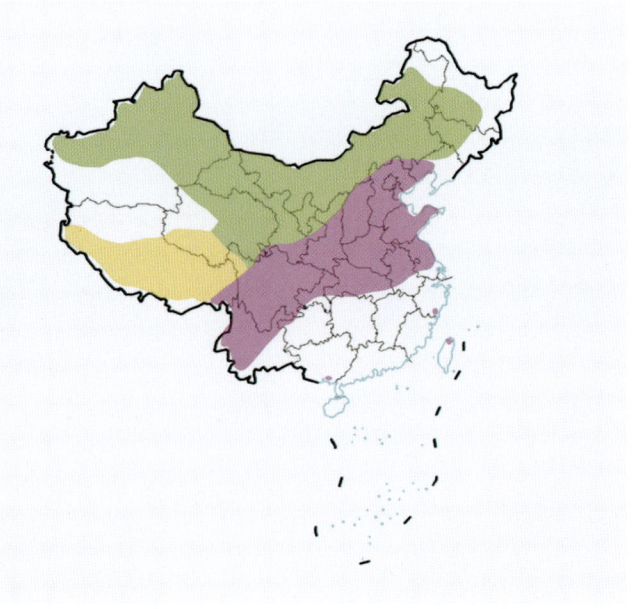

国内分布图
Map of Domestic Distribution

种群 Population

种群数量 Population Size	全球种群数量大于 10,000 只 (Ferguson-Lees and Christie，2001)，国内种群数量估计 3,000 ~ 15,000 只 (Brazil，2009) / The global population size more than 10,000 individuals (Ferguson-Lees and Christie, 2001), and the population size in China is estimated to be 3,000 ~ 15,000 individuals (Brazil，2009)
种群趋势 Population Trend	下降 / Decreasing

生境与生态系统 Habitat (s) and Ecosystem (s)

生　　境 Habitat(s)	主要栖息于山地和草原地区，冬季也常出现在丘陵、农田、沼泽、村庄，甚至是城市附近 / *B. hemilasius* inhabits in mountain areas and grasslands, and is seen in hills, farmlands, marshes, villages and even close to cities
生态系统 Ecosystem(s)	陆地生态系统 / Terrestrial Ecosystem

威胁 Threat (s)

主要威胁 Major Threat(s)	在繁殖地，农业开荒和放牧导致的草原退化是其主要威胁因子，在越冬地，适宜栖息地利用方式的改变，以及杀虫剂和鼠药的使用，是种群数量下降的主要原因 / Major threats that might reduce the population size include deterioration of grasslands owing to agricultural reclamation and grazing in breeding sites; and change of land use and over use of pesticides in wintering sites

保护级别与保护行动 Protection Category and Conservation Action (s)

IUCN 红色名录 (2016) IUCN Red List (2016)	无危 / LC
保护行动 Conservation Action(s)	列入国家 II 级重点保护野生动物和 CITES 公约附录 II / It has been listed as the second class in National Key Protected Wild Animal List and the Appendix II of the CITES

大鵟 *Buteo hemilasius*　　　　　　　　　　　　　　　　　刘怡 摄　By Yi Liu

四川林鸮
Strix davidi

易危 VU C2a(i)

| 数据缺乏 DD | 无危 LC | 近危 NT | 易危 VU | 濒危 EN | 极危 CR | 区域灭绝 RE | 野外灭绝 EW | 灭绝 EX |

分类地位 Taxonomic Status

动物界 Animalia	脊索动物门 Chordata	鸟纲 Aves	鸮形目 Strigiformes	鸱鸮科 Strigidae

学名 Scientific Name	*Strix davidi*
命名人 Species Authority	Sharpe, 1875
英文名 English Name(s)	Sichuan Wood Owl
同物异名 Synonym(s)	*Strix uralensis davidi*
种下单元评估 Infra-specific Taxa Assessed	除羽色略有差异外，与长尾林鸮 (*Strix uralensis*) 鸣声并无差异，有人认为四川林鸮是本种的一个亚种 / It is regarded by some scientists as subspecies of *S. uralensis* due to the same calling songs, although there are a few differences in plumage color

评估信息 Assessment Information

评估年份 Year Assessed	2016
评定人 Assessor(s)	孙悦华 / Yuehua Sun
其他贡献人 Other Contributor(s)	无 / None

理由 Justification: 四川林鸮分布区狭窄，种群密度低，栖息地不断缩减。因此，列为易危等级 / The distribution area of *Strix davidi* is small, the population density is low in nature and the habitats are continually decreasing. Thus, it is listed as Vulnerable

评估历史 Assessment History: 《国家重点保护野生动物名录》：II (1989)，I (2021)；《中国物种红色名录：第二卷 脊椎动物 下册》(2009)：易危 / *National Key Protected Wild Animal List*: II (1989), I (2021); *China Species Red List*: *Vol. II Vertebrates Part 2* (2009): VU

地理分布 Geographical Distribution

国内分布 Domestic Distribution
仅分布于青海东南部、四川西部、西藏东南部和北部及甘肃南部 / It only occurs in southeast Qinghai, west Sichuan, southeast and north Tibet (Xizang) and south Gansu
分布标注 Distribution Note
特有种 / Endemic

国内分布图
Map of Domestic Distribution

种群 Population

种群数量 Population Size	未有评估，但根据甘肃南部的研究，种群密度极低（汤宋华，2008；石美，2012；Fang *et al.*, 2009）/ There is no assessment on its population size. The population density is low in nature based on the study in south Gansu (Tang, 2008; Shi, 2012; Fang *et al.*, 2009)
种群趋势 Population Trend	下降 / Decreasing

生境与生态系统 Habitat (s) and Ecosystem (s)

生　　境 Habitat(s)	栖息于发育成熟的高山针叶林及针阔混交林（石美，2012）/ *S. davidi* inhabits in mature coniferous and mixed coniferous broad-leaved forest in high mountains (Shi, 2012)
生态系统 Ecosystem(s)	陆地生态系统 / Terrestrial Ecosystem

威胁 Threat (s)

主要威胁 Major Threat(s)	栖息地的丧失和破碎化是主要原因，对老龄树的择伐减少了潜在巢址也是重要因素之一 / Major threats include habitat loss and fragmentation, and decrease of potential nests as a result of deforestation of large trees

保护级别与保护行动 Protection Category and Conservation Action (s)

IUCN 红色名录 (2016) IUCN Red List (2016)	IUCN 将该物种视为 *S. uralensis* 的一个亚种，对 *S. uralensis* 的评估为无危 LC (2016) / It is regarded as subspecies of *S. uralensis* by IUCN, and *Strix uralensis* was assessed as LC
保护行动 Conservation Action(s)	列入国家 I 级重点保护野生动物和 CITES 公约附录 II。在甘肃莲花山自然保护区对其繁殖生物学有了初步研究，部分分布区位于保护区范围内，人工巢箱悬挂（Fang *et al.*, 2009）有利于其繁殖。需要对其分布范围和种群数量进行准确的估计，同时加强对栖息地的保护，完善其生态学资料 / It has been listed as the first class in National Key Protected Wild Animal List, and in the Appendix II of the CITES. Primary studies on breeding ecology has been conducted in Lianhua Shan Nature Reserve of Gansu. The nature reserves cover some of the distribution area, and the artificial nests is beneficial for its breeding (Fang *et al.*, 2009). Surveys on its distribution area and population size are urgently required, and improvement of ecological study and habitat conservation are suggested

四川林鸮 *Strix davidi*　　　　唐万玲 摄　By Wanling Tang

鬼鸮
Aegolius funereus
易危 VU C2a(i)

| 数据缺乏 DD | 无危 LC | 近危 NT | 易危 VU | 濒危 EN | 极危 CR | 区域灭绝 RE | 野外灭绝 EW | 灭绝 EX |

分类地位 Taxonomic Status

动物界 Animalia	脊索动物门 Chordata	鸟纲 Aves	鸮形目 Strigiformes	鸱鸮科 Strigidae
学名 Scientific Name		*Aegolius funereus*		
命名人 Species Authority		Linnaeus, 1758		
英文名 English Name(s)		Boreal Owl		
同物异名 Synonym(s)		*Cryptoglaux funerea*		
种下单元评估 Infra-specific Taxa Assessed		*A. f. beickianus* 亚种为中国特有，分布范围狭窄 / *A. f. beickianus* is endemic to China, and the distribution area is small		

评估信息 Assessment Information

评估年份 Year Assessed	2016
评定人 Assessor(s)	孙悦华 / Yuehua Sun
其他贡献人 Other Contributor(s)	无 / None

理由 Justification: 鬼鸮分布范围较小，分布区相互隔离，自然条件下种群密度低，受到栖息地丧失的威胁。因此，列为易危等级 / The distribution areas of *Aegolius funereus* are small and fragmented. The population density is low in nature and threatened by habitat loss. Thus, it is listed as Vulnerable

评估历史 Assessment History:《国家重点保护野生动物名录》：II (1989, 2021)；《中国濒危动物红皮书·鸟类》(1998)：未定；《中国物种红色名录：第二卷 脊椎动物 下册》(2009)：无危 / *National Key Protected Wild Animal List*: II (1989, 2021); *China Red Data Book of Endangered Animals: Aves* (1998): I; *China Species Red List: Vol. II Vertebrates Part 2* (2009): LC

地理分布 Geographical Distribution

国内分布 Domestic Distribution
为陕西、甘肃南部、青海东部、云南西北部、四川、新疆西北部、黑龙江、吉林和内蒙古东北部的留鸟 / It's resident in Shaanxi, south Gansu, east Qinghai, northwest Yunnan, Sichuan, northwest Xinjiang, Heilongjiang, Jilin and northeast Inner Mongolia (Nei Mongol)
分布标注 Distribution Note
非特有种 / Non-endemic

国内分布图
Map of Domestic Distribution

种群 Population

种群数量 Population Size	未评估，但根据甘肃南部对亚种 *A. f. beickianus* 的研究，种群密度较低 (汤宋华，2008) / There is no assessment of population size, while it is estimated that the population density is low in nature based on the studies of *A. f. beickianus* in south Gansu (Tang, 2008)
种群趋势 Population Trend	下降 / Decreasing

生境与生态系统 Habitat (s) and Ecosystem (s)

生　　境 Habitat(s)	栖息于针叶林和针阔混交林，白天躲藏于林中枝叶茂密处，繁殖期 4 ～ 7 月，窝卵数 3 ～ 6 枚，孵卵期 25 ～ 27 天 / *A. funereus* inhabits in mixed coniferous broad-leaved forests, and roosts in the dense forests during days. It breeds in April to July, with clutch of 3 ～ 6 eggs, and breeding period of 25 ～ 27 days
生态系统 Ecosystem(s)	陆地生态系统 / Terrestrial Ecosystem

威胁 Threat (s)

主要威胁 Major Threat(s)	栖息地的丧失和破碎化是该物种主要受威胁因子，对老龄树的择伐减少了其潜在巢址也是其中的重要因素之一 / The major threats are habitat loss and fragmentation. Besides, deforestation of large trees might reduce the potential nests

保护级别与保护行动 Protection Category and Conservation Action (s)

IUCN 红色名录 (2016) IUCN Red List (2016)	无危 / LC
保护行动 Conservation Action(s)	列入国家 II 级重点保护野生动物和 CITES 公约附录 II。部分分布区位于保护区内，在甘肃莲花山自然保护区进行了人工巢箱悬挂，并开展了对亚种 *A. f. beickianus* 的基础生物学研究，发现挂置巢箱对其种群有一定的增加作用 (Fang *et al.*, 2009)。需加强对栖息地的保护，评估种群数量和分布范围 / It has been listed as the second class in National Key Protected Wild Animal List and Appendix II of the CITES. Systematic studies on the species have been conducted globally. In China, there are several nature reserves in its distribution area. Artificial nests have been used in Lianhua Shan Nature Reserve, and it is proved to be effective in population conservation (Fang *et al.*, 2009). Habitat conservation, assessment of population size and distribution area are suggested

鬼鸮 *Aegolius funereus*　　　　许传辉 摄　By Chuanhui Xu

白喉犀鸟 *Anorrhinus austeni*

易危 VU C1; D

| 数据缺乏 DD | 无危 LC | 近危 NT | 易危 VU | 濒危 EN | 极危 CR | 区域灭绝 RE | 野外灭绝 EW | 灭绝 EX |

分类地位 Taxonomic Status

动物界 Animalia	脊索动物门 Chordata	鸟纲 Aves	犀鸟目 Bucerotiformes	犀鸟科 Bucerotidae
学名 Scientific Name		*Anorrhinus austeni*		
命名人 Species Authority		Jerdon, 1872		
英文名 English Name(s)		Austen's Brown Hornbill		
同物异名 Synonym(s)		*Ptilolaemus austeni*		
种下单元评估 Infra-specific Taxa Assessed		无 / None		

评估信息 Assessment Information

评估年份 Year Assessed	2016
评定人 Assessor(s)	杨晓君 / Xiaojun Yang
其他贡献人 Other Contributor(s)	吴飞、常云艳等 / Fei Wu, Yunyan Chang *et al.*

理由 Justification: 在局部区域可以见到一定数量的白喉犀鸟。调查显示由于生境破坏和片断化，种群数量下降明显。该物种在中国境内的具体种群数量以及种群数量下降幅度等相关信息，需要进一步的调查和研究。因此，列为易危等级 / Some individuals of *Anorrhinus austeni* could be seen in some local areas. Based on surveys, its population is undergoing a significant decline as a result of habitat loss and fragmentation. Monitoring and studies on population size and trend are to be improved. Thus, it is listed as vulnerable

评估历史 Assessment History: 《国家重点保护野生动物名录》：Ⅱ (1989)，Ⅰ (2021)；《中国濒危动物红皮书·鸟类》(1998)：稀有；《中国物种红色名录：第二卷 脊椎动物 下册》(2009)：易危 / *National Key Protected Wild Animal List*: Ⅱ (1989), Ⅰ (2021); *China Red Data Book of Endangered Animals* (1998): R; *China Species Red List: Vol. II Vertebrates Part 2* (2009): VU

地理分布 Geographical Distribution

国内分布 Domestic Distribution

云南省西双版纳傣族自治州的勐海和勐腊，可能分布于云南德宏州盈江县（杨岚等，2006）以及普洱市澜沧县 / It inhabits in Menghai and Mengla of Xishuangbanna, and might occurs in Yingjiang of Dehong (Yang *et al.*, 2006) and Lancang of Puer, Yunnan

分布标注 Distribution Note

非特有种 / Non-endemic

国内分布图
Map of Domestic Distribution

种群 Population

种群数量 Population Size	在中国境内确切的种群数量缺乏系统调查，局部地区可见到超过 10 只的种群 / No systematic survey has been conducted in China, and a group of more than ten individuals could be seen in some local areas
种群趋势 Population Trend	下降 / Decreasing

生境与生态系统 Habitat (s) and Ecosystem (s)

生　　境 Habitat(s)	分布于海拔 600～1,500 m 的热带雨林 / *A. austeni* inhabits in tropical rainforests at 600～1,500 m
生态系统 Ecosystem(s)	陆地生态系统 / Terrestrial Ecosystem

威胁 Threat (s)

主要威胁 Major Threat(s)	栖息的热带雨林破坏和片断化、捕猎 / Loss and fragmentation of tropical rainforests; illegal hunting

保护级别与保护行动 Protection Category and Conservation Action (s)

IUCN 红色名录 (2016) IUCN Red List (2016)	近危 / NT
保护行动 Conservation Action(s)	列入国家Ⅰ级重点保护野生动物和 CITES 公约附录Ⅱ，分布区主要在自然保护区内 / It has been listed as the first class in National Key Protected Wild Animal List and Appendix Ⅱ of the CITES. Nature reserves cover most of its distribution areas

白喉犀鸟 *Anorrhinus austeni* 　　　　　　　　高正华 摄　By Zhenghua Gao

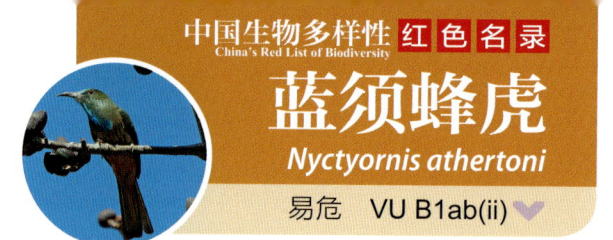

蓝须蜂虎
Nyctyornis athertoni
易危 VU B1ab(ii)

| 数据缺乏 DD | 无危 LC | 近危 NT | 易危 VU | 濒危 EN | 极危 CR | 区域灭绝 RE | 野外灭绝 EW | 灭绝 EX |

分类地位 Taxonomic Status

动物界 Animalia	脊索动物门 Chordata	鸟纲 Aves	佛法僧目 Coraciiformes	蜂虎科 Meropidae
学 名 Scientific Name		*Nyctyornis athertoni*		
命 名 人 Species Authority		Jardine and Selby, 1828		
英 文 名 English Name(s)		Blue-bearded Bee-eater		
同物异名 Synonym(s)		*Merops athertoni*		
种下单元评估 Infra-specific Taxa Assessed		无 / None		

评估信息 Assessment Information

评 估 年 份 Year Assessed	2016
评 定 人 Assessor(s)	梁伟 / Wei Liang
其他贡献人 Other Contributor(s)	杨灿朝 / Canchao Yang

理由 Justification: 蓝须蜂虎分布于热带雨林, 分布区较小。因此, 列为易危等级 / *Nyctyornis athertoni* inhabits in tropical rainforests, and the distribution area is small. Thus, it is listed as Vulnerable

评估历史 Assessment History: 《国家重点保护野生动物名录》: Ⅱ (2021); 《中国物种红色名录: 第二卷 脊椎动物 下册》(2009): 无危 / *National Key Protected Wild Animal List*: Ⅱ (2021); *China Species Red List*: *Vol. Ⅱ Vertebrates Part 2* (2009): LC

地理分布 Geographical Distribution

国内分布 Domestic Distribution
分布于广西、云南和海南 / It occurs in Guangxi, Yunnan and Hainan
分布标注 Distribution Note
非特有种 / Non-endemic

国内分布图
Map of Domestic Distribution

种群 Population

种群数量 Population Size	未评估 / No assessment
种群趋势 Population Trend	下降 / Decreasing

生境与生态系统 Habitat(s) and Ecosystem(s)

生　　境 Habitat(s)	为热带天然林的不常见留鸟 / *N. athertoni* is rarely seen resident in tropical rain forests
生态系统 Ecosystem(s)	陆地生态系统 / Terrestrial Ecosystem

威胁 Threat(s)

主要威胁 Major Threat(s)	天然林面积的持续下降、人工林的大量种植和栖息地质量的降低 / Habitat loss and deterioration owing to the decline of natural forests and increase of secondary and artificial forests

保护级别与保护行动 Protection Category and Conservation Action(s)

IUCN 红色名录 (2016) IUCN Red List (2016)	无危 / LC
保护行动 Conservation Action(s)	列入国家Ⅰ级重点保护野生动物。建议对该物种的种群数量、栖息地和生活史进行深入研究，加大对天然雨林的保护 / It has been listed as the first class in National Key Protected Wild Animal List. Studies on population size, habitat and life history are suggested, and conservation of natural rainforests are to be improved

蓝须蜂虎 *Nyctyornis athertoni*　　　　田穗兴 摄　By Suixing Tian

中国生物多样性红色名录

斑头大翠鸟
Alcedo hercules

易危 VU C1+2a(i)

| DD 数据缺乏 | LC 无危 | NT 近危 | **VU 易危** | EN 濒危 | CR 极危 | RE 区域灭绝 | EW 野外灭绝 | EX 灭绝 |

分类地位 Taxonomic Status

| 动物界 Animalia | 脊索动物门 Chordata | 鸟纲 Aves | 佛法僧目 Coraciiformes | 翠鸟科 Alcedinidae |

学名 Scientific Name	*Alcedo hercules*
命名人 Species Authority	Laubmann, 1917
英文名 English Name(s)	Blyth's Kingfisher
同物异名 Synonym(s)	无 / None
种下单元评估 Infra-specific Taxa Assessed	无 / None

评估信息 Assessment Information

评估年份 Year Assessed	2016
评定人 Assessor(s)	周放 / Fang Zhou
其他贡献人 Other Contributor(s)	舒晓莲 / Xiaolian Shu

理由 Justification: 斑头大翠鸟为狭域分布种，种群数量很少，在大多数分布点甚为罕见。主要分布在山区河溪沿岸，种群数量很小。近年来，其分布区内的小水电开发极其严重，对适宜生境的破坏很大，导致种群数量急剧下降。因此，列为易危等级 / The distribution area and population size of *Alcedo hercules* are small. It is rarely seen in most distribution area. It occurs along mountainous rivers, and population size is quite small. The population is undergoing a rapid decline with the habitat loss owing to the small hydroelectricity projects. Thus, it is listed as Vulnerable.

评估历史 Assessment History: 《国家重点保护野生动物名录》：II (2021)；《中国物种红色名录：第二卷 脊椎动物 下册》(2009)：近危 / *National Key Protected Wild Animal List*: II (2021); *China Species Red List: Vol. II Vertebrates Part 2* (2009): NT

地理分布 Geographical Distribution

国内分布 Domestic Distribution

分布于西藏东南部、云南南部和海南、广东北部、江西西南部、福建西北部以及广西的北部和南部，但数量均很稀少 / It occurs in southeast Tibet (Xizang), south Yunnan and Hainan, north Guangdong, southwest Jiangxi, northwest Fujian, north and south Guangxi, while the population size is small

分布标注 Distribution Note

非特有种 / Non-endemic

国内分布图
Map of Domestic Distribution

种群 Population

种群数量 Population Size	Brazil (2009) 估计中国的繁殖种群小于 100 对,实际上应该多于此 / It is estimated to be less than 100 pairs in China by Brazil (2009), and the population size should actually be more than this
种群趋势 Population Trend	下降中 / In decreasing

生境与生态系统 Habitat (s) and Ecosystem (s)

生　　境 Habitat(s)	该物种常见于常绿林中的溪流旁以及毗邻的旷野 (海拔 200 ~ 1,200 m,主要分布在海拔 400 ~ 1,000 m) / *A. hercules* is seen in rivers in evergreen forests and adjacent open areas at 200 ~ 1,200 m. It normally occurs at 400 ~ 1,000 m
生态系统 Ecosystem(s)	陆地和淡水生态系统 /Terrestrial and Freshwater Ecosystem

威胁 Threat (s)

主要威胁 Major Threat(s)　较低的种群密度分布,森林砍伐会使它们的生境缺失和破碎化。人类活动的干扰,如拦河筑坝、溪流污染等都会对它们的营巢巢址和生存繁衍产生威胁 / The population density is low, and major threats include habitat loss and fragmentation owing to deforestation. It is also vulnerable to human disturbance, dam projects, water pollution, *etc.*, which might be threats to its nesting, breeding and survival

保护级别与保护行动 Protection Category and Conservation Action (s)

IUCN 红色名录 (2016) IUCN Red List (2016)	近危 / NT
保护行动 Conservation Action(s)	列入国家Ⅱ级重点保护野生动物、《国家保护的有益的或者有重要经济、科学研究价值的陆生野生动物名录》。需要对该种鸟进行深入调查,以掌握其现有的分布区以及种群状况,评估种群动态。对其生境的影响因子和潜在威胁进行生态学研究。保护其适宜栖息地。科学规划山区水电建设,加强对河流污染的执法力度 / It has been listed as the second class in National Key Protected Wild Animal List, *National Protected List of Terrestrial Wild Animals with Good Benefits or Important Economic and Scientific Values*. Surveys on its distribution area and population dynamics, studies on the habitats and potential threats are suggested. Besides, it is also urgent to conserve the suitable habitat, scientific planning of hydroelectricity projects and promotion of regulation enforcement of river pollution

斑头大翠鸟 *Alcedo hercules*　　　　刘爱华 摄　By Aihua Liu

白腿小隼
Microhierax melanoleucos

易危 VU B1a; D

| 数据缺乏 DD | 无危 LC | 近危 NT | 易危 VU | 濒危 EN | 极危 CR | 区域灭绝 RE | 野外灭绝 EW | 灭绝 EX |

分类地位 Taxonomic Status

动物界 Animalia	脊索动物门 Chordata	鸟纲 Aves	隼形目 Falconiformes	隼科 Falconidae

学 名 Scientific Name	*Microhierax melanoleucos*
命 名 人 Species Authority	Blyth, 1843
英 文 名 English Name(s)	Pied Falconet
同物异名 Synonym(s)	*Ierax melanoleucos*
种下单元评估 Infra-specific Taxa Assessed	无 / None

评估信息 Assessment Information

评估年份 Year Assessed	2016
评定人 Assessor(s)	杨晓君 / Xiaojun Yang
其他贡献人 Other Contributor(s)	孔德军、常云艳等 /Dejun Kong, Yunyan Chang *et al.*

理由 Justification: Brazil (2009) 认为白腿小隼中国种群的数量不足 100 个繁殖对。目前在江西婺源的有效种群数量大于 40 个繁殖对，总数量约 120 只 (何芬奇等，2014)，该物种分布范围虽然广泛，但是分布区面积小于 20,000 km²，并且分布区严重隔离。种群的成熟个体不足 1,000 只，易受到人类活动的影响，可能在极短时间出现灭绝风险。因此，列为易危等级 / The population size of *Microhierax melanoleucos* is estimated to be less than 100 pairs in China (Brazil, 2009). The breeding population in Wuyuan of Jiangxi is more than 40 pairs, equivalent to 120 individuals (He *et al.*, 2014). It is widely distributed, while the distribution area is less than 20,000 km², and fragmented. The population size is less than 1,000 individuals and easily disturbed by human beings. It might become extremely endangered or even extinct in a short time. Thus, it is listed as Vulnerable

评估历史 Assessment History: 《国家重点保护野生动物名录》：Ⅱ (1989，2021)；《中国物种红色名录：第二卷 脊椎动物 下册》(2009)：无危 / *National Key Protected Wild Animal List*: Ⅱ (1989, 2021); *China Species Red List: Vol. II Vertebrates Part 2* (2009): LC

地理分布 Geographical Distribution

国内分布 Domestic Distribution
为河南、云南、贵州、安徽、江西、江苏、浙江、福建、广东、广西等地的留鸟 / It's resident in Henan, Yunnan, Guizhou, Anhui, Jiangxi, Jiangsu, Zhejiang, Fujian, Guangdong, Guangxi
分布标注 Distribution Note
非特有种 / Non-endemic

国内分布图
Map of Domestic Distribution

种群 Population

种群数量 Population Size	全球种群数量 1,000～10,000 只 (Ferguson-Lees and Christie，2001)，我国的种群数量不足 100 个繁殖对 (Brazil，2009) / The global population size is estimated to be 1,000～10,000 individuals, and the population size in China is less than 100 pairs (Brazil, 2009)
种群趋势 Population Trend	下降 / Decreasing

生境与生态系统 Habitat (s) and Ecosystem (s)

生　　境 Habitat(s)	栖息于亚热带常绿阔叶林中，有时也在林缘耕地边的灌丛上栖息，常见单个或两个一起活动。栖息地海拔 160～500 m 重点保护野生动物 / *M. melanoleucos* inhabits in subtropical evergreen broad-leaved forests, and sometimes also occurs in shrubs along the farmlands at the edges of forests. It is seen in one or two individuals
生态系统 Ecosystem(s)	陆地生态系统 / Terrestrial Ecosystem

威胁 Threat (s)

主要威胁 Major Threat(s)	为林栖型物种，森林的砍伐和栖息地减少或质量退化将构成威胁 / It inhabits in forests, and threatened by habitat loss and deterioration owing to deforestation

保护级别与保护行动 Protection Category and Conservation Action (s)

IUCN 红色名录 (2016) IUCN Red List (2016)	无危 / LC
保护行动 Conservation Action(s)	列入国家 II 级重点保护野生动物和 CITES 公约附录 II，部分分布区在自然保护区内，从而得到保护 / It has been listed as the second class in National Key Protected Wild Animal List and Appendix II of the CITES. Several nature reserves have covered some of its distribution area and it is under protection in this way

白腿小隼 *Microhierax melanoleucos*　　　　王吉义 摄　By Jiyi Wang

黄爪隼
Falco naumanni

易危　VU A2bcde+3bcde+4bcde; C2a(i)

| 数据缺乏 DD | 无危 LC | 近危 NT | 易危 VU | 濒危 EN | 极危 CR | 区域灭绝 RE | 野外灭绝 EW | 灭绝 EX |

分类地位 Taxonomic Status

动物界 Animalia	脊索动物门 Chordata	鸟纲 Aves	隼形目 Falconiformes	隼科 Falconidae

学　名 Scientific Name	*Falco naumanni*
命名人 Species Authority	Fleischer, 1818
英文名 English Name(s)	Lesser Kestrel
同物异名 Synonym(s)	无 / None
种下单元评估 Infra-specific Taxa Assessed	无 / None

评估信息 Assessment Information

评估年份 Year Assessed	2016
评定人 Assessor(s)	马鸣 / Ming Ma
其他贡献人 Other Contributor(s)	杜利民、徐峰 / Limin Du, Feng Xu

理由 Justification: 黄爪隼在中国比较罕见，种群数量不足3,000只。主要分布于西北地区，多出现在荒无人烟的地方，如新疆、甘肃、内蒙古等，也偶然出现在东北和华北等地 (郑光美，2017)。由于草原灭蝗、灭鼠，滥用杀虫剂、毒鼠药、阻燃剂等，在西部地区种群数量呈较大波动。因此，列为易危等级 / *Falco naumanni* is rarely seen in China, and the population size is less than 3,000 individuals. It occurs in northwest China, mostly desolate and uninhabited, *e.g.* Xinjiang, Gansu, Inner Mongolia (Nei Mongol), and seldom seen in Jilin, Liaoning, Hebei, Beijing, Tianjin, Shandong, Henan, Shanxi, Hubei, Sichuan, Yunnan (Zheng, 2017). The population in the west is fluctuating owing to the over use of pesticide during pest control and fire retardant. Thus, it is listed as Vulnerable

评估历史 Assessment History: 《国家重点保护野生动物名录》：II (1989，2021)；《中国物种红色名录：第二卷 脊椎动物 下册》(2009)：易危 / *National Key Protected Wild Animal List*: II (1989, 2021); *China Species Red List: Vol. II Vertebrates Part 2* (2009): VU

地理分布 Geographical Distribution

国内分布 Domestic Distribution

繁殖于内蒙古、甘肃北部、新疆北部，越冬于南部沿海和云南，迁徙经过中东部各省 / It breeds in Inner Mongolia (Nei Mongol), north of Gansu and north Xinjiang, and winters in Yunnan and coastal of south China, and stopover central and east provinces of China

分布标注 Distribution Note

非特有种 / Non-endemic

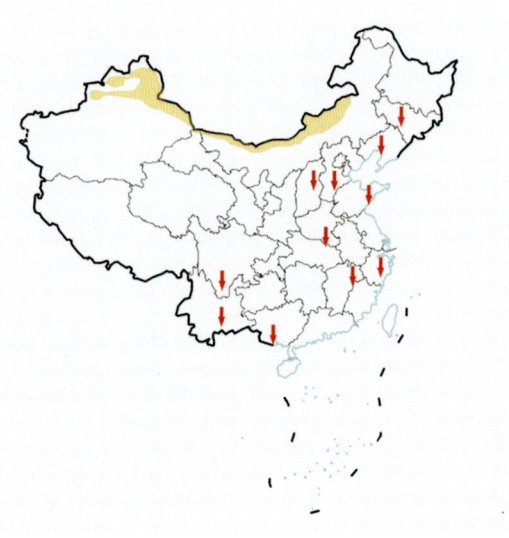

国内分布图
Map of Domestic Distribution

种群 Population

种群数量 Population Size 初步估计新疆北部有黄爪隼 850～1,100 只（繁殖或迁徙路过）；内蒙古有 150～200 只；甘肃有 110～170 只（迁徙）；其他省区有 100～180 只；总计 1,200～1,600 只 / It is estimated that the population size is 850～1,100 individuals in north Xinjiang (breeding or migrating); 150～200 individuals in Inner Mongolia (Nei Mongol); 110～170 individuals in Gansu (migrating); and 100～180 individuals in other provinces, which is 1,200～1,600 individuals in total

种群趋势 Population Trend 在西北部数量变化比较大，有的年份数量剧增，随处可见；而有的年份渺无踪迹。繁殖地也很不固定（杜利民和马鸣，2013）。喜集群繁殖。近 10 年，数量明显增加，新疆北部（包括天山北麓）至少有 6～8 个地点被确定为黄爪隼的繁殖地或迁徙通道。种群数量不稳定，变化比较大。总体的数量不足 3,000 只。过去 60 年，黄爪隼的种群数量急剧下降，种群数量每 10 年递减约 46%，累计减少了 95% / The population of *F. naumanni* in northwest is fluctuating: it is increasing sharply in some years; while hardly seen in some years. And the breeding sites are changeable (Du and Ma, 2013). It breeds in groups, and the population is increasing in recent ten years. There are 6～8 sites in north Xinjiang used by *F. naumanni* for breeding or migration. The population is not stable and the population size is less than 3,000 individuals. The population is undergoing a rapid decline in past 60 years, it decreased by 46% at each 10 years, and 95% in total

生境与生态系统 Habitat (s) and Ecosystem (s)

生境 Habitat(s) 黄爪隼体型较小，栖息于开阔而干热的半荒漠和草原。以采食直翅目昆虫、蜈蚣等无脊椎动物为主（占 95.3%），偶然捕食小雀、啮齿类、蜥蜴等。新疆伊犁谷地，秋季会出现数百只的集群，数量比较大。在乌鲁木齐近郊，黄爪隼捕捉的猎物数虽然众多，但绝大多数是青虫、蚤斯、蜈蚣、螳螂、蝼蛄、飞蛾等无脊椎动物（杜利民和马鸣，2013）/ *F. naumanni* is small. It inhabits in open areas in hot and dry climates. It feeds on insects (orthoptera) and invertebrates (centipedes), which take up 95.3%, and occasionally takes small sparrows, rodents, and lizards. It could be seen in large group of several hundreds of individuals in autumn in Yili Valley of Xinjiang. In suburban area of Urumqi, most food resources are insects, katydid, centipedes, mantis, mole crickets, *etc.* (Du and Ma, 2013)

生态系统 Ecosystem(s) 陆地生态系统 / Terrestrial Ecosystem

威胁 Threat (s)

主要威胁 Major Threat(s) 持续几十年的草原灭蝗、灭鼠、过度放牧、西部大开发及采矿所导致的栖息地丧失等，对黄爪隼的生存威胁很大。农药及杀虫剂污染、繁殖成功率低、食物资源匮乏、栖息地丧失、工业添加剂污染以及全球气候变化等（马鸣，2011a）/ *F. naumanni* is vulnerable to the habitat loss owing to over use of pesticide during pest control, over foraging, development of the west and mining. Major threats include pollution from pesticide and addictive, low breeding success, food shortage, habitat loss, and global climate change (Ma, 2011a)

保护级别与保护行动 Protection Category and Conservation Action (s)

IUCN 红色名录 (2016)
IUCN Red List (2016) 无危 / LC

保护行动 Conservation Action(s) 列入国家 II 级重点保护野生动物、CITES 公约附录 II 和迁徙物种保护公约 (CMS) 附录 I。要加强对本种的繁殖、迁徙习性、种群数量、气候及食物周期性变化的研究 / It has been listed as the second class in the National Key Protected Wild Animal List, Appendix II of the CITES and Appendix I of the Convention on Migratory Species. Studies on breeding, migration, population size, response to climate change and periodical change of food are to be improved

黄爪隼 *Falco naumanni* 孙少海 摄 By Shaohai Sun

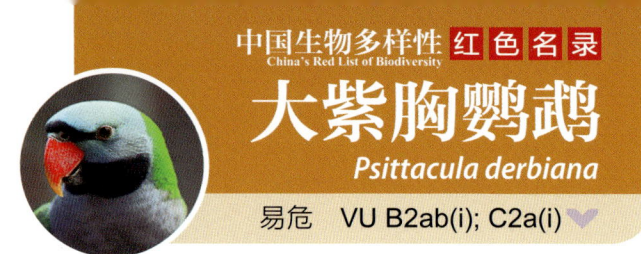

大紫胸鹦鹉
Psittacula derbiana

易危 VU B2ab(i); C2a(i)

| 数据缺乏 DD | 无危 LC | 近危 NT | 易危 VU | 濒危 EN | 极危 CR | 区域灭绝 RE | 野外灭绝 EW | 灭绝 EX |

分类地位 Taxonomic Status

动物界 Animalia	脊索动物门 Chordata	鸟纲 Aves	鹦形目 Psittaciformes	鹦鹉科 Psittacidae

学名 Scientific Name	*Psittacula derbiana*
命名人 Species Authority	Fraser, 1852
英文名 English Name(s)	Lord Derby's Parakeet
同物异名 Synonym(s)	*Palaeornis derbianus*
种下单元评估 Infra-specific Taxa Assessed	无 / None

评估信息 Assessment Information

评估年份 Year Assessed	2016
评定人 Assessor(s)	冉江洪 / Jianghong Ran
其他贡献人 Other Contributor(s)	窦亮等 / Liang Dou *et al.*

理由 Justification: 由于大面积森林采伐，大紫胸鹦鹉营巢栖息地面积大量减少，加上盗捕严重，野外数量呈现逐渐下降的趋势。对四川省原有记录的宝兴县和美姑县的调查显示，已经多年没有发现其踪迹，估计已经绝灭。对丹巴、雅江和木里县的调查结果显示，其种群数量和分布范围比以前都有大幅度的下降。因此，列为易危等级 / The logging resulted in the loss of much breeding habitat of *Psittacula derbiana*, and the heavy trapping pressure made the wild population decreased. No wild observation in Baoxing County and Meigu County, and it is historically distributed in Sichuan Province, where local extinction is estimated. Its population and distribution is undergoing a rapid decline in Danba, Yajiang and Muli County. Thus, it is listed as Vulnerable

评估历史 Assessment History:《国家重点保护野生动物名录》: II (1989, 2021);《中国濒危动物红皮书·鸟类》(1998): 易危;《中国物种红色名录: 第二卷 脊椎动物 下册》(2009): 近危 / *National Key Protected Wild Animal List*: II (1989, 2021); *China Red Data Book of Endangered Animals*: *Aves* (1998): V; *China Species Red List*: *Vol. II Vertebrates Part 2* (2009): NT

地理分布 Geographical Distribution

国内分布 Domestic Distribution
分布于四川西南部、云南西部、西藏东部、广西西部等地 / It occurs in southwest Sichuan, west Yunnan, east Tibet (Xizang) and west Guangxi
分布标注 Distribution Note
非特有种 / Non-endemic

国内分布图
Map of Domestic Distribution

种群 Population

种群数量 Population Size	野外种群数量不明。云南中甸县（香格里拉县）估计为 0.1 只/km² （周林，2004）。四川雅江县格西沟自然保护区的密度为 0.5～2.5 只/km² （2005～2007 年）/ Wild population is unknown. The density is estimated as 0.1 individuals/km² (Zhou, 2004) and 0.5～2.5/km² (2005～2007) in Zhongdian (Shangri-la) County in Yunnan and Yajiang Gexigou Nature Reserve in Sichuan, respectively
种群趋势 Population Trend	下降 / Decreasing

生境与生态系统 Habitat (s) and Ecosystem (s)

生境 Habitat(s)	主要栖息于海拔 2,000～4,000 m 的山地，阔叶林到针叶林区都是其活动范围，随季节变化作垂直迁徙，有时会前往农耕区进行觅食 / *P. derbiana* inhabits coniferous and broad-leaved forests, ranging from 2,000 to 4,000 m. It undertakes some altitudinal movements, and sometimes feeds in croplands
生态系统 Ecosystem(s)	森林生态系统 / Forest Ecosystem

威胁 Threat (s)

主要威胁 Major Threat(s)	栖息地破坏和人类盗捕等干扰严重，栖息地面积减少，栖息地破碎化严重 / Heavy trapping pressure, human disturbance, and habitat loss

保护级别与保护行动 Protection Category and Conservation Action (s)

IUCN 红色名录 (2016) IUCN Red List (2016)	近危 / NT
保护行动 Conservation Action(s)	列入国家 II 级重点保护野生动物。大多数栖息地未受到保护，尚无以该物种为目标保护对象的保护区，四川格西沟国家级自然保护区内有一定数量的大紫胸鹦鹉分布。本物种的栖息地和种群数量急剧减少，盗猎严重，亟待加强对该种的保护 / It has been listed as the second class in National Key Protected Wild Animal List. Most habitats are unprotected and no nature reserve was set up with *P. derbiana* as conservation target, while some presence in Sichuan Gexigou National Nature Reserve. The decreasing population, habitat loss and heavy trapping pressure make the protection of the species urgent

大紫胸鹦鹉 *Psittacula derbiana*　　　　张铭 摄　By Ming Zhang

绯胸鹦鹉
Psittacula alexandri
易危 VU B1ab(iii)

| 数据缺乏 DD | 无危 LC | 近危 NT | 易危 VU | 濒危 EN | 极危 CR | 区域灭绝 RE | 野外灭绝 EW | 灭绝 EX |

分类地位 Taxonomic Status

动物界 Animalia	脊索动物门 Chordata	鸟纲 Aves	鹦形目 Psittaciformes	鹦鹉科 Psittacidae
学名 Scientific Name		*Psittacula alexandri*		
命名人 Species Authority		Linnaeus, 1758		
英文名 English Name(s)		Red-breasted Parakeet		
同物异名 Synonym(s)		无 / None		
种下单元评估 Infra-specific Taxa Assessed		无 / None		

评估信息 Assessment Information

评估年份 Year Assessed	2016
评定人 Assessor(s)	杨灿朝 / Canchao Yang
其他贡献人 Other Contributor(s)	梁伟 / Wei Liang

理由 Justification: 绯胸鹦鹉分布于热带雨林。由于非法捕猎和贸易使得该种的种群数量急剧下降，并导致地区性灭绝 (Juniper and Parr, 1998)。因此，列为易危等级 / *Psittacula alexandri* distributes in tropical rain forest. The population is suspected to be in rapid decline and even local extinction owing to illegal hunting and trade. Thus, it is listed as Vulnerable

评估历史 Assessment History: 《国家重点保护野生动物名录》: II (1989, 2021);《中国濒危动物红皮书·鸟类》(1998): 易危;《中国物种红色名录: 第二卷 脊椎动物 下册》(2009): 近危 / *National Key Protected Wild Animal List*: II (1989, 2021); *China Red Data Book of Endangered Animals*: *Aves* (1998): V; *China Species Red List*: Vol. II Vertebrates Part 2 (2009): NT

地理分布 Geographical Distribution

国内分布 Domestic Distribution
分布于西藏东南部、云南南部、广西西部和南部、海南 / It occurs in southeast Tibet (Xizang), south Yunnan, west and south Guangxi, Hainan
分布标注 Distribution Note
非特有种 / Non-endemic

国内分布图
Map of Domestic Distribution

种群 Population

种群数量 Population Size	未评估。在许多分布区野生种群难以见到 / Not assessed. Wild population is difficult to observe in its distribution area
种群趋势 Population Trend	下降 / Decreasing

生境与生态系统 Habitat (s) and Ecosystem (s)

生　　境 Habitat(s)	天然原始林 / Natural primitive forest
生态系统 Ecosystem(s)	森林生态系统 / Forest Ecosystem

威胁 Threat (s)

主要威胁 Major Threat(s)	由于非法捕猎和贸易使得该种的种群数量急剧下降，造成地区性灭绝 (Juniper and Parr, 1998)。另外，生境破坏也是重要的威胁因素，例如，天然林面积的持续下降和栖息地质量的降低 / The population is suspected to be in rapid decline and even local extinction owing to illegal hunting and trade (Juniper and Parr, 1998). Besides, habitat loss and habitat quality decline are also threats

保护级别与保护行动 Protection Category and Conservation Action (s)

IUCN 红色名录 (2016) IUCN Red List (2016)	近危 / NT
保护行动 Conservation Action(s)	列入国家 II 级重点保护野生动物和 CITES 公约附录 II。建议对该物种的种群数量、栖息地和狩猎状况等进行深入调查；加大对天然林的保护 / It has been listed as the second class in National Key Protected Wild Animal List and Appendix II of the CITES. Surveys in population, habitats and hunting, as well as improvement of the conservation of natural forest are suggested

绯胸鹦鹉 *Psittacula alexandri*　　　　　童光琦 摄　By Guangqi Tong

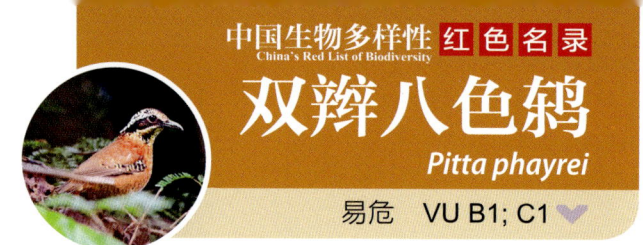

双辫八色鸫

Pitta phayrei

易危　VU B1; C1

| 数据缺乏 DD | 无危 LC | 近危 NT | 易危 VU | 濒危 EN | 极危 CR | 区域灭绝 RE | 野外灭绝 EW | 灭绝 EX |

分类地位 Taxonomic Status

动物界 Animalia	脊索动物门 Chordata	鸟纲 Aves	雀形目 Passeriformes	八色鸫科 Pittidae
学名 Scientific Name		*Pitta phayrei*		
命名人 Species Authority		Blyth, 1862		
英文名 English Name(s)		Eared Pitta		
同物异名 Synonym(s)		*Anthocincla phayrei*		
种下单元评估 Infra-specific Taxa Assessed		无 / None		

评估信息 Assessment Information

评估年份 Year Assessed	2016
评定人 Assessor(s)	周放 / Fang Zhou
其他贡献人 Other Contributor(s)	余丽江 / Lijiang Yu

理由 Justification: 双辫八色鸫在中国的分布区小，仅见于云南西双版纳，推测分布面积小于 20,000 km²。分布区天然森林面积持续减少和破碎化。种群小，罕见。因此，列为易危等级 / The distributed area of *Pitta phayrei* in China is small (estimated less than 20,000 km²), only Xishuangbanna, Yunnan Province. The natural forest is decreased and fragmented. The population of *Pitta phayrei* is small. Thus, it is listed as Vulnerable

评估历史 Assessment History: 《国家重点保护野生动物名录》：II (1989，2021)；《中国濒危动物红皮书·鸟类》(1998)：稀有；《中国物种红色名录：第二卷 脊椎动物 下册》(2009)：近危（几近易危）/ *National Key Protected Wild Animal List*: II (1989, 2021); *China Red Data Book of Endangered Animals*: Aves (1998): R; *China Species Red List*: Vol. II Vertebrates Part 2 (2009): NT (nearly VU)

地理分布 Geographical Distribution

国内分布 Domestic Distribution
仅分布于云南西双版纳 / It only occurs in Xishuangbanna, Yunnan
分布标注 Distribution Note
非特有种 / Non-endemic

国内分布图
Map of Domestic Distribution

种群 Population

种群数量 Population Size	在中国甚罕见。种群数量不明，近10年仅有1例来自红外相机的监测记录（张明霞等，2014）和1例云南西双版纳的观鸟记录 / This species is seldom seen in China and the population is unknown, with only one record from infrared camera in recent 10 years (Zhang *et al.*, 2014) and one birding record in Xishuangbanna, Yunnan
种群趋势 Population Trend	下降 / Decreasing

生境与生态系统 Habitat(s) and Ecosystem(s)

生境 Habitat(s)	栖息于海拔800 m以上的热带-亚热带常绿林 / *P. phayrei* distributes in tropical and subtropical evergreen forests above 800 m
生态系统 Ecosystem(s)	陆地生态系统 / Terrestrial Ecosystem

威胁 Threat(s)

主要威胁 Major Threat(s)	西双版纳的热带雨林曾遭严重砍伐而造成栖息地丧失 (Li *et al.*, 2009) / Habitat loss owing to heavy logging in tropical rain forest in Xishuangbanna (Li *et al.*, 2009)

保护级别与保护行动 Protection Category and Conservation Action(s)

IUCN 红色名录 (2016) IUCN Red List (2016)	无危 / LC
保护行动 Conservation Action(s)	列入国家II级重点保护野生动物。需加强种群数量的调查和野外研究 / It has been listed as the second class in National Key Protected Wild Animal List, and population survey and field research are to be improved

双辫八色鸫 *Pitta phayrei* 　　　　　林树成 摄　By Shucheng Lin

蓝枕八色鸫
Pitta nipalensis
易危 VU C1

| 数据缺乏 DD | 无危 LC | 近危 NT | 易危 VU | 濒危 EN | 极危 CR | 区域灭绝 RE | 野外灭绝 EW | 灭绝 EX |

分类地位 Taxonomic Status

动物界 Animalia	脊索动物门 Chordata	鸟纲 Aves	雀形目 Passeriformes	八色鸫科 Pittidae
学名 Scientific Name		*Pitta nipalensis*		
命名人 Species Authority		Hodgson, 1837		
英文名 English Name(s)		Blue-naped Pitta		
同物异名 Synonym(s)		*Paludicola nipalensis*		
种下单元评估 Infra-specific Taxa Assessed		无 / None		

评估信息 Assessment Information

评估年份 Year Assessed	2016
评定人 Assessor(s)	周放 / Fang Zhou
其他贡献人 Other Contributor(s)	陆舟 / Zhou Lu

理由 Justification: 蓝枕八色鸫见于云南南部及东南部、西藏东南部的墨脱、广西西南部及西北部保存完好的常绿阔叶林中，分布区较为狭窄。由于植被丧失和破碎化而导致种群数量下降。特别是华南地区近年来推广速生桉种植，使其栖息地快速丧失（蒋明康等，2006）。近几年来，在野外仅有少量有关蓝枕八色鸫的观察记录，推测种群数极为稀少。因此，列为易危等级 / *Pitta nipalensis* occurs in well reserved evergreen broad-leaved forest in south and southeast Yunnan, Motuo in southeast Tibet, southwest and northwest Guangxi. Its distribution is small and population is undergoing a decline due to subtropical forest loss and fragmentation, especially habitat loss as a result of the planation of *Eucalyptus* in southern China. The assessment of in situ conservation is "less conservation" (Jiang *et al.*, 2006). In recent years, only a few records of *Pitta nipalensis* in the wild and the population is estimated small. Thus, it is listed as Vulnerable

评估历史 Assessment History: 《国家重点保护野生动物名录》：II（1989，2021）；《中国物种红色名录：第二卷 脊椎动物 下册》(2009)：近危（近乎易危）/ *National Key Protected Wild Animal List*: II (1989, 2021); *China Species Red List: Vol. II Vertebrates Part 2* (2009): NT (nearly VU)

地理分布 Geographical Distribution

国内分布 Domestic Distribution
分布于云南南部、西藏东南部、广西西部和北部 / It occurs in south Yunnan, southeast Tibet (Xizang), west and north Guangxi
分布标注 Distribution Note
非特有种 / Non-endemic

国内分布图
Map of Domestic Distribution

种群 Population

种群数量 Population Size	种群数量十分稀少 / The population size is quite small
种群趋势 Population Trend	下降中 / In decreasing

生境与生态系统 Habitat(s) and Ecosystem(s)

生 境 Habitat(s)	热带雨林、亚热带常绿阔叶林 / Tropical and subtropical evergreen broad-leaved forest
生态系统 Ecosystem(s)	陆地生态系统 / Terrestrial Ecosystem

威胁 Threat(s)

主要威胁 Major Threat(s)	原生的常绿阔叶林被大量破坏，导致栖息地减少和质量下降可能是导致蓝枕八色鸫受威胁的主要原因 / Habitat loss and quality reduction due to the destroy of natural evergreen broad-leaved forest might be the major threats of *Pitta nipalensis*

保护级别与保护行动 Protection Category and Conservation Action(s)

IUCN 红色名录 (2016) IUCN Red List (2016)	无危 / LC
保护行动 Conservation Action(s)	列入国家II级重点保护野生动物。应加大对蓝枕八色鸫的研究和调查力度，继续开展分布区、生境选择和种群现状调查。以准确获得其种群数量和栖息地数据资料 / It has been listed as the second class in National Key Protected Wild Animal List. The research and survey should be improved, especially in distribution, habitat selection and population assessment

蓝枕八色鸫 *Pitta nipalensis*　　　　　　　　　　　　　　　　　　　马耀基 摄　By Yaoji Ma

栗头八色鸫
Pitta oatesi
易危 VU C1

| 数据缺乏 DD | 无危 LC | 近危 NT | 易危 VU | 濒危 EN | 极危 CR | 区域灭绝 RE | 野外灭绝 EW | 灭绝 EX |

分类地位 Taxonomic Status

动物界 Animalia	脊索动物门 Chordata	鸟纲 Aves	雀形目 Passeriformes	八色鸫科 Pittidae
学名 Scientific Name		*Pitta oatesi*		
命名人 Species Authority		Hume, 1873		
英文名 English Name(s)		Rusty-naped Pitta		
同物异名 Synonym(s)		*Hydrornis oatesi*		
种下单元评估 Infra-specific Taxa Assessed		无 / None		

评估信息 Assessment Information

评估年份 Year Assessed	2016
评定人 Assessor(s)	周放 / Fang Zhou
其他贡献人 Other Contributor(s)	余丽江 / Lijiang Yu

理由 Justification: 栗头八色鸫在中国罕见于云南西部和东南部，分布区小。近10年来的野外直接观察记录极少，估计种群数量非常稀少。分布区天然森林面积持续减少和破碎化。因此，列为易危等级 / *Pitta oatesi* is rarely seen in west and southeast Yunnan, with very small distribution area in China. There are very few observation records in the wild and the population is estimated very small. The area of natural forest in distributed area is undergoing a decline and fragmentation. Thus, it is listed as Vulnerable

评估历史 Assessment History:《国家重点保护野生动物名录》：II (1989，2021)；《中国物种红色名录：第二卷 脊椎动物 下册》(2009)：无危 / *National Key Protected Wild Animal List*: II (1989, 2021); *China Species Red List: Vol. II Vertebrates Part 2* (2009): LC

地理分布 Geographical Distribution

国内分布 Domestic Distribution
仅分布于云南西南部，留鸟 / It only occurs in southwest Yunnan as resident
分布标注 Distribution Note
非特有种 / Non-endemic

国内分布图
Map of Domestic Distribution

种群 Population

种群数量 Population Size	未知 / Unknown
种群趋势 Population Trend	下降 / Decreasing

生境与生态系统 Habitat (s) and Ecosystem (s)

生　　境 Habitat(s)	栖息于海拔 1,800 m 以下的热带、亚热带地区，在茂密的常绿阔叶林下的阴湿处活动觅食 / *P. oatesi* uses dense evergreen broad-leaved forest in tropical and subtropical area for roosting and foraging
生态系统 Ecosystem(s)	陆地生态系统 / Terrestrial Ecosystem

威胁 Threat (s)

主要威胁 Major Threat(s)	热带雨林和季雨林遭砍伐而造成栖息地丧失 / The habitat loss owing to the destroy of rain and monsoon forest in tropical area

保护级别与保护行动 Protection Category and Conservation Action (s)

IUCN 红色名录 (2016) IUCN Red List (2016)	无危 / LC
保护行动 Conservation Action(s)	列入国家 II 级重点保护野生动物，建议加强种群数量和生境选择的调查研究，明确种群现状，以便开展针对性的保护措施 / It has been listed as the second class in National Key Protected Wild Animal List. The survey and research on population and habitat selection is suggested to get better knowledge and make specific conservation

栗头八色鸫 *Pitta oatesi* 　　　　　董炜 摄　By Wei Dong

绿胸八色鸫
Pitta sordida
易危 VU C1

| 数据缺乏 DD | 无危 LC | 近危 NT | 易危 VU | 濒危 EN | 极危 CR | 区域灭绝 RE | 野外灭绝 EW | 灭绝 EX |

分类地位 Taxonomic Status

动物界 Animalia	脊索动物门 Chordata	鸟纲 Aves	雀形目 Passeriformes	八色鸫科 Pittidae
学名 Scientific Name		*Pitta sordida*		
命名人 Species Authority		Müller, 1776		
英文名 English Name(s)		Hooded Pitta		
同物异名 Synonym(s)		*Turdus sordidus*		
种下单元评估 Infra-specific Taxa Assessed		无 / None		

评估信息 Assessment Information

评估年份 Year Assessed	2016
评定人 Assessor(s)	周放 / Fang Zhou
其他贡献人 Other Contributor(s)	余丽江 / Lijiang Yu

理由 Justification: 绿胸八色鸫在中国的分布区小，仅见于云南南部和中部，西藏东南部及四川西北部，其中云南哀牢山自然保护区的徐家坝和四川阿坝州是本物种近20年内的最新野外记录地点（Wang et al., 2000；王育章等，2006）。分布区天然森林面积持续减少和破碎化。因此，列为易危等级 / The distribution area of *Pitta sordida* is small: only south, and middle Yunnan, southeast Tibet (Xizang) and northwest Sichuan. Two new wild records were at Xujiaba in Yunnan Ailao Shan nature reserve and Aba in Sichuan (Wang et al., 2000; Wang et al., 2006). The area of natural forest is undergoing a decline and fragmentation. Thus, it is listed as Vulnerable

评估历史 Assessment History: 《国家重点保护野生动物名录》(1989，2021): II；《中国物种红色名录：第二卷 脊椎动物 下册》(2009): 近危 / National Key Protected Wild Animal List (1989, 2021): II; China Species Red List: Vol. II Vertebrates Part 2 (2009): NT

地理分布 Geographical Distribution

国内分布 Domestic Distribution
为云南南部和中部、西藏东南部的留鸟 / It is resident in south and middle Yunnan, southeast Tibet (Xizang)
分布标注 Distribution Note
非特有种 / Non-endemic

国内分布图
Map of Domestic Distribution

种群 Population

种群数量 Population Size	种群数量不明。近 20 年内的记录甚少，在哀牢山自然保护区北段徐家坝进行长达十几年的鸟类调查中曾记录到该鸟，但未提及数量 (Wang *et al.*, 2000)；四川省阿坝藏族羌族自治州理县古尔沟曾采到 1 号标本 (王育章等，2006)。分布区雨林遭受砍伐，生境质量退化，推测呈下降趋势 / The population size of *P. sordida* is unclear. There is very few records in recent 20 years, with only one record at Xujiaba in Ailao Shan Nature Reserve in a long time monitoring for more than 10 years, while no information on its population size (Wang *et al.*, 2000). One specimen was gotten at Guergou in Aba, Sichuan (Wang *et al.*, 2006). The destroy of rain forest and deterioration of habitat quality might drive the population decline
种群趋势 Population Trend	下降 / Decreasing

生境与生态系统 Habitat (s) and Ecosystem (s)

生　　境 Habitat(s)	栖息于热带雨林、季雨林和湿性常绿阔叶林区，高可至海拔 2,000 m。多见单个或成对在林下或林缘沟谷地带觅食 / *P. sordida* uses tropical rainforest, monsoon forest, and evergreen broad-leaved forest, which could be as high as 2000 m. Single or pairs might forage in the underwood or valleys at the edge of the forests
生态系统 Ecosystem(s)	陆地生态系统 / Terrestrial Ecosystem

威胁 Threat (s)

主要威胁 Major Threat(s)	雨林遭砍伐而造成栖息地丧失，盗猎和贸易等威胁因素也可能导致其种群减少 / Habitat loss owing to the destroy of rain forest might be a major threat. The illegal hunting and trading might also led to the population decline

保护级别与保护行动 Protection Category and Conservation Action (s)

IUCN 红色名录 (2016) IUCN Red List (2016)	无危 / LC
保护行动 Conservation Action(s)	列入国家 II 级重点保护鸟类。分布于保护区之外的种群缺乏保护措施。建议加强种群数量和生境选择的调查研究，明确种群现状以便开展针对性的保护措施。加强保护区之外的保护 / It has been listed as the second class in National Key Protected Wild Animal List. There is no conservation actions in the areas outside the nature reserves. The surveys and researches on population size and habitat selection are suggested to get more clear knowledge and make more specific conservation. Besides, the wildlife outside nature reserves should also be protected

绿胸八色鸫 *Pitta sordida* 　　　　　　　　　　　　　　　郑鹏 摄　By Peng Zheng

仙八色鸫
Pitta nympha

易危　VU A2cd+3cd+4cd

| 数据缺乏 DD | 无危 LC | 近危 NT | 易危 VU | 濒危 EN | 极危 CR | 区域灭绝 RE | 野外灭绝 EW | 灭绝 EX |

分类地位 Taxonomic Status

动物界 Animalia	脊索动物门 Chordata	鸟 纲 Aves	雀形目 Passeriformes	八色鸫科 Pittidae
学 名 Scientific Name		*Pitta nympha*		
命 名 人 Species Authority		Temminck and Schlegel, 1850		
英 文 名 English Name(s)		Fairy Pitta		
同物异名 Synonym(s)		无 / None		
种下单元评估 Infra-specific Taxa Assessed		无 / None		

评估信息 Assessment Information

评估年份 Year Assessed	2016
评定人 Assessor(s)	张正旺 / Zhengwang Zhang
其他贡献人 Other Contributor(s)	无 / None

理由 Justification: 仙八色鸫野生种群密度非常低, 种群数量估计少于10,000只; 由于其繁殖区的森林采伐以及一些地区将其作为笼养鸟而非法捕捉和贩卖, 野生种群数量呈快速下降趋势。因此, 列为易危等级 / The density of *Pitta nympha* of wild populations is very low, and the population is estimated to be less than 10,000. Due to deforestation in their breeding habitats and illegal capture for caged birds in some areas, the number of wild populations has shown a rapid decline. Thus, it is listed as Vulnerable

评估历史 Assessment History: 《国家重点保护野生动物名录》: Ⅱ (1989, 2021); 《中国濒危动物红皮书·鸟类》(1998): 稀有; 《中国物种红色名录: 第二卷 脊椎动物 下册》(2009): 易危 / *National Key Protected Wild Animal List*: Ⅱ (1989, 2021); *China Red Data Book of Endangered Animals*: Aves (1998): R; *China Species Red List*: Vol. Ⅱ Vertebrates Part 2 (2009): VU

地理分布 Geographical Distribution

国内分布 Domestic Distribution

为中国东部地区的夏候鸟和旅鸟, 分布于河北、天津、山东、河南南部、甘肃、云南、贵州、湖北、湖南、安徽、江西、江苏、上海、浙江、福建、广东、香港、澳门、广西、海南、台湾 / It is summer visitor and passage migrant in Eastern China. It occurs in Hebei, Tianjin, Shandong, south Henan, Gansu, Yunnan, Guizhou, Hubei, Hunan, Anhui, Jiangxi, Jiangsu, Shanghai, Zhejiang, Fujian, Guangdong, Hong Kong, Macao, Guangxi, Hainan, Taiwan

分布标注 Distribution Note

非特有种 / Non-endemic

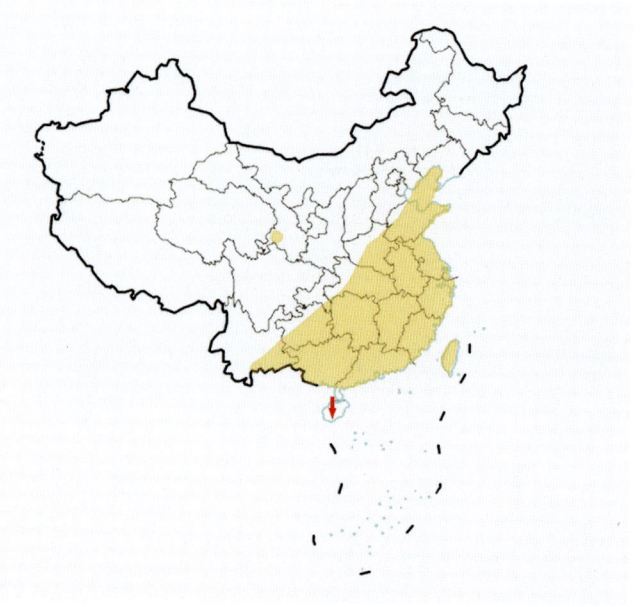

国内分布图
Map of Domestic Distribution

种群 Population

种群数量 Population Size	在台湾繁殖种群达 2,000 只。在江西、海南等地均有一定的数量 / There is a breeding population of 2,000 in Taiwan, and some are also found in Jiangxi, Hainan, *etc.*
种群趋势 Population Trend	下降 / Decreasing

生境与生态系统 Habitat (s) and Ecosystem (s)

生　　境 Habitat(s)	属于典型的森林鸟类，繁殖区在亚热带森林内 / *P. nympha* is typical forest bird, and uses subtropical forests as its breeding sites
生态系统 Ecosystem(s)	陆地生态系统 / Terrestrial Ecosystem

威胁 Threat (s)

主要威胁 Major Threat(s)	主要是繁殖区森林的丧失和片断化。在我国东南沿海地区，由于木材生产和林地转变成其他类型的土地，导致天然森林被大面积砍伐。非法猎捕和人为活动的干扰是另外一个威胁因素 / The loss and fragmentation of the forest in breeding area is the major threat. Lots of forest are destroyed due to logging and land-use change. Illegal trapping and human disturbance are also threats

保护级别与保护行动 Protection Category and Conservation Action (s)

IUCN 红色名录 (2016) IUCN Red List (2016)	易危 / VU A2cd+3cd+4cd
保护行动 Conservation Action(s)	列入国家 II 级重点保护野生动物和 CITES 公约附录 II。有关该物种的生态学资料还不多，应开展其种群分布、数量的调查以及相关的生态生物学研究，明确其栖息地需求和种群现状，以便为保护工作提供依据 / It has been listed as the second class of National Key Protected Wild Animal List and the Appendix II of the CITES. The biology information is far less clear, therefore the survey of distribution, population size and biological research is suggested, to get more knowledge of its population status and habitat need and to be a scientific support

仙八色鸫 *Pitta nympha*　　　　戎景山 摄　By Jingshan Rong

大盘尾 *Dicrurus paradiseus*

易危 VU A1c; C1

| 数据缺乏 DD | 无危 LC | 近危 NT | **易危 VU** | 濒危 EN | 极危 CR | 区域灭绝 RE | 野外灭绝 EW | 灭绝 EX |

分类地位 Taxonomic Status

动物界 Animalia	脊索动物门 Chordata	鸟纲 Aves	雀形目 Passeriformes	卷尾科 Dicruridae

学名 Scientific Name	*Dicrurus paradiseus*
命名人 Species Authority	Linnaeus, 1766
英文名 English Name(s)	Greater Racket-tailed Drongo
同物异名 Synonym(s)	*Cuculus paradiseus*
种下单元评估 Infra-specific Taxa Assessed	无 / None

评估信息 Assessment Information

评估年份 Year Assessed	2016
评定人 Assessor(s)	邓文洪 / Wenhong Deng
其他贡献人 Other Contributor(s)	无 / None

理由 Justification: 大盘尾野外种群数量稀少，极难发现。在我国大盘尾的适宜栖息地锐减，栖息地片断化严重，在野外极难发现。因此，列为易危等级 / The wild population of *Dicrurus paradiseus* is quite small. The suitable habitats are undergoing a rapid decline and fragmentation, which make it quite difficult to be seen in the wild. Thus, it is listed as Vulnerable

评估历史 Assessment History: 《国家重点保护野生动物名录》：Ⅱ (2021)；《中国物种红色名录：第二卷 脊椎动物 下册》(2009)：海南亚种近危 (近乎易危) / *National Key Protected Wild Animal List*: Ⅱ (2021); *China Species Red List*: *Vol. II Vertebrates Part 2* (2009): *D. P. johni* NT (nearly VU)

地理分布 Geographical Distribution

国内分布 Domestic Distribution	
为云南西部夏候鸟和海南的留鸟 / It occurs in west Yunnan as summer visitor, and in Hainan as resident	
分布标注 Distribution Note	
非特有种 / Non-endemic	

国内分布图
Map of Domestic Distribution

种群 Population

种群数量 Population Size	种群数量不足 20,000 / The population size is estimated to be less than 20,000
种群趋势 Population Trend	下降 / Decreasing

生境与生态系统 Habitat(s) and Ecosystem(s)

生　　境 Habitat(s)	栖息于热带地区的雨林及季雨林中 / *D. paradiseus* inhabits rainforest and monsoon forest in tropical area
生态系统 Ecosystem(s)	陆地生态系统 / Terrestrial Ecosystem

威胁 Threat(s)

主要威胁 Major Threat(s)	栖息地的丧失与片断化；栖息地的人为干扰 / Habitat loss and fragmentation and human disturbance might be the major threats

保护级别与保护行动 Protection Category and Conservation Action(s)

IUCN 红色名录 (2016) IUCN Red List (2016)	无危 / LC
保护行动 Conservation Action(s)	列入国家 II 级重点保护野生动物、《国家保护的有益的或者有重要经济、科学研究价值的陆生野生动物名录》/ It has been listed as the second class in National Key Protected Wild Animal List, *National Protected List of Terrestrial Wild Animals with Good Benefits or Important Economic and Scientific Values*

大盘尾 *Dicrurus paradiseus* 　　裘世雄 摄　By Shixiong Qiu

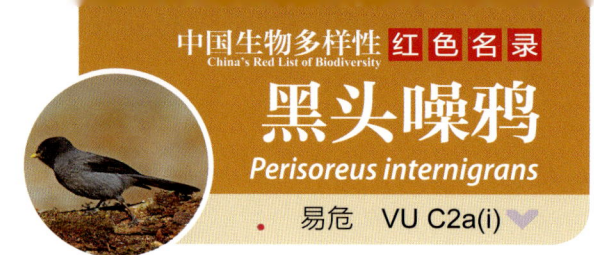

黑头噪鸦
Perisoreus internigrans

易危 VU C2a(i)

| 数据缺乏 DD | 无危 LC | 近危 NT | **易危 VU** | 濒危 EN | 极危 CR | 区域灭绝 RE | 野外灭绝 EW | 灭绝 EX |

分类地位 Taxonomic Status

动物界 Animalia	脊索动物门 Chordata	鸟纲 Aves	雀形目 Passeriformes	鸦科 Corvidae

学名 Scientific Name	*Perisoreus internigrans*
命名人 Species Authority	Thayer & Bangs, 1912
英文名 English Name(s)	Sichuan Jay
同物异名 Synonym(s)	*Boanerges internigrans*
种下单元评估 Infra-specific Taxa Assessed	无 / None

评估信息 Assessment Information

评估年份 Year Assessed	2016
评定人 Assessor(s)	孙悦华 / Yuehua Sun
其他贡献人 Other Contributor(s)	无 / None

理由 Justification: 黑头噪鸦分布范围较小，种群密度低，种群相互隔离且种群数量持续下降。因此，列为易危等级 / The distribution area of *Perisoreus internigrans* is small and the density in the wild is low. Several populations are isolated from each other and the populations are undergoing a decline. Thus, it is listed as Vulnerable

评估历史 Assessment History: 《国家重点保护野生动物名录》：Ⅰ (2021)；《中国物种红色名录：第二卷 脊椎动物 下册》(2009)：易危 / *National Key Protected Wild Animal List*: Ⅰ (2021); *China Species Red List: Vol. II Vertebrates Part 2* (2009): VU

地理分布 Geographical Distribution

国内分布 Domestic Distribution

分布于甘肃南部、西藏东部、青海东南部、四川西部 / It occurs in south Gansu, east Tibet (Xizang), southeast Qinghai and west Sichuan

分布标注 Distribution Note

特有种 / Endemic

国内分布图
Map of Domestic Distribution

种群 Population

种群数量 Population Size	估计种群数量 2,500～10,000 只 / The population size is estimated to be 2,500～10,000 individuals
种群趋势 Population Trend	下降 / Decreasing

生境与生态系统 Habitat (s) and Ecosystem (s)

生　　境 Habitat(s)	主要栖息于海拔 3,000～4,500 m 的高山暗针叶林中，树栖，以家族为单位合作繁殖，通常集 3～4 只的小群活动。在主要分布区的调查表明，活动区约 42 hm², 巢间距为 2～3 km / *P. internigrans* uses conifer forest in mountain areas with high altitude of 3,000～4,500 m. it roosts in the tree, and forms small flocks, and the mean flock size is 3～4 individuals, with the home range of 42 hm² and nest distance of 2～3 km
生态系统 Ecosystem(s)	陆地生态系统 / Terrestrial Ecosystem

威胁 Threat (s)

主要威胁 Major Threat(s)	森林的砍伐及气候变化导致的栖息地丧失和破碎化 (Lu *et al.*, 2012) / The destroy of forest and the habitat loss and fragmentation due to climate change (Lu *et al.*, 2012)

保护级别与保护行动 Protection Category and Conservation Action (s)

IUCN 红色名录 (2016) IUCN Red List (2016)	易危 / VU C2a (i)
保护行动 Conservation Action(s)	列入国家 I 级重点保护野生动物、《国家保护的有益的或者有重要经济、科学研究价值的陆生野生动物名录》。关于栖息地、繁殖和行为已有初步研究，部分分布区涵盖在保护区内 / It has been listed as the first class in National Key Protected Wild Animal List, *National Protected List of Terrestrial Wild Animals with Good Benefits or Important Economic and Scientific Values*. There have been some preliminary researches on habitat, breeding and behavior. Nature reserves cover some of its distribution

黑头噪鸦 *Perisoreus internigrans*　　　　井夫 摄　By Jingfu

黑尾地鸦
Podoces hendersoni

易危 VU C2(i); D1

| 数据缺乏 DD | 无危 LC | 近危 NT | 易危 VU | 濒危 EN | 极危 CR | 区域灭绝 RE | 野外灭绝 EW | 灭绝 EX |

分类地位 Taxonomic Status

动物界 Animalia	脊索动物门 Chordata	鸟纲 Aves	雀形目 Passeriformes	鸦科 Corvidae
学　名 Scientific Name		*Podoces hendersoni*		
命名人 Species Authority		Hume, 1871		
英文名 English Name(s)		Mongolian Ground Jay		
同物异名 Synonym(s)		无 / None		
种下单元评估 Infra-specific Taxa Assessed		对新疆、青海、甘肃和内蒙古黑尾地鸦分布数量与种群生存力分析显示，未来20年的灭绝概率小于10%（马鸣等，2006）/ The analysis of population viability of *Podoces hendersoni* in Xinjiang, Qinghai, Gansu, and Inner Mongolia (Nei Mongol) indicated that the probability of extinction in 20 years is less than 10% (Ma *et al.*, 2006)		

评估信息 Assessment Information

评估年份 Year Assessed	2016
评定人 Assessor(s)	马鸣 / Ming Ma
其他贡献人 Other Contributor(s)	徐峰、雷富民、吴道宁、丁鹏等 / Feng Xu, Fumin Lei, Daoning Wu, Peng Ding *et al.*

理由 Justification: 黑尾地鸦分布区域狭窄，种群密度极低。国内仅分布于新疆、甘肃、青海、宁夏、内蒙古等地，属于极端干旱地区物种。因此，列为易危等级 / *Podoces hendersoni* occurs in Xinjiang, Gansu, Qinghai, Ningxia, and Inner Mongolia (Nei Mongol) as resident. *Podoces hendersoni* appears in extremely dry area, and the population size is very small. Thus, it is listed as Vulnerable

评估历史 Assessment History:《国家重点保护野生动物名录》：II（2021）；《中国物种红色名录：第二卷 脊椎动物 下册》（2009）：近危 / *National Key Protected Wild Animal List*: II (2021); *China Species Red List: Vol. II Vertebrates Part 2* (2009): NT

地理分布 Geographical Distribution

国内分布 Domestic Distribution
主要分布于新疆的南部和北部（马鸣，2011b）、甘肃西北部、宁夏、青海北部、内蒙古西部 / It mainly distributes in south and north Xinjiang (Ma, 2011b), and also occurs in northwest Gansu, Ningxia, north Qinghai, west Inner Mongolia (Nei Mongol)
分布标注 Distribution Note
非特有种 / Non-endemic

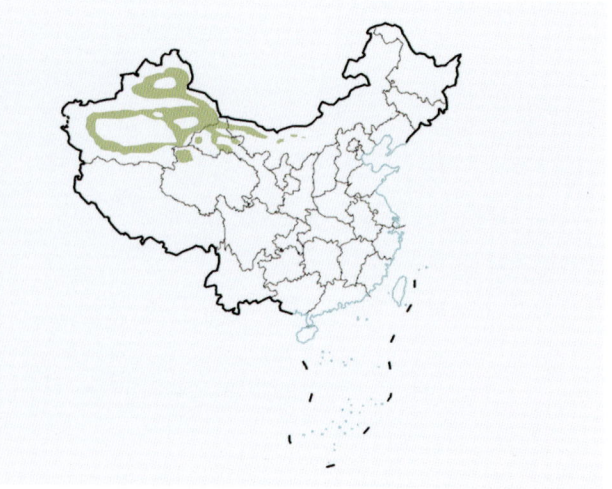

国内分布图 Map of Domestic Distribution

种群 Population

种群数量 Population Size	在新疆卡拉麦里保护区，种群密度为 5～8 只 /100 km² (马鸣等，2006) / The population density is 5～8 individuals per 100 km² in Kalamaili Nature Reserve (Ma *et al.*, 2006)
种群趋势 Population Trend	稳定 / Stable

生境与生态系统 Habitat (s) and Ecosystem (s)

生　　境 Habitat(s)	栖息于荒漠之中，生活环境极其特化，年降水量仅 80～140 mm。喜欢植被稀疏的戈壁 (赵正阶，2001；马鸣等，2006)。栖息地海拔 300～3,400 m，适应性较强。杂食性，以昆虫、蜥蜴、植物种子为主食，也采食植物的根、茎、叶片、花朵、公路上散落的谷物及人类丢弃的垃圾等 / *P. hendersoni* uses the deserts, with precipitation of 80～140 mm. It prefers hard Gobi Desert with sparse vegetation (Zhao, 2001; Ma *et al.*, 2006). It shows strong adaptability with distributed altitude of 300～3,400 m. It mainly feeds on insects, lizards, seeds, and also takes flowers, leaves and roots, and grain scattered in the road, or even garbage left by human beings
生态系统 Ecosystem(s)	陆地生态系统 / Terrestrial Ecosystem

威胁 Threat (s)

主要威胁 Major Threat(s)	过度放牧、薪柴砍伐和垦荒造成的栖息地与营巢地丧失。植被退化、土地盐渍化、荒漠化、全球气候变化及环境条件持续恶化等 (柯杨等，2010) / Habitat and breeding sites loss due to over grazing, firewood cutting and reclamation. Long time development of oasis and agriculture may lead to the vegetation degeneration, soil salinization, and desertification, global climate change, and environment deterioration, *etc*. Besides, the survey is difficult and the research is not so deep (Ke *et al.*, 2010)

保护级别与保护行动 Protection Category and Conservation Action (s)

IUCN 红色名录 (2016) **IUCN Red List (2016)**	无危 / LC
保护行动 **Conservation Action(s)**	列入国家 II 级重点保护野生动物、《国家保护的有益的或者有重要经济、科学研究价值的陆生野生动物名录》。部分分布区位于保护区内。公众对其知之甚少，需要宣传。同时加强科研与保护投入，改善其生存现状 / It has been listed as the second class in National Key Protected Wild Animal List, *National Protected List of Terrestrial Wild Animals with Good Benefits or Important Economic and Scientific Values*. The information of *Podoces hendersoni* is quite few for the public. The input of scientific research and conservation is suggested to improve its status

黑尾地鸦 *Podoces hendersoni* 　　　　俞亮 摄　By Liang Yu

白尾地鸦
Podoces biddulphi

易危 VU C2(i); D1

| 数据缺乏 DD | 无危 LC | 近危 NT | 易危 VU | 濒危 EN | 极危 CR | 区域灭绝 RE | 野外灭绝 EW | 灭绝 EX |

分类地位 Taxonomic Status

动物界 Animalia	脊索动物门 Chordata	鸟纲 Aves	雀形目 Passeriformes	鸦科 Corvidae
学名 Scientific Name		*Podoces biddulphi*		
命名人 Species Authority		Hume, 1874		
英文名 English Name(s)		Xinjiang Ground Jay		
同物异名 Synonym(s)		无 / None		
种下单元评估 Infra-specific Taxa Assessed		新疆塔克拉玛干沙漠白尾地鸦种群未来20年的灭绝概率小于10% (Ma, 2011) / The analysis of population viability of *Podoces biddulphi* in Taklimakan Desert indicated that the probability of extinction in 20 years is less than 10% (Ma, 2011)		

评估信息 Assessment Information

评估年份 Year Assessed	2016
评定人 Assessor(s)	马鸣 / Ming Ma
其他贡献人 Other Contributor(s)	徐峰、吴逸群、王传波、巴图尔汗、才代、贾泽信等 / Feng Xu, Yiqun Wu, Chuanbo Wang, Baturhan, Dai Cai, Zexin Jia *et al.*

理由 Justification: 白尾地鸦为新疆特有种，分布区域狭窄，种群密度极低，环境条件恶劣，仅见于塔克拉玛干沙漠及其边缘地区，属于极端干旱地区。种群数量不足10,000只。因此，列为易危等级 / *Podoces biddulphi* is an endemic species restricted in Xinjiang, with narrow distribution and extremely low density. It only occurs in extremely dry area in Taklimakan Desert and the edge area. The population size is less than 10,000 individuals. Thus, it is listed as Vulnerable

评估历史 Assessment History: 《国家重点保护野生动物名录》：II (2021)；《中国物种红色名录：第二卷 脊椎动物 下册》(2009)：近危 / *National Key Protected Wild Animal List*: II (2021); *China Species Red List: Vol. II Vertebrates Part 2* (2009): NT

地理分布 Geographical Distribution

国内分布 Domestic Distribution
主要分布于新疆塔克拉玛干沙漠区域 (马鸣, 2001b)。在甘肃和青海有偶然记录 / It mainly occurs in Taklimakan Desert in Xinjiang, and seldom appears in Gansu and Qinghai (Ma, 2001b)
分布标注 Distribution Note
特有种 / Endemic

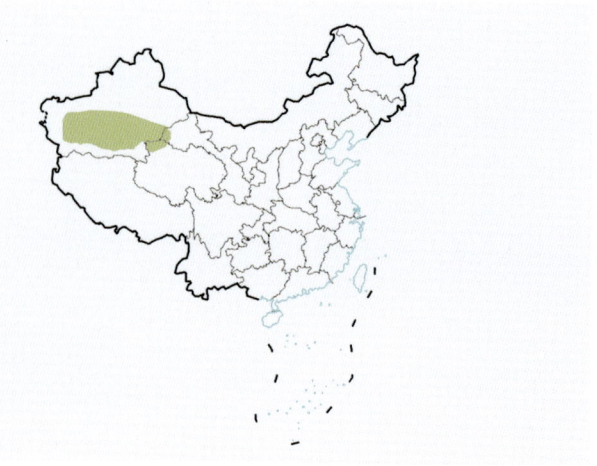

国内分布图
Map of Domestic Distribution

种群 Population

种群数量 Population Size	初步估计新疆有 4,100 ~ 6,700 只（留鸟），在甘肃和青海不足 100 只 / It's estimated that 4,100 ~ 6,700 individuals in Xinjiang (resident) and less than 100 individuals in Gansu and Qinghai
种群趋势 Population Trend	稳定（Ma，1999；马鸣等，2004b）/ Stable (Ma, 1999; Ma *et al.*, 2004b)

生境与生态系统 Habitat(s) and Ecosystem(s)

生境 Habitat(s)	栖息于塔克拉玛干大沙漠之中，生活环境极其特化，年降水量仅 50 mm。多营巢于低矮的灌丛中（马鸣等，2004b）。栖息地海拔 800 ~ 1,200 m。以昆虫、蜥蜴、植物种子为主食，也采食植物的根、茎、叶片、谷物及人类丢弃的垃圾等 / *P. biddulphi* uses the Taklimakan Desert with precipitation of 50 mm. It commonly nests in low shrubs (Ma *et al.*, 2004). The altitude of habitats covers 800 ~ 1200 m. It mainly feeds on insects, lizards, seeds, and also takes leaves, roots, and grain, or even garbage left by human beings
生态系统 Ecosystem(s)	陆地生态系统 / Terrestrial Ecosystem

威胁 Threat(s)

主要威胁 Major Threat(s)	农业垦荒造成的栖息地和营巢地丧失、塔里木河断流、地下水位降低、植被退化、土地盐渍化，还有油田建设、高速公路撞击、非法捕捉等因素 / Habitat and breeding sites loss due to reclamation, the cutoff of Tarim River, the lowering of groundwater, the vegetation degeneration, soil salinization, development of oil field and highway and illegal trapping

保护级别与保护行动 Protection Category and Conservation Action(s)

IUCN 红色名录 (2016) IUCN Red List (2016)	近危 / NT
保护行动 Conservation Action(s)	列入国家 II 级重点保护野生动物。白尾地鸦是中国特有种，需要加强科研投入，了解其生存现状 / It has been listed as the second class in National Key Protected Wild Animal List. *P. biddulphi* is endemic species of China. Input of scientific research is needed to get better knowledge of its status

白尾地鸦 *Podoces biddulphi* 马鸣 摄 By Ming Ma

歌百灵
Mirafra javanica

易危 VU A2abcd; B1b(ii,iii)

| 数据缺乏 DD | 无危 LC | 近危 NT | 易危 VU | 濒危 EN | 极危 CR | 区域灭绝 RE | 野外灭绝 EW | 灭绝 EX |

分类地位 Taxonomic Status

动物界 Animalia	脊索动物门 Chordata	鸟纲 Aves	雀形目 Passeriformes	百灵科 Alaudidae
学 名 Scientific Name		*Mirafra javanica*		
命 名 人 Species Authority		Horsfield, 1821		
英 文 名 English Name(s)		Horsfield's Bush Lark		
同物异名 Synonym(s)		无 / None		
种下单元评估 Infra-specific Taxa Assessed		无 / None		

评估信息 Assessment Information

评 估 年 份 Year Assessed	2016
评 定 人 Assessor(s)	邓文洪 / Wenhong Deng
其他贡献人 Other Contributor(s)	无 / None

理由 Justification: 歌百灵仅分布于云南中部、广东、香港、广西的狭小区域内。适宜栖息地缩小，栖息地严重片断化。因此，列为易危等级 / *Mirafra javanica* only occurs in Guangdong, Hong Kong and Guangxi. Its suitable habitats are undergoing a decline and fragmentation. Thus, it is listed as Vulnerable

评估历史 Assessment History: 《国家重点保护野生动物名录》：Ⅱ (2021)；《中国物种红色名录：第二卷 脊椎动物 下册》(2009)：无危 / *National Key Protected Wild Animal List*: Ⅱ (2021); *China Species Red List: Vol. II Vertebrates Part 2* (2009): LC

地理分布 Geographical Distribution

国内分布 Domestic Distribution
仅分布于云南中部、广东、香港、广西等区域，为留鸟 / It only occurs in central of Yunnan, Guangdong, Hong Kong and Guangxi, as resident
分布标注 Distribution Note
非特有种 / Non-endemic

国内分布图
Map of Domestic Distribution

种群 Population

种群数量 Population Size	未知 / Unknown
种群趋势 Population Trend	下降 / Decreasing

生境与生态系统 Habitat(s) and Ecosystem(s)

生　　境 Habitat(s)	主要栖息于开阔而平坦的草地、农田、牧场、旷野等低地，也出现于林缘疏林草地和农田 / *M. javanica* uses open and flat lower land, *e.g.* grasslands, farmlands, meadows, *etc*. It also occurs in grassland and farmland at the edge of woods
生态系统 Ecosystem(s)	草原生态系统 / Grassland Ecosystem

威胁 Threat(s)

主要威胁 Major Threat(s)	分布区狭窄，适宜栖息地缩小，栖息地片断化 / Its distribution area is narrow and its suitable habitats are undergoing a decline and fragmentation

保护级别与保护行动 Protection Category and Conservation Action(s)

IUCN 红色名录 (2016) IUCN Red List (2016)	无危 / LC
保护行动 Conservation Action(s)	列入国家 II 级重点保护野生动物、《国家保护的有益的或者有重要经济、科学研究价值的陆生野生动物名录》/ It has been listed as the second class in National Key Protected Wild Animal List, *National Protected List of Terrestrial Wild Animals with Good Benefits or Important Economic and Scientific Values*

歌百灵 *Mirafra javanica*　　　　　莫建平 摄　By Jianping Mo

蒙古百灵
Melanocorypha mongolica

易危 VU A2abcd+B1b(ii,iii)

| 数据缺乏 DD | 无危 LC | 近危 NT | 易危 VU | 濒危 EN | 极危 CR | 区域灭绝 RE | 野外灭绝 EW | 灭绝 EX |

分类地位 Taxonomic Status

动物界 Animalia	脊索动物门 Chordata	鸟纲 Aves	雀形目 Passeriformes	百灵科 Alaudidae
学名 Scientific Name		*Melanocorypha mongolica*		
命名人 Species Authority		Pallas,1776		
英文名 English Name(s)		Mongolian Lark		
同物异名 Synonym(s)		*Alauda mongolica*		
种下单元评估 Infra-specific Taxa Assessed		内蒙古种群的生存力分析显示，种群未来20年的灭绝概率小于10% / The analysis of population viability of *Melanocorypha mongolica* indicated that the probability of extinction in 20 years is less than 10%		

评估信息 Assessment Information

评估年份 Year Assessed	2016
评定人 Assessor(s)	邢莲莲 / Lianlian Xing
其他贡献人 Other Contributor(s)	杨贵生、宋丽军等 / Guisheng Yang, Lijun Song *et al.*

理由 Justification: 蒙古百灵分布较广，特别是内蒙古锡林郭勒盟数量很大，但是每年的繁殖季节被大量抓捕，损失量多达十几万只，导致种群数量呈急剧下降趋势。因此，列为易危等级 / *Melanocorypha mongolica* is a representative species in grassland in China. It has an extremely large range, especially in Xilingol in Inner Mongolia (Nei Mongol), with large population. While in breeding season, more than 100,000 birds might be trapped illegally or dead, which makes the population drop dramatically. Thus, it is listed as Vulnerable

评估历史 Assessment History: 《国家重点保护野生动物名录》：II (2021)；《中国物种红色名录：第二卷 脊椎动物 下册》(2009)：近危 / National Key Protected Wild Animal List: II (2021); China Species Red List: Vol. II Vertebrates Part 2 (2009): NT

地理分布 Geographical Distribution

国内分布 Domestic Distribution

分布于黑龙江西南部、吉林西部、北京、天津、河北北部、山东、陕西北部、内蒙古、宁夏、甘肃西部、青海东部、云南 / It occurs in southwest Heilongjiang, west Jilin, Beijing, Tianjin, north Hebei, Shandong, north Shaanxi, Inner Mongolia (Nei Mongol), Ningxia, west Gansu, east Qinghai and Yunnan

分布标注 Distribution Note

非特有种 / Non-endemic

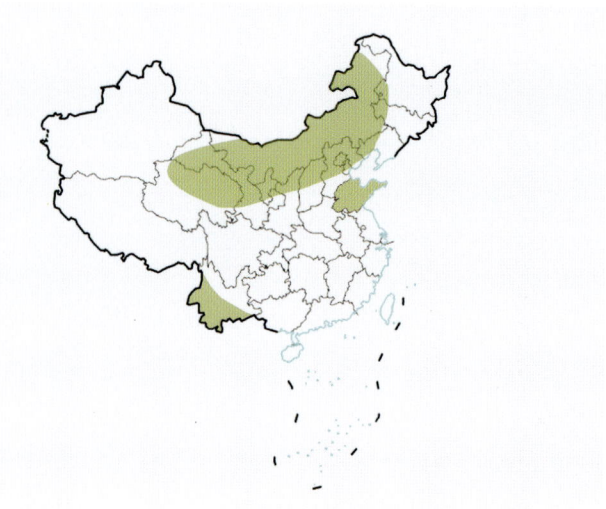

国内分布图
Map of Domestic Distribution

种群 Population

种群数量 Population Size	种群数量约 30 万只 / The population size is about 300,000 individuals
种群趋势 Population Trend	急剧下降 / Decreasing dramatically

生境与生态系统 Habitat (s) and Ecosystem (s)

生　　境 Habitat(s)	我国各种类型草原，包括典型草原及荒漠草原 / All types of grassland, including typical grassland and desert grassland
生态系统 Ecosystem(s)	草原生态系统 / Grassland Ecosystem

威胁 Threat (s)

主要威胁 Major Threat(s)	捕猎，草原退化及过度放牧使巢区条件受损，天敌、人等对雏鸟和卵的破坏增加等 / The destroy of nest by illegal trapping, the deterioration of grassland and over grazing are the major threats. Besides, the loss of subadult and eggs by human beings aggravates the situation

保护级别与保护行动 Protection Category and Conservation Action (s)

IUCN 红色名录 (2016) IUCN Red List (2016)	无危 / LC
保护行动 Conservation Action(s)	列入国家 II 级重点保护野生动物、《国家保护的有益的或者有重要经济、科学研究价值的陆生野生动物名录》。1994 年被定为内蒙古自治区区鸟，需进一步加强对抓捕雏幼鸟的执法力度 / It has been listed as the second class in National Key Protected Wild Animal List, *National Protected List of Terrestrial Wild Animals with Good Benefits or Important Economic and Scientific Values*. *Melanocorypha mongolica* was designated as provincial bird in Inner Mongolia (Nei Mongol). The efforts of law enforcement need to be improved

蒙古百灵 *Melanocorypha mongolica*　　　　　　　　　　　谢金平 摄　By Jinping Xie

远东苇莺
Acrocephalus tangorum

易危 VU C2a(ii)

| 数据缺乏 DD | 无危 LC | 近危 NT | **易危 VU** | 濒危 EN | 极危 CR | 区域灭绝 RE | 野外灭绝 EW | 灭绝 EX |

分类地位 Taxonomic Status

动物界 Animalia	脊索动物门 Chordata	鸟纲 Aves	雀形目 Passeriformes	苇莺科 Acrocephalidae

学 名 Scientific Name	*Acrocephalus tangorum*
命 名 人 Species Authority	La Touche, 1912
英 文 名 English Name(s)	Manchurian Reed Warbler
同物异名 Synonym(s)	无 / None
种下单元评估 Infra-specific Taxa Assessed	无 / None

评估信息 Assessment Information

评 估 年 份 Year Assessed	2016
评 定 人 Assessor(s)	董路 / Lu Dong
其他贡献人 Other Contributor(s)	无 / None

理由 Justification: 远东苇莺的繁殖区仅限于我国东北及邻近的俄罗斯地区，种群数量不足 10,000 只，且大多个体存在于同一个种群中。因此，列为易危等级 / *Acrocephalus tangorum* only breeds in Northeast China and closed area in Russia. The population size is less than 10,000, most in one subspecies. Thus, it is listed as Vulnerable

评估历史 Assessment History: 《中国物种红色名录：第二卷 脊椎动物 下册》(2009)：易危 / *China Species Red List: Vol. II Vertebrates Part 2* (2009): VU

地理分布 Geographical Distribution

国内分布 Domestic Distribution

繁殖于黑龙江、吉林，迁徙经过辽宁、北京、天津、河北、山东、内蒙古东北部、上海、香港、广西 / It breeds in Heilongjiang and Jilin, and migrates across in Liaoning, Beijing, Tianjin, Hebei, Shandong, northeast Inner Mongolia (Nei Mongol), Shanghai, Hong Kong and Guangxi

分布标注 Distribution Note

非特有种 / Non-endemic

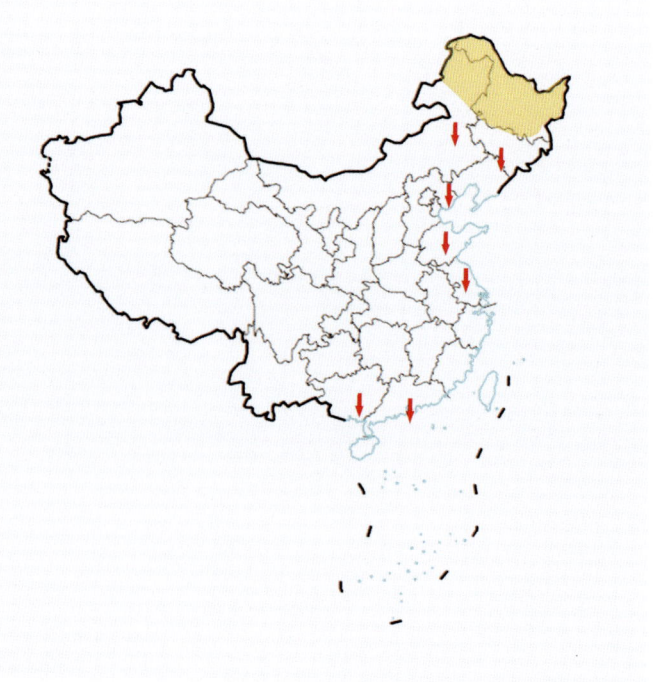

国内分布图
Map of Domestic Distribution

种群 Population

种群数量 Population Size	本物种可繁殖个体数量 2,500 ～ 10,000 只，且基本存在于同一个种群中 / The population size is estimated to be 2,500 ～ 10,000 mature individuals, almost in one population
种群趋势 Population Trend	下降 / Decreasing

生境与生态系统 Habitat(s) and Ecosystem(s)

生境 Habitat(s)	在繁殖地，主要栖息于湿地附近的芦苇灌丛中。越冬期可见于湿地周边或干旱草地及灌丛 / *A. tangorum* uses reeds adjacent to wetlands when breeding and occurs in areas surrounding wetlands or grassland and shrubs
生态系统 Ecosystem(s)	陆地生态系统 / Terrestrial Ecosystem

威胁 Threat(s)

主要威胁 Major Threat(s)	越冬区适宜栖息地面积的减少，是本物种的主要威胁因子。由于农业开发与城市化以及对草地资源的过度利用，导致湿地等适宜栖息地面积快速下降，是重要影响因素 / The loss of suitable wintering habitat due to development of agriculture, urbanization, and over use of grassland is the major threat

保护级别与保护行动 Protection Category and Conservation Action(s)

IUCN 红色名录 (2016) IUCN Red List (2016)	易危 / VU C2a (ii)
保护行动 Conservation Action(s)	部分分布区位于保护区内，需开展对其繁殖区适宜栖息地变化及种群动态研究 / The study on change of suitable breeding habitats and population dynamics is needed

远东苇莺 *Acrocephalus tangorum*   唐海全江 摄   By Tanghaiquanjiang

东亚蝗莺 *Locustella pleskei*

易危 VU C2a(i)

| 数据缺乏 DD | 无危 LC | 近危 NT | 易危 VU | 濒危 EN | 极危 CR | 区域灭绝 RE | 野外灭绝 EW | 灭绝 EX |

分类地位 Taxonomic Status

动物界 Animalia	脊索动物门 Chordata	鸟纲 Aves	雀形目 Passeriformes	蝗莺科 Locustellidae
学名 Scientific Name		*Locustella pleskei*		
命名人 Species Authority		Taczanowski, 1890		
英文名 English Name(s)		Pleske's Warbler		
同物异名 Synonym(s)		无 / None		
种下单元评估 Infra-specific Taxa Assessed		无 / None		

评估信息 Assessment Information

评估年份 Year Assessed	2016
评定人 Assessor(s)	刘阳 / Yang Liu
其他贡献人 Other Contributor(s)	黄秦 / Qin Huang

理由 Justification: 东亚蝗莺的中国繁殖种群可能受到诸多因素的影响而减少。中国东部沿海是其主要迁徙路线、华南地区是其主要越冬地，这些地区均面临着生境退化及非法捕猎的风险，对种群造成直接威胁。因此，列为易危等级 / The breeding population of *Locustella pleskei* in China might be undergoing a decline due to several factors. It uses the east coastal when migration and South China as wintering sites, while these areas are under risk of habitat deterioration and illegal hunting. Thus, it is listed as Vulnerable

评估历史 Assessment History:《中国物种红色名录：第二卷 脊椎动物 下册》(2009)：易危 / *China Species Red List: Vol. II Vertebrates Part 2* (2009): VU

地理分布 Geographical Distribution

国内分布 Domestic Distribution

迁徙时偶见于东部沿海地区，在黄海外海岛屿繁殖 (Qiao *et al.*, 2006)。迁徙时经过华东沿海江苏、上海、浙江、福建、台湾，越冬地在中国华南沿海省份广东湿地 / It uses the east coastal when migration, *e.g.* Jiangsu, Shanghai, Zhejiang, Fujian, Taiwan, and wetlands of Guangdong in South China as wintering sites. There are also a few breeding records on islands off the coast of the Yellow Sea (Qiao *et al.*, 2006)

分布标注 Distribution Note

非特有种 / Non-endemic

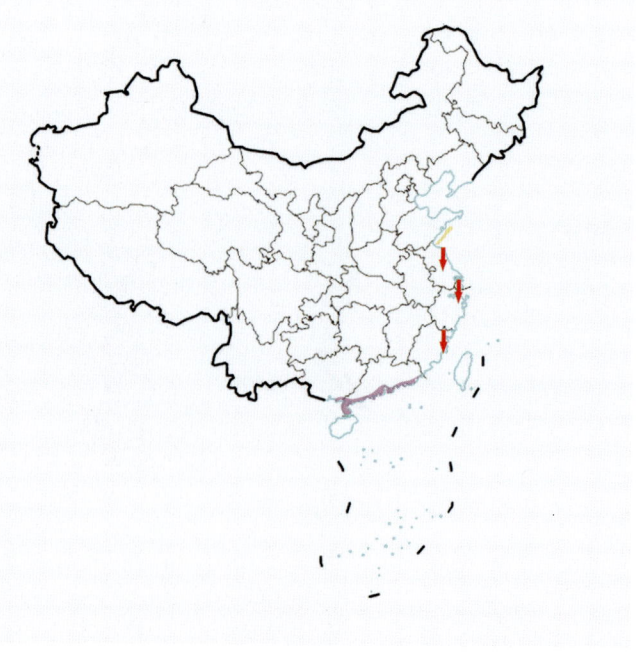

国内分布图
Map of Domestic Distribution

种群 Population

种群数量 Population Size	种群数量 1,000 ~ 1,500 只 / The population size is estimated to be 1,000 ~ 1,500 individuals
种群趋势 Population Trend	下降 / Decreasing

生境与生态系统 Habitat (s) and Ecosystem (s)

生　　境 Habitat(s)	栖息于低山和岛屿的灌丛中，越冬时栖息于海边的红树林和芦苇灌丛中 / *L. pleskei* inhabits in bushes in low mountains and islands. In its wintering range, birds have been found in extensive reedbeds, low shrubs near reedbeds and mangroves
生态系统 Ecosystem(s)	湿地生态系统 / Wetland Ecosystem

威胁 Threat (s)

主要威胁 Major Threat(s)	繁殖区可能受到岛屿引入鼠类、火山爆发、人为干扰等威胁，海岛的开发亦有可能破坏繁殖地的植被。非繁殖季节沿海湿地生境的丧失退化，

以及非法猎捕等 (BirdLife International, 2014) / The introduction of rodent, the volcanic eruption, and human disturbance are the major threats in breeding sites, while the degradation of wintering habitat in coastal wetlands and illegal trapping might be threats in wintering and stopover areas (BirdLife International, 2014)

保护级别与保护行动 Protection Category and Conservation Action (s)

IUCN 红色名录 (2016) IUCN Red List (2016)	易危 / VU C2a (i)
保护行动 Conservation Action(s)	暂无 / None

东亚蝗莺 *Locustella pleskei* 　　　　　　　薛琳 摄　By Lin Xue

台湾鹎
Pycnonotus taivanus

易危　VU A2ce+3ce+4ce

| 数据缺乏 DD | 无危 LC | 近危 NT | 易危 VU | 濒危 EN | 极危 CR | 区域灭绝 RE | 野外灭绝 EW | 灭绝 EX |

分类地位 Taxonomic Status

动物界 Animalia	脊索动物门 Chordata	鸟纲 Aves	雀形目 Passeriformes	鹎科 Pycnonotidae

学名 Scientific Name	*Pycnonotus taivanus*
命名人 Species Authority	Styan, 1893
英文名 English Name(s)	Styan's Bulbul
同物异名 Synonym(s)	无 / None
种下单元评估 Infra-specific Taxa Assessed	无 / None

评估信息 Assessment Information

评估年份 Year Assessed	2016
评定人 Assessor(s)	丁平 / Ping Ding
其他贡献人 Other Contributor(s)	无 / None

理由 Justification: 台湾鹎与白头鹎分布区重叠；相互之间缺少生殖隔离，导致基因渐渗引起的遗传多样性丧失，正处于种群快速缩减的状况下，同时也承受着栖息地丧失的威胁，种群数量至少减少30%，20年内有绝灭风险。因而，列为易危等级 / The distribution of *Pycnonotus taivanus* is overlapped with *P. sinensis*, and there is no reproductive isolation between the two species. With the risk of habitat loss, the population of *P. taivanus* is undergoing a rapid decline of at least 30%, and is estimated to suffer the risk of extinction in 20 years (Wang, 2009). Thus, it is listed as Vulnerable

评估历史 Assessment History: 《国家重点保护野生动物名录》：II (2021)；《中国物种红色名录：第二卷 脊椎动物 下册》(2009)：易危 / *National Key Protected Wild Animal List*: II (2021); *China Species Red List: Vol. II Vertebrates Part 2* (2009): VU

地理分布 Geographical Distribution

国内分布 Domestic Distribution

在台湾东部、南部和西南部的沿海平原以及山地均有发现。但由于其种群与白头鹎杂交，基因完整性较高的种群分布范围已经缩小到仅在台东县的一些区域 / It occurs in coastal plain and mountains in east, south and southwest Taiwan. Due to the interbreeding of *P. taivanus* and *P. sinensis*, the population of pure *P. taivanus* has declined and only occurs in limited area in Taidong County

分布标注 Distribution Note

特有种 / Endemic

国内分布图
Map of Domestic Distribution

种群 Population

种群数量 Population Size	种群数量 10,000～100,000 个繁殖对 (Brazil，2009) / The population size is estimated to be 10,000～100,000 pairs
种群趋势 Population Trend	下降 / Decreasing

生境与生态系统 Habitat(s) and Ecosystem(s)

生　　境 Habitat(s)	主要栖息于低海拔丘陵和山脚平原地区的次生阔叶林和疏林灌丛中，也出现于村庄、农田及路边的丛林中，有时甚至出现于城镇公园。经常小群活动 / *P. taivanus* mainly uses the secondary broad-leaved forests and shrubs in hills and plains, while also occurs in village, farmland, woods, and even parks in the town. It usually occurs in small groups
生态系统 Ecosystem(s)	森林生态系统 / Forest Ecosystem

威胁 Threat(s)

主要威胁 Major Threat(s)	台湾鹎最主要的威胁是与其近亲白头鹎进行杂交，这很大程度上是由于生境的改变。这两种生物重叠的生境越来越多，杂交的机会也不断增加。杂交的后代与纯种具有相同的生殖能力。因而台湾鹎面临同化的风险 (卢汰春，2010) / The main risk of *P. taivanus* is interbreeding with *P. sinensis* due to the change of habitat. The increase of habitat overlap of the two species brings more chance for interbreeding. The cross and the pure offspring have the same breeding ability. *P. taivanus* is probably confronted with the risk of assimilation (Lu, 2010)

保护级别与保护行动 Protection Category and Conservation Action(s)

IUCN 红色名录 (2016) IUCN Red List (2016)	易危 / VU A2ce+3ce+4ce
保护行动 Conservation Action(s)	列入国家 II 级重点保护野生动物、《国家保护的有益的或者有重要经济、科学研究价值的陆生野生动物名录》。台湾地区 1985 年对其进行积极保护，在垦丁自然保护区常见 / It has been listed as the second class in National Key Protected Wild Animal List, *National Protected List of Terrestrial Wild Animals with Good Benefits or Important Economic and Scientific Values*. It is protected in Taiwan since 1985, and is usually seen in Kending Nature Reserve

台湾鹎 *Pycnonotus taivanus*　　　　　王智斌 摄　By Zhibin Wang

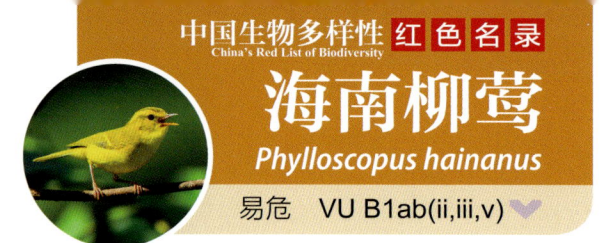

海南柳莺
Phylloscopus hainanus
易危 VU B1ab(ii,iii,v)

| 数据缺乏 DD | 无危 LC | 近危 NT | **易危 VU** | 濒危 EN | 极危 CR | 区域灭绝 RE | 野外灭绝 EW | 灭绝 EX |

分类地位 Taxonomic Status

动物界 Animalia	脊索动物门 Chordata	鸟纲 Aves	雀形目 Passeriformes	柳莺科 Phylloscopidae
学 名 Scientific Name		*Phylloscopus hainanus*		
命 名 人 Species Authority		Olsson, Alström & Colston, 1993		
英 文 名 English Name(s)		Hainan Leaf Warbler		
同物异名 Synonym(s)		*Seicercus hainanus*		
种下单元评估 Infra-specific Taxa Assessed		无 / None		

评估信息 Assessment Information

评估年份 Year Assessed	2016
评定人 Assessor(s)	梁伟 / Wei Liang
其他贡献人 Other Contributor(s)	杨灿朝 / Canchao Yang

理由 Justification：海南柳莺为海南特有种。由于砍伐、种植橡胶林等因素，海南的天然林面积持续锐减。因此，列为易危等级 / *Phylloscopus hainanus* is endemic to Hainan. Due to the deforestation and rubber plantation, the area of natural forest is undergoing a rapid decline and fragmentation. Thus, it is listed as Vulnerable

评估历史 Assessment History：《中国物种红色名录：第二卷 脊椎动物 下册》(2009)：易危 / *China Species Red List: Vol. II Vertebrates Part 2* (2009): VU

地理分布 Geographical Distribution

国内分布 Domestic Distribution
仅见于海南 / It only occurs in Hainan
分布标注 Distribution Note
特有种 / Endemic

国内分布图
Map of Domestic Distribution

种群 Population

种群数量 Population Size	种群成熟个体 1,500～7,000 只 / The population size is estimated to be 1,500～7,000 mature individuals
种群趋势 Population Trend	下降 / Decreasing

生境与生态系统 Habitat (s) and Ecosystem (s)

生　　境 Habitat(s)	天然原始林 / Nature forest
生态系统 Ecosystem(s)	陆地生态系统（热带森林生态系统）/ Terrestrial Ecosystem (Tropical Forest Ecosystem)

威胁 Threat (s)

主要威胁 Major Threat(s)	分布区狭小，仅分布于海南的天然林。由于砍伐、种植橡胶林和农地开垦，海南的天然林面积从 1943 年的 16,920 km^2 下降至 1994 年的 3,000 km^2 / *P. hainanus* only occurs in the natural forest in Hainan. Due to the deforestation, rubber plantation and slash-and-burn cultivation, the natural forest has decreased from 16,920 km^2 in 1943 to 3,000 km^2 in 1994

保护级别与保护行动 Protection Category and Conservation Action (s)

IUCN 红色名录 (2016) IUCN Red List (2016)	易危 / VU B1ab (ii,iii,v)
保护行动 Conservation Action(s)	列入《国家保护的有益的或者有重要经济、科学研究价值的陆生野生动物名录》。建议对该物种的种群数量、栖息地和繁殖生态等进行深入研究；加大对天然林和栖息地质量的保护 / It was listed as *National Protected List of Terrestrial Wild Animals with Good Benefits or Important Economic and Scientific Values*. The study on the population size, the habitat and breeding ecology of *P. hainanus* is suggested. Meanwhile, the natural forest and habitat quality should be protected

海南柳莺 *Phylloscopus hainanus*　　　　　　　　　　　许传辉 摄　By Chuanhui Xu

暗色鸦雀
Sinosuthora zappeyi

易危 VU B1ab(i,ii,iii); C2a(i)

中国生物多样性红色名录 / China's Red List of Biodiversity

| 数据缺乏 DD | 无危 LC | 近危 NT | 易危 VU | 濒危 EN | 极危 CR | 区域灭绝 RE | 野外灭绝 EW | 灭绝 EX |

分类地位 Taxonomic Status

动物界 Animalia	脊索动物门 Chordata	鸟纲 Aves	雀形目 Passeriformes	莺鹛科 Sylviidae
学名 Scientific Name		*Sinosuthora zappeyi*		
命名人 Species Authority		Thayer and Bangs, 1912		
英文名 English Name(s)		Grey-hooded Parrotbill		
同物异名 Synonym(s)		*Paradoxornis zappeyi*		
种下单元评估 Infra-specific Taxa Assessed		无 / None		

评估信息 Assessment Information

评估年份 Year Assessed	2016
评定人 Assessor(s)	孙悦华 / Yuehua Sun
其他贡献人 Other Contributor(s)	无 / None

理由 Justification: 暗色鸦雀分布范围较小，数个小种群间彼此隔离，同时受到旅游等人为活动的干扰。因此列为易危等级 / The distribution area of *Sinosuthora zappeyi* is small, and populations are isolated, meanwhile are under high pressure of human disturbance. Thus, it is listed as Vulnerable

评估历史 Assessment History: 《国家重点保护野生动物名录》：II (2021)；《中国濒危动物红皮书·鸟类》(1998)：稀有；《中国物种红色名录：第二卷 脊椎动物 下册》(2009)：易危 / *National Key Protected Wild Animal List*: II (2021); *China Red Data Book of Endangered Animals*: *Aves* (1998): R; *China Species Red List*: *Vol. II Vertebrates Part 2* (2009): VU

地理分布 Geographical Distribution

国内分布 Domestic Distribution
分布于四川、贵州西北部和云南东北部 / It occurs in Sichuan, northwest Guizhou and northeast Yunnan
分布标注 Distribution Note
特有种 / Endemic

国内分布图
Map of Domestic Distribution

种群 Population

种群数量 Population Size	未知 / Unknown
种群趋势 Population Trend	下降 / Decreasing

生境与生态系统 Habitat(s) and Ecosystem(s)

生境 Habitat(s)	栖息于海拔 2,300～3,200 m 的高山和高原地带的箭竹丛和灌丛中，尤以开阔的湖边和溪流沿岸灌丛和高草丛中较常见。除繁殖期间成对或单独活动外，其他季节多成群活动。营巢于竹丛 (Jiang *et al.*, 2009) / *S. zappeyi* uses the bamboos and shrubs in mountain and plateau areas with elevation of 2,300～3,200 m, especially around the open lake, or along the river. It forms groups, breeding season excepted, and nests in the bamboos (Jiang *et al.*, 2009)
生态系统 Ecosystem(s)	陆地生态系统 / Terrestrial Ecosystem

威胁 Threat(s)

主要威胁 Major Threat(s)	栖息地的丧失和破碎化，景区的旅游开发带来的人为干扰和破坏，所依赖的竹丛大规模开花可能也是影响其数量的因素之一 / The major threats of *S. zappeyi* are habitat loss and fragmentation. Besides, the human disturbance of tourism, and the blooming of bamboo are also threats to the population

保护级别与保护行动 Protection Category and Conservation Action(s)

IUCN 红色名录 (2016) IUCN Red List (2016)	易危 / VU C2a (i)
保护行动 Conservation Action(s)	列入国家 II 级重点保护野生动物、《国家保护的有益的或者有重要经济、科学研究价值的陆生野生动物名录》。部分分布区位于保护区内，开展了初步的生活史研究，需要对其数量和分布进行更准确的估计，加强保护区间的连接度，评估旅游带来的风险 / It has been listed as the second class in National Key Protected Wild Animal List, *National Protected List of Terrestrial Wild Animals with Good Benefits or Important Economic and Scientific Values*. Some distribution areas are located in the nature reserves, and studies on life history has started. The estimates of population and distribution, as well as network of nature reserves, and assessment of tourism are needed

暗色鸦雀 *Sinosuthora zappeyi* 　　　　王文娟 摄　By Wenjuan Wang

中国生物多样性红色名录

金额雀鹛
Schoeniparus variegaticeps

易危　VU B1ab(ii,iv); C2a(i)

| 数据缺乏 DD | 无危 LC | 近危 NT | 易危 VU | 濒危 EN | 极危 CR | 区域灭绝 RE | 野外灭绝 EW | 灭绝 EX |

分类地位 Taxonomic Status

| 动物界 Animalia | 脊索动物门 Chordata | 鸟纲 Aves | 雀形目 Passeriformes | 幽鹛科 Pellorneidae |

学名 Scientific Name	*Schoeniparus variegaticeps*
命名人 Species Authority	Yen, 1932
英文名 English Name(s)	Golden-fronted Fulvetta
同物异名 Synonym(s)	*Alcippe variegaticeps*
种下单元评估 Infra-specific Taxa Assessed	无 / None

评估信息 Assessment Information

评估年份 Year Assessed	2016
评定人 Assessor(s)	周放 / Fang Zhou
其他贡献人 Other Contributor(s)	蒋爱伍 / Aiwu Jiang

理由 Justification: 金额雀鹛分布区极为狭窄且不连续，仅见于广西中北部和四川西南部的高海拔森林中，估计实际栖息地面积小于 1,000 km²。近年来，在分布区内很少有金额雀鹛的观察记录，估计种群数量已经极为稀少。因此，列为易危等级 / The distribution area of *Schoeniparus variegaticeps* is small and fragmented, which is estimated to be less than 1,000 km². It only occurs in forests at high altitude in middle and north Guangxi, and southwest Sichuan. There are very few records, and the population size is estimated to be quite small. Thus, it is listed as Vulnerable

评估历史 Assessment History: 《国家重点保护野生动物名录》: Ⅰ (2021); 《中国物种红色名录: 第二卷 脊椎动物 下册》(2009): 易危 / National Key Protected Wild Animal List: I (2021); China Species Red List: Vol. II Vertebrates Part 2 (2009): VU

地理分布 Geographical Distribution

国内分布 Domestic Distribution	为四川中部和广西北部的留鸟 / It is resident in central Sichuan and north Guangxi
分布标注 Distribution Note	特有种 / Endemic

国内分布图
Map of Domestic Distribution

种群 Population

种群数量 Population Size	估计种群数量不足 2,000 对 / The population size is estimated to be less than 2,000 pairs
种群趋势 Population Trend	下降 / Decreasing

生境与生态系统 Habitat(s) and Ecosystem(s)

生　　境 Habitat(s)	海拔 1,500 m 以上的常绿阔叶林、针阔混交林和竹林中 / *S. variegaticeps* uses evergreen broad-leaved forests, mixed coniferous broad-leaved forests and bamboos above 1,500 m
生态系统 Ecosystem(s)	陆地生态系统 / Terrestrial Ecosystem

威胁 Threat(s)

主要威胁 Major Threat(s)	原生的常绿阔叶林被大量破坏，导致栖息地减少和质量下降可能是金额雀鹛受威胁的主要原因。另外，由于分布区不连续，栖息地片断化也是金额雀鹛受威胁的原因之一 / The major threats of the species are habitat loss and deterioration due to destroy of natural evergreen broad-leaved forests. Besides, the habitat fragmentation is also a threat

保护级别与保护行动 Protection Category and Conservation Action(s)

IUCN 红色名录 (2016) IUCN Red List (2016)	易危 / VU B1ab (ii,iii,iv,v); C2a (i)
保护行动 Conservation Action(s)	列入国家 I 级重点保护野生动物、《国家保护的有益的或者有重要经济、科学研究价值的陆生野生动物名录》。建议加大对其种群数量和栖息地需求的研究 / It has been listed as the first class in National Key Protected Wild Animal List, *National Protected List of Terrestrial Wild Animals with Good Benefits or Important Economic and Scientific Values*. It is necessary to improve the study on population size and habitat requirements are suggested

金额雀鹛 *Schoeniparus variegaticeps*　　　　王文娟 摄　By Wenjuan Wang

黑额山噪鹛
Garrulax sukatschewi

易危 VU B1ab(i,ii,iii); C2a(i)

| 数据缺乏 DD | 无危 LC | 近危 NT | 易危 VU | 濒危 EN | 极危 CR | 区域灭绝 RE | 野外灭绝 EW | 灭绝 EX |

分类地位 Taxonomic Status

动物界 Animalia	脊索动物门 Chordata	鸟 纲 Aves	雀形目 Passeriformes	噪眉科 Leiothrichidae
学 名 Scientific Name		*Garrulax sukatschewi*		
命 名 人 Species Authority		Berezowski and Bianchi, 1891		
英 文 名 English Name(s)		Snowy-cheeked Laughingthrush		
同物异名 Synonym(s)		*Trochalopteron sukatschewi*		
种下单元评估 Infra-specific Taxa Assessed		无 / None		

评估信息 Assessment Information

评估年份 Year Assessed	2016
评 定 人 Assessor(s)	孙悦华 / Yuehua Sun
其他贡献人 Other Contributor(s)	无 / None

理由 Justification：黑额山噪鹛种群数量较小；分布区狭窄，占有面积少于 2,000 km²，且受到栖息地丧失和破碎化的威胁。因此，列为易危等级 / The population size of *Garrulax sukatschewi* is small and the distribution area is small, which is less than 2,000 km². The species is under the threats of habitat loss and fragmentation. Thus, it is listed as Vulnerable

评估历史 Assessment History：《国家重点保护野生动物名录》：Ⅰ (2021)；《中国濒危动物红皮书·鸟类》(1998)：稀有；《中国物种红色名录：第二卷 脊椎动物 下册》(2009)：易危 / *National Key Protected Wild Animal List*: Ⅰ (2021); *China Red Data Book of Endangered Animals*: *Aves* (1998): R; *China Species Red List*: *Vol. II Vertebrates Part 2* (2009): VU

地理分布 Geographical Distribution

国内分布 Domestic Distribution
为甘肃南部和四川北部的留鸟 / It is resident in south Gansu and north Sichuan
分布标注 Distribution Note
特有种 / Endemic

国内分布图
Map of Domestic Distribution

种群 Population

种群数量 Population Size	种群数量 2,500 ～ 9,999 只 / The population size is roughly estimated to be 2,500 ～ 9,999 individuals
种群趋势 Population Trend	下降 / Decreasing

生境与生态系统 Habitat (s) and Ecosystem (s)

生　　境 Habitat(s)	栖息于海拔 2,300 ～ 3,200 m 的高山和高原地带的箭竹丛和灌丛中，尤以开阔的湖边和溪流沿岸灌丛和高草丛中较常见，营巢于竹丛 (Jiang et al., 2009) / G. sukatschewi uses the bamboos and shrubs in mountain and plateau area at 2,300 ～ 3,200 m, especially the shrubs and grasses in the open area adjacent to the lake or the banks along the stream. The nest is founded in the bamboos (Jiang et al., 2009)
生态系统 Ecosystem(s)	陆地生态系统 / Terrestrial Ecosystem

威胁 Threat (s)

主要威胁 Major Threat(s)	森林栖息地的丧失和破碎化是影响其生存的最主要因素，林下植被的清理也对其生存产生不利影响 (Wang et al., 2011) / The major threats are habitat loss and fragmentation. The clean of the vegetation at the bottom of the forests is also negative (Wang et al., 2011)

保护级别与保护行动 Protection Category and Conservation Action (s)

IUCN 红色名录 (2016) IUCN Red List (2016)	易危 / VU C2a (i)
保护行动 Conservation Action(s)	列入国家 I 级重点保护野生动物、《国家保护的有益的或者有重要经济、科学研究价值的陆生野生动物名录》。对栖息地、繁殖、行为等进行了初步研究，获得了一些基础生活史资料 (Bi et al., 2003; Wang et al., 2011); 保护区涵盖了其大部分分布范围。种群数量仍需估计，其生活史资料需要进一步完善 / It has been listed as the first class in National Key Protected Wild Animal List, National Protected List of Terrestrial Wild Animals with Good Benefits or Important Economic and Scientific Values. There are some primary studies on habitat, breeding and behavior (Bi et al., 2003; Wang et al., 2011). Some nature reserves have been set up to cover most of its distribution area. Studies on the population size and life history

黑额山噪鹛 Garrulax sukatschewi　　　　　　　　　　　　　　　　　　　FK1187 摄　By FK1187

白点噪鹛
Garrulax bieti
易危 VU C1

| 数据缺乏 DD | 无危 LC | 近危 NT | 易危 VU | 濒危 EN | 极危 CR | 区域灭绝 RE | 野外灭绝 EW | 灭绝 EX |

分类地位 Taxonomic Status

动物界 Animalia	脊索动物门 Chordata	鸟纲 Aves	雀形目 Passeriformes	噪鹛科 Leiothrichidae

学名 Scientific Name	*Garrulax bieti*
命名人 Species Authority	Oustalet, 1897
英文名 English Name(s)	White-speckled Laughingthrush
同物异名 Synonym(s)	*Ianthocincla Bieti*
种下单元评估 Infra-specific Taxa Assessed	无 / None

评估信息 Assessment Information

评估年份 Year Assessed	2016
评定人 Assessor(s)	董路 / Lu Dong
其他贡献人 Other Contributor(s)	无 / None

理由 Justification: 白点噪鹛的种群数量有限，分布区狭窄且生境破碎化严重，局域地理种群数量不超过1,000只，且呈下降趋势。林业和农业的发展导致其栖息的亚热带山地森林被破坏。因此，列为易危等级 / The population size and distribution area of *Garrulax bieti* are small and its habitat is heavily fragmented. The local population is less than 1,000 individuals and is undergoing a decline. The subtropical forests have been destroyed due to the development of forestry and agriculture. Thus, it is listed as Vulnerable

评估历史 Assessment History: 《国家重点保护野生动物名录》：Ⅰ (2021)；《中国物种红色名录：第二卷 脊椎动物 下册》(2009)：易危 / *National Key Protected Wild Animal List*: Ⅰ (2021); *China Species Red List: Vol. II Vertebrates Part 2* (2009): VU

地理分布 Geographical Distribution

国内分布 Domestic Distribution
分布于云南西北部和四川西南部 / It occurs in northwest Yunnan and southwest Sichuan
分布标注 Distribution Note
特有种 / Endemic

国内分布图
Map of Domestic Distribution

种群 Population

种群数量 Population Size	种群数量 2,500～9,999 只 / The population size is estimated to be 2,500～9,999 individuals
种群趋势 Population Trend	下降 / Decreasing

生境与生态系统 Habitat (s) and Ecosystem (s)

生　　境 Habitat(s)	本物种适宜的生境尚缺乏了解。主要分布于海拔 2,500～4,500 m 植被发育良好且人为干扰较少的针阔混交林、针叶林和杜鹃花林，尤其偏好竹灌层丰富的森林 / There is very few information on its suitable habitat. It mainly occurs in the mixed coniferous broad-leaved forest, coniferous forest and azalea shrubs at 2,500～4,500 m without human disturbance, especially the forests with bamboos and shrubs
生态系统 Ecosystem(s)	陆地生态系统 / Terrestrial Ecosystem

威胁 Threat (s)

主要威胁 Major Threat(s)	森林面积的减少和适宜生境的破碎化是其主要威胁因子。云南的森林覆盖率在近 40 年内至少下降了 50%，由于伐木，四川的一些分布点生境已经被完全破坏。非法捕猎也是威胁因子之一 / The loss of forest and the fragmentation of suitable habitat are the major threats. The forest coverage rate has declined by 50% in 40 years in Yunnan. And some habitats in Sichuan have been totally destroyed owing to logging. The illegal trapping is another threat

保护级别与保护行动 Protection Category and Conservation Action (s)

IUCN 红色名录 (2016) IUCN Red List (2016)	易危 / VU C2a (i)
保护行动 Conservation Action(s)	列入国家 I 级重点保护野生动物、《国家保护的有益的或者有重要经济、科学研究价值的陆生野生动物名录》。分布区内已建立了多个保护区，如白马雪山国家级自然保护区、哈巴雪山自然保护区和玉龙雪山自然保护区等 / It has been listed as the first class in National Key Protected Wild Animal List, *National Protected List of Terrestrial Wild Animals with Good Benefits or Important Economic and Scientific Values*. Several nature reserves have been set up in its distribution area, *e.g.* Baima Jokul National Nature Reserve, Haba Jokul Nature Reserve, and Yulong Jokul Nature Reserve, *etc.*

白点噪鹛 *Garrulax bieti*　　　　　　　　　邹渝 摄　By Yu Zou

中国生物多样性红色名录

灰胸薮鹛
Liocichla omeiensis

易危　VU B1ab(i-v); C2a(i,ii)

| 数据缺乏 DD | 无危 LC | 近危 NT | 易危 VU | 濒危 EN | 极危 CR | 区域灭绝 RE | 野外灭绝 EW | 灭绝 EX |

分类地位 Taxonomic Status

动物界 Animalia	脊索动物门 Chordata	鸟纲 Aves	雀形目 Passeriformes	噪鹛科 Leiothrichidae

学名 Scientific Name	*Liocichla omeiensis*
命名人 Species Authority	Riley, 1926
英文名 English Name(s)	Emei Shan Liocichla
同物异名 Synonym(s)	无 / None
种下单元评估 Infra-specific Taxa Assessed	无 / None

评估信息 Assessment Information

评估年份 Year Assessed	2016
评定人 Assessor(s)	卢欣 / Xin Lu
其他贡献人 Other Contributor(s)	无 / None

理由 Justification: 灰胸薮鹛分布区狭窄，由于亚热带森林被破坏，栖息地片断化，估计分布区少于 20,000 km²，所有种群数目可能不足 10,000 个，成熟个体只有 1,667～6,666 个，种群可能随着栖息地丧失而正在逐渐下降。因此，列为易危等级 / The distribution area of *Liocichla omeiensis* is small and estimated to be less than 20,000 km² owing to the destroy of subtropical forest and fragmentation of habitat. The population size is estimated to be less than 10,000 with 1,667～6,666 mature individuals. Its population and distribution are undergoing a rapid decline with the habitat loss. Thus, it is listed as Vulnerable

评估历史 Assessment History: 《国家重点保护野生动物名录》：Ⅰ (2021)；《中国物种红色名录：第二卷 脊椎动物 下册》(2009)：易危 / *National Key Protected Wild Animal List*: I (2021); *China Species Red List: Vol. II Vertebrates Part 2* (2009): VU

地理分布 Geographical Distribution

国内分布 Domestic Distribution
分布于四川南部及云南东北部 / It occurs in south Sichuan and northeast Yunnan

分布标注 Distribution Note
特有种 / Endemic

国内分布图
Map of Domestic Distribution

种群 Population

种群数量 Population Size	估计种群数量不足 10,000 只 / The population size is estimated to be less than 10,000 individuals
种群趋势 Population Trend	下降 / Decreasing

生境与生态系统 Habitat (s) and Ecosystem (s)

生　　境 Habitat(s)	夏季多分布于海拔 1,400～2,400 m 的亚热带及温带山地常绿阔叶林、次生林、竹林和林缘灌丛区域，冬季一般在海拔 500～1,400 m 的地带 / *L. omeiensis* uses subtropical and temperate evergreen broad-leaved forests, secondary forests, bamboos and shrubs at 1,400～2,400 m in summer, and 500～1,400 m in winter
生态系统 Ecosystem(s)	陆地生态系统 / Terrestrial Ecosystem

威胁 Threat (s)

主要威胁 Major Threat(s)	主要的威胁源于森林的片断化和丧失，伐木以及农业活动造成的森林的逐渐消失。竹子等其他森林产品被收集用来交易，以及放牧。猎捕以及交易也是其威胁来源之一 / The major threats are habitat loss and fragmentation owing to the logging and agriculture activities. The forest products (bamboo, *etc.*) are collected for trading. Illegal trapping and pet bird trading are also threats

保护级别与保护行动 Protection Category and Conservation Action (s)

IUCN 红色名录 (2016) IUCN Red List (2016)	易危 / VU B1ab (i-v)
保护行动 Conservation Action(s)	列入国家 I 级重点保护野生动物、《国家保护的有益的或者有重要经济、科学研究价值的陆生野生动物名录》。四川省将其列入重点保护野生动物名录。建议建立该种在峨眉山的保护小区 / It has been listed as the first class in National Key Protected Wild Animal List, *National Protected List of Terrestrial Wild Animals with Good Benefits or Important Economic and Scientific Values*. Nature reserve in Emei Shan should be set up, and including the species into Sichuan Key Protected Wild Animal List is suggested

灰胸薮鹛 *Liocichla omeiensis*　　　　　　陈孝齐 摄　By Xiaoqi Chen

四川旋木雀
Certhia tianquanensis

易危　VU C2a(i)

| 数据缺乏 DD | 无危 LC | 近危 NT | **易危 VU** | 濒危 EN | 极危 CR | 区域灭绝 RE | 野外灭绝 EW | 灭绝 EX |

分类地位 Taxonomic Status

动物界 Animalia	脊索动物门 Chordata	鸟纲 Aves	雀形目 Passeriformes	旋木雀科 Certhiidae
学名 Scientific Name		*Certhia tianquanensis*		
命名人 Species Authority		Li, 1995		
英文名 English Name(s)		Sichuan Treecreeper		
同物异名 Synonym(s)		无 / None		
种下单元评估 Infra-specific Taxa Assessed		无 / None		

评估信息 Assessment Information

评估年份 Year Assessed	2016
评定人 Assessor(s)	孙悦华 / Yuehua Sun
其他贡献人 Other Contributor(s)	张雁云 / Yanyun Zhang

理由 Justification: 四川旋木雀分布区相对较小且集中，人类活动对其干扰较大。因此，列为易危等级 / The distribution area of *Certhia tianquanensis* is relatively small and concentrated, and human activities are more disturbing. Thus, it is listed as Vulnerable

评估历史 Assessment History: 《国家重点保护野生动物名录》：Ⅱ (2021)；《中国物种红色名录：第二卷 脊椎动物 下册》(2009)：易危 / *National Key Protected Wild Animal List*: Ⅱ (2021); *China Species Red List: Vol. Ⅱ Vertebrates Part 2* (2009): VU

地理分布 Geographical Distribution

国内分布 Domestic Distribution
为陕西南部、四川中部和北部地区的留鸟 / It is resident in south Shaanxi, central and north of Sichuan

分布标注 Distribution Note
特有种 / Endemic

国内分布图
Map of Domestic Distribution

种群 Population

种群数量 Population Size	数据不详。在四川瓦屋山峨眉冷杉林中有较高密度的繁殖个体 (Sun et al., 2009) / No specific information while there are relatively high density of breeding birds in the *Abies* forests in Wawu Shan in Sichuan
种群趋势 Population Trend	下降中 / In decreasing

生境与生态系统 Habitat (s) and Ecosystem (s)

生 境 Habitat(s)	栖息和繁殖于冷杉与糙皮桦混交林，林下有浓密的竹林灌丛 (Martens et al., 2002)，巢址与冷杉林中的桦树枯木相关 (Sun et al., 2009)，冬季可以从海拔 2,500 m 以上的区域迁移到海拔 1,600 m 的区域觅食 / *C. tianquanensis* uses the mixed *Abies* forests and *Betula* forests with dense bamboo shrubs for breeding and roosting (Martens et al., 2002). The nests are significantly related to the dead woods of *Betula* (Sun et al., 2009). It migrates from the areas above 2,500 m to 1,600 m for foraging
生态系统 Ecosystem(s)	陆地生态系统 / Terrestrial Ecosystem

威胁 Threat (s)

主要威胁 Major Threat(s)	分布区缩小，栖息地退化 / The major threat is the decline of distribution and habitat degeneration

保护级别与保护行动 Protection Category and Conservation Action (s)

IUCN 红色名录 (2016) IUCN Red List (2016)	近危 / NT
保护行动 Conservation Action(s)	列入国家 II 级重点保护野生动物。部分分布区位于太白山自然保护区，卧龙自然保护区和喇叭河自然保护区。需进一步加强其原生生境冷杉林的保护，应保护好其生境中的枯立木 / It has been listed as the second class in National Key Protected Wild Animal List. Part of its distribution areas are located in Taibai Shan Nature Reserve, Wolong Nature Reserve and Labahe Nature Reserve. The protection of natural *Abies* forests should be improved and the dead stand wood should also be protected

四川旋木雀 *Certhia tianquanensis*　　胡敬林 摄　By Jinglin Hu

滇䴓
Sitta yunnanensis

易危 VU A3bcd+4bcd; B2b(I,ii,iii)

| 数据缺乏 DD | 无危 LC | 近危 NT | **易危 VU** | 濒危 EN | 极危 CR | 区域灭绝 RE | 野外灭绝 EW | 灭绝 EX |

分类地位 Taxonomic Status

动物界 Animalia	脊索动物门 Chordata	鸟纲 Aves	雀形目 Passeriformes	䴓科 Sittidae
学　　名 Scientific Name		*Sitta yunnanensis*		
命　名　人 Species Authority		Ogilvie-Grant, 1900		
英　文　名 English Name(s)		Yunnan Nuthatch		
同物异名 Synonym(s)		无 / None		
种下单元评估 Infra-specific Taxa Assessed		无 / None		

评估信息 Assessment Information

评 估 年 份 Year Assessed	2016
评 定 人 Assessor(s)	杨晓君 / Xiaojun Yang
其他贡献人 Other Contributor(s)	吴飞、常云艳等 /Fei Wu, Yunyan Chang *et al.*

理由 Justification: 滇䴓分布区相对较小，生境破坏，种群数量有下降趋势。因此，列为易危等级 / The population of *Sitta yunnanensis* is undergoing a decline due to the small distribution and destroy of habitat. Thus, it is listed as Vulnerable

评估历史 Assessment History: 《国家重点保护野生动物名录》：Ⅱ (2021)；《中国物种红色名录：第二卷 脊椎动物 下册》(2009)：近危 / *National Key Protected Wild Animal List*: Ⅱ (2021); *China Species Red List*: *Vol. II Vertebrates Part 2* (2009): NT

地理分布 Geographical Distribution

国内分布 Domestic Distribution
分布于西藏东南部、云南、四川西部和贵州西部 / It occurs in southeast Tibet (Xizang), Yunnan, west Sichuan and west Guizhou
分布标注 Distribution Note
特有种 / Endemic

国内分布图
Map of Domestic Distribution

种群 Population

种群数量 Population Size	局部常见，但由于分布区相对较小，以及生境破坏等原因，种群数量有下降趋势 / The population is undergoing a decline owing to the limited distribution and habitat loss
种群趋势 Population Trend	下降 / Decreasing

生境与生态系统 Habitat (s) and Ecosystem (s)

生　　境 Habitat(s)	栖息于中山和高山沟谷林、山坡针叶林或针阔混交林 / *S. yunnanensis* uses woods in valleys of middle and high mountains, and coniferous forest or mixed coniferous broad-leaved forest at mountainside
生态系统 Ecosystem(s)	陆地生态系统 / Terrestrial Ecosystem

威胁 Threat (s)

主要威胁 Major Threat(s)	栖息地破坏 / Destroy of habitat

保护级别与保护行动 Protection Category and Conservation Action (s)

IUCN 红色名录 (2016) IUCN Red List (2016)	近危 / NT
保护行动 Conservation Action(s)	列入国家 II 级重点保护野生动物、《国家保护的有益的或者有重要经济、科学研究价值的陆生野生动物名录》/ It has been listed as the second class in National Key Protected Wild Animal List, *National Protected List of Terrestrial Wild Animals with Good Benefits or Important Economic and Scientific Values*

滇䴓 *Sitta yunnanensis*　　　　李建东 摄　By Jiandong Li

中国生物多样性 红色名录
China's Red List of Biodiversity

淡紫鸭
Sitta solangiae

易危 VU B1ab(iii)

数据缺乏 DD	无危 LC	近危 NT	易危 VU	濒危 EN	极危 CR	区域灭绝 RE	野外灭绝 EW	灭绝 EX

分类地位 Taxonomic Status

动物界 Animalia	脊索动物门 Chordata	鸟纲 Aves	雀形目 Passeriformes	鸭科 Sittidae

学 名 Scientific Name	*Sitta solangiae*
命 名 人 Species Authority	Delacour & Jabouille, 1930
英 文 名 English Name(s)	Yellow-billed Nuthatch
同物异名 Synonym(s)	无 / None
种下单元评估 Infra-specific Taxa Assessed	无 / None

评估信息 Assessment Information

评 估 年 份 Year Assessed	2016
评 定 人 Assessor(s)	梁伟 / Wei Liang
其他贡献人 Other Contributor(s)	杨灿朝 / Canchao Yang

理由 Justification: 淡紫鸭分布面积少于 20,000 km², 分布区小且分割成 3～4 个区域。部分保护区存在不同程度的旅游和开发等, 导致栖息地质量持续衰退。因此, 列为易危等级 / The distribution area of *Sitta solangiae* is small (less than 20,000 km²) and splitted into 3～4 parts. The habitat quality is undergoing a deterioration due to the development and tourism of some nature reserves. Thus, it is listed as Vulnerable

评估历史 Assessment History: 《中国物种红色名录: 第二卷 脊椎动物 下册》(2009): 近危 / *China Species Red List*: *Vol. II Vertebrates Part 2* (2009): NT

地理分布 Geographical Distribution

国内分布 Domestic Distribution
分布于海南 / It occurs in Hainan
分布标注 Distribution Note
非特有种 / Non-endemic

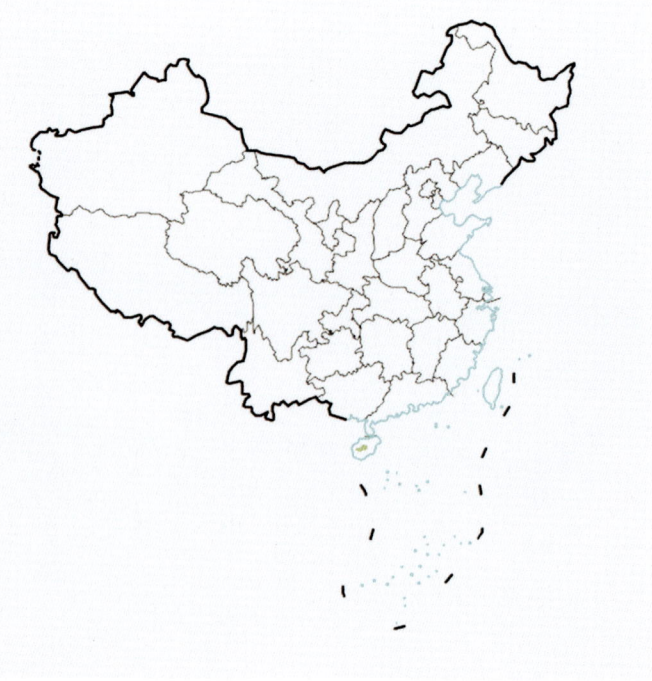

国内分布图
Map of Domestic Distribution

种群 Population

种群数量 Population Size	未评估。在海南各主要林区的天然林中尚可见 / The population size has not been assessed. *Sitta solangiae* could be seen in natural forests in Hainan
种群趋势 Population Trend	下降 / Decreasing

生境与生态系统 Habitat (s) and Ecosystem (s)

生　　境 Habitat(s)	天然原生林 / Natural forests
生态系统 Ecosystem(s)	陆地生态系统 / Terrestrial Ecosystem

威胁 Threat (s)

主要威胁 Major Threat(s)	海南天然林面积的持续下降和栖息地质量的降低 / The major threats are the decline of the area of natural forests and the deterioration of habitat quality

保护级别与保护行动 Protection Category and Conservation Action (s)

IUCN 红色名录 (2016) IUCN Red List (2016)	近危 / NT
保护行动 Conservation Action(s)	列入《国家保护的有益的或者有重要经济、科学研究价值的陆生野生动物名录》。其分布区内已经有多个国家级和省级自然保护区，应加大对天然栖息地质量的保护。建议对该物种的种群数量、栖息地、生活史等进行深入研究 / It has been listed as *National Protected List of Terrestrial Wild Animals with Good Benefits or Important Economic and Scientific Values*. Studies on population size, habitat, home range, breeding, behavior, and food pattern are suggested to get knowledge on its life history. National and provincial nature reserve should be set up and the protection of natural habitat should be improved

淡紫鳾 *Sitta solangiae*　　田穗兴 摄　By Suixing Tian

鹩哥
Gracula religiosa

易危　VU A2acd

| 数据缺乏 DD | 无危 LC | 近危 NT | 易危 VU | 濒危 EN | 极危 CR | 区域灭绝 RE | 野外灭绝 EW | 灭绝 EX |

分类地位 Taxonomic Status

动物界 Animalia	脊索动物门 Chordata	鸟纲 Aves	雀形目 Passeriformes	椋鸟科 Sturnidae
学名 Scientific Name		*Gracula religiosa*		
命名人 Species Authority		Linnaeus, 1758		
英文名 English Name(s)		Hill Myna		
同物异名 Synonym(s)		无 / None		
种下单元评估 Infra-specific Taxa Assessed		无 / None		

评估信息 Assessment Information

评估年份 Year Assessed	2016
评定人 Assessor(s)	邓文洪 / Wenhong Deng
其他贡献人 Other Contributor(s)	无 / None

理由 Justification: 在中国鹩哥的非法捕猎压力较大，野外较难见到。因此，列为易危等级 / *Gracula religiosa* is under high pressure of illegal trapping for pet bird trading, and seldom seen in the wild. Thus, it is listed as Vulnerable

评估历史 Assessment History: 《国家重点保护野生动物名录》：II (2021)；《中国濒危动物红皮书·鸟类》(1998)：易危；《中国物种红色名录：第二卷 脊椎动物 下册》：濒危 / *National Key Protected Wild Animal List*: II (2021); *China Red Data Book of Endangered Animals*: *Aves* (1998): V; (2009) *China Species Red List*: *Vol. II Vertebrates Part 2* (2009): EN

地理分布 Geographical Distribution

国内分布 Domestic Distribution
分布于云南西部和南部、广东、澳门、广西西南部和海南 / It occurs in west and south Yunnan, Guangdong, Macao, southwest Guangxi and Hainan
分布标注 Distribution Note
非特有种 / Non-endemic

国内分布图
Map of Domestic Distribution

种群 Population

种群数量 Population Size	估计种群数量不足 20,000 只 / The population size is estimated to be less than 20,000 individuals
种群趋势 Population Trend	下降 / Decreasing

生境与生态系统 Habitat (s) and Ecosystem (s)

生　　境 Habitat(s)	栖息于低山丘陵和山脚平原地区的次生林、常绿阔叶林、落叶阔叶林、竹林和混交林中，尤以林缘疏林地区较常见，也见于耕地、旷野和村寨附近的小块树林中 / *G. religiosa* uses secondary forests, evergreen broad-leaved forests, mixed deciduous broad-leaved forest with bamboos, especially the edge of forests. It also occurs in the farmland, the open land, the small patches of woods adjacent to the village
生态系统 Ecosystem(s)	陆地生态系统 / Terrestrial Ecosystem

威胁 Threat (s)

主要威胁 Major Threat(s)	由于过度捕捉和环境条件恶化，致使种群数量日趋减少。在广西、云南极为少见，海南的种群数量也比较稀少 / The population size is undergoing a decline due to the over trapping and the deterioration of environment. It's rarely seen in Guangxi and Yunnan and the population in Hainan is also small

保护级别与保护行动 Protection Category and Conservation Action (s)

IUCN 红色名录 (2016) IUCN Red List (2016)	无危 / LC
保护行动 Conservation Action(s)	列入国家 II 级重点保护野生动物、《国家保护的有益的或者有重要经济、科学研究价值的陆生野生动物名录》和 CITES 公约附录 II / It has been listed as the second class in National Key Protected Wild Animal List, *National Protected List of Terrestrial Wild Animals with Good Benefits or Important Economic and Scientific Values* and Appendix II of the CITES

鹩哥 *Gracula religiosa*　　　　　　　　　　　　　　　　　　　林杰 摄　By Jie Lin

褐头鸫 *Turdus feae*

中国生物多样性红色名录

易危 VU C2a(ii)

| 数据缺乏 DD | 无危 LC | 近危 NT | **易危 VU** | 濒危 EN | 极危 CR | 区域灭绝 RE | 野外灭绝 EW | 灭绝 EX |

分类地位 Taxonomic Status

动物界 Animalia	脊索动物门 Chordata	鸟纲 Aves	雀形目 Passeriformes	鸫科 Turdidae
学名 Scientific Name		*Turdus feae*		
命名人 Species Authority		Salvadori, 1887		
英文名 English Name(s)		Grey-sided Thrush		
同物异名 Synonym(s)		*Merula feae*		
种下单元评估 Infra-specific Taxa Assessed		无 / None		

评估信息 Assessment Information

评估年份 Year Assessed	2016
评定人 Assessor(s)	张正旺 / Zhengwang Zhang
其他贡献人 Other Contributor(s)	董路 / Lu Dong

理由 Justification: 褐头鸫在中国的繁殖分布区局限于华北较高海拔的山地，分布面积较小；种群密度非常低，种群数量估计少于 10,000 只；推测其繁殖种群数量处于持续下降中。因此，列为易危等级 / *Turdus feae* only breeds in the high mountains in North China and the distribution area is small. The population density is quite low and the population size is estimated to be less than 10,000 individuals. The breeding population is presumed to be undergoing a decline. Thus, it is listed as Vulnerable

评估历史 Assessment History:《国家重点保护野生动物名录》：II (2021)；《中国物种红色名录：第二卷 脊椎动物 下册》(2009)：无危 / *National Key Protected Wild Animal List*: II (2021); *China Species Red List*: *Vol. II Vertebrates Part 2* (2009): LC

地理分布 Geographical Distribution

国内分布 Domestic Distribution
繁殖于北京、河北、山西、山东，迁徙期见于四川和贵州 / It breeds in Beijing, Hebei, Shanxi, Shandong, and occurs in Sichuan and Guizhou when migrating
分布标注 Distribution Note
特有种 / Endemic

国内分布图
Map of Domestic Distribution

种群 Population

种群数量 Population Size	种群数量 1,500～7,000 只成鸟 / The population size is estimated to be 1,500～7,000 mature individuals
种群趋势 Population Trend	下降 / Decreasing

生境与生态系统 Habitat(s) and Ecosystem(s)

生　　境 Habitat(s)	属于典型的森林鸟类，在海拔 1,000～1,900 m 的温带阔叶林和油松林中繁殖。其巢址平均海拔 1,805 m (1,730～1,930 m)，巢树均为华北落叶松，巢树平均高 11 m (8～14 m) / T. feae is typical forest species, and breeds in the temperate broad-leaved forests and pine forests at 1,000～1,900 m. The average height of the nests is 1,805 m (1,730～1,930 m), and all the trees for nests are *Larix principis-rupprechtii*, with an average height of 11 m (8～14 m)
生态系统 Ecosystem(s)	陆地生态系统 / Terrestrial Ecosystem

威胁 Threat(s)

主要威胁 Major Threat(s)	主要是繁殖区栖息地的丧失和片断化。在河北和山西，天然阔叶林已经减少到原来面积的 20% 以下。此外，非法猎捕是另外一个威胁因素 / The major threats are breeding habitat loss and fragmentation. The natural broad-leaved forest has declined by more than 80%. Besides, illegal trapping is another threat

保护级别与保护行动 Protection Category and Conservation Action(s)

IUCN 红色名录 (2016) IUCN Red List (2016)	易危 / VU C2a (ii)
保护行动 Conservation Action(s)	列入国家Ⅱ级重点保护野生动物、《国家保护的有益的或者有重要经济、科学研究价值的陆生野生动物名录》和迁徙物种保护公约 (CMS) 附录Ⅱ。在其繁殖区已经建立了山西庞泉沟、北京百花山、河北雾灵山、小五台山、老岭等自然保护区 / It has been listed as the second class in National Key Protected Wild Animal List, *National Protected List of Terrestrial Wild Animals with Good Benefits or Important Economic and Scientific Values* and Appendix II of the Convention on Migratory Species. Shanxi Pangquangou National Nature Reserve, Beijing Baihua Shan National Nature Reserve, Hebei Wuling Shan National Nature Reserve, Xiaowutai Shan National Nature Reserve, Laoling Nature Reserve have been set up in its breeding area

褐头鸫 *Turdus feae*　　　　　　　　　　　　　　　　　　　许崇山 摄　By Chongshan Xu

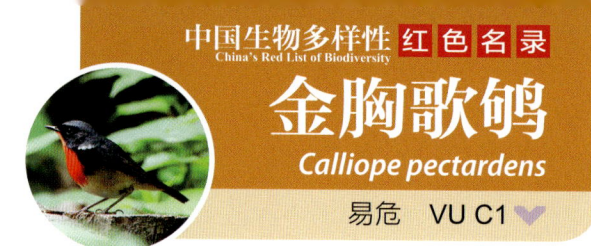

金胸歌鸲
Calliope pectardens
易危 VU C1

| 数据缺乏 DD | 无危 LC | 近危 NT | 易危 VU | 濒危 EN | 极危 CR | 区域灭绝 RE | 野外灭绝 EW | 灭绝 EX |

分类地位 Taxonomic Status

动物界 Animalia	脊索动物门 Chordata	鸟纲 Aves	雀形目 Passeriformes	鹟科 Muscicapidae
学 名 Scientific Name		*Calliope pectardens*		
命 名 人 Species Authority		David, 1877		
英 文 名 English Name(s)		Firethroat		
同物异名 Synonym(s)		*Luscinia pectardens*, *Calliope pectardens*		
种下单元评估 Infra-specific Taxa Assessed		无 / None		

评估信息 Assessment Information

评 估 年 份 Year Assessed	2016
评 定 人 Assessor(s)	丁平 / Ping Ding
其他贡献人 Other Contributor(s)	无 / None

理由 Justification: 迄今只有少量的金胸歌鸲繁殖记录，没有越冬地记录，对该物种了解甚少。可能由于其较小的种群数量以及活动区域的限制，面临生存威胁，加之其栖息地人迹罕至，缺乏详细的调查记录。因此，列为易危等级 / The data of *Calliope pectardens* is very limited yet, with only a few breeding records and no wintering record. It might face threats to subsistence owing to the small population and distribution area. Thus, it is listed as Vulnerable

评估历史 Assessment History: 《国家重点保护野生动物名录》：II (2021)；《中国物种红色名录：第二卷 脊椎动物 下册》(2009)：近危 / *National Key Protected Wild Animal List*: II (2021); *China Species Red List*: *Vol. II Vertebrates Part 2* (2009): NT

地理分布 Geographical Distribution

国内分布 Domestic Distribution
繁殖于陕西南部、西藏南部、云南北部、四川、重庆 / It breeds in south Shaanxi, south Tibet (Xizang), north Yunnan, Sichuan and Chongqing
分布标注 Distribution Note
特有种 / Endemic

国内分布图
Map of Domestic Distribution

种群 Population

种群数量 Population Size	关于该物种的描述极少,仅最近在中国卧龙国家级自然保护区有该物种的记录,数量极为稀少,估计数量可能有 10,000 ~ 19,999 只,成年个体大致为 6,000 ~ 15,000 只 (BirdLife International,2014) / The knowledge of *Calliope pectardens* is very little and the latest record was in Sichuan Wolong national Nature Reserve. The population size is quite small, and estimated to be 10,000 ~ 19,999 individuals, with 6,000 ~ 15,000 mature individuals (BirdLife International, 2014)
种群趋势 Population Trend	下降 / Decreasing

生境与生态系统 Habitat (s) and Ecosystem (s)

生境 Habitat(s)	生活于山林中、沟谷和溪沟边的稠密灌丛间,海拔 2,800 ~ 3,700 m 的森林里。于森林地面取食昆虫 / *C. pectardens* uses the forests, dense shrubs in the valleys and along the stream, at 2,800 ~ 3,700 m
生态系统 Ecosystem(s)	陆地生态系统 / Terrestrial Ecosystem

威胁 Threat (s)

主要威胁 Major Threat(s)	由于森林的砍伐以及农田的扩增,该物种的繁殖地和越冬地很有可能面临着片断化风险 / The breeding and wintering sites might be fragmented owing to the loss of forest and the increase of cropland

保护级别与保护行动 Protection Category and Conservation Action (s)

IUCN 红色名录 (2016) IUCN Red List (2016)	近危 / NT
保护行动 Conservation Action(s)	列入国家 II 级重点保护野生动物和《国家保护的有益的或者有重要经济、科学研究价值的陆生野生动物名录》/ It has been listed as the second class in National Key Protected Wild Animal List, *National Protected List of Terrestrial Wild Animals with Good Benefits or Important Economic and Scientific Values*

金胸歌鸲 *Calliope pectardens*　　　　　　　　　　　　　曲大勇 摄　By Dayong Qu

白喉林鹟
Cyornis brunneatus
易危 VU C1

| 数据缺乏 DD | 无危 LC | 近危 NT | 易危 VU | 濒危 EN | 极危 CR | 区域灭绝 RE | 野外灭绝 EW | 灭绝 EX |

分类地位 Taxonomic Status

动物界 Animalia	脊索动物门 Chordata	鸟纲 Aves	雀形目 Passeriformes	鹟科 Muscicapidae
学名 Scientific Name		*Cyornis brunneatus*		
命名人 Species Authority		Slater, 1897		
英文名 English Name(s)		Brown-chested Jungle Flycatcher		
同物异名 Synonym(s)		*Rhinomyias brunneatus*, *Siphia brunneata*		
种下单元评估 Infra-specific Taxa Assessed		无 / None		

评估信息 Assessment Information

评估年份 Year Assessed	2016
评定人 Assessor(s)	周放 / Fang Zhou
其他贡献人 Other Contributor(s)	无 / None

理由 Justification: 白喉林鹟栖息于亚热带山地森林，植被丧失和破碎化导致其种群数量下降。低海拔地区森林的繁殖地和越冬地不断遭到破坏，特别是华南地区近年来推广速生桉种植，使其栖息地快速丧失，部分以往活动较多的地区已少有发现。因此，列为易危等级 / The population of *Cyornis brunneatus* is undergoing a decline due to the loss and fragmentation of habituated subtropical mountain forest. The destroy of forests in breeding and wintering sites, especial the plantation of eucalyptus, has led to the loss of habitat. *Cyornis brunneatus* is rarely observed. Thus, it is listed as Vulnerable

评估历史 Assessment History: 《国家重点保护野生动物名录》：II (2021)；《中国物种红色名录：第二卷 脊椎动物 下册》(2009)：易危 / *National Key Protected Wild Animal List*: II (2021); *China Species Red List: Vol. II Vertebrates Part 2* (2009): VU

地理分布 Geographical Distribution

国内分布 Domestic Distribution

分布于中国东南部地区，为河南、云南、贵州、湖北、湖南、安徽、江西、江苏、上海、浙江、福建、广东、香港、广西、台湾的夏候鸟，但各地的记录多见于有限的分布点；在广西南部越冬 (赵东东等，2013) / It breeds in Southeast China, *e.g.* limited sites in Henan, Yunnan, Guizhou, Hubei, Hunan, Anhui, Jiangxi, Jiangsu, Shanghai, Zhejiang, Fujian, Guangdong, Hong Kong, Guangxi, Taiwan, and there is also wintering population in south Guangxi (Zhao *et al.*, 2013)

分布标注 Distribution Note

非特有种 / Non-endemic

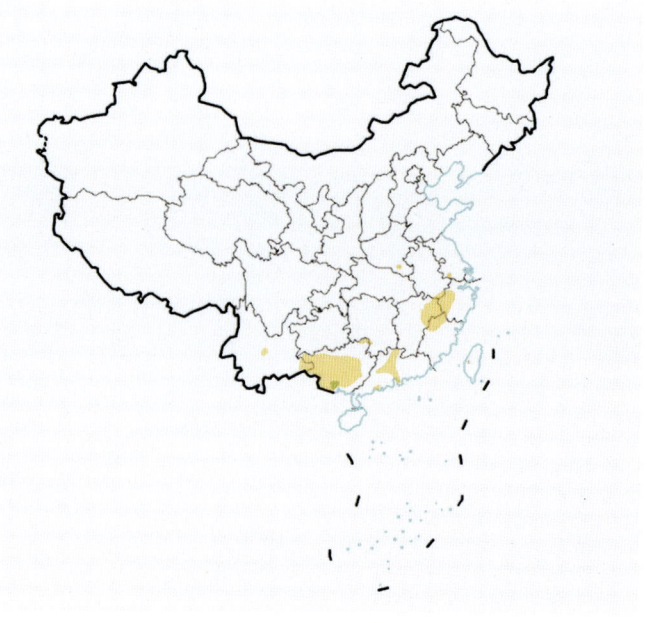

国内分布图 Map of Domestic Distribution

种群 Population

种群数量 Population Size	种群数量不足 10,000 对 / The population size is estimated to be less than 10,000 pairs
种群趋势 Population Trend	下降 / Decreasing

生境与生态系统 Habitat (s) and Ecosystem (s)

生　　境 Habitat(s)	常栖于低海拔地区的茂密灌木丛、亚热带常绿阔叶林及季雨林中，高可至海拔 1,600 m 的林缘下层。在毗邻天然森林的茂密竹丛、次生林及人工林内亦有活动 / C. brunneatus uses dense shrubs, subtropical evergreen broad-leaved forest and rainforest at low latitude, with elevation up to 1,600 m. It also occurs in the adjacent dense bamboos, secondary forests and planted forest
生态系统 Ecosystem(s)	陆地生态系统 / Terrestrial Ecosystem

威胁 Threat (s)

主要威胁 Major Threat(s)	中国东南部地区的天然林和次生林逐渐改变成人工用材林，使白喉林鹟的栖息地受损严重，栖息地的持续丧失和破碎化范围扩大，是白喉林鹟受到的主要威胁 / The habitat of Cyornis brunneatus has been destroyed by the change of natural and secondary forest into planted forest in southeast China. Habitat loss and fragmentation are the major threats of it

保护级别与保护行动 Protection Category and Conservation Action (s)

IUCN 红色名录 (2016) IUCN Red List (2016)	易危 / VU C2a (ii)
保护行动 Conservation Action(s)	列入国家 II 级重点保护野生动物、《国家保护的有益的或者有重要经济、科学研究价值的陆生野生动物名录》。建议加强对正在变更的森林类型的监测和管理，进一步开展白喉林鹟分布和种群数量调查 / It has been listed as the second class in National Key Protected Wild Animal List, National Protected List of Terrestrial Wild Animals with Good Benefits or Important Economic and Scientific Values. Improvement of monitoring and management of the change of forest, and survey on distribution and population size of C. brunneatus are suggested

白喉林鹟 Cyornis brunneatus　　　　　　　　　　陈锋 摄　By Feng Chen

贺兰山岩鹨
Prunella koslowi
易危 VU C2a(i)

| 数据缺乏 DD | 无危 LC | 近危 NT | 易危 VU | 濒危 EN | 极危 CR | 区域灭绝 RE | 野外灭绝 EW | 灭绝 EX |

分类地位 Taxonomic Status

动物界 Animalia	脊索动物门 Chordata	鸟纲 Aves	雀形目 Passeriformes	岩鹨科 Prunellidae

学 名 Scientific Name	*Prunella koslowi*
命 名 人 Species Authority	Przewalski, 1887
英 文 名 English Name(s)	Mongolian Accentor
同物异名 Synonym(s)	*Accentor koslowi*
种下单元评估 Infra-specific Taxa Assessed	无 / None

评估信息 Assessment Information

评 估 年 份 Year Assessed	2016
评 定 人 Assessor(s)	张雁云 / Yanyun Zhang
其他贡献人 Other Contributor(s)	张立勋 / Lixun Zhang

理由 Justification: 贺兰山岩鹨在越冬分布区有一定的数量，近期缺乏全面调查。但多次调查中，均未发现数量超过 1,000 只个体的种群。因此，列为易危等级 / There are a number of *Prunella koslowi* in desert in winter. No comprehensive investigation has been conducted recently, and no population above 1,000 individuals was recorded in each survey. Thus, it is listed as Vulnerable

评估历史 Assessment History: 《国家重点保护野生动物名录》：II (2021)；《中国物种红色名录：第二卷 脊椎动物 下册》(2009)：近危 (近乎易危) / *National Key Protected Wild Animal List*: II (2021); *China Species Red List: Vol. II Vertebrates Part 2* (2009): NT (nearly VU)

地理分布 Geographical Distribution

国内分布 Domestic Distribution
分布于内蒙古西部、宁夏和甘肃中部 / It occurs in west Inner Mongolia (Nei Mongol), Ningxia and central Gansu
分布标注 Distribution Note
特有种 / Endemic

国内分布图
Map of Domestic Distribution

种群 Population

种群数量 Population Size	缺乏系统的数量调查，在其越冬分布区有一定的数量，1987 年在宁夏中卫县迎水桥，曾记录到 34 只个体 (常家传等，1991)，1987 年至 1989 年 10 月在阿拉善高原 (腾格里沙漠、乌兰布和沙漠) 和鄂尔多斯高原 (毛乌素沙地) 的 5 个地点采到 18 只标本 (常家传等，1991)。2009 年 2 月至 2010 年 12 月在贺兰山国家级自然保护区调查结果显示为稀有种 (李元刚等，2012) / There has been no systematic survey on population size. There are a number of *P. koslowi* in desert in winter: 34 individuals were recorded in Yingshui Bridge in Zhongwei County, Ningxia in 1987 (Chang *et al.*, 1991); 18 specimens were gotten in 5 sites in Alashan Plateau (Tengger Desert and Ulan Buh Desert) and Ordos Plateau (Maowusu Sandy Land) in 1987 ~ 1989 (Chang *et al.*, 1991). It is a rare species in the survey of Helan Shan National Nature Reserve from February 2009 to December 2012 (Li *et al.*, 2012)
种群趋势 Population Trend	下降 / Decreasing

生境与生态系统 Habitat (s) and Ecosystem (s)

生　　境 Habitat(s)	栖息在阿拉善高原和鄂尔多斯高原的荒漠、半荒漠的沙枣、小叶杨、红柳、沙蒿、油蒿、沙米等灌丛中 (常家传等，1991；陈服官等，1998)。主要取食沙枣、沙米、杂草种子、柽柳芽等植物性食物，以及昆虫等动物性食物 (常家传等，1991；王香亭，1990) / *P. koslowi* uses shrubs of *Elaeagnus angustifolia*, *Populus simonii*, *Tamarix ramosissima*, *Artemisia desertorum*, *Artemisia ordosica*, *Agriophyllum squarrosum* in desert and semi-desert on the Alashan Plateau and Ordos Plateau (Chang *et al.*, 1991; Chen *et al.*, 1998). It feeds on seeds of *Elaeagnus angustifolia*, *Agriophyllum squarrosum* and other grasses, and shoots of *Tamarix ramosissima etc.*, and insects (Chang *et al.*, 1991; Wang, 1990)
生态系统 Ecosystem(s)	陆地生态系统、荒漠生态系统 / Terrestrial Ecosystem, Desert Ecosystem

威胁 Threat (s)

主要威胁 Major Threat(s)	荒漠化进一步加剧导致的栖息地退化是其受威胁的主要因子 / The major threat is habitat deterioration owing to desertification

保护级别与保护行动 Protection Category and Conservation Action (s)

IUCN 红色名录 (2016) IUCN Red List (2016)	无危 / LC
保护行动 Conservation Action(s)	列入国家 II 级重点保护野生动物、《国家保护的有益的或者有重要经济、科学研究价值的陆生野生动物名录》。建议加强分布区和种群现状等生态生物学研究 / It has been listed as the second class in National Key Protected Wild Animal List, *National Protected List of Terrestrial Wild Animals with Good Benefits or Important Economic and Scientific Values*. Biological and ecological studies on population status and distribution are needed

贺兰山岩鹨 *Prunella koslowi*　　　　　　　　　　　　　　　关克 摄　By Ke Guan

禾雀
Lonchura oryzivora

易危 VU A2bde+3bde+4bde

| 数据缺乏 DD | 无危 LC | 近危 NT | 易危 VU | 濒危 EN | 极危 CR | 区域灭绝 RE | 野外灭绝 EW | 灭绝 EX |

分类地位 Taxonomic Status

| 动物界 Animalia | 脊索动物门 Chordata | 鸟纲 Aves | 雀形目 Passeriformes | 梅花雀科 Estrildidae |

学 名 Scientific Name	*Lonchura oryzivora*
命 名 人 Species Authority	Linnaeus, 1758
英 文 名 English Name(s)	Java Sparrow
同物异名 Synonym(s)	*Loxia oryzivora*
种下单元评估 Infra-specific Taxa Assessed	无 / None

评估信息 Assessment Information

评估年份 Year Assessed	2016
评 定 人 Assessor(s)	丁平 / Ping Ding
其他贡献人 Other Contributor(s)	无 / None

理由 Justification: 禾雀是一种传统的观赏鸟类。极其猖獗的非法捕捉导致禾雀的数量锐减。因此，列为易危等级 / *Padda oryzivora* is a famous pet birds, and the serious illegal trapping has caused the rapid decline. Thus, it is listed as Vulnerable

评估历史 Assessment History: 《中国物种红色名录：第二卷 脊椎动物 下册》(2009)：无危 / China Species Red List: *Vol. II Vertebrates Part 2* (2009): LC

地理分布 Geographical Distribution

国内分布 Domestic Distribution

禾雀的数量曾经较丰富，如今急剧减少。明朝时禾雀引入到中国，在江苏、浙江、福建、广东、香港、广西及台湾等地有分布 / There were a large number of *Padda oryzivora*, while the population size is undergoing a rapid decline. It was introduced into China in Ming Dynasty, and occurs in Jiangsu, Zhejiang, Fujian, Guangdong, Hong Kong, Guangxi and Taiwan

分布标注 Distribution Note

非特有种 / Non-endemic

国内分布图
Map of Domestic Distribution

种群 Population

种群数量 Population Size	中国大陆估计有 100～10,000 繁殖对 (Brazil, 2009) / The population size is about 100～10,000 pairs in Chinese mainland (Brazil, 2009)
种群趋势 Population Trend	下降 / Decreasing

生境与生态系统 Habitat (s) and Ecosystem (s)

生　　境 Habitat(s)	禾雀是群居鸟类，尤其在非繁殖期结成大群。栖息于乡村、耕地（尤其是稻田）、灌木丛、矮树，以及城镇林园 / *L. oryzivora* forms large groups, especially during non-breeding period. It uses villages, croplands, shrubs, short trees and woods in town
生态系统 Ecosystem(s)	热带、亚热带森林生态系统 / Tropical and Subtropical Forest Ecosystem

威胁 Threat (s)

主要威胁 Major Threat(s)	非法贸易是禾雀数量锐减的主要原因。禾雀的集群活动使它们更容易受到大规模的集中捕杀。栖息地的开发也使其数量不断减少 / The population size is undergoing a rapid decline due to the illegal trading. The ecology of group activity makes it vulnerable to the extensive trapping. The development activities in its habitat is another threat

保护级别与保护行动 Protection Category and Conservation Action (s)

IUCN 红色名录 (2016) IUCN Red List (2016)	易危 / VU A2bde+3bde+4bde
保护行动 Conservation Action(s)	列入 CITES 公约附录 II。国内目前尚未实施具体的保护行动 / It has been listed as the Appendix II of the CITES, while there is no conservation action in China

禾雀 *Lonchura oryzivora*　　　　　　　　　　　　　　　　　　　　侯志刚 摄　By Zhigang Hou

藏雀
Carpodacus roborowskii

易危 VU B2b(i-iv); C2(i, ii); D2

| 数据缺乏 DD | 无危 LC | 近危 NT | 易危 VU | 濒危 EN | 极危 CR | 区域灭绝 RE | 野外灭绝 EW | 灭绝 EX |

分类地位 Taxonomic Status

动物界 Animalia	脊索动物门 Chordata	鸟纲 Aves	雀形目 Passeriformes	燕雀科 Fringillidae
学名 Scientific Name		*Carpodacus roborowskii*		
命名人 Species Authority		Przewalski, 1887		
英文名 English Name(s)		Tibetan Rosefinch		
同物异名 Synonym(s)		*Kozlowia roborowskii*		
种下单元评估 Infra-specific Taxa Assessed		无 / None		

评估信息 Assessment Information

评估年份 Year Assessed	2016
评定人 Assessor(s)	刘阳 / Yang Liu
其他贡献人 Other Contributor(s)	黄秦 / Qin Huang

理由 Justification: 虽然藏雀分布局限在青藏高原东北部，但是分布区基本连续，分布区的面积估计达到 194,000 km² (BirdLife International，2014)。分布区没有缩小、分割和破碎化的迹象。其种群数量虽然缺少估计，但是没有急剧减少或者波动的迹象，总体被认为很稳定 (BirdLife International，2014)。因此，列为易危等级 / The distribution area of *Carpodacus roborowskii* is restricted in northeast Qinghai-Tibet (Xizang) Plateau, and estimated to be 194,000 km² (BirdLife International, 2014), which is not in decline or fragmentation. The population size has not been estimated, while it is suspected to be stable in the absence of evidence for any declines or fluctuation(BirdLife International，2014). Thus, it is listed as Vulnerable

评估历史 Assessment History: 《国家重点保护野生动物名录》：Ⅱ (2021)；《中国濒危动物红皮书·鸟类》(1998)：稀有；《中国物种红色名录：第二卷 脊椎动物 下册》(2009)：近危 / *National Key Protected Wild Animal List*: Ⅱ (2021); *China Red Data Book of Endangered Animals*: Aves (1998): R; *China Species Red List*: Vol. Ⅱ Vertebrates Part 2 (2009): NT

地理分布 Geographical Distribution

国内分布 Domestic Distribution
分布于西藏东北部、青海和新疆东南部 / It occurs in northeast Tibet (Xizang), Qinghai and southeast Xinjiang
分布标注 Distribution Note
特有种 / Endemic

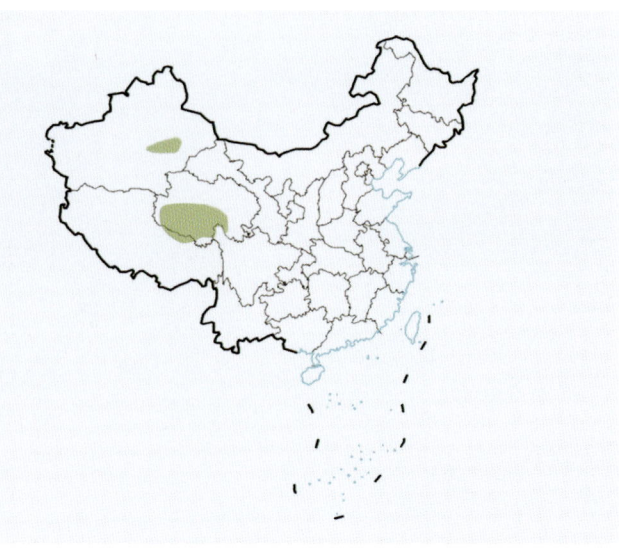

国内分布图
Map of Domestic Distribution

种群 Population

种群数量 Population Size	未知 / Unknown
种群趋势 Population Trend	稳定 / Stable

生境与生态系统 Habitat(s) and Ecosystem(s)

生　　境 Habitat(s)	栖息于海拔 4,500～5,400 m 的多岩石的高山草甸 / *C. roborowskii* uses rocky alpine meadow at 4,500～5,400 m
生态系统 Ecosystem(s)	高山草甸(陆地)生态系统 / Alpine Meadow (Terrestrial) Ecosystem

威胁 Threat(s)

主要威胁 Major Threat(s)	分布区处于人迹罕至的高海拔的地带，没有明显的致危因素 / It distributed in sparsely populated high-altitude areas, and there is no evidence for any substantial threats

保护级别与保护行动 Protection Category and Conservation Action(s)

IUCN 红色名录 (2016) IUCN Red List (2016)	无危 / LC
保护行动 Conservation Action(s)	列入国家 II 级重点保护野生动物、《国家保护的有益的或者有重要经济、科学研究价值的陆生野生动物名录》/ It has been listed as the second class in National Key Protected Wild Animal List, *National Protected List of Terrestrial Wild Animals with Good Benefits or Important Economic and Scientific Values*

藏雀 *Carpodacus roborowskii*　　　　董江天 摄　By Jiangtian Dong

藏鹀
Emberiza koslowi

易危 VU B1b(i-iv); C1

| 数据缺乏 DD | 无危 LC | 近危 NT | 易危 VU | 濒危 EN | 极危 CR | 区域灭绝 RE | 野外灭绝 EW | 灭绝 EX |

分类地位 Taxonomic Status

动物界 Animalia	脊索动物门 Chordata	鸟纲 Aves	雀形目 Passeriformes	鹀科 Emberizidae
学名 Scientific Name		*Emberiza koslowi*		
命名人 Species Authority		Bianchi, 1904		
英文名 English Name(s)		Tibetan Bunting		
同物异名 Synonym(s)		无 / None		
种下单元评估 Infra-specific Taxa Assessed		无 / None		

评估信息 Assessment Information

评估年份 Year Assessed	2016
评定人 Assessor(s)	卢欣 / Xin Lu
其他贡献人 Other Contributor(s)	无 / None

理由 Justification: 对藏鹀了解较少，其具有一个相对稳定的小种群。当前并未显示种群有明显下降或有实质性的威胁，但对其分布区大小和潜在的威胁因子需要进一步研究和评估。估计有 3,000～7,500 对成熟个体，数据可能有所高估。为特有种，分布区狭窄。因此，列为易危等级 / The knowledge of *Emberiza koslowi* is limited, while the population is suspected to be stable in the absence of evidence for any declines or threats. Studies and assessment on the distribution area and the potential threats are needed. It is rare and the population size is estimated to be 3,000～7,500 breeding pairs, which might be overestimated. Thus, it is listed as Vulnerable

评估历史 Assessment History: 《国家重点保护野生动物名录》：II (2021)；《中国濒危动物红皮书·鸟类》(1998)：稀有；《中国物种红色名录：第二卷 脊椎动物 下册》(2009)：近危 / *National Key Protected Wild Animal List*: II (2021); *China Red Data Book of Endangered Animals*: *Aves* (1998): R; *China Species Red List*: *Vol. II Vertebrates Part 2* (2009): NT

地理分布 Geographical Distribution

国内分布 Domestic Distribution

为青海南部、甘肃西南部、四川东北部和青海南部的留鸟 / It is resident in south Qinghai, southwest Gansu, northeast Sichuan and south Qinghai

分布标注 Distribution Note

特有种 / Endemic

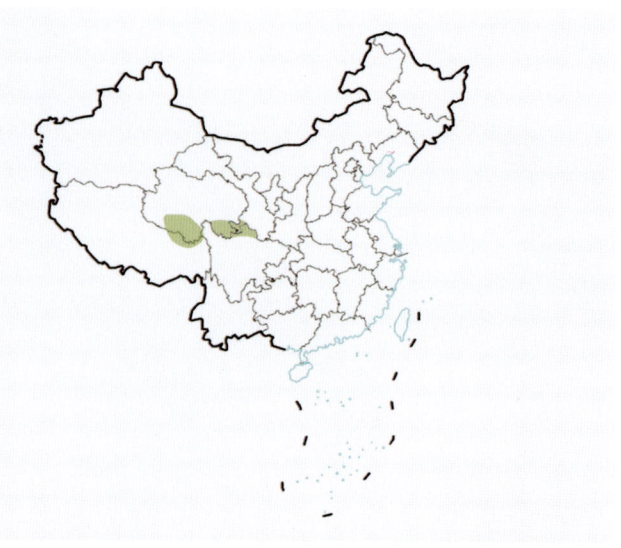

国内分布图
Map of Domestic Distribution

种群 Population

种群数量 Population Size	估计有 3,000～7,500 对成熟个体 / The population size is estimated to be 3,000～7,500 breeding pairs
种群趋势 Population Trend	稳定 / Stable

生境与生态系统 Habitat(s) and Ecosystem(s)

生　　境 Habitat(s)	海拔 3,600～4,600 m，高山草甸、草原和灌丛，青藏高原山柳灌丛地带等。林线以上的开阔而荒瘠的高山灌丛、矮小桧树丛、杜鹃林及裸露地面 (Thewlis and Martins，2000) / *E. koslowi* uses alpine meadow, grassland and shrubs at 3,600～4,600 m. It inhabits barren areas, juniper and rhododendron shrub on valley slopes above the treeline (Thewlis and Martins, 2000)
生态系统 Ecosystem(s)	高原灌丛生态系统 / Plateau Shrub Ecosystem

威胁 Threat(s)

主要威胁 Major Threat(s)	栖息地减少和质量下降是影响因素之一，但目前并未构成严重威胁 / Habitat decline might influence the population, while it has not been a threat yet

保护级别与保护行动 Protection Category and Conservation Action(s)

IUCN 红色名录 (2016) IUCN Red List (2016)	近危 / NT
保护行动 Conservation Action(s)	列入国家 II 级重点保护野生动物、《国家保护的有益的或者有重要经济、科学研究价值的陆生野生动物名录》。以青海省果洛州白玉乡为中心的小范围分布受到当地藏族喇嘛以及牧民保护。还需要对当前分布和栖息地趋势做更进一步的调查和监测，精确认识其栖息地需求，保护并扩大适宜栖息地以应对未来可能出现的威胁 / It has been listed as the second class in National Key Protected Wild Animal List, *National Protected List of Terrestrial Wild Animals with Good Benefits or Important Economic and Scientific Values*. A restricted area of Baiyu in Guoluo is protected by local Tibetan Lama and herdsman. Surveys and monitoring on distribution and habitat trend are required to determine its habitat requirements, and to protect and enlarge suitable habitat in order to cope with the threats in future

藏鹀 *Emberiza koslowi*　　　　　　　　　　　　　　　　　　　　　张永 摄　By Yong Zhang

硫黄鹀
Emberiza sulphurata

易危 VU C2a(ii)

| 数据缺乏 DD | 无危 LC | 近危 NT | 易危 VU | 濒危 EN | 极危 CR | 区域灭绝 RE | 野外灭绝 EW | 灭绝 EX |

分类地位 Taxonomic Status

动物界 Animalia	脊索动物门 Chordata	鸟纲 Aves	雀形目 Passeriformes	鹀科 Emberizidae

学名 Scientific Name	*Emberiza sulphurata*
命名人 Species Authority	Temminck & Schlegel, 1848
英文名 English Name(s)	Yellow Bunting
同物异名 Synonym(s)	*Schoeniclus sulphuratus*
种下单元评估 Infra-specific Taxa Assessed	无 / None

评估信息 Assessment Information

评估年份 Year Assessed	2016
评定人 Assessor(s)	刘阳 / Yang Liu
其他贡献人 Other Contributor(s)	黄秦 / Qin Huang

理由 Justification: 硫黄鹀的繁殖地区狭窄，全球评估为 3,750～14,999 只。过境时罕见于东部沿海省区和台湾，在南方省区为罕见冬候鸟。繁殖和越冬地都面临栖息地丧失和破碎化的威胁。农药的使用和非法捕猎亦导致本种数量下降。因此，列为易危等级 / The breeding area of *Emberiza sulphurata* is restricted and the global population is estimated to be 3,750～14,999 individuals. It is rarely seen in east coast and Taiwan when migration. And it rarely occurs in south China in winter. Both breeding and wintering sites are under the threats of habitat loss and fragmentation. The population is undergoing a decline owing to the abuse of pesticide and illegal trapping. Thus, it is listed as Vulnerable

评估历史 Assessment History: 《中国物种红色名录：第二卷 脊椎动物 下册》(2009)：易危 / *China Species Red List: Vol. II Vertebrates Part 2* (2009): VU

地理分布 Geographical Distribution

国内分布 Domestic Distribution
为山东、江西、江苏东部、上海、浙江、福建、广东、香港、广西、台湾不常见旅鸟 / It is uncommon passage migrate in the coast of Shandong, Jiangxi, eastern Jiangsu, Shanghai, Zhejiang, Fujian, Guangdong, Hong Kong, Guangxi, and Taiwan
分布标注 Distribution Note
非特有种 / Non-endemic

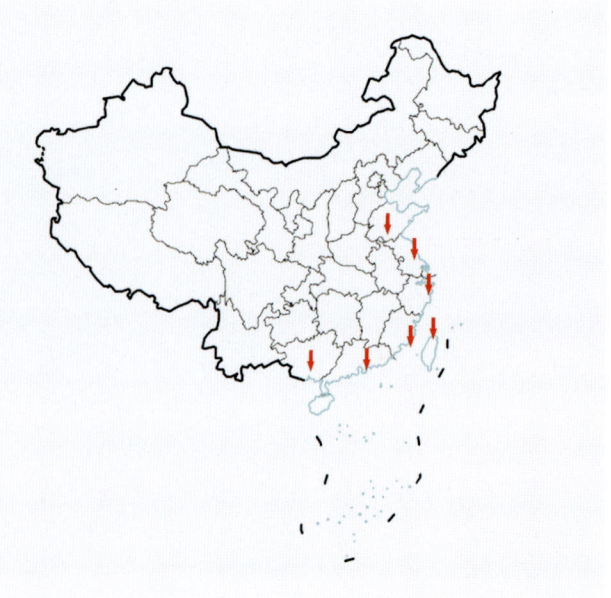

国内分布图
Map of Domestic Distribution

种群 Population

种群数量 Population Size	种群数量 500～1,000 只 / The population size is 500～1,000 individuals
种群趋势 Population Trend	下降 / Decreasing

生境与生态系统 Habitat (s) and Ecosystem (s)

生　　境 Habitat(s)	繁殖于山麓地带的落叶林或混交林及次生植被。迁徙期间见于开阔生境下多灌丛的植被 / *E. sulphurata* breeds in the deciduous forest, mixed forest and secondary vegetation in mountain areas. It is seen in the open area with shrubs during migrating
生态系统 Ecosystem(s)	温带森林生态系统 / Temperate Forest Ecosystem

威胁 Threat (s)

主要威胁 Major Threat(s)	繁殖栖息地的丧失和破碎化，农药的使用导致繁殖成功率降低，迁徙路线上的栖息地丧失和非法捕猎，致使该物种数量显著下降 / The population is undergoing a decline owing to the breeding habitat loss and fragmentation, abuse of pesticide in breeding sites and low reproductive success, and habitat loss and illegal trapping in the migratory route

保护级别与保护行动 Protection Category and Conservation Action (s)

IUCN 红色名录 (2016) IUCN Red List (2016)	易危 / VU C2a (ii)
保护行动 Conservation Action(s)	列入《国家保护的有益的或者有重要经济、科学研究价值的陆生野生动物名录》/ It has been listed as *National Protected List of Terrestrial Wild Animals with Good Benefits or Important Economic and Scientific Values*

硫黄鹀 *Emberiza sulphurata*　　　　　　　　　　　　　　　　　　　陈光辉 摄　By Guanghui Chen

参考文献
References

白哈斯, 高玮, 周道玮. 2003. 火烧对栗斑腹鹀巢址选择的影响. 东北师大学报(自然科学版), 35(1): 60-65. Bai H S, Gao W, Zhou D W. 2003. Impacts of fire on the nest site selection of *Embreriza jankowskii* taczanovski. *Journal of Northeast Normal University* (Natural Science Edition), 35(1): 60-65. (In Chinese with English Abstract).

鲍伟东, 李晓京, 史阳. 2005. 北京地区隼形目鸟类物种多样性现状调查. 四川动物, 24(4): 557-558. Bao W D, Li X J, Shi Y. 2005. A recent survey on bird species diversity of Falconiformes in Beijing region. *Sichuan Journal of Zoology*, 24(4): 557-558. (In Chinese with English Abstract)

Barter M, 陈立伟, 曹垒, 等. 2004. 长江中下游水鸟调查报告(2004年1-2月). 北京: 中国林业出版社. Barter M, Chen L W, Cao L, *et al.* 2004. Waterbird Survey of the Middle and Lower Changjiang River Floodplain (January-February 2004). Beijing: China Forestry Publishing House. (In Chinese and English)

Barter M, 雷刚, 曹垒. 2005. 长江中下游水鸟调查报告 (2005年2月). 北京: 中国林业出版社. Barter M, Lei G, Cao L. 2005. Waterbird Survey of the Middle and Lower Changjiang River Floodplain (February 2005). Beijing: China Forestry Publishing House. (In Chinese and English)

才代, 马鸣, 原洪, 等. 1997. 天山巴音布鲁克湿地的黑鹳(*Ciconia nigra*). 地方病通报, 12(增刊): 21-22. Cai D, Ma M, Yuan H, *et al.* 1997. Black Stork in Bayinbulak. *Endemic Diseases Bulletin*, 12(Sup): 21-22. (In Chinese with English Abstract)

仓决卓玛, 杨乐, 李建川, 等. 2005. 西藏林周县澎波河谷冬春季鸟类调查. 西藏科技, 45(12): 24, 30. Tsamchue D M, Yang L, Li J C, *et al.* 2005. Pengbo river valley winter spring birds investigation in Lin Zhou county of Tibet. *Tibet's Science & Technology*, 45(12): 24, 30. (In Chinese)

曹玉萍, 于德海, 刘玉铉. 1989. 波斑鸨在河北省的首次发现与饲养. 野生动物, 11(4): 32. Cao Y P, Yu D H, Liu Y X. 1989. The first discovery and raise of the Houbara Bustard in Hebei province. *Chinese Journal Wildlife*, 11(4): 32. (In Chinese)

常家传, 杭福兰, 卢文喜, 等. 1991. 贺兰山岩鹨(漠岩鹨)的研究. 野生动物, 13(3): 22-24. Chang J C, Hang F L, Lu W X, *et al.* 1991. A study on Mongolian Accentor. *Chinese Journal Wildlife*, 13(3): 22-24. (In Chinese with English Abstract)

常家传, 杭馥兰. 1983. 黑龙江省鱼鸮. 野生动物, 5(2): 53. Chang J C, Hang F L. 1983. The identification of fish owl in Heilongjiang province. *Chinese Journal Wildlife*, 5(2): 53. (In Chinese)

陈服官, 罗时有, 郑光美, 等. 1998. 中国动物志 鸟纲 第九卷 雀形目: 太平鸟科-岩鹨科. 北京: 科学出版社. Chen F G, Luo S Y, Zheng G M, *et al.* 1998. Fauna Sinica Aves. Vol. 9. Passerformes: Bombycillidae- Prunellidae. Beijing: Science Press. (In Chinese)

陈克林. 2006. 黄渤海湿地与迁徙水鸟研究. 北京: 中国林业出版社. Chen K L. 2006. Study on Migratory Waterbirds and Wetlands in the Yellow Sea. Beijing: China Forestry Publishing House. (In Chinese)

陈水华, 范忠勇, 陆祎玮. 2014. 中华凤头燕鸥育雏数量创新纪录——人工引导鸟类选择繁殖地试验再获成功. 浙江林业, 24(10): 18-19. Chen S H, Fan Z Y, Lu Y W. 2014. A successful trial of artificially guided to select breeding sites of bird - A new record number of chicks of Chinese crested tern. *Zhejiang Forestry*, 24(10): 18-19. (In Chinese)

陈水华, 范忠勇. 2013. 中华凤头燕鸥. 动物学杂志, 48(5): 788-789. Chen S H, Fan Z Y. 2013. Chinese Crested Tern. *Chinese Journal of Zoology*, 48(5): 788-789. (In Chinese with English Abstract)

陈水华, 颜重威, 范忠勇, 等. 2005. 浙江韭山列岛的黑嘴端凤头燕鸥繁殖群调查初报. 动物学杂志, 40(1): 96-97. Chen S H, Yan C W, Fan Z Y, *et al.* 2005. The breeding colony of Chinese Crested Tern at Jiushan Islands in Zhejiang. *Chinese Journal of Zoology*, 40(1): 96-97. (In Chinese with English Abstract)

陈天波, 蒙渊君, 蒋爱伍. 2007. 弄岗自然保护区冠斑犀鸟的分布与种群数量. 广西农业生物科学, 26(1): 20-23. Chen T B, Meng Y J, Jiang A W. 2007. Distribution and population size of Oriental Pied Hornbill in Nonggang Nature Reserve. *Journal of Guangxi Agricultural and Biological Sciences*, 26(1): 20-23.

程瑾瑞, 高玮, 王海涛. 2002. 栗斑腹鹀种群数量变化的分析. 东北师大学报(自然科学版), 34(1): 49-53. Cheng J R, Gao W, Wang H T. 2002. Analyses of variation in quantity of Jankowski's Bunting. *Journal of Northeast Normal University* (Natural Science Edition), 34(1): 49-53. (In Chinese with English Abstract)

程松林, 林剑声. 2011. 江西武夷山国家级自然保护区鸟类多样性调查. 动物学杂志, 46(5): 66-78. Cheng S L, Lin J S. 2011. A Survey on Avian Diversity in Wuyishan National Nature Reserve, Jiangxi. *Chinese Journal of Zoology*, 46(5): 66-78.

邓立斌, 陈端吕, 邓丽群. 2013. 基于层次分析法的自然保护区模糊综合评价研究——以海南霸王岭国家级自然保护区为例. 中国农学通报, 29(29): 118-125. Deng L B, Chen R L, Deng L Q. 2013. Fuzzy comprehensive evaluation on natural reserve based on analytic hierarchy process: a case study on Bawanling National Natural Reserve. *Chinese Agricultural Science Bulletin*, 29(29): 118-125. (In Chinese with English Abstract)

邓章文, 韩联宪, 岩道, 等. 2012. 云南紫溪山巨鸸春季觅食树种选择. 生态学杂志, 47(6): 1-6. Deng Z W, Han L X, Yan D, *et al*. 2012. Tree species preferences of Giant Nuthatch during spring in Zixi Mountain, Yunnan province. *Chinese Journal of Zoology*, 47(6): 1-6. (In Chinese with English Abstract)

丁长青, 郑光美. 1996. 黄腹角雉再引入的初步研究. 动物学报, 42(增刊): 69-73. Ding C Q, Zheng G M. 1996. A primary study on the reintroduction of Cabot's Tragopan. *Acta Zoologica Sinica*, 42(sup): 69-73. (In Chinese with English Abstract)

丁长青, 郑光美. 1997. 黄腹角雉的巢址选择. 动物学报, 43(1): 27-33. Ding C Q, Zheng G M. 1997. The nest site selection of the Yellow-Bellied Tragopan (*Tragopan caboti*). *Acta Zoologica Sinica*, 43(1): 27-33. (In Chinese with English Abstract)

丁长青. 2004. 朱鹮研究. 上海: 上海科技教育出版社. Ding C Q. 2004. Research on the Crested Ibis. Shanghai: Shanghai Scientific and Technology Education Publishing House. (In Chinese)

丁鹏. 2013. 金雕(*Aquila chrvsaetos*)繁殖期行为与巢址选择. 乌鲁木齐: 中国科学院硕士研究生论文. Ding P. 2013. Breeding behavior of Golden Eagle and nest site selection. Ürümqi: Master Dissertation of Chinese Academy of Sciences. (In Chinese with English abstract)

丁平, 姜仕仁, 诸葛阳. 2000. 浙江西部白颈长尾雉栖息地片段化研究. 动物学研究, 21(1): 65-69. Ding P, Jiang S R, Zhu G Y. 2000. The study on fragmentation of habitat used by Elliot's Pheasant in western Zhejiang. *Zoological Research*, 21(1): 65-69. (In Chinese with English Abstract)

丁平, 李智, 姜仕仁, 等. 2002. 白颈长尾雉栖息地小区利用度影响因子研究. 浙江大学学报(理学版), 20(1): 103-108. Ding P, Li Z, Jiang S R, *et al*. 2002. Studies on the factors affecting patch use degree by Elliot's Pheasant. *Journal of Zhejiang University* (Science Edition), 20(1): 103-108. (In Chinese with English Abstract)

丁平, 杨月伟, 李智, 等. 2002. 白颈长尾雉的夜宿地选择研究. 浙江大学学报(理学版), 29(5): 564-568. Ding P, Yang Y W, Li Z, *et al*. 2002. A study on the roost site selection to Elliot's Pheasant. *Journal of Zhejiang University* (Science Edition), 29(5): 564-568. (In Chinese with English Abstract)

丁平, 杨月伟, 李智. 2001. 白颈长尾雉栖息地的植被特征. 浙江大学学报, 28(5): 557-562. Ding P, Yang Y W, Li Z. 2001. Vegetation characteristics of habitats used by Elliot's Pheasant. *Journal of Zhejiang University*, 28(5): 557-562. (In Chinese with English Abstract)

丁平, 杨月伟, 梁伟, 等. 1996. 贵州雷公山自然保护区白颈长尾雉栖息地研究. 动物学报, 42: 62-67. Ding P, Yang Y W, Liang W, *et al*. 1996. The habitat of Elliot's Pheasant in the Leigong Mountain Natural Reserve. *Acta Zoologica Sinica*, 42: 62-67. (In Chinese with English Abstract)

丁平, 诸葛阳. 1988. 白颈长尾雉的生态研究. 生态学报, 8(1): 44-49. Ding P, Zhu G Y. 1988. The ecology of *Syrmaticus ellioti* Swinhoe. *Acta Ecologica Sinica*, 8(1): 44-49. (In Chinese with English Abstract)

丁平, 诸葛阳. 1989. 白颈长尾雉. 动物学杂志, 24(2): 39-43. Ding P, Zhu G Y. 1989. Elliot's Pheasant. *Journal of Zoology*, 24(2): 39-43. (In Chinese with English Abstract)

丁平, 诸葛阳, 张词祖. 1999. 白颈长尾雉繁殖生态的研究. 动物学研究, 55(2): 139-145. Ding P, Zhu G Y, Zhang C Z. 1999. The studies on breeding ecology of *Syrmaticus ellioti* Swinhoe. *Zoology Research*, 55(2): 139-145. (In Chinese with English Abstract)

董磊, 孙悦华. 2007. 四川省唐家河自然保护区发现灰冠鸦雀. 动物学杂志, 42(6): 151. Dong L, Sun Y H. 2007. Rusty-throated Parrotbill found at Tangjiahe Natural Reserve in Sichuan, China. *Chinese Journal of Zoology*, 42(6): 151. (In Chinese with

English Abstract)

杜利民, 马鸣. 2013. 黄爪隼和红隼的繁殖习性记录. 四川动物, 32(5): 766-769. Du L M, Ma M. 2013. Habit difference of Lesser Kestrel and Common Kestrel during breeding season. *Sichuan Journal of Zoology*, 32(5): 766-769. (In Chinese with English Abstract)

樊文颖, 邢小军, 邹小林. 2003. 遗鸥栖息地选择及其保护对策研究. 内蒙古林业科技, 4: 32-33. Fan W Y, Xing X J, Zou X L. 2003. Choice of habitat for *Larus relictus* and protection countermeasures. *Inner Mongolia Forestry Science and Technology*, 4: 32-33. (In Chinese)

樊自立. 1985. 人类活动影响下新疆生态环境的一些变化. 生态学报, 5(4): 291-299. Fan Z L. 1985. Some changes in the ecological environment caused by human activities in Xinjiang. *Acta Ecologica Sinica*, 5(4): 291-299. (In Chinese with English Abstract)

冯理, 闵龙, 韩联宪, 等. 2008. 云南纳帕海自然保护区越冬猛禽考察初报. 四川动物, 27(3): 445-448. Feng L, Min L, Han L X, *et al.* 2008. Wintering observation of raptor in Napahai Natural Reserve, Yunnan. *Sichuan Journal of Zoology*, 27(3): 445-448. (In Chinese with English Abstract)

符建荣, 刘少英, 孙治宇, 等. 2009. 四川长沙贡玛自然保护区的鸟类资源. 四川动物, 28(2): 298-301. Fu J R, Liu S Y, Sun Z Y, *et al.* 2009. Survey on bird resources in Changshagongma Natural Reserve in Sichuan province. *Sichuan Journal of Zoology*, 28(2): 298-301. (In Chinese with English Abstract)

傅承钊. 1986. 新发现的白尾海雕繁殖区. 野生动物, 8(4): 33-34. Fu C Z. 1986. A newly discovered breeding area of White-tailed Eagle. *Chinese Wildlife*, 8(4): 33-34. (In Chinese)

傅桐生, 陈鹏. 1966. 栗斑腹鹀的分布及其繁殖习性. 动物学报, 18(2): 195-198. Fu T S, Chen P. 1996. The distribution and breeding habits of *Emberiza jankowskii*. *Acta Zoologica Sinica*, 18(2): 195-198. (In Chinese with English Abstract)

高继宏, 孙相吾, 陶宇, 等. 1992. 青头潜鸭繁殖研究初报. 野生动物, 14(2): 25-27. Gao J H, Sun X W, Tao Y, *et al.* 1992. A preliminary report on the breeding research of the Baer's Pochard. *Chinese Wildlife*, 14(2): 25-27. (In Chinese)

高立杰, 侯建华, 董建新. 2013. 遗鸥繁殖期新分布——内蒙古袄太湿地. 动物学杂志, 48(1): 141-142. Gao L J, Hou J H, Dong J X. 2013. New distribution of the Relict Gulls in breeding season—Aotai Lake of Inner Mongolia. *Chinese Journal of Zoology*, 48(1): 141-142. (In Chinese with English Abstract)

高铁军, 吴勇, 吴兆军. 1992. 遗鸥在鄂尔多斯中部的分布暨一新群巢的发现. 动物学杂志, 27(5): 31-33. Gao T J, Wu Y, Wu Z J. 1992. New finding of nest colony of the Relict Gull (*Larus relictus*) in Ordos of Inner Mongolia. *Chinese Journal of Zoology*, 27(5): 31-33. (In Chinese with English Abstract)

高玮. 2002a. 栗斑腹鹀 (*Emberiza jankowskii*) 生态学. 长春: 吉林科学技术出版社. Gao W. 2002a. Jankowski's Bunting (*Emberiza jankowskii*) ecology. Changchun: Jilin Science and Technology Press. (In Chinese)

高玮. 2002b. 中国隼形目鸟类生态学. 北京: 科学出版社. Gao W. 2002b. Ecology of Falcon order in China. Beijing: Science Press. (In Chinese)

高玮. 2006. 中国东北地区鸟类及其生态学研究. 北京: 科学出版社. Gao W. 2006. Studies on birds and their ecology in northeast China. Beijing: Science Press. (In Chinese)

高玮, 相桂权, 张凤岭, 等. 1988. 黑嘴松鸡的繁殖习性. 动物学报, 34(2): 194-195. Gao W, Xiang G Q, Zhang F L, *et al.* 1988. On the breeding habit of Black-billed Capercaillie. *Acta Zoologica Sinica*, 34(2): 194-195. (In Chinese with English Abstract)

高行宜, 戴昆. 1994. 新疆北部地区鸨类考察初报. 动物学杂志, 29(2): 52-53. Gao XY, Dai K. 1994. Preliminary report on investigation of bustard in northern Xinjiang. *Chinese Journal of Zoology*, 29(2): 52-53. (In Chinese with English Abstract)

高行宜, 乔德录, 熊义锋, 等. 1996. 新疆昌吉州波斑鸨分布与数量考察初报. 干旱区研究, 13(1): 81-83. Gao X Y, Qiao D L, Xiong Y F, *et al.* 1996. Preliminary investigation report on the distribution and population of Houbara in Changji, Xinjiang. *Arid Zone Research*, 13(1): 81-83. (In Chinese with English Abstract)

高行宜, 杨维康, 乔建芳, 等. 2007. 中国鸨类的分布与现状. 干旱区研究, 24(2): 179-186. Gao X Y, Yang W K, Qiao J F, *et al.* 2007. Distribution and actuality of *Otis tarda* in China. *Arid Zone Research*, 24(2): 179-186. (In Chinese with English Abstract)

高育仁, 余德群. 1990. 海南岛孔雀雉现状. 动物学杂志, 25(4): 42-44. Gao Y R, Yu D Q. 1990. Status of Peacock Pheasant in Hainan Island. *Chinese Journal of Zoology*, 25(4): 42-44. (In Chinese with English Abstract)

高忠燕, 杨晓杰, 王文锋. 2009. 扎龙保护区白枕鹤种群现状及保护对策. 齐齐哈尔师范高等专科学校学报, 29(3): 73-74. Gao Z Y, Yang X J, Wang W F. 2009. Population status and conservation measures of White-naped Cranes in Zhalong

Natural Reserve. *Journal of Science of Teachers' College and University*, 29(3): 73-74. (In Chinese with English Abstract)

葛继稳, 蔡庆华, 胡鸿兴, 等. 2005. 湖北省珍稀濒危保护水禽物种多样性及种群数量. 长江流域资源与环境, 14(1): 50-54. Ge J W, Cai Q H, Hu H X, *et al.* 2005. On species diversity and population size of rare, endangered and national key protected waterfowls in Hubei province, China. *Resources and Environment in the Yangtze Basin*, 14(1): 50-54. (In Chinese with English Abstract)

巩会生, 曾治高, 王西峰, 等. 2009. 陕西省国家重点保护野生动物的现状及分析. 陕西师范大学学报(自然科学版), 37(1): 52-59. Gong H S, Zeng Z G, Wang X F, *et al.* 2009. Current status and analysis of national protected animal species in Shaanxi province of China. *Journal of Shaanxi Normal University* (Natural Science Edition), 37(1): 52-59. (In Chinese with English Abstract)

苟军, 张耀东. 2007. 新疆北部发现白头硬尾鸭繁殖. 动物学杂志, 42(6): 52. Gou J, Zhang Y D. 2007. White-headed Duck was found breeding in northern Xinjiang. *Journal of Zoology*, 42(6): 52. (In Chinese with English Abstract)

关贯勋, 梁之华, 郭汉佳, 等. 2010. 澳门鸟类资源调查报告. 四川动物, 29(1): 91-98. Guan G X, Liang Z H, Guo H J, *et al.* 2010. Macao birds count report. *Sichuan Journal of Zoology*, 29(1): 91-98. (In Chinese with English Abstract)

郭玉民, 刘相林, 徐纯柱, 等. 2005. 小兴安岭白头鹤繁殖地种群数量初步调查. 动物学杂志, 40(4): 51-54. Guo Y M, Liu X L, Xu C Z, *et al.* 2005. A preliminary census of Hooded Crane population in the breeding area of Lesser Xingan Mountains. *Chinese Journal of Zoology*, 40(4): 51-54. (In Chinese with English Abstract)

郭玉民, 文丞, 林剑声, 等. 2016. 青头潜鸭(*Aythya baeri*)在中国的近期分布. 野生动物学报, 37(4): 382-385. Guo Y M, Wen C, Lin J S. 2016. Current appearance of Baer's Pochard in China. *Chinese Journal of Wildlife*, 37(4): 382-385. (In Chinese with English Abstract)

国家林业局. 2009. 中国重点陆生野生动物资源调查. 北京: 中国林业出版社. State Forestry Administration. 2009. Survey of key terrestrial wildlife resources in China. Beijing: China Forestry Press. (In Chinese)

韩晓东, 吴景才, 赵正阶. 1994. 大王家岛附近越冬长尾鸭的初步观察. 动物学杂志, 29(4): 22-24. Han X D, Wu J C, Zhao Z J. 1994. Preliminary observation of Long-tailed Duck overwintering around Dawangjia Island. *Chinese Journal of Zoology*, 29(4): 22-24. (In Chinese with English Abstract)

何芬奇, 林剑声, 王英永, 等. 2014. 婺源鸟类纪录解析. 动物学杂志, 49(2): 170-184. He F Q, Lin J S, Wang Y Y, *et al.* 2014. Bird records from Wuyuan, NE Jiangxi of SE China. *Chinese Journal of Zoology*, 49(2): 170-184. (In Chinese with English Abstract)

何芬奇, 林植, 江航东, 等. 2007. 白眉山鹧鸪在福建的分布. 动物学杂志, 42(4): 149-150. He F Q, Lin Z, Jiang H D, *et al.* 2007. The occurrence of the White-necked Partridge in Fujian, SE China. *Chinese Journal of Zoology*, 42(4): 149-150. (In Chinese with English Abstract)

何芬奇, 林植, 江航东. 2013. 中国的紫水鸡——其分布与种下分类问题的回顾与探讨. 动物学杂志, 48(3): 490-496. He F Q, Lin Z, Jiang H D. 2013. The Purple Gallinule in China—a review on its distribution and the subspecies. *Chinese Journal of Zoology*, 48(3): 490-496. (In Chinese with English Abstract)

何芬奇, 卢汰春, 芦春雷. 1986. 绿尾虹雉的繁殖生态研究. 生态学报, 6(2): 186-192. He F Q, Lu T C, Lu C L. 1986. Study on the breeding ecology of the Chinese Monal. *Acta Ecologica Sinica*, 6(2): 186-192. (In Chinese with English Abstract)

何芬奇, Melville D, 邢小军, 等. 2002. 遗鸥研究概述. 动物学杂志, 7(3): 65-70. He F Q, Melville D, Xing X J, *et al.* 2002. A review on studies of the Relict Gull *Larus relictus*. *Chinese Journal of Zoology*, 7(3): 65-70. (In Chinese with English Abstract)

何芬奇, 田秀华, 于海玲, 等. 2008. 略论东方白鹳的繁殖分布区域的扩展. 动物学杂志, 43(6): 154-157. He F Q, Tian X H, Yu H L, *et al.* 2008. The new breeding range and subpopulations of the Oriental Stork. *Chinese Journal of Zoology*, 43(6): 154-157. (In Chinese with English Abstract)

何芬奇, 张荫荪. 1994. 遗鸥鄂尔多斯种群研究的最新报道. 生物多样性, 2(2): 88-90. He F Q, Zhang Y S. 1994. The latest news on the Relict Gull Ordos population. *Biodiversity Science*, 2(2): 88-90. (In Chinese with English Abstract)

何芬奇, 张荫荪. 1998. 有关棕头鸥和遗鸥两近似种的分类与分布问题研究. 动物分类学报, 23(1): 105-111. He F Q, Zhang Y S. 1998. About some uncertainty of the distribution and taxonomy of *Larus brunnicephalus* and *Larus relictus*. *Acta Zootaxonomica Sinica*, 23(1): 105-111. (In Chinese with English Abstract)

何晓瑞. 1994. 云南历史上梅花鹿、犀牛和赤颈鹤的分布及其绝迹原因的研究. 云南大学学报 (自然科学版), 16(3): 294-298. He X R. 1994. Studies on geographic distribution and disappearance causes of sika deer, ehinoceroses and Sarus Crane in the history of Yunnan. *Journal of Yunnan University* (Natural Sciences Edition), 16(3): 294-298. (In Chinese with English

Abstract)

何业恒, 何文君. 1990. 试论褐马鸡地理分布的历史变迁. 湖南师范大学自然科学学报, 13(3): 275-280. He Y H, He W J. 1990. On the Geographical Distribution and historical re-distribution of *Crossoptilon mantchuricum*. *Journal of Natural Science of Hunan Normal University*, 13(3): 275-280. (In Chinese with English Abstract)

贺福银, 王建强, 田家龙, 等. 2004. 大兴安岭地区松鸡科鸟类. 中国林副特产, 19(5): 42-43. He F Y, Wang J Q, Tian J L, et al. 2004. The bird species of Tetraonidae family in Daxing'anling area. *Forest By-product and Speciality in China*, 19(5): 42-43. (In Chinese)

黑龙江省林业厅. 1990. 国际鹤类保护与研究. 北京: 中国林业出版社. Heilongjiang Forestry Department. 1990. International Crane Protection and Research. Beijing: China Forestry Press. (In Chinese)

洪孝宇. 2007a. 黄脚渔鸮在台湾的分布模式. 屏东: 屏东科技大学硕士学位论文. Hong X Y. 2007a. The Distribution Pattern of Tawny Fish Owl in Taiwan. Pingtung: Master Dissertation of Pingtung University of Science and Technology. (In Chinese with English Abstract)

洪孝宇. 2007b. 黄脚渔鸮的分布现况与族群概估. 野生动物保育汇报及通讯, 11(1): 19-26. Hong X Y. 2007b. The distribution and the estimated population of Tawny Fish Owl. *Wildlife Conservation Report and Communication*, 11(1): 19-26. (In Chinese with English Abstract)

胡鸿兴, 康洪莉, 贡国鸿, 等. 2005. 湖北省湿地冬季水鸟多样性研究. 长江流域资源与环境, 14(4): 424-428. Hu H X, Kang H L, Gong G H, et al. 2005. Biodiversity of winter waterbirds in Hubei, China. *Resources and Environment in the Yangtze Basin*, 14(4): 424-428. (In Chinese with English Abstract)

胡军华, 胡慧建, 杨道德, 等. 2007. 广东海丰紫水鸡种群密度调查. 动物学杂志, 42(1): 107-111. Hu J H, Hu H J, Yang D D, et al. 2007. Population estimation of Purple Swamphen (*Porphyrio porphyrio*) in Haifeng, Guangdong. *Chinese Journal of Zoology*, 42(1): 107-111. (In Chinese with English Abstract)

黄海魁. 2000. 云南野生动植物地理分布特点和退化原因分析. 云南地理环境研究, 12(1): 80-87. Huang H K. 2000. The conservation and the distributed feature Yunnan wild fauna. *Yunnan Geographic Environment Research*, 12(1): 80-87. (In Chinese with English Abstract)

黄人鑫, 米尔曼, 邵红光. 1992. 中国鸟类新纪录——阿尔泰雪鸡. 动物分类学报, 17(4): 501-502. Huang R X, Milman, Shao H G. 1992. A new record of birds from China—Altai Snowcock. *Acta Zootaxonomica Sinica*, 17(4): 501-502. (In Chinese with English Abstract)

黄人鑫, 邵红光, 米尔曼. 1991. 阿尔泰雪鸡生态的初步观察. 四川动物, 10(3): 36. Huang R X, Shao H G, Milman. 1991. Preliminary research of ecology of Altai Snowcock. *Sichuan Journal of Zoology*, 10(3): 36. (In Chinese)

黄松, 方秀峰, 黄接棠. 2003. 白颈长尾雉(*Syrmaticus ellioti*)的人工繁殖. 安徽大学学报(自然科学版), 27(3): 98-102. Huang S, Fang X F, Huang J T. 2003. Artificial propagation on *Syrmaticus ellioti*. *Journal of Anhui University* (Natural Science Edition), 27(3): 98-102. (In Chinese with English Abstract)

黄族豪, 柯坫华, 乐枫玲. 2013. 江西省猛禽鸟类多样性研究. 井冈山大学学报(自然科学版), 34(6): 96-99. Huang Z H, Ke D H, Le F L. 2013. Raptor diversity of Jiangxi province. *Journal of Jinggangshan University* (Natural Science), 34(6): 96-99. (In Chinese with English Abstract)

江航东, 林清贤, 林植, 等. 2005. 福建沿海岛屿水鸟考察报告. 动物分类学报, 30(4): 852-856. Jiang H D, Lin Q X, Lin Z, et al. 2005. Report on the waterbirds occurring on/around the offshore islands in Fujian, sea China. *Acta Zootaxonomica Sinica*, 30(4): 852-856. (In Chinese with English Abstract)

蒋明康, 王智, 秦卫华, 等. 2006. 我国自然保护区内国家重点保护物种保护成效评价. 生态与农村环境学报, 22(4): 35-38. Jiang M K, Wang Z, Qin W H, et al. 2006. Effectiveness of national priority wildlife protection in natural reserves. *Journal of Ecology and Rural Environment*, 22(4): 35-38. (In Chinese with English Abstract)

经宇, 孙悦华, 方昀. 2003. 黑头噪鸦的繁殖及生活史特征. 动物学杂志, 38: 91-92. Jing Y, Sun Y H, Fang Y. 2003. Notes on the natural history of the Sichuan Jay (*Perisoreus internigrans*). *Chinese Journal of Zoology*, 38: 91-92. (In Chinese with English Abstract)

柯杨, 黄原, 雷富民, 等. 2010. 黑尾地鸦线粒体基因组序列测定与分析. 遗传, 32(9): 951-960. Ke Y, Huang Y, Lei F M, et al. 2010. Sequencing and analysis of the complete mitochondrial genome of *Podoces hendersoni* (Ave, Corvidae). *Hereditas*, 32(9): 951-960. (In Chinese with English Abstract)

雷富民, 卢汰春. 2006. 中国鸟类特有种. 北京: 科学出版社. Lei F M, Lu T C. 2006. The endemic birds of China. Beijing: Science press. (In Chinese)

雷忻, 王文强, 廉振民. 2011. 陕西红碱淖遗鸥研究现状分析. 延安大学学报(自然科学版), 30(4): 93-96. Lei X, Wang W Q, Lian Z M. 2011. A Review of Studies on the status of *Larus relictus* in Hongjiannao lake, Shaanxi Province. *Journal of Yanan University* (Natural Science Edition), 30(4): 93-96. (In Chinese with English Abstract)

黎德武, 吴发清, 何定富, 等. 1985. 神农架地区的鸟类区系. 华中师范学报(自然科学版), (2): 81-89. Li D W, Wu F Q, He D F, *et al.* 1985. An avifauna survey of the Shennongjia area in Hubei province. *Journal of Central China Teachers College* (Natural Sciences Edition), (2): 81-89. (In Chinese with English Abstract)

李成龄, 江永生. 1990. 洪湖野鸭越冬种群的数量调查. 野生动物, 12(1): 9-10. Li C L, Jiang Y S. 1990. Quantitative investigation on population of wintering wild ducks in Honghu Lake. *Chinese Wildlife*, 12(1): 9-10. (In Chinese with English Abstract)

李德品, 唐涛, 蔡庆华, 等. 2011. 滇西北紫水鸡的分布及其种群现状. 四川动物, 30(4): 644-648. Li D P, Tang T, Cai Q H, *et al.* 2011. Distribution and population status of *Porphyrio porphyrio* in northwest Yunnan, China. *Sichuan Journal of Zoology*, 30(4): 644-648. (In Chinese with English Abstract)

李东明, 孙立汉, 高士平, 等. 2004. 基于RS和GIS技术的河北省鸡形目鸟类分布格局. 动物学研究, 25(4): 304-310. Li D M, Sun L H, Gao S P, *et al.* 2004. Galliformes distribution in Hebei Province based on RS and GIS techniques. *Zoological Research*, 25(4): 304-310. (In Chinese with English Abstract)

李凤山, 刘观华, 吴建东, 等. 2011. 鄱阳湖湿地和水鸟的生态研究. 北京: 科学普及出版社. Li F S, Liu G H, Wu J D, *et al.* 2011. Ecological Study of Wetlands and Waterbirds at Poyang Lake. Beijing: Popular Science Press. (In Chinese)

李凤山, 杨芳. 2003. 云贵高原黑颈鹤的种群数量和分布. 动物学杂志, 38(3): 43-46. Li F S, Yang F. 2003. Population numbers and distribution of Black-necked Cranes (*Grus nigricollis*) in the Yungui Gaoyuan plateau. *Chinese Journal of Zoology*, 38(3): 43-46. (In Chinese with English Abstract)

李桂垣. 1995. 普通旋木雀*Certhia familiaris* (Passeriformes: Certhiidae)的新亚种. 动物分类学报, 20(3): 373-377. Li G Y. 1995. New subspecies of Eurasian Treecreeper (*Certhia familiaris*) (Passeriformes: Certhiidae). *Acta Zoologica Sinica*, 20(3): 373-377. (In Chinese with English Abstract)

李海清, 马秀杰, 贾艳玲, 等. 2006. 内蒙古图牧吉国家级自然保护区珍稀濒危鸟类种类、分布特点及保护措施. 内蒙古林业调查设计, 29(6): 57-59. Li H Q, Ma X J, Jia Y L, *et al.* 2006. Species distribution characteristics and protective measures of rare and endangered birds in national preservation area of Tumuji in Inner Mongolia. *Inner Mongolia Forestry Investigation and Design*, 29(6): 57-59. (In Chinese with English Abstract)

李晶晶, 韩联宪, 曹宏芬, 等. 2013. 珠穆朗玛峰国家级自然保护区鸟类区系及其垂直分布特征. 动物学研究, 34(6): 531-548. Li J J, Han L X, Cao H F, *et al.* 2013. The fauna and vertical distribution of birds in Mount Qomolangma National Natural Reserve. *Zoological Research*, 34(6): 531-548. (In Chinese with English Abstract)

李景瑞. 2005. 海南孔雀雉的活动区和栖息地利用研究. 北京: 北京师范大学硕士学位论文. Li J R. 2005. The home range and habitat use of Hainan Peacock Pheasant. Beijing: Master Dissertation of Beijing Normal University. (In Chinese with English Abstract)

李仕宁, 谢林顺, 杨琼荣. 2012. 茄新省级森林经营所鸟类资源调查. 热带林业, 40(4): 42-46. Li S N, Xie L S, Yang Q R. 2012. Avifauna of Jiaxin Provincial Natural Reserve, Hainan. *Tropical Forestry*, 40(4): 42-46. (In Chinese with English Abstract)

李文发, 赵和生, 王景权, 等. 1994. 兴凯湖自然保护区的鸟类资源. 国土与自然资源研究, (2): 47-50. Li W F, Zhao H S, Wang J Q, *et al.* 1994. Bird Resources in Xingkai Lake Nature Reserve. *Territory & Natural Resources Study*, (2): 47-50. (In Chinese with English Abstract)

李小惠, 谭洪治, 陈采安, 等. 1990. 白额山鹧鸪的生态观察. 动物学杂志, 25(1): 25-27, 67. Li X H, Tan H Z, Chen C A, *et al.* 1990. A preliminary study on White-necklaced partridge. *Chinese Journal of Zoology*, 25(1): 25-27, 67. (In Chinese with English Abstract)

李小燕, 杨磊, 李东洋, 等. 2012. 广东莲花山白盆珠省级自然保护区鸟类资源调查. 四川动物, 31(4): 650-654. Li X Y, Yang L, Li D Y, *et al.* 2012. Survey of bird resources in Guangdong Lianhuashan Baipenzhu Provincial Natural Reserve. *Sichuan Journal of Zoology*, 31(4): 650-654. (In Chinese with English Abstract)

李雪艳, 梁璐, 宫鹏, 等. 2012. 中国观鸟数据揭示鸟类分布变化. 科学通报, 57(31): 2956-2963. Li X Y, Liang L, Gong P, *et al.* 2012. Bird watching in China reveals bird distribution changes. *Chinese Science Bulletin*, 57(31): 2956-2963. (In Chinese with English Abstract)

李杨, 袁梨, 史洋, 等. 2015. 北京地区珍稀鸟类生态分布的GIS分析. 北京林业大学学报, (5): 119-125. Li Y, Yuan L, Shi Y, *et al.* 2015. GIS analysis of the distribution dynamics of rare birds in Beijing. *Journal of Beijing Forestry University*, (5): 119-

125. (In Chinese with English Abstract)

李元刚, 李志刚, 胡天华. 2012. 宁夏贺兰山国家级自然保护区鸟类区系组成及其特征研究. 西北林学院学报, 27(1): 109-115. Li Y G, Li Z G, Hu T H. 2012. Structure and characteristic of Avifuna in Helan Mountain National Nature Reserve. *Journal of Northwest Forestry University*, 27(1): 109-115. (In Chinese with English Abstract)

梁斌, 陈水华, 王忠德. 2007. 浙江五峙山列岛黄嘴白鹭的巢位选择研究. 生物多样性, 15(1): 92-96. Liang B, Chen S H, Wang Z D. 2007. Nest selection of Chinese Egret (*Egretta eulophotes*) in Wuzhishan Archipelago, Zhejiang. *Biodiversity Science*, 15(1): 92-96. (In Chinese with English Abstract)

梁军, 张明明, 李路云, 等. 2013. 贺兰山白尾海雕的越冬种群数量和分布研究简报. 四川动物, 32(5): 780-782. Liang J, Zhang M M, Li L Y, et al. 2013. Population numbers and distribution of winter *Haliaeetus albicilla* in Helan mountains. *Sichuan Journal of Zoology*, 32(5): 780-782. (In Chinese with English Abstract)

林芳君, 蒋萍萍, 丁平. 2010. 白颈长尾雉微卫星多态性的遗传学分析. 动物学研究, 31(5): 461-468. Lin F J, Jiang P P, Ding P. 2012. Genetic analysis of microsatellite polymorphism in the Elliot's Pheasant (*Syrmaticus ellioti*) in China. *Zoological Research*, 31(5): 461-468. (In Chinese with English Abstract)

林剑声, 刘伟民, 何芬奇. 2005. 云南思茅莱阳河保护区发现蓝腰短尾鹦鹉*Psittinus cyanurus*. 动物学研究, 26(3): 321. Lin J S, Liu W M, He F Q. 2005. The Blue-rumped Parrot *Psittinus cyanurus* found at Laiyanghe NR in Simao of S Yunnan, China. *Zoological Research*, 26(3): 321. (In Chinese with English Abstract)

林植, 叶振伟, 何芬奇. 2012. 关于厦门的紫水鸡. 动物学杂志, 47(6): 125-127. Lin Z, Ye Z W, He F Q. 2012. Finding the colony of Purple Swamphen (*Porphyrio porphyrio*) at Amoy of SE China. *Chinese Journal of Zoology*, 47(6): 125-127. (In Chinese with English Abstract)

刘伯文. 1996. 内蒙古兴安盟图牧吉地区鸟类资源的考察. 东北林业大学学报, 5(24): 92-100. Liu B W. 1996. A report on bird resource in Tumuji region, Xing'an League, Inner Mongolia. *Journal of Northeast Forestry University*, 5(24): 92-100. (In Chinese with English Abstract)

刘伯文. 1997. 东北地区大鸨的现状和保护. 国土与自然资源研究, (4): 61-63. Liu B W. 1997. The status and conservation of Great Bustard in Northeast China. *Land and Natural Resources Research*, (4): 61-63. (In Chinese with English Abstract)

刘春悦, 江红星, 孙效维, 等. 2013. 白鹤中途停歇地主要食物藨草球茎密度的空间插值方法研究. 动物学杂志, 48(3): 382-390. Liu C Y, Jiang H X, Sun X W, et al. 2013. Comparison of the spatial interpolation methods for the tuber density of two Scirpus species: main food of Siberian Cranes at the stopover site. *Chinese Journal of Zoology*, 48(3): 382-390. (In Chinese with English Abstract)

刘焕金, 申守义, 王全喜. 1989. 乌雕冬季的生态观察. 动物学杂志, 24(4): 14-16. Liu H J, Shen S Y, Wang Q X. 1989. A research on winter ecology of the Greater Spotted Eagle. *Chinese Journal of Zoology*, 24(4): 14-16. (In Chinese)

刘焕金, 苏化龙, 冯敬义, 等. 1985. 山西省黑鹳的数量分布. 生态学报, 5(2): 193-194. Liu H J, Su H L, Feng J Y, et al. 1985. The numerical distribution of the Black Stork in Shanxi province. *Acta Ecologica Sinica*, 5(2): 193-194. (In Chinese with English Abstract)

刘焕金, 苏化龙, 申守义, 等. 1986. 陕西省金雕的地理分布. 国土与自然资源研究, (3): 36-45. Liu H J, Su H L, Shen S Y, et al. 1986. Geographical distribution of Golden Eagle in Shaanxi. *Land and Natural Resources Research*, (3): 36-45. (In Chinese with English Abstract)

刘迺发, 包新康, 廖继承. 2013. 青藏高原鸟类分类与分布. 北京: 科学出版社. Liu N F, Bao X K, Liao J C. 2013. The classification and distribution of the birds in Qingzang plateau. Beijing: Science Press. (In Chinese)

刘迺发, 黄族豪, 文陇英. 2004. 西部荒漠地区的湿地和水禽多样性. 湿地科学, 2(4): 259-266. Liu N F, Huang Z H, Wen L Y. 2004. Wetland and water bird diversity in desert area of the western China. *Wetland Science*, 2(4): 259-266. (In Chinese with English Abstract)

刘荣, 王本瑞, 侯永平. 1995. 金雕种群密度的调查. 动物学杂志, 30(2): 23-24. Liu R, Wang B R, Hou Y P. 1995. Survey of population density of Golden Eagle. *Journal of Zoology*, 30(2): 23-24. (In Chinese)

刘少初, 次仁. 1993. 西藏的雉鹑. 野生动物, 15(2): 18-21. Liu S C, Ci Ren. 1993. Pheasant in Tibet. *Journal of Wildlife*, 15(2): 18-21. (In Chinese)

刘文盈, 张秋良, 邢小军, 等. 2008. 鄂尔多斯高原盐沼湿地底栖动物多样性特征与遗鸥繁殖期觅食的相关性研究. 干旱区资源与环境, 22(3): 185-192. Liu W Y, Zhang Q L, Xing X J, et al. 2008. The relationship between the zoobenthos and *Larus relictu*'s foraging behavior in saline-alkaline wetland of Ordos plateau. *Journal of Arid Land Resources and Environment*, 22(3): 185-192. (In Chinese with English Abstract)

刘小如, 丁宗苏, 方伟宏, 等. 2010. 台湾鸟类志(上中下). 台北: 农委会林务局. Liu X R, Ding Z S, Fang W H, *et al.* 2010. The avifauna of Taiwan (Vol. 1-3). Taipei: Forest Bureau, Council of Agriculture. (In Chinese)

刘小如, 丁宗苏, 方伟宏, 等. 2012. 台湾鸟类志(上中下). 第二版. 台北: "中央研究院"生物多样性研究中心. Severinghaus L L, Ding T S, Fang W H, *et al.* 2012. The avifauna of Taiwan (2nd ed). (Vol. 1-3). Taipei: Academia Sinica Biodiversity Research Center. (In Chinese)

刘宇, 杨志杰, 左斌, 等. 2008. 中华秋沙鸭(*Mergus squamatus*)在江西省的越冬分布及种群数量调查. 东北师范大学学报, 40(1): 111-115. Liu Y, Yang Z J, Zuo B, *et al.* 2008. Wintering distribution and population size of Scaly-sided Merganser *Mergus squamatus* in Jiangxi Province. *Journal of Northeast Normal University*, 40(1): 111-115. (In Chinese with English Abstract)

隆廷伦, 郭耕. 1998. 绿尾虹雉冬季生态的跟踪观测研究. 四川动物, 17(3): 104-105. Long T L, Guo G. 1998. Tracing study on ecology in winter of *Lophophorus lhuysii*. *Sichuan Journal of Zoology*, 17(3): 104-105. (In Chinese)

卢汰春. 1991. 中国珍稀濒危野生鸡类. 福州: 福建科学技术出版社. Lu T C. 1991. Rare and endangered wild pheasant in China. Fuzhou: Fujian Science and Technology Press. (In Chinese)

卢汰春. 2010. 台湾鹎——即将消失的中国特有种. 大自然, 31(5): 64-65. Lu T C. 2010. Taiwan Bulbul-an endangered Chinese endemic species. *Journal of China Nature*, 31(5): 64-65. (In Chinese with English Abstract)

卢汰春, 刘如笋, 何芬奇. 1986. 绿尾虹雉生态学研究. 动物学报, 32(3): 273-279. Lu T C, Liu R S, He F Q. 1986. Ecological Studies on Chinese Monal. *Acta Zoologica Sinica*, 32(3): 273-279. (In Chinese with English Abstract)

吕士成. 2009. 盐城沿海丹顶鹤种群动态与湿地环境变迁的关系. 南京师大学报(自然科学版), 32(4): 89-93. Lyu S C. 2009. Relationship between the population dynamics of the wintering red-crowned crane in natural reserve for rare birds in beach region and wetland environment variance of Yancheng. *Journal of Nanjing Normal University* (Natural Science Edition), 32(4): 89-93. (In Chinese with English Abstract)

罗益奎, 蒋爱伍, 陈辈乐, 等. 2013. 广西冠斑犀鸟的种群数量及分布状况. 生物多样性, 21(3): 352-358. Luo Y K, Jiang A W, Chen B L, *et al.* 2013. Population number and distribution of oriental pied hornbill in Guangxi, China. *Biodiversity Science*, 21(3): 352-358. (In Chinese with English Abstract)

马福, 张建龙. 2009. 中国重点陆生野生动物资源调查. 北京: 中国林业出版社. Ma F, Zhang J L. 2009. Survey of key terrestrial wildlife resources in China. Beijing: China Forestry Press. (In Chinese)

马国瑞. 1989. 绿尾虹雉在甘肃的分布和生态初步观察. 天水师专学报(混合版), 9(1): 101-104. Ma G R. 1989. Preliminary observation of the distribution and ecology of Luweihong in Gansu. *Journal of Tianshui Teachers College* (mixed edition), 9(1): 101-104. (In Chinese with English Abstract)

马鸣. 1994. 黑鹳(*Ciconia nigra*)在塔里木盆地的分布与繁殖. 见: 中国鸟类学会水鸟组. 中国水鸟研究. 上海: 华东师范大学出版社. Ma M. 1994. The distribution and reproduction of Black Stork (*Ciconia nigra*) in the Tarim Basin. *In*: Waterbird Group, China Ornithological Society From the Chinese Waterbird Research. Shanghai: East China Normal University Press. (In Chinese)

马鸣. 2001a. 塔克拉玛干沙漠白尾地鸦的分布与生态习性. 干旱区研究, 18(3): 29-35. Ma M. 2001. The distribution and ecological habits and characteristics of Xinjiang Ground Jays in Taklamakan Desert. *Arid Zone Research*, 18(3): 29-35. (In Chinese with English Abstract)

马鸣. 2001b. 新疆白鹳(*Ciconia ciconia*)种群的历史分布区考证及其绝迹原因分析. 中国学术期刊文摘(科技快报), 7(6): 734-736. Ma M. 2001. The criticism on the historical distribution and the population disappeared of White Stork (*Ciconia ciconia* asiatica) in Xinjiang, China. *Chinese Science Abstracts*, 7(6): 734-736. (In Chinese with English Abstract)

马鸣. 2002. 历史上新疆白鹳的地理分布区域考证. 干旱区地理, 25(2): 139-142. Ma M. 2002. A textual research on the geographical habitation of White Storks (*Ciconia ciconia* asiatica). *Arid Land Geography*, 25(2): 139-142. (In Chinese with English Abstract)

马鸣. 2009. 新疆白头硬尾鸭现状. 中国鹤类通讯, 13(1): 43-45. Ma M. 2009. The current status of White-headed Duck in Xinjiang. *China Crane Newsletter*, 13(1): 43-45. (In Chinese with English Abstract)

马鸣. 2011a. 为新疆的猛禽深深悲哀. 中国鸟类观察, 11(5): 10-13. Ma M. 2011a. Sorrow for raptors of Xinjiang. *China Bird Watch*, 11(5): 10-13. (In Chinese with English Abstract)

马鸣. 2011b. 新疆鸟类分布名录. 第二版. 北京: 科学出版社. Ma M. 2011b. A checklist on the distribution of the birds in Xinjiang (2nd ed). Beijing: Science Press. (In Chinese)

马鸣, 巴吐尔汗, 才代, 等. 1997a. 新疆白鹳(*Ciconia ciconia*)与黑鹳(*Ciconia nigra*)资源状况及生态研究. 地方病通报, 12(增刊): 33-35. Ma M, Batuerhan, Cai D, *et al.* 1997. Resource status and ecological research of White Stork (*Ciconia ciconia*) and Black Stork (*Ciconia nigra*). *Endemic Disease Bulletin*, 12(Suppl): 33-35. (In Chinese)

马鸣, 巴吐尔汗, 贾泽新, 等. 1992a. 新疆鸟类新纪录两种——夜鹭和白尾海雕. 干旱区研究, 9(1): 61-62. Ma M, Batuerhan, Jia Z X, *et al.* 1992a. Two new bird records in Xinjiang-the Black—crowned Night Heron and the White-tailed Sea Eagle. *Arid Zone Research*, 9(1): 61-62. (In Chinese)

马鸣, 巴吐尔汗, 陆健健. 1993. 新疆南部地区黑鹳(*Ciconia nigra*)种群密度及其繁殖. 动物学研究, 14(4): 374, 382-383. Ma M, Batuerhan, Lu J J. 1993. The population density and breeding of Black Stork (*Ciconia nigra*) in southern Xinjiang. *Zoological Research*, 14(4): 374, 382-383. (In Chinese)

马鸣, 胡宝文, 克德尔汗·巴亚恒, 等. 2010a. 新疆艾比湖遗鸥和细嘴鸥的数量现状. 动物学杂志, 45(1): 45-49. Ma M, Hu B W, Kedeerhan B, *et al.* 2010a. The population status Relict Gull and Slender-billed Gull in Aibi Lake in Xinjiang. *Journal of Zoology*, 45(1): 45-49. (In Chinese with English Abstract)

马鸣, 贾泽新. 1993. 黑鹳(*Ciconia nigra*)诸鸟的繁殖地新报. 动物学杂志, 28(1): 41-42. Ma M, Jia Z X. 1993. A report on the new breeding sites of Black Stork (*Ciconia nigra*). *Journal of Zoology*, 28(1): 41-42. (In Chinese)

马鸣, 克德尔汗·巴亚恒, 李飞, 等. 2010b. 新疆艾比湖湿地自然保护区鸟类清单及秋季迁徙数量统计. 四川动物, 29(6): 912-918. Ma M, Kedeerhan B, Li F, *et al.* 2010b. List of birds and count of autumn migration in Ebinur Wetland Natural Reserve. *Sichuan Journal of Zoology*, 29(6): 912-918. (In Chinese with English Abstract)

马鸣, 罗宁, 贾泽信. 1992b. 塔克拉玛干沙漠腹地动物调查. 动物学杂志, 27(5): 43. Ma M, Luo N, Jia Z X. 1992b. Investigation of Taklimakan Desert hinterland Animals. *Journal of Zoology*, 27(5): 43. (In Chinese)

马鸣, 梅宇, Potapov E, 等. 2007. 中国西部地区猎隼(*Falco cherrug*)繁殖生物学与保护. 干旱区地理, 30(5): 654-659. Ma M, Mei Y, Potapov E, *et al.* 2007. Breeding biology of Saker Falcon *Falco cherrug* and protection plan in Tibet, Qinghai and Xinjiang, the west of China. *Arid Land Geography*, 30(5): 654-659. (In Chinese with English Abstract)

马鸣, 梅宇. 2007. 在新疆发现两种珍稀野鸭. 动物学研究, 28(6): 673-674. Ma M, Mei Y. 2007. White-headed Ducks and Long-tailed Ducks were recorded in Xinjiang, China. *Zoological Research*, 28(6): 673-674. (In Chinese with English Abstract)

马鸣, 欧咏, 段刚. 1997b. 中日塔克拉玛干沙漠徒步科学探险报告. 干旱区研究, 14(3): 55-58. Ma M, Ou Y, Duan G. 1997. The biology report of Sino-Nipponian scientific exploration of foot in Taklimakan Desert, 1997. *Arid Zone Research*, 14(3): 55-58. (In Chinese)

马鸣, Potapov E, 殷守敬, 等. 2005. 新疆、青海、西藏猎隼(*Falco cherrug*)生存状况与繁殖生态. 见: 中国鸟类学会. 第八届中国动物学会鸟类学分会全国代表大会暨第六届海峡两岸鸟类学研讨会论文集: 307-313. Ma M, Potapov E, Yin S J. 2005. The status and breeding ecology of *Falco cherrug* in Xinjiang, Qinghai and Tibet. *In*: China Ornithological Society. Studies on Chinese Ornithology: 307-313. (In Chinese)

马鸣, 王岐山. 2000. 长脚秧鸡重新在新疆发现. 动物学研究, 21(5): 348. Ma M, Wang Q S. 2000. Re-discovery of Corncrake *Crex crex* in Xinjiang, China. *Zoological Research*, 21(5): 348. (In Chinese with English Abstract)

马鸣, 魏顺德, 程军. 2004a. 卫星跟踪下的黑鹳迁徙. 动物学杂志, 39(2): 102. Ma M, Wei S D, Cheng J. 2004a. Satellite tracking on Black Stork migration under. *Journal of Zoology*, 39(2): 102. (In Chinese with English Abstract)

马鸣, 徐峰, 才代, 等. 2004b. 塔克拉玛干沙漠特有物种——白尾地鸦. 乌鲁木齐: 新疆科学技术出版社: 1-131. Ma M, Xu F, Cai D, *et al.* 2004b. Xinjiang Ground Jay *Podoces biddulphi*, an endemic species in Taklimakan Desert. Ürümqi: Xinjiang Science and Technology Publishing House. (In Chinese)

马鸣, 殷守敬, 徐峰, 等. 2006. 新疆黑尾地鸦初步调查. 动物学杂志, 41(2): 135. Ma M, Yin S J, Xu F, *et al.* 2006. Preliminary investigation of Mongolian Ground Jay in Xinjiang. *Journal of Zoology*, 41(2): 135. (In Chinese with English Abstract)

马鸣, 张同, 徐峰. 2014. 新疆南部黑颈鹤种群分布及数量. 动物学研究, 35(S1): 105-110. Ma M, Zhang T, Xu F. 2014. Numbers and distribution of the black-necked crane (Grus nigricollis) in the southern Xinjiang. *Zoological Research*, 35(S1): 105-110. (In Chinese with English Abstract)

马逸清. 1986. 中国鹤类研究文集. 哈尔滨: 黑龙江教育出版社. Ma Y Q. 1986. Collection of Chinese Crane Research. Harbin: Heilongjiang Education Press. (In Chinese)

梅宇, 马鸣, Dixon A, 等. 2008. 中国西部电网电击猛禽致死事故调查. 动物学杂志, 43(4): 114-117. Mei Y, Ma M, Dixon A, *et al.* 2008. Investigation on raptor of electrocution along power lines in the Western China. *Chinese Journal of Zoology*, 43(4): 114-117. (In Chinese with English Abstract)

彭波涌, 胡军华, 胡慧建. 2006. 西洞庭湖鸟类物种多样性分析. 四川动物, 25(4): 850-854. Peng B Y, Hu J H, Hu H J. 2006.

Analysis of bird diversity in West Dongting Lake, Hunan Province. *Sichuan Journal of Zoology*, 25(4): 850-854. (In Chinese with English Abstract)

彭岩波, 丁平. 2005. 白颈长尾雉春季扩散活动的影响因子. 动物学研究, 26(4): 373-378. Peng Y B, Ding P. 2005. Factors affecting movement of spring dispersal of Elliot's Pheasants. *Zoological Research*, 26(4): 373-378. (In Chinese with English Abstract)

钱法文. 2005. 中国鹤类保护现状调查. 森林与人类, 25(5): 31-38. Qian F W. 2005. Survey of the status of crane conservation in China. *Forest & Humankind*, 25(5): 31-38. (In Chinese with English Abstract)

钱燕文, 张洁, 汪松, 等. 1965. 新疆南部的鸟兽. 北京: 科学出版社: 298-299. Qian Y W, Zhang J, Wang S, *et al.* 1965. Birds and Mammals in Southern Xinjiang. Beijing: Science Press. (In Chinese)

乔建芳, 杨维康, Combreau O, 等. 2003. 新疆木垒波斑鸨的繁殖成功率. 动物学报, 49(3): 310-317. Qiao J F, Yang W K, Combreau O, *et al.* 2003. Breeding success of Houbara Bustard (*Chlamydotis undulata macqueenii*) population in Mulei, Xinjiang. *Current Zoology*, 49(3): 310-317. (In Chinese with English Abstract)

邱冠华. 2009. 甘肃敦煌西湖湿地鸟类调查及其栖息地重要性评价. 北京: 北京林业大学硕士学位论文. Qiu G H. 2009. Wetland birds and their habitats importance evaluation in Dunhuang West Lake Area. Beijing: Master Dissertation of Beijing Forestry University. (In Chinese)

屈文政, 杨岚. 1998. 犀鸟科中国新纪录——花冠皱盔犀鸟. 动物分类学报, 23(2): 222-223. Qu W Z, Yang L. 1998. Wreathed Hornbill (*Aceros* (Rhyticeros) *undulatus*), a new record of bucerotidae from China. *Acta Zootaxonomica Sinica*, 23(2): 222-223. (In Chinese)

任巍, 张志麟, 米小其, 等. 2008. 中华秋沙鸭在湖南的新分布. 动物学杂志, 43(4): 19. Ren W, Zhang Z L, Mi X Q, *et al.* 2008. New distribution record of Scaly-sided Merganser in Hunan. *Chinese Journal of Zoology*, 43(4): 19. (In Chinese with English Abstract)

任文博, 张建云, 袁朝晖. 等. 2014. 秦岭南麓黄脚渔鸮新分布观察初报. 陕西林业科技, 41(3): 63-64. Ren W B, Zhang J Y, Yuan C H, *et al.* 2014. Report on the new distribution of *ketupa flavipes* in the south slope of Qinling Mountains. *Shaanxi Forest Science and Technology*, 41(3): 63-64. (In Chinese with English Abstract)

沈钧, 余新华. 1988. 白颈长尾雉的饲养. 动物学杂志, 23(6): 37-38. Shen J, Yu X H. 1988. Breeding of Elliot's Pheasant in captivity. *Chinese Journal of Zoology*, 23(6): 37-38. (In Chinese)

石建斌, 郑光美. 1995. 白颈长尾雉的活动区. 北京师范大学学报(自然科学版), 31(4): 513-519. Shi J B, Zheng G M. 1995. The home-range of Elliot's Pheasant. *Journal of Beijing Normal University* (Natural Science), 31(4): 513-519. (In Chinese with English Abstract)

石美. 2012. 甘肃莲花山斑尾榛鸡 (*Tetrastes sewerzowi*) 春季领域保护行为和四川林鸮繁殖生态观察. 北京: 中国科学院硕士学位论文. Shi M. 2012. Observation on territory behavior of the Chinese Grouse (*Tetrastes sewerzowi*) and reproductive ecology of Sichuan Wood Owl (*Strix davidi*) at Lianhuashan, Gansu, China. Beijing: Dissertation of Chinese Academy of Sciences. (In Chinese with English Abstract)

苏化龙. 1988. 金雕——处于濒危状态中的大型猛禽. 动物学杂志, 23(5): 36-40. Su H L. 1988. Golden Eagle—a large and endangered raptor species. *Chinese Journal of Zoology*, 23(5): 36-40. (In Chinese)

苏化龙, 林英华, 李迪强, 等. 2000. 中国鹤类现状及其保护对策. 生物多样性, 8(2): 180-191. Su H L, Lin Y H, Li D Q, *et al.* 2000. Status of Chinese cranes and their conservation strategies. *Biodiversity Science*, 8(2): 180-191. (In Chinese with English Abstract)

粟通萍, 王绍能, 蒋爱伍. 2012. 广西猫儿山地区鸟类组成及垂直分布格局. 动物学杂志, 47(6): 54-65. Su T P, Wang S N, Jiang A W. 2012. Species composition and vertical distribution pattern of the birds in Mao'ershan Region, Northeastern Guangxi. *Chinese Journal of Zoology*, 47(6): 54-65. (In Chinese with English Abstract)

孙悦华, Martens J. 2005. 陕西秦岭发现四川旋木雀. 动物学杂志, 40(4): 33. Sun Y H, Martens J. 2005. New distribution of the Sichuan Treecreeper in the Qinling Mountains, Shaanxi. *Chinese Journal of Zoology*, 40(4): 33. (In Chinese with English Abstract)

孙悦华, 郑光美. 1992. 黄腹角雉活动区的无线电遥测研究. 动物学报, 38(4): 385-392. Sun Y H, Zhen G M. 1992. The home range of Cabot's Targopans by radiotracking. *Acta Zoologica Sinica*, 38(4): 385-392. (In Chinese with English Abstract)

谭耀匡, 关贯勋. 2003. 中国动物志 鸟纲 第七卷 夜鹰目 雨燕目 咬鹃目 佛法僧目 鴷形目. 北京: 科学出版社. Tan Y K, Guan G X. 2003. Fauna Sinica·Aves. Vol. 7. Caprimulgiformes, Apodiformes, Trogoniformes, Coraciiformes and Piciformes. Beijing: Science Press. (In Chinese)

汤宋华. 2008. 莲花山鬼鸮(*Aegolius funereus beickianus*)和四川林鸮(*Strix davidi*)的繁殖生态与育雏行为. 北京: 中国科学院硕士学位论文. Tang S H. 2008. Reproductive Ecology and Brooding Behaviors of *Aegolius funereus* Beickianus and *Strix davidi* in Sichuan. Beijing: Master Dissertation of Chinese Academy of Sciences. (In Chinese with English Abstract)

田桷, 薛文, 马俊, 等. 1998. 遗鸥(*Larus relictus*)繁殖群在内蒙古东部的新发现. 内蒙古大学学报(自然科学版), 29(5): 694-696. Tian L, Xue W, Ma J, *et al*. 1998. The discovery of the Relict Gull's breeding population in the east of Inner Mongolia. *Journal of Inner Mongolia University* (Natural Science Edition), 29(5): 694-696. (In Chinese with English Abstract)

田秀华, 王进军. 2001. 中国大鸨. 长春: 东北林业大学出版社. Tian X H, Wang J J. 2001. The Great Bustard in China. Changchun: Northeast Forestry University Press. (In Chinese)

佟富春, 高玮, 肖以华, 等. 2002a. 吉林白城地区草原栗斑腹鹀窝卵数、营巢成功率和繁殖成功率的研究. 应用生态学报, 13(3): 281-284. Tong F C, Gao W, Xiao Y H, *et al*. 2002a. Clutch size, nesting success and breeding success rate in *Emberiza jankowskii* in the grassland at Baicheng in Jilin Province. *Chinese Journal of Applied Ecology*, 13(3): 281-284. (In Chinese with English Abstract)

佟富春, 肖以华, 白哈斯, 等. 2002b. 吉林省白城地区干草原栗斑腹鹀的繁殖生态. 生态学报, 22(9): 1485-1490. Tong F C, Xiao Y H, Bai H S, *et al*. 2002b. Breeding ecology of Jankowski's Bunting in the Dry Grassland in Baicheng, Jilin Province. *Acta Ecologica Sinica*, 22(9): 1485-1490. (In Chinese with English Abstract)

涂业苟, 俞长好, 黄晓凤, 等. 2009. 鄱阳湖区域越冬雁鸭类分布与数量. 江西农业大学学报, 31(4): 460-764. Tu Y G, Yu C H, Huang X F, *et al*. 2009. Distribution and population of the overwintering Anatidae Waterfowl in the Poyang Lake. *Acta Agriculturae Universitatis Jiangxiensis*, 31(4): 460-764. (In Chinese with English Abstract)

万冬梅, 高玮. 赵匠, 等. 2002. 辽宁猛禽迁徙规律的研究. 东北师大学报(自然科学版), 34(2): 78-83. Wan D M, Gao W, Zhao J, *et al*. 2002. On the migration law of raptor in Liaoning Province. *Journal of Northeast Normal University* (Natural Science Edition), 34(2): 78-83. (In Chinese with English Abstract)

汪继超, 史海涛. 2002. 海南孔雀雉. 生物学通报, 37(11): 24-25. Wang J C, Shi H T. 2002. *Polyplectron katsumatae*. *Bulletin of Biology*, 37(11): 24-25. (In Chinese)

汪清雄, 杨超, 刘铮, 等. 2013. 红碱淖遗鸥孵卵行为. 生态学杂志, 2(2): 375-379. Wang Q X, Yang C, Liu Z, *et al*. 2013. Hatching behavior of relict gull *Larus relictus* in Hongjiannao of Shannxi Province, Northwest China. *Chinese Journal of Ecology*, 2(2): 375-379. (In Chinese with English Abstract)

汪松, 谢焱. 2004. 中国物种红色名录(第一卷·红色名录). 北京: 高等教育出版社. Wang S, Xie Y. 2004. China Species Red List (Vol. 1). Beijing: Higher Education Press. (In Chinese and English)

汪松, 谢焱. 2009. 中国物种红色名录(第二卷·脊椎动物·下册). 北京: 高等教育出版社. Wang S, Xie Y. 2009. China Species Red List (Vol. 2 Veterbrate. Part 2). Beijing: Higher Education Press. (In Chinese and English)

汪志如, 单继红, 李言阔, 等. 2010. 江西省中华秋沙鸭越冬种群现状调查及胁迫因素分析. 四川动物, 29(4): 597-600. Wang Z R, Shan J H, Li Y K, *et al*. 2010. Winter population status and endangered factors of Scaly-sided Merganser (*Mergus squamatus*) in Jiangxi Province. *Sichuan Journal of Zoology*, 29(4): 597-600. (In Chinese with English Abstract)

王超, 刘冬平, 庆保平. 2014. 野生朱鹮的种群数量和分布现状. 动物学杂志, 49(5): 666-671. Wang C, Liu D P, Qing B P. 2014. The current population and distribution of wild Crested Ibis *Nipponia nippon*. *Chinese Journal of Zoology*, 49(5): 666-671. (In Chinese with English Abstract)

王凤昆, 于文涛, 刘化金. 2007. 中俄兴凯湖湿地水鸟迁徙调查. 野生动物, 28(2): 17-19. Wang F K, Yu W T, Liu H J. 2007. Survey on Migration of Waterfowl in Xingkai Lake between China and Russia. *Chinese Journal of Wildlife*, 28(2): 17-19. (In Chinese with English Abstract)

王家骏. 1984. 世界猛禽. 上海: 上海科学技术出版社. Wang J J. 1984. Raptor of the world. Shanghai: Shanghai Scientific & Technical Publishers. (In Chinese)

王岐山, 马鸣, 高育仁. 2006. 中国动物志 鸟纲 第五卷 鹤形目 鸻形目 鸥形目. 北京: 科学出版社. Wang Q S, Ma M, Gao Y R. 2006. Fauna Sinica. Aves. Vol. 5. Gruiformes, Charadriiformes and Lariformes. Beijing: Science Press. (In Chinese)

王天厚, 钱国桢. 1988. 长江口杭州湾鸻形目鸟类. 上海: 华东师范大学出版社. Wang T H, Qian G Z. 1988. Waders (*Charadiiformes*) community of the Chang-jiang River Estuary and Hang-zhou Bay. Shanghai: East China Normal University Press. (In Chinese)

王香亭. 1990. 宁夏脊椎动物志. 银川: 宁夏人民出版社. Wang X T. 1990. Vertebrate Fauna of Ninxia. Yinchuan: Ningxia People's Publishing House. (In Chinese)

王香亭. 1991. 甘肃脊椎动物志. 兰州: 甘肃科学技术出版社. Wang X T. 1991. Vertebrate Fauna of Gansu. Lanzhou: Gansu

Science & Technology Press. (In Chinese)

王鑫. 2013. 食物对越冬小白额雁分布、能量与氮平衡以及行为的影响. 合肥: 中国科学技术大学博士学位论文. Wang X. 2013. The effect of diet on distribution, energy and nitrogen budget, and behavior of wintering Lesser White-fronted Geese. Hefei: Doctor Thesis of University of Science and Technology of China. (In Chinese)

王育章, 胡锦矗, 吴天勋, 等. 2006. 四川鸟类一新记录-绿胸八色鸫. 四川动物, 26(3): 537. Wang Y Z, Hu J C, Wu T X, *et al.* 2006. A new bird record in Sichuan-*Pitta sordida*. *Sichuan Journal of Zoology*, 26(3): 537. (In Chinese with English Abstract)

王直军, 李国锋, 曹敏, 等. 2001. 西双版纳勐宋轮歇演替区鸟类多样性及食果鸟研究. 动物学研究, 22(3): 205-210. Wang Z J, Li G F, Cao M, *et al.* 2001. Study on bird diversity and frugivorous birds in fallow succession forest regions of Mengsong, Xishuangbanna. *Zoological Research*, 22(3): 205-210. (In Chinese with English Abstract)

王紫江, 赵雪冰, 罗康. 2015. 昆明地区鸟类50年的变化. 四川动物, 34(4): 599-613. Wang Z J, Zhao X B, Luo K. 2015. Avian Changes in Fifty Years (1963-2013) in Kunming, China. *Sichuan Journal of Zoology*, 34(4): 599-613. (In Chinese with English Abstract)

魏顺德, 马鸣, 陈怀玉, 等. 1990. 塔里木盆地黑鹳(*Ciconia nigra*)的分布与繁殖. 八一农学院学报, 13(1): 55-58. Wei S D, Ma M, Chen H Y, *et al.* 1990. On the Distribution and Reproduction of Black Stork in Tarim Basin. *Journal of Xinjiang Agricultural University*, 13(1): 55-58. (In Chinese)

温立嘉, 时坤, 黄建, 等. 2014. 西藏墨脱鸟兽红外相机监测初报. 生物多样性, 22(6): 798-799. Wen L J, Shi K, Huang J, *et al.* 2014. Preliminary analysis of mammal and bird diversity monitored with camera traps in Medog, Tibet. *Biodiversity Science*, 22(6): 798-799. (In Chinese with English Abstract)

温战强, 郑光美. 1998. 黄腹角雉的饲养繁殖. 动物学杂志, 33(3): 22-27. Wen Z Q, Zheng G M. 1998. Artificial raising and breeding of Cabot's Tragopan (*Tragopan Caboti*). *Chinese Journal of Zoology*, 33(3): 22-27. (In Chinese)

温战强, 郑光美, 陈大元, 等. 1997. 黄腹角雉精子超微结构的研究. 动物学报, 43(2): 127-132. Wen Z Q, Zheng G M, Chen D Y, *et al.* 1997. Ultrastructure of Spermatozoon of Cabot's Tragopan (*Tragopan Caboti*). *Current Zoology*, 43(2): 127-132. (In Chinese with English Abstract)

文陇英, 弓加文, 刘迺发. 2008. 四川茂县雉鹑习性的调查. 动物学杂志, 43(2): 73-76. Wen L Y, Gong J W, Liu N F. 2008. Habit of *Tetraophasis obscurus* in Mao County, Sichuan. *Chinese Journal of Zoology*, 43(2): 73-76. (In Chinese with English Abstract)

文贤继, 杨晓君, 韩联宪. 1995. 绿孔雀在中国的分布现状调查. 生物多样性, 3(1): 46-51. Wen X J, Yang X J, Han L X. 1995. Investigations on the current status of the distribution of Green Peafowl in China. *Biodiversity Science*, 3(1): 46-51. (In Chinese with English Abstract)

乌力吉. 1996. 呼伦贝尔草原上的大鸨. 野生动物, (6): 11. Wu L J. 1996. Great Bustard on Hulunbeier Grassland. *Chinese Journal of Wildlife*, (6): 11. (In Chinese)

吴逸群, 马鸣, 徐峰, 等. 2006. 新疆准葛尔盆地猎隼(*Falco cherrug*)繁殖期食性及其对鼠类的防控. 新疆农业大学学报, 29(2): 13-16. Wu Y Q, Ma M, Xu F, *et al.* 2006. Feeding habits analysis of Saker Falcon and the effect of raptor on prevention mouse population. *Journal of Xinjiang Agricultural University*, 29(2): 13-16. (In Chinese)

武明录, 陈立根, 牛浩. 1999. 白肩雕繁殖地新记录. 野生动物, (5): 45-48. Wu M L, Chen L G, Niu H. 1999. New breeding record of Imperial Eagle. *Chinese Journal of Wildlife*, (5): 45-48. (In Chinese)

肖红, 王开锋, 冯宁. 2013. 陕西定边苟池湿地发现遗鸥繁殖群分布. 动物学杂志, 48(5): 776-777. Xiao H, Wang K F, Feng N. 2013. New breeding distribution of Relict Gulls in Gouchi Wetland, Dingbian, Shaanxi Province. *Chinese Journal of Zoology*, 48(5): 776-777. (In Chinese with English Abstract)

邢莲莲. 1996. 乌梁素海鸟类志. 呼和浩特: 内蒙古大学出版社. Xing L L. 1996. The Avifauna of Wuliangsuhai Inner Mongolia. Hohhot: Inner Mongolia University Press. (In Chinese)

邢莲莲. 2013. 达里诺尔野鸟. 北京: 中国大百科全书出版社. Xing L L. 2013. Wild Bird in Darinor. Beijing: Encyclopedia of China Publishing House. (In Chinese)

邢莲莲, 杨贵生, 张永让, 等. 1988. 内蒙古乌梁素海鸟类区系及生态分布的研究. 内蒙古大学学报(自然科学版), 19(3): 524-534. Xing L L, Yang G S, Zhang Y R, *et al.* 1988. Studies of faunai and ecology distribution on birds in Wuliangsuhai region, Nei Mongol. *Acta Scientiarum Naturalium Universitatis Neimongol* (Natural Science), 19(3): 524-534. (In Chinese)

徐华林. 2013. 深圳湾水鸟生物多样性初步研究. 野生动物, 34(5): 291-295. Xu H L. 2013. Research on the waterfowl biodiversity in Shenzhen Bay. *Chinese Journal of Wildlife*, 34(5): 291-295. (In Chinese with English Abstract)

徐剑. 1997. 粤北地区保护动物现状及对策. 韶关大学学报, 18(4): 70-75. Xu J. 1997. Status and management of the protected animal in the north of Guangding. *Journal of Shaoguan University*, 18(4): 70-75. (In Chinese with English Abstract)

徐利, 姜兆文, 马逸清, 等. 1996. 紫貂冬季食性的分析. 兽类学报, 16(4): 272-277. Xu L, Jiang Z W, Ma Y Q, *et al.* 1996. Winter food habits of Sable (*Martes zibellina*) in Daxinganling Mountains, China. *Acta Theriologica Sinica*, 16(4): 272-277. (In Chinese with English Abstract)

徐言朋, 郑家文, 丁平, 等. 2007. 官山白颈长尾雉活动区域海拔高度的季节变化及其影响因素. 生物多样性, 15(4): 337-343. Xu Y P, Zhen J W, Ding P, *et al.* 2007. Seasonal change in ranging of Elliot's pheasant and its determining factors in Guanshan National Nature Reserve, Jiangxi. *Biodiversity Science*, 15(4): 337-343. (In Chinese with English Abstract)

徐雨. 2012. 四川雉鹑(*Tetraophasis szechenyii*)繁殖期夜栖地及夜栖行为的研究. 成都: 四川大学博士学位论文. Xu Y. 2012. Study on the roost habitat and behavior of Sichuan Pheasant (*Tetraophasis szechenyii*). Chengdu: Dissertation of Sichuan University. (In Chinese with English Abstract)

徐振武, 冯宁, 王中强, 等. 2006. 陕北红碱淖湿地遗鸥资源分布与保护管理对策. 西北林学院学报, 21(2): 126-129. Xu Z W, Feng N, Wang Z Q, *et al.* 2006. *Larus relictus* resource distribution at Hongjiannao Lake in the North of Shaanxi and protecting countermeasures. *Journal of Northwest Forestry University*, 21(2): 126-129. (In Chinese with English Abstract)

许维枢. 1995. 中国猛禽——鹰隼类. 北京: 中国林业出版社. Xu W S. 1995. Chinese Raptor—hawk and falcon, Beijing: China Forestry Publishing. (In Chinese)

杨陈, 周立志, 朱文中. 2007. 越冬地东方白鹳繁殖生物学的初步研究. 动物学报, 53(2): 215-226. Yang C, Zhou L Z, Zhu W Z. 2007. A preliminary study on the breeding biology of the oriental white stork *Ciconia boyciana* in its wintering area. *Current Zoology*, 53(2): 215-226. (In Chinese with English Abstract)

杨道德, 吴宏道, 朱德冲, 等. 2001. 广东象头山保护区野生动物资源调查和保护对策. 中南林学院学报, 21(1): 69-73. Yang D D, Wu H D, Zhu D C, *et al.* 2001. Investigation and protective strategies of wildlife resources of Xiangtoushan Natural Reserve in Guangdong Province. *Journal of Central South Forestry University*, 21(1): 69-73. (In Chinese with English Abstract)

杨贵生, 邢莲莲. 1998. 内蒙古脊椎动物名录及分布. 呼和浩特: 内蒙古大学出版社. Yang G S, Xing L L. 1998. A checklist and distribution of vertebrates in Inner Mongolia. Hohhot: Inner Mongolia University Press. (In Chinese)

杨洪燕, 张正旺. 2006. 渤海湾地区红腹滨鹬迁徙动态的初步研究. 动物学杂志, 41(3): 85-89. Yang H Y, Zhang Z W. 2006. Migrating dynamic of Red Knots in Bohai Bay. *Chinese Journal of Zoology*, 41(3): 85-89. (In Chinese with English Abstract)

杨岚. 1983. 中国八色鸫科鸟类的分类研究. 动物学研究, 4(3): 219-226. Yang L. 1983. A taxonomic study on Chinese birds of the *Genus Pitta* (Pittidae). *Zoological Research*, 4(3): 219-226. (In Chinese)

杨岚. 1987. 赤颈鹤在云南分布的现状. 动物学研究, 8(3): 17. Yang L. 1987. An investigation of the distribution of Sarus Crane in Yunnan province, China. *Zoological Research*, 8(3): 17. (In Chinese)

杨岚, 等. 1995. 云南鸟类志 上卷: 非雀形目. 昆明: 云南科技出版社. Yang L. 1995. The Avifauna of Yunnan China, Vol 1: Non-Passeriformes. Kunming: Yunnan Science and Technology Press. (In Chinese)

杨岚, 屈文政, 李强. 2006. 鸟类. 见: 杨宇明, 杜凡. 云南铜壁关自然保护区科学考察研究. 昆明: 云南科技出版社: 237-257. Yang L, Qu W Z, Li Q. 2006. Aves. *In*: Yang Y M and Du F. Intergrated scientific studies of Yunnan Tongbiguan nature reserve. Kunming: Yunnan Science and Technology Press. (In Chinese)

杨岚, 杨晓君. 1997. 云南发现中华秋沙鸭和黑嘴鸥. 动物学研究, 18(4): 388. Yang L, Yang X J. 1997. Chinese Merganser and Saunder's Gull watched in Yunnan, China. *Zoological Research*, 18(4): 388. (In Chinese)

杨岚, 杨晓君. 2004. 云南鸟类志 下卷: 雀形目. 昆明: 云南科技出版社. Yang L, Yang X J. 2004. The Avifauna of Yunnan China (Vol. 2: Passeriformes). Kunming: Yunnan Science and Technology Press. (In Chinese)

杨楠, 徐雨, 冉江洪, 等. 2009. 四川雉鹑繁殖习性初报. 动物学杂志, 44(2): 48-51. Yang N, Xu Y, Ran J H, *et al.* 2009. Notes on the breeding habits of the Buff-throated Partridge. *Chinese Journal of Zoology*, 44(2): 48-51. (In Chinese with English Abstract)

杨维康, 乔建芳, 高行宜, 等. 2005. 波斑鸨的生态生物学研究现状. 干旱区研究, 22(2): 205-210. Yang W K, Qiao J F, Gao X Y, *et al.* 2005. Actuality of the study on ecological biology of *Chlamydotis undulata*. *Arid Zone Research*, 22(2): 205-210. (In Chinese with English Abstract)

杨晓君, 常云艳. 2014. 云南鹤类与研究现状. 动物学研究, 35(S1): 51-60. Yang X J, Chang Y Y. 2014. Cranes and their research in Yunnan Province. *Zoological Research*, 35(S1): 51-60. (In Chinese with English Abstract)

杨晓君, 钱法文, 李凤山, 等. 2005. 中国首次卫星跟踪黑颈鹤研究初报. 动物学研究, 26(6): 657-658. Yang X J, Qian F W, Li F S, et al. 2005. First Satellite Tracking of Black-necked Cranes in China. *Zoological Research*, 26(6): 657-658. (In Chinese with English Abstract)

杨晓君, 文贤继, 杨岚. 1997. 云南东南部和西北部绿孔雀分布的调查. 动物学研究, 18(1): 12. Yang X J, Wen J X, Yang L. 1997. The range of Green Peafowl (*Pavo muticus imperator*) in Southeast and northwest Yunnan province. *China Zoological Research*, 18(1): 12. (In Chinese)

杨兴家, 金玄善. 2000. 图们江下游湿地水鸟及其生态分布. 动物学杂志, 35(5): 26-30. Yang X J, Jin X S. 2000. The water birds and their ecology distribution along the lower reaches of Tumen River. *Chinese Journal of Zoology*, 35(5): 26-30. (In Chinese with English Abstract)

杨元昌, 张庆, 何纪昌. 1989. 西双版纳鸟类区系. 西南林学院学报, 9(1): 70-83. Yang Y C, Zhang Q, He J C. 1989. Avian fauna of Xishuangbanna. *Journal of Southwest Forestry College*, 9(1): 70-83. (In Chinese with English Abstract)

杨月伟, 丁平, 姜仕仁, 等. 1999. 针阔混交林内白颈长尾雉栖息地利用的影响因子研究. 动物学报, 45(3): 279-286. Yang Y W, Ding P, Jiang S R, et al. 1999. Factors affecting habitat used by Elliot's Pheasant (*Symaticus ellioti*) in mixed coniferous and broadleaf forests. *Current Zoology*, 45(3): 279-286. (In Chinese with English Abstract)

杨志杰, 罗维桢, 易国栋, 等. 2003. 中华秋沙鸭的繁殖生态研究. 东北师大学报(自然科学版), 35(2): 123-124. Yang Z J, Luo W Z, Yi G D, et al. 2003. Study on the breeding ecology of Scaly-sided Merganser. *Journal of Northeast Normal University* (Natural Science Edition), 35(2): 123-124. (In Chinese with English Abstract)

姚孝原. 1996. 大鸨的越冬数量及分布. 四川动物, 15(4): 169-169. Yao X Y. 1996. The population size and distribution of the overwintering Great Bustard. *Sichuan Journal of Zoology*, 15(4): 169-169. (In Chinese)

尹秉高, 刘务林. 1993. 西藏珍稀野生动物与保护. 北京: 中国林业出版社. Yin B G, Liu W L. 1993. Rare wildlife and protection in Tibet. Beijing: China Forestry Publishing House. (In Chinese)

尹远新, 葛东宁, 关学敏, 等. 2009. 我国东北地区黑嘴松鸡的种群及栖息地现状. 国土与自然资源研究, (2): 90-91. Yin Y X, Ge D N, Guan X M, et al. 2009. Population and habitat status of Black-billed Capercailiie in the Northeast of China. *Territory & Natural Resources Study*, (2): 90-91. (In Chinese with English Abstract)

尹祚华, 雷富民. 1999. 中国首次发现黑脸琵鹭的繁殖地. 动物学杂志, 34(6): 30-31. Yin Z H, Lei F M. 1999. The new discovery of breeding site of the Black-faced Spoonbill (*Platalea minor*) in China. *Chinese Journal of Zoology*, 34(6): 30-31. (In Chinese)

尹祚华, 雷富民. 2000. 长山列岛发现黄嘴白鹭的繁殖种群. 动物学杂志, 35(5): 39-41. Yin Z H, Lei F M. 2000. A breeding population of the Chinese Egret on the Changshan Islands. *Chinese Journal of Zoology*, 35(5): 39-41. (In Chinese)

余丽江, 陆舟, 舒晓莲, 等. 2015. 广西西南喀斯特地区的鸟类多样性. 广西师范大学学报(自然科学版), 33(2): 103-108. Yu L J, Lu Z, Shu X L, et al. 2015. Bird diversity in the Karst region of southwest Guangxi. *Journal of Guangxi Normal University* (Natural Science), 33(2): 103-108. (In Chinese with English Abstract)

袁国映. 1989. 新疆脊椎动物简志. 乌鲁木齐: 新疆人民出版社. Yuan G Y. 1989. A primary fauna of vertebrates in Xinjiang. Ürümqi: Xinjiang People's Publishing House. (In Chinese)

曾翌硕, 林文隆. 2010. 台湾的猫头鹰. 台中: 台中县野鸟救伤保育学会. Zeng Y S, Lin W L. 2010. The Owls of Taiwan. Taichung: Wild Bird Rescue and Conservation Society of Taichung County. (In Chinese)

张国钢, 郑光美, 张正旺. 2004. 山西五台山地区褐马鸡的再引入. 动物学报, 50(1): 126-132. Zhang G G, Zhen G M, Zhang Z W. 2004. Reintroduction of Brown Eared pheasant *Crossoptilon mantchuricum* in Wutaishan Mountains of Shanxi, China. *Acta Zoologica Sinica*, 50(1): 126-132. (In Chinese with English Abstract)

张锦辉, 符英丽. 2002. 海南野生动物资源与保护对策. 海南师范学院学报(自然科学版), 15(1): 84-88. Zhang J H, Fu Y L. 2002. Wildlife resources in Hainan and their protection. *Journal of Hainan Normal University* (Natural Science), 15(1): 84-88. (In Chinese with English Abstract)

张军平, 郑光美. 1990. 黄腹角雉的种群数量及其结构研究. 动物学研究, 11(4): 291-297. Zhang J P, Zhen G M. 1990. The studies of the population number and structure of Cabot's Tragopan (*Tragopan caboti*). *Zoological Research*, 11(4): 291-297. (In Chinese)

张青霞. 1996. 历山自然保护区乌雕冬季生态观察. 四川动物, 16(4): 170-172. Zhang Q X. 1996. Preliminary study on winter ecology of Greater Spotted Eagle in Lishan Nature Reserve. *Sichuan Journal of Zoology*, 16(4): 170-172. (In Chinese)

张明霞, 曹林, 权锐昌. 2014. 利用红外相机监测西双版纳森林动态样地的野生动物多样性. 生物多样性, 22(6): 830-832. Zhang M X, Cao L, Quan R C. 2014. Camera trap survey of animals in Xishuangbanna Forest Dynamics Plot, Yunnan.

Biodiversity Science, 22(6): 830-832. (In Chinese with English Abstract)

张琼. 2012. 甘肃白水江灰冠鸦雀栖息地研究. 甘肃科技, 28(12): 139-140. Zhang Q. 2012. Study on habitat of Rusty-throated Parrotbill in Baishui River, Gansu. *Gansu Science and Technology*, 28(12): 139-140. (In Chinese)

张荣祖. 1999. 中国动物地理. 北京: 科学出版社. Zhang R Z. 1999. Chinese Animal Geography. Beijing: Science Press. (In Chinese)

张荣祖, 赵肯堂. 1978. 关于《中国动物地理区划》的修改. 动物学报, 24(2): 196-202. Zhang R Z, Zhao K T. 1978. On the zoogeographical regions of China. *Current Zoology*, 24(2): 196-202. (In Chinese with English Abstract)

张涛. 1995. 甘肃白水江自然保护区绿尾虹雉分布与生态的初步观察. 动物学杂志, 30(4): 25-28. Zhang T. 1995. Distribution and ecology of Chinese Monals in Baishuijiang Nature Reserve Gansu. *Chinese Journal of Zoology*, 30(4): 25-28. (In Chinese)

张荫荪, 何芬奇. 1994. 对遗鸥分类问题的进一步论证. 动物分类学报, 19(3): 378-382. Zhang Y S, He F Q. 1994. On the taxonomic status of the Relict Gull (*Larus relictus*). *Acta Zootaxonomic Sinica*, 19(3): 378-382. (In Chinese with English Abstract)

张荫荪, 何芬奇, 陈荣伯. 1993. 遗鸥繁殖生境选择及其繁殖地湿地鸟类群落研究. 动物学研究, 14(2): 128-135. Zhang Y S, He F Q, Chen R B. 1993. Breeding habitat selection of the Relict Gull and the wetland bird community around its breeding sites. *Zoological Research*, 14(2): 128-135. (In Chinese)

张永文, 张晓峰. 2012. 陕西秦岭地区黑喉歌鸲繁殖行为监测简报. 陕西林业科技, 40(6): 54-55. Zhang Y W, Zhang X F. 2012. Discovery of the propagation site of rare bird *Luscinia obscura* Distributed in Qinling Mountains. *Shaanxi Forest Science and Technology*, 40(6): 54-55. (In Chinese with English Abstract)

张正旺, 丁长青, 丁平, 等. 2003. 中国鸡形目鸟类的现状与保护对策. 生物多样性, 11(5): 414-421. Zhang Z W, Ding C Q, Ding P, *et al*. 2003. The current status and a conservation strategy for species of Galliformes in China. *Biodiversity Science*, 11(5): 414-421. (In Chinese with English Abstract)

张正旺, 尹荣伦, 郑光美. 1989. 笼养黄腹角雉繁殖期取食活动性的研究. 动物学研究, 10(4): 333-339. Zhang Z W, Yin R L, Zhen G M. 1989. Feeding activity of the Cabot's Tragopan during the breeding season in captivity. *Zoological Research*, 10(4): 333-339. (In Chinese)

赵序茅, 马鸣, 丁鹏, 等. 2013b. 金雕巢期行为谱及时间分配. 干旱区地理, 36(6): 1084-1089. Zhao X M, Ma M, Ding P, *et al*. 2013. Ethogram and time budget of golden eagle during breeding season in Tianshan Mts. *Arid Land Geography*, 36(6): 1084-1089. (In Chinese with English Abstract)

赵序茅, 马鸣, 邢睿. 2012. 不是所有的金雕都有机会翱翔蓝天——小金雕成长记. 大自然, 33(4): 48-50. Zhao XM, Ma M, Xing R. 2012. Not all Golden Eagles have the opportunity to fly in the blue sky. *China Nature*, 2012. 33(4): 48-50. (In Chinese)

赵序茅, 马鸣, 张同, 等. 2013a. 白头硬尾鸭行为时间分配及日活动节律. 生态学杂志, 32(9): 2439-2443. Zhao X M, Ma M, Zhang T, *et al*. 2013a. Behavioral time budget and diurnal rhythm of White-headed Duck in Northwest China. *Chinese Journal of Ecology*, 32(9): 2439-2443. (In Chinese with English Abstract)

赵正阶. 1985. 长白山鸟类志. 长春: 吉林科学技术出版社. Zhao Z J. 1985. The Avifauna of Changbai Mountain. Changchun: Jilin Science & Technology Publishing House. (In Chinese)

赵正阶. 1993. 中华秋沙鸭的迁徙和越冬. 野生动物, (2): 22-23. Zhao Z J. 1993. Migration and overwintering of Scaly-sided Merganser. *Chinese Journal of Wildlife*, (2): 22-23. (In Chinese)

赵正阶. 1995. 中国鸟类志·上卷·非雀形目. 长春: 吉林科学技术出版社. Zhao Z J. 1995. The Avifauna of China. Vol.1 Non-Passeriformes. Changchun: Jilin Science & Technology Publishing House. (In Chinese)

赵正阶. 2001. 中国鸟类志·下卷·雀形目. 长春: 吉林科学技术出版社. Zhao Z J. 2001. The Avifauna of China. Vol. 2 Passeriformes. Changchun: Jilin Science & Technology Publishing House. (In Chinese)

赵正阶, 韩晓冬, 吴景才, 等. 1994. 中华秋沙鸭的繁群数量、分布和保护策略. 见: 中国鸟类学会水鸟组, 中国水鸟研究. 上海: 华东师范大学出版社. Zhao Z J, Han X D, Wu J C, *et al*. 1994. The population, distribution and protection strategies of Scaly-sided Merganser. *In*: Waterbird team of China Ornithological Society. the Chinese Waterbird Research. Shanghai: East China Normal University Press. (In Chinese)

赵正阶, 吴景才, 张淑华, 等. 1993. 中华秋沙鸭在长白山地区的分布和种群数量. 动物学研究, 14(3): 221-225. Zhao Z J, Wu J C, Zhang S H, *et al*. 1993. Breeding population density of Chinese Merganser in Changbai Mountain. *Zoological Research*, 14(3): 221-225. (In Chinese)

赵忠, 何毅, 杨鹏翼, 等. 2011. 肃南肃北草原野生动物资源调查研究. 草业学报, 20(2): 67-75. Zhao Z, He Y, Yang P Y, *et al*. 2011. Investigation on wild animal resource in Sunan and Subei counties. *Acta Prataculturae Sinica*, 20(2): 67-75. (In Chinese with English Abstract)

赵作审, 赵中琴. 1996. 黑龙江省鸡类资源. 动物学报, 42(增刊): 45-48. Zhao Z S, Zhao Z Q. 1996. The resource of Heilongjiang pheasants. *Acta Zoologica Sinica*, 42(supplement): 45-48. (In Chinese with English Abstract)

郑宝赉, 杨岚, 杨德华, 等. 1985. 中国动物志 鸟纲 第八卷 雀形目: 阔嘴鸟科-和平鸟科. 北京: 科学出版社. Zhen B L, Yang L, Yang D H, *et al*. 1985. Fauna Sinica Aves. Vol. 8. Passeriformes: Eurylaimidae-Irenidae. Beijing: Science Press. (In Chinese)

郑光美. 2011. 中国鸟类分类与分布名录. 第2版. 北京: 科学出版社. Zhen G M. 2011. A Checklist on the Classification and Distribution of the Birds of China. 2nd ed. Beijing: Science Press. (In Chinese and English)

郑光美. 2012. 鸟类学. 北京: 北京师范大学出版社. Zheng G M. 2012. Ornithology. Beijing: Beijing Normal University Press. (In Chinese)

郑光美. 2015. 中国雉类. 北京: 高等教育出版社. Zheng G M. 2015. Pheasants in China. Beijing: Higher Education Press. (In Chinese)

郑光美. 2017. 中国鸟类分类与分布名录. 第3版. 北京: 科学出版社. Zheng G M. 2017. A Checklist on the Classification and Distribution of the Birds of China. 3rd ed. Beijing: Science Press. (In Chinese and English)

郑光美, 王岐山. 1998. 中国濒危动物红皮书鸟类. 北京: 科学出版社. Zheng G M, Wang Q S. 1998. China Red Data Book of Endangered Animals (Aves). Beijing: Science Press. (In Chinese and English)

郑光美, 赵欣如, 宋杰, 等. 1985. 黄腹角雉的繁殖生态研究. 生态学报, 5(4): 91-97, 104. Zheng G M, Zhao X R, Song J, *et al*. 1985. The breeding ecology of Cabot's Tragopan. *Acta Ecological Sinica*, 5(4): 91-97, 104. (In Chinese with English Abstract)

郑光美, 赵欣如, 宋杰, 等. 1986. 黄腹角雉的食性研究. 生态学报, 6(3): 283-288. Zheng G M, Zhao X R, Song J, *et al*. 1986. Feeding ecology of the Cabot's Tragopan (*Tragopan Caboti*). *Acta Ecologica Sinica*, 6(3): 283-288. (In Chinese with English Abstract)

郑作新. 1976. 中国鸟类分布名录. 第2版. 北京: 科学出版社. Zheng Z X. 1976. A Checklist of the bird distribution in China. Beijing: Science Press. (In Chinese)

郑作新, 李桂垣, 张清茂. 1983. 暗色鸦雀的一新亚种——二郎山亚种. 动物分类学报, (3): 106-108. Zheng Z X, Li G Y, Zhang Q M. 1983. A new subspecies of *Paradoxornis zappeyi*—*P. z. erlanshanicus*. *Sinica Acta Zootaxonomic*, (3): 106-108. (In Chinese with English Abstract)

郑作新, 谭耀匡, 卢汰春, 等. 1978. 中国动物志 鸟纲 第四卷 鸡形目. 北京: 科学出版社. Zheng Z X, Tan Y K, Lu T C, *et al*. 1978. Fauna Sinica. Aves. Vol. 4. Galliformes. Beijing: Science Press. (In Chinese)

郑作新, 冼耀华, 关贯勋, 等. 1991. 中国动物志 鸟纲 第六卷 鸽形目 鹦形目 鹃形目 鸮形目. 北京: 科学出版社. Zheng Z X, Xian Y H, Guan G X, *et al*. 1991. Fauna Sinica. Aves. Vol. 6. Columbiformes, Psittaciformes, Cuculiformes and Strigiformes. Beijing: Science Press. (In Chinese)

郑作新, 张荣祖. 1956. 中国动物地理区域. 地理学报, 22(1): 93-109. Zheng Z X, Zhang R Z. 1956. On tentative scheme for dividing zoogeographical regions of China. *Acta Geographica Sinic*, 22(1): 93-109. (In Chinese with English Abstract)

中国动物学会鸟类学分会. 2005. 中国观鸟年报2005. China Ornithological Society. 2005. China Bird Report 2005. (In Chinese)

中国沿海水鸟同步调查项目组. 2009. 中国沿海水鸟同步调查报告(2005年9月-2007年12月). 香港: 香港观鸟会有限公司. China Coastal Waterbird Census Group. 2009. China coastal waterbird census Report (September 2005-December 2007). Hong Kong: Hong Kong Bird Watching Association Limited. (In Chinese)

中国沿海水鸟同步调查项目组. 2011. 中国沿海水鸟同步调查报告(2008年1月-2009年12月). 香港: 香港观鸟会有限公司. China Coastal Waterbird Investigation Project Team. 2011. China coastal waterbird census Report (January 2008-December 2009). Hong Kong: Hong Kong Bird Watching Association Limited. (In Chinese)

钟福生, 董婉未, 李威娜, 等. 2012. 梅江流域鸟类群落结构及其多样性. 生态环境学报, 21(5): 825-833. Zhong F S, Dong W W, Li W N, *et al*. 2012. Community structure and diversity of birds in Meijiang River Valley. *Ecology and Environment Sciences*, 21(5): 825-833. (In Chinese with English Abstract)

钟福生, 颜亨梅, 李丽平, 等. 2007. 东洞庭湖湿地鸟类群落结构及其多样性. 生态学杂志, 26(12): 1959-1968. Zhong F S, Yan H M, Li L P, *et al*. 2007. Community structure and diversity of birds on east Dongting Lake wetland of Hunan Province.

Chinese Journal of Ecology, 26(12): 1959-1968. (In Chinese with English Abstract)

周放. 2011. 广西陆生脊椎动物分布名录. 北京: 中国林业出版社. Zhou F. 2011. A Checklist of the Distribution of Terrestrial Vertebrates in Guangxi. Beijing: China Forestry Publishing. (In Chinese)

周放, 余丽江, 陆舟, 等. 2005. 海南鸦巢址选择的初步调查. 动物学杂志, 40(1): 54-58. Zhou F, Yu L J, Lu Z, et al. 2005. A primary investigation of nest selection of White-eared Night Heron. *Chinese Journal of Zoology*, 40(1): 54-58. (In Chinese with English Abstract)

周放, 周解. 2004. 十万大山地区野生动物研究与保护. 北京: 中国林业出版社. Zhou F, Zhou J. 2004. Study and Conservation of Wildlife in Shiwandashan Region. Beijing: China Forestry Publishing House. (In Chinese)

周福璋, 丁文宁. 1982. 白鹤越冬习性. 动物学杂志, 26(4): 19-21. Zhou F Z, Ding W N. 1982. The overwinter behavior of Siberian Crane. *Chinese Journal of Zoology*, 26(4): 19-21. (In Chinese with English Abstract)

周福璋, 丁文宁, 王子玉. 1981. 发现大群白鹤在中国越冬. 动物学报, 27(2): 179. Zhou F Z, Ding W N, Wang Z Y. 1981. A large flock of White Cranes (*Grus leucogeranus*) wintering in China. *Current Zoology*, 27(2): 179. (In Chinese)

周纪刚, 郑洲翔, 牛晓楠, 等. 2013. 惠州鸟类资源调查. 安徽农业科学, 41(14): 6285-6287. Zhou J G, Zheng Z X, Niu X N, et al. 2013. Survey of bird resources in Huizhou. *Journal of Anhui Agricultural Sciences*, 41(14): 6285-6287. (In Chinese with English Abstract)

周林. 2004. 大绯胸鹦鹉及其保护与利用. 野生动物, (1): 30-31. Zhou L. 2004. The protection and utilization of Great Cockatoo. *Chinese Journal of Wildlife*, (1): 30-31. (In Chinese with English Abstract)

周晓, 陈东东, Kress S W, 等. 2017. 海鸟种群的人工招引与恢复技术及其应用. 生物多样性, 25(4): 364-371. Zhou X, Chen D D, Kress S W, et al. 2017. A review of the use of active seabird restoration techniques. *Biodiversity Science*, 25(4): 364-371. (In Chinese with English Abstract)

周宇垣. 1955. 广东大绿鸠的一个新亚种. 中山大学学报(自然科学版), 1(3): 128-130. Zhou Y Y. 1955. On a new Green Imperial Pigeon from the district of Lofaoshan, Kwangtung, China. *Journal of sun yat-sen university* (Natural Science), 1(3): 128-130. (In Chinese)

朱磊, 孙悦华, 胡锦矗. 2012. 中国鸮形目鸟类分类现状. 四川动物, 31(1): 170-175. Zhu L, Sun Y H, Hu J C. 2012. The taxonomic status of Chinese Owl (Strigifromes). *Sichuan Journal of Zoology*, 31(1): 170-175. (In Chinese with English Abstract)

邹发生, 叶冠锋. 2016. 广东陆生脊椎动物分布名录. 广州: 广东科技出版社. Zou F S, Ye G F. 2016. A Checklist of the Distribution of Terrestrial Vertebrates in Guangzhou. Guangzhou: Guangdong Science and Technology Press. (In Chinese)

Bai Q, Chen J, Chen Z, et al. 2015. Identification of coastal wetlands of international importance for waterbirds: A review of China Coastal Waterbird Surveys 2005-2013. *Avian Research*, 6: 12.

Balachandran S, Sakthivel R. 1994. Site-fidelity to the unusual nesting site of Brahminy Kite *Haliastur indus* (Boddaert). *Journal of Bombay Natural History Society*, 91(1): 139.

Barter M, Cao L, Chen L, et al. 2005. Results of a survey for waterbirds in the lower Yangtze floodplain, China, in January-February 2004. *Forktail*, 21: 1.

Barter M, Zhuang X L, Wang X, et al. 2014. Abundance and distribution of wintering Scaly-sided Mergansers *Mergus squamatus* in China: Where are the missing birds? *Bird Conservation International*, 24(4): 406-415.

Batbayar N, Takekawa J Y, Newman S H, et al. 2011. Migration strategies of Swan Geese *Anser cygnoides* from northeast Mongolia. *Wildfowl*, 61: 90-109.

Beton D, Snape R, Saydam B. 2013. Status and ecology of the Bonelli's Eagle, *Aquila fasciatus*, in the Pentadaktylos mountain range, Cyprus. *Zoology in the Middle East*, 59(2): 123-130.

Bi Z L, Gu Y, Jia C X, et al. 2003. Nests, eggs, and nestling behavior of the Snowy-cheeked Laughingthrush (*Garrulax sukatschewi*) at Lianhuashan Natural Reserve, Gansu, China. *Wilson Bulletin*, 115(4): 474-477.

BirdLife International. 2013. http: //www.birdlife.org.[2019-7-9]

Bishop M A, Tsamchu D, Li F. 2012. Number and distribution of Black-necked Cranes wintering in Zhigatse Prefecture, Tibet. *Chinese Birds*, 3: 191-198.

Bradter U, Gombobaatar S, Uuganbayar C, et al. 2005. Reproductive performance and nest-site selection of White-naped Cranes *Grus vipio* in the Ulz river valley, north-eastern Mongolia. *Bird Conservation International*, 15(4): 313-326.

Bradter U, Gombobaatar S, Uuganbayar C, et al. 2007. Time budgets and habitat use of White-naped Cranes *Grus vipio* in the Ulz river valley, north-eastern Mongolia during the breeding season. *Bird Conservation International*, 17: 259-271.

Brazil M. 2009. Birds of East Asia. London: A and C Black.

Brickle N W, Duckworth J W, Tordoff A W, et al. 2008. The status and conservation of Galliformes in Cambodia, Laos and Vietnam. *Biodiversity and Conservation*, 17(6): 1393-1427.

Brickle N W. 2002. Habitat use, predicted distribution and conservation of Green peafowl (*Pavo muticus*) in Dak Lak Province, Vietnam. *Biological Conservation*, 105(2): 189-197.

Byers C, Olsson U, Curson J. 1995. Buntings and Sparrows: A Guide to the Buntings and North American Sparrows. Robertsbridge: Pica Press.

Cao L, Barter M, Lei G. 2008a. New Anatidae population estimates for eastern China: Implications for current flyway estimates. *Biological Conservation*, 141(9): 2301-2309.

Cao L, Pang Y L, Liu N F. 2007. Waterbirds of the Xi Sha Archipelago, South China Sea. *Waterbirds*, 30(2): 296-300.

Cao L, Wang X, Wang Q S, et al. 2008b. Wintering Anatidae in China: A preliminary analysis. *Casarca*, 11: 161-180.

Cao L, Zhang Y, Barter M, et al. 2010. Anatidae in eastern China during the non-breeding season: Geographical distributions and protection status. *Biological Conservation*, 143(3): 650-659.

Carrascal L M, Seoane J. 2009. Factors affecting large-scale distribution of the Bonelli's Eagle *Aquila fasciata* in Spain. *Ecological Research*, 24(3): 565-573.

Carrete M, Sánchez-Zapata J A, Martínez J E, et al. 2002. Factors influencing the decline of a Bonelli's Eagle *Hieraaetus fasciatus* population in SE Spain: demography, habitat or competition? *Biodiversity and Conservation*, 11(6): 975-985.

Chan S. 2004. Yellow-breasted Bunting *Emberiza aureola*. Birding ASIA, 16-17.

Chan S, Fang W H, Lee K S, et al. 2010. International single species action plan for the conservation of the Black-faced Spoonbill (*Platalea minor*). Tokyo and Bonn, BirdLife International Asia Division and CMS Secretariat.

Chang J, Wang B, Zhang Y Y, et al. 2008. Molecular evidence for species status of the endangered Hainan Peacock Pheasant. *Zoological Science*, 25(1): 30-35.

Chen S H, Chang S H, Liu Y, et al. 2009. Low population and severe threats: Status of the critically endangered Chinese Crested Tern *Sterna bernsteini*. *Oryx*, 43(2): 209-212.

Chen S H, Fan Z Y, Chen C S, et al. 2010. A new breeding site of the Critically Endangered Chinese Crested Tern *Sterna bernsteini* in the Wuzhishan Archipelago, eastern China. *Forktail*, 26(26): 132-134.

Chen S H, Fan Z Y, Chen C S, et al. 2011. The Breeding Biology of Chinese Crested Terns in Mixed Species Colonies in Eastern China. *Bird Conservation International*, 21(3): 266-273.

Chen S H, Fan Z Y, Roby D D, et al. 2015. Human harvest climate change and their synergistic effects drove the Chinese Crested Tern to the brink of extinction. *Global Ecology and Conservation*, 4: 137-145.

Chen S H, Huang Q, Fan Z Y, et al. 2012. The update of Zhejiang bird checklist. *Chinese Birds*, 3(2): 118-136.

Cheng T H. 1987. A synopsis of the avifauna of China. Beijing: Science Press.

Choudhury A. 2003. Birds of Eaglenest Wildlife Sanctuary and Sessa Orchid Sanctuary. *Forktail*, 19: 1-13.

Choudhury A. 2009a. Birds of south Asia-the Ripley guide: book review. *Newsletter and Journal of the Rhino Foundation for Nature in North-East India*, 8: 16-24.

Choudhury A. 2009b. Mrs Hume's Pheasant in northeastern India. *Tigerpaper*, 36(2): 4-10.

Christie D A, de Juana E. 2016. Handbook of the Birds of the World Alive. Barcelona: Lynx Edicions. (Retrieved from http://www.hbw.com/node/55025 on 15 April 2016).

Collar N J, Andreev A V, Chan S, et al. 2001. Threatened birds of Asia, the BirdLife International red data books. Part A. Cambridge: BridLife International.

Collar N J, Butchart S H M. 2013. Conservation breeding and avian diversity: Chances and challenges. *International Zoo Yearbook*, 48(1): 7-28.

Collar N J, Robson C. 2007. Family Timaliidae (Babblers). *In*: del Hoyo J, Elliott A, Christie D A. Handbook of birds of the world, vol. 12: Picathartes to Tits and Chickadees. Barcelona: Lynx Edicions.

Combreau O. 2007. Arabic falconry and the illegal Houbara trade in Arabia. Falco: 16-17.

Combreau O, Launay F, Lawerence M. 2001. An assessment of annual mortality rates in adult sized migrant Houbara Bustards *Chlamydotis macqueenii*. *Animal Conservation*, 4(2): 133-141.

Combreau O, Qiao J, Lawerence M, et al. 2002. Breeding success in a Houbara Bustard *Chlamydotis macqueenii* population on the eastern fringe of the Jungar Basin, People's Republic of China. *Ibis*, 144(2): E45-E56.

Cramp S, Simmons K E L, Perrins C M. 1994. The Birds of the Western Palearctic. Oxford: Oxford University Press.

Crivelli A J. 1996. Action plan for the Dalmatian Pelican (*Pelecanus crispus*). *In*: Heredia B, Rose L, Painter M. Globally threatened birds in Europe: Action plans. Strasbourg: Council of Europe and BirdLife International: 53-66

Crivelli A J, Catsadorakis G, Hatzilacou D, *et al.* 1997. *Pelicanus crispus* Dalmatian Pelican. *Birds of the Western Palearctic Update*, 1(3): 149-153.

Crivelli A J, Catsadorakis G, Hatzilacou D, *et al.* 2000. Status and population development of Great White Pelican *Pelecanus onocrotalus* and Dalmatian Pelican *P. crispus* breeding in the Palearctic. *In*: Yesou P, Sultana J. Monitoring and conservation of birds, mammals and sea turtles in the Mediterranean and Black Seas: Proceedings of the 5th Medmaravis Symposium, Gozo, Malta, 29 September-3 October 1998, pp. 38-46. Valetta: Environment Protection Department.

Crivelli A J, Hatzilacou D, Catsadorakis G. 1998. The breeding biology of the Dalmatian Pelican *Pelecanus crispus*. *Ibis*, 140(3): 472-481.

Crivelli A J, Krivenko V G, Vinogradov V G. 1994. Pelicans in the former USSR. Slimbridge: International Waterfowl and Wetlands Research Bureau.

Crivelli A J, Marsili L, Focardi S, *et al.* 1999. Organochlorine compounds in pelicans (*Pelecanus crispus* and *Pelecanus onocrotalus*) nesting at Lake Mikri Prespa, north-western Greece. *Bulletin of Environmental Contamination and Toxicology*, 62(4): 383-389.

Cuthbert R, Green R E, Ranade S, *et al.* 2006. Rapid population declines of Egyptian Vulture (*Neophron percnopterus*) and Red-headed Vulture (*Sarcogyps calvus*) in India. *Animal Conservation*, 9(3): 349-354.

Dai B, Dowell S, Garson P J, *et al.* 2009. Habitat utilization by the threatened Sichuan Partridge *Arborophila rufipectus*: Consequences for managing newly protected areas in southern China. *Bird Conservation International*, 19(2): 187-198.

Das A, Green R E, Taggart M. 2011. Are conservation actions reducing the threat to India's vulture populations? *Current Science*, 101(11): 1480-1481.

Davies E. 2011. Rare robin breeding sites found. BBC Nature. Available at: file:///J:/SPI/Science/6b%20Red%20List%20new%20info/Asia/Luscinia%20obscura%20BBC%20Dec11.htm.

del Hoyo J, Collar N J, Christie D A, *et al.* 2014. HBW and BirdLife international illustrated checklist of the birds of the world. Barcelona: Lynx Edicions and BirdLife International.

del Hoyo J, Elliott A, Christie D. 2009. Handbook of the Birds of the World, volume 14: Bush-shrikes to Old Word Sparrows. Barcelona: Lynx Edicions.

del Hoyo J, Elliot A, Sargatal J. 1992. Handbook of the Birds of the World, volume 1: Ostrich to Ducks. Barcelona: Lynx Edicions.

del Hoyo J, Elliott A, Sargatal J. 1994. Handbook of the Birds of the World, volume 2: New World Vultures to Guineafowl. Barcelona: Lynx Edicions.

del Hoyo J, Elliott A, Sargatal J. 1996. Handbook of the Birds of the World, volume 3: Hoatzin to Auks. Barcelona: Lynx Edicions.

del Hoyo J, Elliott A, Sargatal J. 1997. Handbook of the Birds of the World, volume 4: Sandgrouse to Cuckoos. Barcelona: Lynx Edicions.

del Hoyo J, Elliott A, Sargatal J. 2001. Handbook of the Birds of the World, volume 6: Mousebirds to Hornbills. Barcelona: Lynx Edicions.

Deng W H, Zheng G M, Zhang Z W, *et al.* 2005. Providing artificial nest platforms for Cabot's Tragopan *Tragopan caboti* (Aves: Galliformes): A useful conservation tool? *Oryx*, 39(2): 158-163.

Dennis T E, Baxter C I. 2006. The Status of the White-bellied Sea-eagle and Osprey on Kangaroo Island in 2005. *South Australian Ornithologist*, 35(1-2): 47-51.

Dennis T E, McIntosh R R, Shaughnessy P D. 2011. Effects of human disturbance on productivity of White-bellied Sea-Eagles (*Haliaeetus leucogaster*). *Emu*, 111(2): 179-185.

Ding C Q. 2010. Crested Ibis. *Chinese Birds*, (1): 156-162.

Ding P, Ma M, Bayaheng K, *et al.* 2013. Golden Eagle *Aquila chrysaetos* in Xinjiang: Nest-site selection in different reproductive areas. *Acta Ecologica Sinica*, 33(3): 11-19.

Dodsworth P T L. 1912. Extension of the habitat of the Brahminy Kite *Haliastur indus*. *Journal of Bombay Natural History Society*, 21: 665-666.

Dong L, Zhang J, Sun Y, *et al.* 2011. Phylogeographic patterns and conservation units of a vulnerable species, Cabot's tragopan (*Tragopan caboti*), endemic to southeast China. *Conservation Genetics*, 11(6): 2231-2242.

Evans T, Clements T. 2004. Current status and future monitoring of Green Peafowl in southern Mondulkiri. *Cambodia Bird News*, 12: 18-20.

Fang J Y, Wang Z H, Zhao S Q, *et al.* 2006. Biodiversity changes in the lakes of the central Yangtze. *Frontiers in Ecology and the Environment*, 4: 369-377.

Fang Y, Tang S H, Gu Y, *et al.* 2009. Conservation of Tengmalm's Owl and Sichuan Wood Owl in Lianhuashan Mountain, Gansu, China. *Ardea*, 97: 649.

Farm K, Garden B. 2008. Survivorship and dispersal ability of a rehabilitated Brown Fish Owl (*Ketupa zeylonensis*) released more than a decade after admission to a wildlife rescue centre in Hong Kong SAR China. Hong Kong, Kadoorie Farm and Botanic Garden Publication Series No. 3.

Fellows J R, Zhou F, Lee K W. 2001. Status update on White-eared Night Heron Gorsachius magnificus in South China. *Bird Conservation International*, 11(2): 101-111.

Ferguson-Lees J, Christie D A. 2001. Raptors of the World. London: Christopher Helm.

Fijn R C, Hiemstra D, Phillips R A, *et al.* 2013. Arctic Terns *Sterna paradisaea* from the Netherlands migrate record distances across three oceans to Wilkes Land, East Antarctica. *Ardea*, 101 (1): 3-12.

Fox A D, Hearn R D, Cao L, *et al.* 2013. Preliminary observations of diurnal feeding patterns of Swan Geese *Anser cygnoides* using two different habitats at Shengjin Lake, Anhui Province, China. *Wildfowl*, 58: 20-30.

Franco A M A, Marques J T, Sutherland W J. 2005. Is nest-site availability limiting Lesser Kestrel populations? A multiple scale approach. *Ibis*, 147(4): 657-666.

Fujimaki Y. 1987. Joint survey report of Japan and USSR on Steller's Sea Eagle. *In*: The Third Japan-USSR Bird Protection Symposium. Tokyo: Wild Bird Society of Japan.

Fuller R A, Carroll J P, McGowan P J K. 2000. Partridges, Quails, Francolins, Snowcocks, Guineafowl, and Turkeys. Status survey and conservation action plan 2000-2004. Gland, IUCN and World Pheasant Association.

Galligan T H, Amano T, Prakash V M, *et al.* 2014. Have population declines in Egyptian Vulture and Red-headed Vulture in India slowed since the 2006 ban on veterinary diclofenac? *Bird Conservation International*, 24(3): 272-281.

Gao Y R. 1999. Conservation status of endemic Galliformes on Hainan Island, China. *Bird Conservation International*, 9: 411-416.

Gee N G, Moffett L I, Wilder G D. 1926. A Tentative List of Chinese Birds. The Peking Society of Natural History. viii , 144

Gilbert M, Gombobataar S. 2009. The Status and Distribution of Pallas's Fish Eagle in Mongolia: a report on field surveys June-August 2009. Boise: The Peregrine Fund.

Gilbert M, Tingay R, Losolmaa J, *et al.* 2014. Distribution and status of the Pallas's Fish Eagle *Haliaeetus leucoryphus* in Mongolia: A cause for conservation concern? *Bird Conservation International*, 24(3): 379-388.

Gill F, Donsker D, Rasmussen P. 2020. IOC World Bird List (v10.1). doi: 10.14344/IOC.ML.10.1.

Gokula V. 2011. An ethogram of Spot-billed Pelican (*Pelecanus philippensis*). *Chinese Birds*, 2(4): 183-192.

Goriup P. 1997. The world status of the Houbara Bustard Chlamydotis undulata. *Bird Conservation International*, 7(4): 373-397.

Grimmett R. 1991. Little known oriental bird: Biddulph's Ground Jay. *Oriental Bird Club Bulletin*, 13: 26-29.

Han L, Lu Y, Han H. 2009. The status and distribution of Green Peafowl *Pavo muticus* in Yunnan Province, China. *International Journal of Galliformes Conservation*, 1: 29-31.

Handbook of the Birds of the World and BirdLife International. 2019. Handbook of the Birds of the World and BirdLife International digital checklist of the birds of the world. Version 4. Available at: http://datazone.birdlife.org/userfiles/file/Species/Taxonomy/HBW-BirdLife_Checklist_v4_Dec19.zip.

Harris J. 2012. Cranes, Agriculture, and Climate Change Proceedings of a Workshop Organized by the International Crane Foundation and Muraviovka Park for Sustainable Land Use. Wisconsin: International Crane Foundation Baraboo, Wisconsin.

Harris J, Mirande C. 2013. A global overview of Cranes: Status, threats and conservation priorities. *Chinese Birds*, 4(3): 189-209.

Hasegawa H, De Gange A R. 1982. The Short-tailed Albatross, *Diomedea albtrus*, its status, distribution and natural history. *American Birds*, 36(5): 806-814.

Hawkes L A, Balachandran S, Batbayar N, *et al.* 2013. The paradox of extreme high-altitude migration in Bar-headed geese *Anser indicus. Proceedings of the Royal Society*, 280(1750): 20122114.

Higuchi H, Nagendran M, Darman Y, *et al.* 2000. Migration and habitat use of oriental White Storks from satellite tracking studies. *Global Environment Research*, 156(4): 169-182.

Higuchi H, Nagendran M, Pierre J P. 2006. Satellite-tracking the migration of cranes and storks. *Acta Zoologica Sinica*, 52: 206-210.

Higuchi H, Pierre J P, Krever V, *et al.* 2004. Using a remote technology in conservation: Satellite tracking White-naped Cranes in Russia and Asia. *Conservation Biology*, 18(1): 136-147.

Higuchi H, Shibaev Y, Minton J, *et al.* 1998. Satellite tracking of the migration of the Red-Crowned Crane *Grus japonensis*. *Ecological Research*, 13(3): 273-282.

Holt D W, Berkley R, Deppe C, *et al.* 2016. Blakiston's Eagle-owl (*Bubo blakistoni*). *In*: del Hoyo J, Elliott A, Sargatal J, *et al.* Handbook of the Birds of the World Alive. Barcelona: Lynx Edicions.

Holt P. 2008. Eight new ornithological records from Xinjiang, China. *Arid Land Geography*, 31(2): 243-248.

Hong Kong Bird Watching Society. 2007. Proceedings: keeping Asia's spoonbills airborne: International symposium on research and conservation of the Black-faced Spoonbill, Hong Kong, 16-18 January 2006. Hong Kong: Hong Kong Bird Watching Society.

Hong Kong Bird Watching Society. 2014. 2014 Black-faced Spoonbill: Results of international census. Hong Kong: Hong Kong Bird Watching Society.

Hughes B, Robinson J A, Green A J, *et al.* 2006. International single species action plan for the conservation of the White-headed Duck Oxyura leucocephala. Technical Series, CMS (No. 13) and AEWA (No. 8).

Iamsiri A, Gale G. 2004. Hume's pheasant in Thailand. Tragopan, 20/21: 25.

IBC. 2014. Sarus Crane (*Grus antigone*) the internet bird collection. Available at: http://ibc. lynxeds. com/species/sarus-crane-grus-antigone.

Inskipp C, Baral H S. 2011. Potential impacts of agriculture on Nepal birds. *Our Nature*, 8(1): 270-312.

Inskipp T, Lindsey N, Duckworth W. 1996. An Annotated Checklist of the Birds of the Oriental Region (1st ed). Oriental Bird Club.

IUCN. 2012. IUCN Red List of Threatened Species (ver. 2012. 1). Available at: http://www. iucnred list.org.

IUCN. 2013. IUCN Red List of Threatened Species (ver. 2013. 2). Available at: http://www. iucnredlist.org.

IUCN. 2014. IUCN Red List of Threatened Species (ver. 2014. 2). Available at: http://www. iucnredlist.org.

IUCN. 2015. The IUCN Red List of Threatened Species (ver. 2015. 4). Available at:www.iucnredlist.org.

IUCN. 2019. IUCN Redlist summary statistics. Available at: https://www.iucnredlist.org/resources/summary-statistics#Summary%20Tables.

James D A, Kannan R. 2009. Nesting habitat of the Great Hornbill (*Buceros bicornis*) in the Anaimalai Hills of southern India. *Wilson Journal of Ornithology*, 121(3): 485-492.

Jiang H X, Hou Y Q, Chu G Z, *et al.* 2010. Breeding population dynamics and habitat transition of Saunders's Gull *Larus saundersi* in Yancheng National Nature Reserve, China. *Bird Conservation International*, 20(1): 13-24.

Jiang Y, Sun Y H, Lv N, *et al.* 2009. Breeding biology of the Grey-hooded Parrotbill (*Paradoxornis zappeyi*) at Wawushan, Sichuan, China. *The Wilson Journal of Ornithology*, 121(4), 800-803.

Jiang Y L, Gao W, Lei F, *et al.* 2008. Nesting biology and population dynamics of endangered Jankowski's Bunting (*Emberiza jankowskii*) in Western Jilin, China. *Bird Conservation International*, 18(2): 153-163.

Jing Y, Fang Y, Strickland D, *et al.* 2009. Alloparenting in the rare Sichuan Jay (*Perisoreus internigrans*). *Condor*, 111: 662-667.

Jing Y, Lyu N, Fang Y, *et al.* 2011. Home range, population density, and habitat utilization of the Sichuan Jay. *Chinese Birds*, 2(2): 94-100.

Judas J, Lawrence M, Combreau O. 2009. High mortality of Asian Houbara *Chlamydotis macqueenii* in Iran. *Falco*, 14-15.

Juniper T, Parr M. 1998. A Guide to the Parrots of the World. Sussex: Pica Press.

Kanai Y, Ueta M, Germogenov N, *et al.* 2002. Migration routes and important resting areas of Siberian Cranes (*Grus leucogeranus*) between northeastern Siberia and China as revealed by satellite tracking. *Biological Conservation*, 106(3): 339-346.

Karyakin I V, Nikolenko E G, Barashkova A N, *et al.* 2010. Golden Eagle in the Altai-Sayan Region, Russia. *Raptors Conservation*, 18: 82-152.

Keane A M, Carroll J P, Fuller R A, et al. 2005. Partridges, Quails, Francolins, Snowcocks, Guineafowl and Turkeys: Status Survey and Conservation Action Plan 2005-2009. Gland, IUCN and WPA.

Khachar S, Mundkur T. 1989. Status and distribution of the King Vulture *Sarcogyps calvus* (Scopoli) in Gujarat: Results of a recent enquiry. *Journal of Bombay Natural History Society*, 86: 360-362.

Kim E Y, Goto R, Iwata H, et al. 1999. Preliminary survey of lead poisoning of Steller's Sea Eagle (*Haliaeetus pelagicus*) and White-tailed Sea Eagle (*Haliaeetus albicilla*) in Hokkaido, Japan. *Environmental Toxicology and Chemistry*, 18(3): 448-451.

Kong D J, Wu F, Shan P F, et al. 2018. Status and distribution changes of the endangered Green Peafowl (*Pavo muticus*) in China over the past three decades (1990s-2017). *Avian Research*, 9: 18.

Kong D J, Yang X J, Liu Q, et al. 2011. Winter habitat selection of vulnerable black-necked cranes *Grus nigricollis*: Implications for determining effective conservation actions. *Oryx*, 45(2): 258-264.

König C, Weick F. 2008. Owls of the World. 2nd ed. London: Christopher Helm.

Lee K S, Lau M W N, Fellowes J R, et al. 2006. Forest bird fauna of south China: Notes on current distribution and status. *Forktail*, 22: 23-38.

Lewthwaite R E, Zou F S. 2015. A checklist of the birds of guangdong with notes on its ornithological exploration. *Journal of Chinese Zoology*, 50(4): 499-517.

Li B C, Jiang P P, Ding P. 2007. First breeding observations and a new locality record of White-eared Night-heron Gorsachius magnificus in Southeast China. *Waterbirds*, 30: 301-304.

Li C, Xiong K N, Wu G M. 2013. Process of biodiversity research of Karst areas in China. *Acta Ecologica Sinica*, 33(4): 192-200.

Li F S, Wu J D, Harris J, et al. 2012. Number and distribution of cranes wintering at Poyang Lake, China during 2011-2012. *Chinese Birds*, 3(3): 180-190.

Li H M, Ma Y X, Liu W J, et al. 2009. Clearance and fragmentation of tropical rain forest in Xishuangbanna, SW, China. *Biodiversity and Conservation*, 18(13): 3421-3440.

Li N, Zhou W, Yang Y Y, et al. 2008. Microscopic analysis on the winter diet of *Syrmaticus humiae* in Dazhongshan Nature Reserve, Yunnan. *Newsletter of China Ornithological Society*, 17: 28.

Li X T. 2004. Raptors of China. Beijing: China Forestry Publishing House.

Li X T, Liu R S. 1993. The Brown Eared Pheasant. Beijing: International Academic Publishers.

Li Z W, Mundkur T. 2003. Status Overview and Recommendations for Conservation of the White-Headed Duck *Oxyura leucocephala* in Central Asia. Kuala Lumpur: Wetlands International.

Liang C T, Chang S H, Fang W H. 2000. Little known oriental bird: Discovery of a breeding colony of Chinese Crested Tern. *OBC Bulletin*, 32: 18.

Liang W, Cai Y, Yang C C. 2013. Extreme levels of hunting of birds in a remote village of Hainan Island, China. *Bird Conservation International*, 23(1): 45-52.

Liang W, Zhang Z W. 2011. Hainan Peacock Pheasant (*Polyplectron katsumatae*): An endangered and rare tropical forest bird. *Chinese Birds*, 2(2): 111-116.

Liao W B, Fuller R A, Hu J C, et al. 2008. Habitat use by endangered Sichuan Partridges *Arborophila rufipectus* during the breeding season. *Acta Ornithologica*, 43(2): 179-184.

Liao W B, Hu J C, Li C, et al. 2008. Roosting behaviour of the endangered Sichuan Hill-partridge *Arborophila rufipectus* during the breeding season. *Bird Conservation International*, 18(3): 260-266.

Liu B Y, Li L, Huw L, et al. 2016. Comparing post-release survival and habitat use by captive-bred Cabot's Tragopan (*Tragopan caboti*) in an experimental test of soft-release reintroduction strategies. *Avian Research*, 7: 19.

Liu P Q, Li F, Song H D, et al. 2010. A survey to the distribution of the Scaly-sided Merganser (*Mergus squamatus*) in Changbai Mountain range (China side). *Chinese Birds*, 1(2): 148-155.

Liu Q, Li F, Buzzard P, et al. 2012. Migration routes and new breeding areas of Black-necked Cranes. *Wilson Journal of Ornithology*, 124(4): 704-712.

Liu Y, Guo D S, Qiao Y L, et al. 2009. Regional extirpation of the critically endangered Chinese Crested Tern from Shandong Coast, China? *Waterbirds*, 32(4): 597-599.

Liu Y, Zhang Z W, Li J Q, et al. 2008. A survey of the birds of the Dabie Shan range, central China. *Forktail*, 24: 80-91.

Liu Z, Zhou W, Zhang Q, et al. 2008. The characteristics of plant community and selection of foraging sites in Hume's Pheasant

(*Syrmaticus humiae*) in Nanhua part of Ailaoshan National Nature Reserve. *Newsletter of China Ornithological Society*, 17(2): 26-27.

Londei T. 2011. Podoces ground-jays and roads: Observations from Taklimakan Desert, China. *Forktail*, 27: 109-111.

Londei T. 2013. About the geographic distribution of the Xinjiang Ground Jay. *Chinese Birds*, 4: 184-186.

López-López P, Sarà M, Di V M. 2012. Living on the edge: assessing the extinction risk of critically endangered Bonelli's Eagle in Italy. PLoS One, 7(5): e37862. doi: 10.1371/journal.pone.0037862.

Lu H F, Campbell D, Chen J, et al. 2007.Conservation and economic viability of nature reserves: An emergy evaluation of the Yancheng Biosphere Reserve. *Biological Conservation*, 139(3-4): 415-423.

Ludlow F, Kinnear N B. 1933. A contribution to the ornithology of Chinese Turkestan. *Ibis*, 75: 240-259.

Lyu N, Jing Y, Lloyd H, et al. 2012. Assessing the distributions and potential risks from climate change for the Sichuan Jay (*Perisoreus internigrans*). *Condor*, 114(2): 365-376.

Ma M. 1998. Xinjiang Ground Jay in the Taklimakan desert. *Oriental Bird Club Bulletin*, 27: 57-58.

Ma M. 1999. Saker smugglers target western China. O*riental Bird Club Bulletin*, 29: 17.

Ma M. 2000. Important bird areas (IBAs) with globally threatened birds of Xinjiang, China. *Arid Zone Research*, 10: 281-284.

Ma M. 2004. Recent data on Saker smuggling in China. *Falco*, 23: 17-18.

Ma M. 2011. Status of the Xinjiang Ground Jay: population, breeding ecology and conservation. *Chinese Birds*, 2(1): 59-62.

Ma M. 2013. Government-sponsored falconry practices, rodenticides, and land development jeopardize Golden Eagles (*Aquila chrysaetos*) in western China. *Journal of Raptor Research*, 47(1): 76-79.

Ma M, Cai D. 2002. The fate of the White Stork (Ciconia ciconia asiatica) in Xinjiang, China. Abstract of 23rd International Ornithological Congress, Beijing.

Ma M, Ka K H. 2004. Records of Xinjiang Ground-jay in Taklimakan Desert, Xinjiang, China. *Forktail*, 20: 121-124.

Ma M, Mei Y, Tian L L, et al. 2006a. The Saker Falcon in the desert of North Xinjiang, China. *Raptors Conservation*, (6): 58-64.

Ma M, Peng D, Li WD, et al. 2010. Breeding ecology and survival status of the Golden Eagle in China. *Raptors Conservation*, (19): 75-87.

Ma M, Tong Z G, Peng D, et al. 2012. Golden Eagle in the Northwestern China. *Raptors Conservation*, (25): 70-78.

Ma M, Wei S D, Xu F, et al. 2006b. Black Stork *Ciconia nigra* in Xinjiang, China. *Biota*, 7: 57-64.

Ma M, Zhang T. 2012. The White-headed Duck *Oxyura leucocephala* inÜrümqi, Xinjiang, China. *Birding ASIA*, 18: 93-96.

Ma M, Zhao X M. 2013. Distribution patterns and ecology of Steppe Eagle in China. *Raptors Conservation*, 27: 172-179.

Ma Z J, Hua N, Peng H B, et al. 2013. Differentiating between stopover and staging sites: functions of the southern and northern Yellow Sea for long-distance migratory shorebirds. *Journal of Avian Biology*, 44(5): 504-512.

Ma Z J, Li B, Li W J, et al. 2009. Conflicts between biodiversity conservation and development in a biosphere reserve. *Journal of Applied Ecology*, 46(3): 527-535.

MacKinnon J, Verkuil Y I, Murray N. 2012. IUCN situation analysis on East and Southeast Asian intertidal habitats, with particular reference to the Yellow Sea (including the Bohai Sea). Occasional Paper of the IUCN Species Survival Commission No. 47. Gland and Cambridge, IUCN.

MacKinnon J R, Phillips K, He F Q. 2000. A Field Guide to the Birds of China. Oxford: Oxford University Press.

Madge S, Burn H. 1994. Crows and Jays. London: Christopher Helm Ltd, A and C Black Ltd: 124-127.

Madge S, McGowan P. 2002. Pheasants, Partridges and Grouse: Including Buttonquails, Sandgrouse and allies. London: Christopher Helm.

Manning T, Ross G A, Symons R. 2008. Environmental Contaminants in White-Bellied Sea-Eagles (*Haliaeetus Leucogaster*) Found in Sydney, Australia. *Australasian Journal of Ecotoxicology*, 14: 21-30.

Martens J, Eck S, Sun Y H. 2002. *Certhia tianquanensis* Li, a treecreeper with relict distribution in Sichuan, China. *Journal of Ornithology*, 143(4): 440-456.

Martens J, Eck S, Sun Y H. 2003. On the discovery of a new treecreeper in China—*Certhia tianquanensis* Li. OBC Bulletin, 37: 65-70.

Martí R, del Moral J C. 2003. Atlas de las Aves Reproductoras de España. Dirección General de Conservación de la Naturaleza-Sociedad Española de Ornithología, Madrid.

Michler I. 2009. Lifting the veil: Wake-up call to counter covert hunting threat. *Africa-Birds and Birding*, 14(3): 58-62.

Mikkola H. 2013. Owls of the World: A Photographic Guide. 2nd ed. London: Christopher Helm.

Minna J, Su H, Lin Y S. 1992. Breeding ecology of Styan's Bulbul *Pycnonotus taivanus* in Taiwan. *Ibis*, 139(3): 518-52.

Mix H M, Bräunlich A. 2000. Dalmatian Pelican. *In*: Reading R P, Miller B. Endangered animals: A reference guide to conflicting issues. London: Greenwood Press: 78-83.

Moores N, Rogers D, Kim R H, *et al.* 2008. The 2006-2008 Saemangeum Shorebird Monitoring Program Report. Busan: Birds Korea Publication.

Morrison W, Rosalind L, Balachandran S. 1992. Unusual nesting site of Brahminy Kite *Haliastur indus*. *Journal of Bombay Natural History Society*, 89: 117-118.

Muñoz A R, Real R, Barbosa A M, *et al.* 2005. Modelling the distribution of Bonelli's Eagle in Spain: Implications for conservation planning. *Diversity and Distributions*, 11(6): 477-486.

Nadeem M S, Asif M, Mahmood T. 2007. Reappearance of Red-headed Vulture *Sarcogyps calvus* in Tharparker, Southeast Pakistan. *Podoces*, 2: 146-148.

Nakagawa H, Lobkov E G, Fujimaki Y. 1987. Winter censuses on *Haliaeetus pelagicus* in Kamchatka and northern Japan in 1985. *Strix*, 6: 14-19.

Naoroji R. 1999. Status of diurnal raptors of Corbett National Park with notes on their ecology and conservation. *Journal of Bombay Natural History Society*, 96: 387-398.

Oaks J L, Gilbert M, Virani M Z, *et al.* 2004. Diclofenac residues as the cause of vulture population decline in Pakistan. *Nature*, 427(6975): 630-633.

Olsson U, Alstrom P, Colston P R. 1993. A new species of *Phylloscopus* warbler from Hainan Island, China. *Ibis*, 135(1): 3-7.

Parnell E. 2016. Brighter future for worlds' rarest tern. *World Birdwatch*, 3: 15-17.

Piatta J F, Wetzela J, Bellb K. 2006. Predictable hotspots and foraging habitat of the endangered Short-tailed Albatross (*Phoebastria albatrus*) in the North Pacific: Implications for conservation. *Deep-Sea Research II: Topical Studies in Oceanography*, 53(3): 387-398.

Potapov E, Ma M. 2004. The highlander: the highest breeding Saker in the World. *Falco*, 23: 10-12.

Potapov E, Mcgrady M J, Utekhina I, *et al.* 2013. Steller's Sea Eagle monitoring at the northern part of the sea of Okhotsk: Birds, People, Technologies. *Raptors Conservation*, 27: 46-57.

Potapov E, Utekhina I, McGrady M, *et al.* 2010. Low breeding success of Steller's Sea Eagles in Magadan district (Russia) in 2009: Start of a decline? *Raptors Conservation*, 18: 163-165.

Poyarkov N D. 2005. Natural history and problems of conservation of the Swan Goose. *Casarca*, (Supl. 1): 139-159.

Qiao Y L, Liu Y, Guo D S, *et al.* 2006. First Chinese breeding record of Pleske's Warbler *Locustella pleskei*, from a small island off Qingdao, Shandong province. *Birding ASIA*, 6: 81-82.

Qin Y, He F Q. 2011. Latest evidence of the existence of the northern flock of the Chinese Crested Tern (*Sterna bernsteini*). *Chinese Birds*, 2(4): 206-207.

Rahmani A R. 1998. A possible decline of vultures in India. *OBC Bulletin*, 28: 40-41.

Rajan S A, Balasubramanian P, Natarajan V. 1992. Eastern Steppe Eagle *Aquila rapax nipalensis* killing mobbing Brahminy Kite *Haliastur indus* (Boddaert) at Pt. Calimere Wildlife Sanctuary, Tamil Nadu. *Journal of Bombay Natural History Society*, 89: 247-248.

Ranjit M, Natarajan V. 1992. Brahminy Kite *Haliastur indus* (Boddaert) preying on bats. *Journal of Bombay Natural History Society*, 89(3): 367.

Rasmussen P C, Anderton J C. 2005. Birds of South Asia: the Ripley Guide. Barcelona: Lynx Edicions.

Rheindt F E. 2004. Notes on the range and ecology of Sichuan Treecreeper *Certhia tianquanensis*. *Forktail*, 20: 141-142.

Robson C. 2007. Birds of South-east Asia. Singapore: Tien Wah Press.

Robson C. 2011. A wintering Chinese Crested Tern *Sterna bernsteini* in eastern Indonesia. *Birding ASIA*, 15: 51.

Rogers D I, Yang H Y, Hassell C J, *et al.* 2010. Red knots (*Calidris canutus piersmai* and *C. c. rogersi*) depend on a small threatened staging area in Bohai Bay, China. *Emu*, 110(4): 307-315.

Roselaar C S. 1992. A new species of mountain finch Leucosticte from western Tibet. *Bull Br Ornithol Club*, 112: 225-231.

Rudnick J. 2006. Conservation of the Eastern Imperial Eagle (*Aquila heliaca*) in central Asia: Using molecular techniques to investigate raptor ecology and noninvasively monitor raptor populations. West Lafayette: Doctoral thesis of Purdue University.

Ryan G E. 2012. Brahminy Kites *Haliastur indus* fishing with Irrawaddy Dolphins *Orcaella brevirostris* in the Mekong River.

Forktail, 28: 161.

Schweizer M, Etzbauer C, Shirihai H, *et al.* 2020. A molecular analysis of the mysterious Vaurie's Nightjar *Caprimulgus centralasicus* yields fresh insight into its taxonomic status. *Journal of Ornithology*, 161: 635-650.

Serrano D, Tella J L, Forero M G, *et al.* 2001. Factors affecting breeding dispersal in the facultatively colonial Lesser Kestrel: Individual experience vs. conspecific cues. *Journal of Animal Ecology*, 70(4): 568-578.

Shaw T. 1938. The avifauna of Tsingtao and neighbouring districts. *Bulletin of the Fan Memorial Institute of Biology* (Zoology Series), 8: 133-222.

Shephard J M, Hughes J M, Catterall C P, *et al.* 2005. Conservation status of the White-bellied Sea-Eagle *Haliaeetus leucogaster* in Australia determined using mtDNA control region sequence data. *Conservation Genetics*, 6(3): 413-429.

Shi H Q, Cao L, Barter M A, *et al.* 2008. Status of the East Asian population of the Dalmatian Pelican *Pelecanus crispus*: The need for urgent conservation action. *Bird Conservation International*, 18(2): 181-193.

Snow D W, Perrins C M. 1998. The Birds of the Western Palearctic (Vol 1): Non-Passerines. Oxford: Oxford University Press.

Solovyeva D V, Liu P Q, Antonov A I, *et al.* 2014. The population size and breeding range of the Scaly-sided Merganser *Mergus squamatus*. *Bird Conservation International*, 24(4): 393-405.

Spierenburg P. 2005. Birds in Bhutan: Status and distribution. Bedford: Oriental Bird Club.

Steve M, Burn H. 1994. Crows and Jays. London: Christopher Helm Ltd, A and C Black Ltd: 124-127.

Su L Y, Zou H F. 2012. Status, threats and conservation needs for the continental population of the Red-crowned Crane. *Chinese Birds*, 3(3): 147-164.

Sun Y, Dong L, Zhang Y Y, *et al.* 2009a. Is a forest road a barrier for the Vulnerable Cabot's tragopan *Tragopan caboti* in Wuyishan, Jiangxi, China? *Oryx*, 43(4): 614-617.

Sun Y H, Jia C X, Fang Y. 2001. The distribution and status of Sichuan Grey Jay (*Perisoreus internigrans*). *Journal fur Ornithologie*, 142(1): 93-98.

Sun Y H, Jiang Y X, Martens J, *et al.* 2009. Notes on the breeding biology of the Sichuan Treecreeper (*Certhia tianquanensis*). *Journal of Ornithology*, 150(4): 925-929.

Sun Y H, Wu H J, Wang Y. 2004. Tawny Fish Owl predation at fish farms in Taiwan. *Journal of Raptor Research*, 38(4): 326-333.

Swinhoe R. 1863. Catalogue of the birds of China, with remarks principally on their geographical distribution. Proceedings of the Zoological Society of London. pp259-339.

Tamada K. 2006. Population change of grassland birds over ten years in Nakashibetsu, eastern Hokkaido. *Ornithological Science*, 5(1): 127-131.

Tamura M, Higuchi H, Shimazaki H, *et al.* 2001. Satellite observation of movements and habitat conditions of Red-Crowned Cranes and Oriental White Storks in East Asia. *Global Environment Research*, 4: 207-218.

Tella J L, Donazar J A, Negro J J, *et al.* 1996. Seasonal and interannual variations in the sex-ratio of Lesser Kestrel *Falco naumanni* broods. *Ibis*, 138(2): 342-345.

Thewlis R M, Martins R P. 2000. Observations of the breeding biology and behaviour of Kozlov's Bunting *Emberiza koslowi*. *Forktail*, 16: 57-59.

Tong M X, Zhang L, Li J, *et al.* 2012. The critical importance of the Rudong mudflats, Jiangsu Province, China in the annual cycle of the Spoon-billed Sandpiper *Calidris pygmeus*. *Wader Study Group Bulletin*, 119(3): 74-77.

Tourenq C, Combreau O, Lawrence M, *et al.* 2005. Alarming Houbara Bustard population trends in Asia. *Biological Conservation*, 121(1): 1-8.

Tourenq C, Combreau O, Pole S B, *et al.* 2004. Monitoring of Asian Houbara Bustard *Chlamydotis macqueenii* populations in Kazakhstan reveals dramatic decline. *Oryx*, 38(1): 62-67.

Ueta M, Melville D S, Wang Y, *et al.* 2002. Discovery of the Breeding Sites and Migration Routes of Black-Faced Spoonbills *Platalea minor*. *Ibis*, 144(2): 340-343.

van Balen B S, Suwelo I S, Hadi D S, *et al.* 1993. The decline of the Brahminy Kite *Haliastur indus* on Java. *Forktail*, 8: 83-88.

Vaurie C. 1960. Systematic notes on Palearctic birds. No. 39. Caprimulgidae: A new species of Caprimulgus. *American Museum Novitates*, (1985): 1-10.

Village A. 1990. The Kestrel. London: Poyser.

Wang H T, Jiang Y L, Gao W. 2010. Jankowski's Bunting (*Emberiza jankowskii*): Current status and conservation. *Chinese Birds*, 1(4): 251-258.

Wang J, Jia C X, Tang S, *et al*. 2011. Breeding biology of the Snowy-cheeked Laughingthrush (*Garrulax sukatschewi*). *The Wilson Journal of Ornithology*, 123(1): 146-150.

Wang X, Barter M, Cao L, *et al*. 2012. Serious contractions in wintering distribution and decline in abundance of Baer's Pochard *Aythya baeri*. *Bird Conservation International*, 22(2): 121-127.

Wang Z J, Carpenter C, Young S S. 2000. Bird distribution and conservation in the Ailao Mountains, Yunnan, China. *Biological Conservation*, 92(1): 45-57.

Watson J. 1997. The Golden Eagle. London: Poyser.

Wetlands International. 2012. Waterfowl Population Estimates. 5th ed. Wageningen: Wetlands International.

Wetlands International. 2014. Waterbird Population Estimates. Retrieved from wpe.wetlands.org on Tuesday 13 May 2014.

Wu F, Kong DJ, Shan PF, *et al*. 2019. Ongoing Green Peafowl protection in China. *Zoological Research*, 40(6): 580-582.

Wu Y Q, Ma M, Xu F, *et al*. 2006. Feeding habits analysis of Saker Falcon and the effect of raptor on prevention mouse population. *Journal of Xinjiang Agricultural University*, 29(2): 13-16.

Xu J L, Zhang Z W, Wang Y, *et al*. 2011. Spatio-temporal responses of male Reeves's Pheasants Syrmaticus reevesii to forest edges in the Dabie Mountains, central China. *Wildlife Biology*, 17(1):16-24.

Yadav B P. 2007. Distribution and Habitat Preferences of Hodgon's Bushchat (*Saxicola insignis*) in Grassland of Suklaphanta Wildlife Reserve of Far-Western Development Region of Nepal. Lodon: Report submitted to OBC.

Yang H Y, Chen B, Barter M, *et al*. 2011. Impacts of tidal land reclamation in Bohai Bay, China: Ongoing losses of critical Yellow Sea waterbird staging and wintering sites. *Bird Conservation International*, 21(3): 241-259.

Yu Y T, Chen Z H. 2008. Dalmatian Pelican *Pelecanus crispus*: The largest waterbird in East Asia, and the rarest? *Birding Asia*, 9: 62-66.

Zhang K J, Yu X, Gan X J, *et al*. 2004. Chinese Crested Tern at Chongming Dao, Shanghai, China. *Birding Asia*, 2: 66.

Zhang Y, Cao L, Barter M, *et al*. 2011. Changing distribution and abundance of Swan Goose *Anser cygnoides* in the Changjiang River floodplain: the likely loss of a very important wintering site. *Bird Conservation International*, 21(1): 36-48.

Zhang Y Y, Zheng G M. 2007. A population viability analysis (PVA) for Cabot's Tragopan (*Tragopan caboti*) in Wuyanling, south-west Ching. *Bird Conservation International*, 17(2): 151-161.

Zhang Z W. 1998. The distributional range of Brown-eared Pheasant *Crossoptilon mantchuricum*. *Tragopan*: 5-7.

Zhao X M, Ma M, Ding P, *et al*. 2013. Chronology of physical and behavior development on the nestlings of Golden Eagle in China. *Selevinia*, 21: 113-118.

Zhao Z J, Nickel H, Groh G. 1994. Vorkommen und gesang der Jankowskiammer (*Emberiza jankowskii*) in der Chinesischen provinz Jilin. *Journal für Ornithologie*, 135: 617-620.

Zhou W, Li N, Deng Z J, *et al*. 2010. Modelling foraging habitats of Hume's Pheasant (*Syrmaticus humiae*) in Dazhong Mountain, Yunnan, southwestern China. *Chinese Birds*, 1(4): 236-243.

Zöckler C, Syroechkovski E E. 2010. Rapid and continued population decline in the Spoon-billed Sandpiper *Eurynorhynchus pygmeus* indicates imminent extinction unless conservation action is taken. *Bird Conservation International*, 20(2): 95-111.

附录 中国鸟类濒危等级评估名录
Appendix Assessment List of Endangered Levels of Chinese Birds

	目	科	物种中文名	物种学名	物种英文名	濒危等级	评估依据	是否特有
0001	GALLIFORMES 鸡形目	Phasianidae 雉科	环颈山鹧鸪	*Arborophila torqueola*	Common Hill Partridge	LC		
0002	GALLIFORMES 鸡形目	Phasianidae 雉科	四川山鹧鸪	*Arborophila rufipectus*	Sichuan Hill Partridge	EN	B2ab(iii), C2a(i)	√
0003	GALLIFORMES 鸡形目	Phasianidae 雉科	红喉山鹧鸪	*Arborophila rufogularis*	Rufous-throated Hill Partridge	LC		
0004	GALLIFORMES 鸡形目	Phasianidae 雉科	白眉山鹧鸪	*Arborophila gingica*	White-necklaced Hill Partridge	VU	C2a(i)	√
0005	GALLIFORMES 鸡形目	Phasianidae 雉科	白颊山鹧鸪	*Arborophila atrogularis*	White-cheeked Hill Partridge	NT		
0006	GALLIFORMES 鸡形目	Phasianidae 雉科	褐胸山鹧鸪	*Arborophila brunneopectus*	Bar-backed Hill Partridge	NT		
0007	GALLIFORMES 鸡形目	Phasianidae 雉科	红胸山鹧鸪	*Arborophila mandellii*	Chestnut-breasted Hill Partridge	VU	C2a(i)	
0008	GALLIFORMES 鸡形目	Phasianidae 雉科	台湾山鹧鸪	*Arborophila crudigularis*	Taiwan Hill Partridge	NT		√
0009	GALLIFORMES 鸡形目	Phasianidae 雉科	海南山鹧鸪	*Arborophila ardens*	Hainan Hill Partridge	EN	B1a+b(iii)	√
0010	GALLIFORMES 鸡形目	Phasianidae 雉科	绿脚树鹧鸪	*Tropicoperdix chloropus*	Green-legged Partridge	NT		
0011	GALLIFORMES 鸡形目	Phasianidae 雉科	花尾榛鸡	*Tetrastes bonasia*	Hazel Grouse	LC		
0012	GALLIFORMES 鸡形目	Phasianidae 雉科	斑尾榛鸡	*Tetrastes sewerzowi*	Chinese Grouse	NT		√
0013	GALLIFORMES 鸡形目	Phasianidae 雉科	镰翅鸡	*Falcipennis falcipennis*	Siberian Grouse	RE		
0014	GALLIFORMES 鸡形目	Phasianidae 雉科	松鸡	*Tetrao urogallus*	Western Capercaillie	EN	C2a(i)	
0015	GALLIFORMES 鸡形目	Phasianidae 雉科	黑嘴松鸡	*Tetrao urogalloides*	Black-billed Capercaillie	EN	A2c+3c+4c	
0016	GALLIFORMES 鸡形目	Phasianidae 雉科	黑琴鸡	*Lyrurus tetrix*	Black Grouse	NT		
0017	GALLIFORMES 鸡形目	Phasianidae 雉科	岩雷鸟	*Lagopus muta*	Rock Ptarmigan	NT		
0018	GALLIFORMES 鸡形目	Phasianidae 雉科	柳雷鸟	*Lagopus lagopus*	Willow Ptarmigan	VU	C2a(i)	
0019	GALLIFORMES 鸡形目	Phasianidae 雉科	雪鹑	*Lerwa lerwa*	Snow Partridge	NT		
0020	GALLIFORMES 鸡形目	Phasianidae 雉科	红喉雉鹑	*Tetraophasis obscurus*	Chestnut-throated Partridge	VU	B2ab(i), C2a(i)	√
0021	GALLIFORMES 鸡形目	Phasianidae 雉科	黄喉雉鹑	*Tetraophasis szechenyii*	Buff-throated Partridge	VU	B2ab(i), C2a(i)	√
0022	GALLIFORMES 鸡形目	Phasianidae 雉科	暗腹雪鸡	*Tetraogallus himalayensis*	Himalayan Snowcock	NT		
0023	GALLIFORMES 鸡形目	Phasianidae 雉科	藏雪鸡	*Tetraogallus tibetanus*	Tibetan Snowcock	NT		
0024	GALLIFORMES 鸡形目	Phasianidae 雉科	阿尔泰雪鸡	*Tetraogallus altaicus*	Altai Snowcock	VU	B1b(ii, iii)+C1	
0025	GALLIFORMES 鸡形目	Phasianidae 雉科	石鸡	*Alectoris chukar*	Chukar Partridge	LC		
0026	GALLIFORMES 鸡形目	Phasianidae 雉科	大石鸡	*Alectoris magna*	Rusty-necklaced Partridge	NT		√

续表

目		科	物种中文名	物种学名	物种英文名	濒危等级	评估依据	是否特有
0027	鸡形目 GALLIFORMES	Phasianidae 雉科	中华鹧鸪	Francolinus pintadeanus	Chinese Francolin	NT		
0028	鸡形目 GALLIFORMES	Phasianidae 雉科	灰山鹑	Perdix perdix	Grey Partridge	LC		
0029	鸡形目 GALLIFORMES	Phasianidae 雉科	斑翅山鹑	Perdix dauurica	Daurian Partridge	LC		
0030	鸡形目 GALLIFORMES	Phasianidae 雉科	高原山鹑	Perdix hodgsoniae	Tibetan Partridge	LC		
0031	鸡形目 GALLIFORMES	Phasianidae 雉科	西鹌鹑	Coturnix coturnix	Common Quail	LC		
0032	鸡形目 GALLIFORMES	Phasianidae 雉科	鹌鹑	Coturnix japonica	Japanese Quail	LC		
0033	鸡形目 GALLIFORMES	Phasianidae 雉科	蓝胸鹑	Synoicus chinensis	Blue-breasted Quail	NT		
0034	鸡形目 GALLIFORMES	Phasianidae 雉科	棕胸竹鸡	Bambusicola fytchii	Mountain Bamboo Partridge	LC		
0035	鸡形目 GALLIFORMES	Phasianidae 雉科	灰胸竹鸡	Bambusicola thoracicus	Chinese Bamboo Partridge	LC		√
0036	鸡形目 GALLIFORMES	Phasianidae 雉科	台湾竹鸡	Bambusicola sonorivox	Taiwan Bamboo Partridge	LC		√
0037	鸡形目 GALLIFORMES	Phasianidae 雉科	血雉	Ithaginis cruentus	Blood Pheasant	NT		
0038	鸡形目 GALLIFORMES	Phasianidae 雉科	黑头角雉	Tragopan melanocephalus	Western Tragopan	DD		
0039	鸡形目 GALLIFORMES	Phasianidae 雉科	红胸角雉	Tragopan satyra	Satyr Tragopan	VU	C2a (i)	
0040	鸡形目 GALLIFORMES	Phasianidae 雉科	灰腹角雉	Tragopan blythii	Blyth's Tragopan	DD		
0041	鸡形目 GALLIFORMES	Phasianidae 雉科	红腹角雉	Tragopan temminckii	Temminck's Tragopan	NT		
0042	鸡形目 GALLIFORMES	Phasianidae 雉科	黄腹角雉	Tragopan caboti	Cabot's Tragopan	EN	B2ab(i)	√
0043	鸡形目 GALLIFORMES	Phasianidae 雉科	勺鸡	Pucrasia macrolopha	Koklass Pheasant	LC		
0044	鸡形目 GALLIFORMES	Phasianidae 雉科	棕尾虹雉	Lophophorus impejanus	Himalayan Monal	NT		
0045	鸡形目 GALLIFORMES	Phasianidae 雉科	白尾梢虹雉	Lophophorus sclateri	Sclater's Monal	EN	C2a(i)	
0046	鸡形目 GALLIFORMES	Phasianidae 雉科	绿尾虹雉	Lophophorus lhuysii	Chinese Monal	EN	B2ab(iii), C2a(i)	√
0047	鸡形目 GALLIFORMES	Phasianidae 雉科	红原鸡	Gallus gallus	Red Junglefowl	NT		
0048	鸡形目 GALLIFORMES	Phasianidae 雉科	黑鹇	Lophura leucomelanos	Kalij Pheasant	NT		
0049	鸡形目 GALLIFORMES	Phasianidae 雉科	白鹇	Lophura nycthemera	Silver Pheasant	LC		
0050	鸡形目 GALLIFORMES	Phasianidae 雉科	蓝腹鹇	Lophura swinhoii	Swinhoe's Pheasant	NT		√
0051	鸡形目 GALLIFORMES	Phasianidae 雉科	白马鸡	Crossoptilon crossoptilon	White Eared Pheasant	NT		√
0052	鸡形目 GALLIFORMES	Phasianidae 雉科	藏马鸡	Crossoptilon harmani	Tibetan Eared Pheasant	NT		√
0053	鸡形目 GALLIFORMES	Phasianidae 雉科	褐马鸡	Crossoptilon mantchuricum	Brown Eared Pheasant	VU	C2a(i)	√

续表

	目		科		物种中文名	物种学名	物种英文名	濒危等级	评估依据	是否特有
0054	鸡形目	GALLIFORMES	雉科	Phasianidae	蓝马鸡	*Crossoptilon auritum*	Blue Eared Pheasant	NT		√
0055	鸡形目	GALLIFORMES	雉科	Phasianidae	白颈长尾雉	*Syrmaticus ellioti*	Elliot's Pheasant	VU	B2ab(iii), C2a(i)	√
0056	鸡形目	GALLIFORMES	雉科	Phasianidae	黑颈长尾雉	*Syrmaticus humiae*	Hume's Pheasant	VU	C1+2a(i)	
0057	鸡形目	GALLIFORMES	雉科	Phasianidae	黑长尾雉	*Syrmaticus mikado*	Mikado Pheasant	NT		√
0058	鸡形目	GALLIFORMES	雉科	Phasianidae	白冠长尾雉	*Syrmaticus reevesii*	Reeves's Pheasant	EN	A2cd+3cd+4cd	√
0059	鸡形目	GALLIFORMES	雉科	Phasianidae	环颈雉	*Phasianus colchicus*	Common Pheasant	LC		
0060	鸡形目	GALLIFORMES	雉科	Phasianidae	红腹锦鸡	*Chrysolophus pictus*	Golden Pheasant	NT		√
0061	鸡形目	GALLIFORMES	雉科	Phasianidae	白腹锦鸡	*Chrysolophus amherstiae*	Lady Amherst's Pheasant	NT		
0062	鸡形目	GALLIFORMES	雉科	Phasianidae	灰孔雀雉	*Polyplectron bicalcaratum*	Grey Peacock Pheasant	EN	C1+2a(i)	
0063	鸡形目	GALLIFORMES	雉科	Phasianidae	海南孔雀雉	*Polyplectron katsumatae*	Hainan Peacock Pheasant	CR	A2cd+3cd+4cd, C2a(i)	√
0064	鸡形目	GALLIFORMES	雉科	Phasianidae	绿孔雀	*Pavo muticus*	Green Peafowl	CR	A2cd+3cd+4cd	
0065	雁形目	ANSERIFORMES	鸭科	Anatidae	栗树鸭	*Dendrocygna javanica*	Lesser Whistling Duck	VU	C2a(i)	
0066	雁形目	ANSERIFORMES	鸭科	Anatidae	鸿雁	*Anser cygnoid*	Swan Goose	VU	A3bcd	
0067	雁形目	ANSERIFORMES	鸭科	Anatidae	豆雁	*Anser fabalis*	Bean Goose	LC		
0068	雁形目	ANSERIFORMES	鸭科	Anatidae	短嘴豆雁	*Anser serrirostris*	Tundra Bean Goose	LC		
0069	雁形目	ANSERIFORMES	鸭科	Anatidae	灰雁	*Anser anser*	Greylag Goose	LC		
0070	雁形目	ANSERIFORMES	鸭科	Anatidae	白额雁	*Anser albifrons*	Greater White-fronted Goose	LC		
0071	雁形目	ANSERIFORMES	鸭科	Anatidae	小白额雁	*Anser erythropus*	Lesser White-fronted Goose	VU	A2abcd	
0072	雁形目	ANSERIFORMES	鸭科	Anatidae	斑头雁	*Anser indicus*	Bar-headed Goose	LC		
0073	雁形目	ANSERIFORMES	鸭科	Anatidae	雪雁	*Anser caerulescens*	Snow Goose	LC		
0074	雁形目	ANSERIFORMES	鸭科	Anatidae	加拿大雁	*Branta canadensis*	Canada Goose	DD(NE)		
0075	雁形目	ANSERIFORMES	鸭科	Anatidae	小美洲黑雁	*Branta hutchinsii*	Cackling Goose	DD(NE)		
0076	雁形目	ANSERIFORMES	鸭科	Anatidae	黑雁	*Branta bernicla*	Brent Goose	DD(NE)		
0077	雁形目	ANSERIFORMES	鸭科	Anatidae	白颊黑雁	*Branta leucopsis*	Barnacle Goose	DD(NE)		
0078	雁形目	ANSERIFORMES	鸭科	Anatidae	红胸黑雁	*Branta ruficollis*	Red-breasted Goose	DD(NE)		
0079	雁形目	ANSERIFORMES	鸭科	Anatidae	疣鼻天鹅	*Cygnus olor*	Mute Swan	NT		

续表

	目	科	物种中文名	物种学名	物种英文名	濒危等级	评估依据	是否特有
0080	ANSERIFORMES	Anatidae	小天鹅	Cygnus columbianus	Tundra Swan	NT		
0081	ANSERIFORMES	Anatidae	大天鹅	Cygnus cygnus	Whooper Swan	NT		
0082	ANSERIFORMES	Anatidae	瘤鸭	Sarkidiornis melanotos	Comb Duck	DD(NE)		
0083	ANSERIFORMES	Anatidae	翘鼻麻鸭	Tadorna tadorna	Common Shelduck	LC		
0084	ANSERIFORMES	Anatidae	赤麻鸭	Tadorna ferruginea	Ruddy Shelduck	LC		
0085	ANSERIFORMES	Anatidae	鸳鸯	Aix galericulata	Mandarin Duck	NT		
0086	ANSERIFORMES	Anatidae	棉凫	Nettapus coromandelianus	Asian Pygmy Goose	EN	C2a(ii)	
0087	ANSERIFORMES	Anatidae	赤膀鸭	Mareca strepera	Gadwall	LC		
0088	ANSERIFORMES	Anatidae	罗纹鸭	Mareca falcata	Falcated Duck	NT		
0089	ANSERIFORMES	Anatidae	赤颈鸭	Mareca penelope	Eurasian Wigeon	LC		
0090	ANSERIFORMES	Anatidae	绿眉鸭	Mareca americana	American Wigeon	DD(NE)		
0091	ANSERIFORMES	Anatidae	绿头鸭	Anas platyrhynchos	Mallard	LC		
0092	ANSERIFORMES	Anatidae	棕颈鸭	Anas luzonica	Philippine Duck	DD(NE)		
0093	ANSERIFORMES	Anatidae	印度斑嘴鸭	Anas poecilorhyncha	Indian Spot-billed Duck	LC		
0094	ANSERIFORMES	Anatidae	斑嘴鸭	Anas zonorhyncha	Spot-billed Duck	LC		
0095	ANSERIFORMES	Anatidae	针尾鸭	Anas acuta	Northern Pintail	LC		
0096	ANSERIFORMES	Anatidae	绿翅鸭	Anas crecca	Green-winged Teal	LC		
0097	ANSERIFORMES	Anatidae	琵嘴鸭	Spatula clypeata	Northern Shoveler	LC		
0098	ANSERIFORMES	Anatidae	白眉鸭	Spatula querquedula	Garganey	LC		
0099	ANSERIFORMES	Anatidae	花脸鸭	Sibirionetta formosa	Baikal Teal	NT		
0100	ANSERIFORMES	Anatidae	云石斑鸭	Marmaronetta angustirostris	Marbled Teal	DD(NE)		
0101	ANSERIFORMES	Anatidae	赤嘴潜鸭	Netta rufina	Red-crested Pochard	LC		
0102	ANSERIFORMES	Anatidae	帆背潜鸭	Aythya valisineria	Canvasback	DD(NE)		
0103	ANSERIFORMES	Anatidae	红头潜鸭	Aythya ferina	Common Pochard	LC		
0104	ANSERIFORMES	Anatidae	青头潜鸭	Aythya baeri	Baer's Pochard	CR	A2cd+3cd+4cd	
0105	ANSERIFORMES	Anatidae	白眼潜鸭	Aythya nyroca	Ferruginous Duck	NT		
0106	ANSERIFORMES	Anatidae	凤头潜鸭	Aythya fuligula	Tufted Duck	LC		

续表

编号	目	科	科	物种中文名	物种学名	物种英文名	濒危等级	评估依据	是否特有
0107	雁形目	ANSERIFORMES	Anatidae	鸭科	斑背潜鸭	*Aythya marila*	Greater Scaup	LC		
0108	雁形目	ANSERIFORMES	Anatidae	鸭科	小绒鸭	*Polysticta stelleri*	Steller's Eider	DD		
0109	雁形目	ANSERIFORMES	Anatidae	鸭科	丑鸭	*Histrionicus histrionicus*	Harlequin Duck	LC		
0110	雁形目	ANSERIFORMES	Anatidae	鸭科	斑脸海番鸭	*Melanitta fusca*	Velvet Scoter	NT		
0111	雁形目	ANSERIFORMES	Anatidae	鸭科	黑海番鸭	*Melanitta americana*	Black Scoter	NT		
0112	雁形目	ANSERIFORMES	Anatidae	鸭科	长尾鸭	*Clangula hyemalis*	Long-tailed Duck	CR	C2a(i)	
0113	雁形目	ANSERIFORMES	Anatidae	鸭科	鹊鸭	*Bucephala clangula*	Common Goldeneye	LC		
0114	雁形目	ANSERIFORMES	Anatidae	鸭科	斑头秋沙鸭	*Mergellus albellus*	Smew	LC		
0115	雁形目	ANSERIFORMES	Anatidae	鸭科	普通秋沙鸭	*Mergus merganser*	Common Merganser	LC		
0116	雁形目	ANSERIFORMES	Anatidae	鸭科	红胸秋沙鸭	*Mergus serrator*	Red-breasted Merganser	LC		
0117	雁形目	ANSERIFORMES	Anatidae	鸭科	中华秋沙鸭	*Mergus squamatus*	Scaly-sided Merganser	EN	C2a(i)	
0118	雁形目	ANSERIFORMES	Anatidae	鸭科	白头硬尾鸭	*Oxyura leucocephala*	White-headed Duck	CR	A2bcde+4bcde, B1ab(i), C2a(i), D	
0119	䴙䴘目	PODICIPEDIFORMES	Podicipedidae	䴙䴘科	小䴙䴘	*Tachybaptus ruficollis*	Little Grebe	LC		
0120	䴙䴘目	PODICIPEDIFORMES	Podicipedidae	䴙䴘科	赤颈䴙䴘	*Podiceps grisegena*	Red-necked Grebe	NT		
0121	䴙䴘目	PODICIPEDIFORMES	Podicipedidae	䴙䴘科	凤头䴙䴘	*Podiceps cristatus*	Great Crested Grebe	LC		
0122	䴙䴘目	PODICIPEDIFORMES	Podicipedidae	䴙䴘科	角䴙䴘	*Podiceps auritus*	Horned Grebe	NT		
0123	䴙䴘目	PODICIPEDIFORMES	Podicipedidae	䴙䴘科	黑颈䴙䴘	*Podiceps nigricollis*	Black-necked Grebe	LC		
0124	红鹳目	PHOENICOPTERIFORMES	Phoenicopteridae	红鹳科	大红鹳	*Phoenicopterus roseus*	Greater Flamingo	NT		
0125	鸽形目	COLUMBIFORMES	Columbidae	鸠鸽科	原鸽	*Columba livia*	Rock Dove	LC		
0126	鸽形目	COLUMBIFORMES	Columbidae	鸠鸽科	岩鸽	*Columba rupestris*	Hill Pigeon	LC		
0127	鸽形目	COLUMBIFORMES	Columbidae	鸠鸽科	雪鸽	*Columba leuconota*	Snow Pigeon	LC		
0128	鸽形目	COLUMBIFORMES	Columbidae	鸠鸽科	欧鸽	*Columba oenas*	Stock Dove	LC		
0129	鸽形目	COLUMBIFORMES	Columbidae	鸠鸽科	中亚鸽	*Columba eversmanni*	Pale-backed Pigeon	DD		
0130	鸽形目	COLUMBIFORMES	Columbidae	鸠鸽科	斑尾林鸽	*Columba palumbus*	Wood Pigeon	LC		
0131	鸽形目	COLUMBIFORMES	Columbidae	鸠鸽科	斑林鸽	*Columba hodgsonii*	Speckled Wood Pigeon	LC		
0132	鸽形目	COLUMBIFORMES	Columbidae	鸠鸽科	灰林鸽	*Columba pulchricollis*	Ashy Wood Pigeon	LC		

附录 中国鸟类濒危等级评估名录

续表

	目	科		物种中文名	物种学名	物种英文名	濒危等级	评估依据	是否特有
0133	鸽形目	COLUMBIFORMES	Columbidae	鸠鸽科	紫林鸽	*Columba punicea*	Pale-capped Pigeon	EN	B1ab(iii), D	
0134	鸽形目	COLUMBIFORMES	Columbidae	鸠鸽科	黑林鸽	*Columba janthina*	Japanese Wood Pigeon	DD		
0135	鸽形目	COLUMBIFORMES	Columbidae	鸠鸽科	欧斑鸠	*Streptopelia turtur*	European Turtle Dove	LC		
0136	鸽形目	COLUMBIFORMES	Columbidae	鸠鸽科	山斑鸠	*Streptopelia orientalis*	Oriental Turtle Dove	LC		
0137	鸽形目	COLUMBIFORMES	Columbidae	鸠鸽科	灰斑鸠	*Streptopelia decaocto*	Eurasian Collared Dove	LC		
0138	鸽形目	COLUMBIFORMES	Columbidae	鸠鸽科	火斑鸠	*Streptopelia tranquebarica*	Red Turtle Dove	LC		
0139	鸽形目	COLUMBIFORMES	Columbidae	鸠鸽科	珠颈斑鸠	*Spilopelia chinensis*	Spotted-necked Dove	LC		
0140	鸽形目	COLUMBIFORMES	Columbidae	鸠鸽科	棕斑鸠	*Streptopelia senegalensis*	Laughing Dove	LC		
0141	鸽形目	COLUMBIFORMES	Columbidae	鸠鸽科	斑尾鹃鸠	*Macropygia unchall*	Barred Cuckoo Dove	NT		
0142	鸽形目	COLUMBIFORMES	Columbidae	鸠鸽科	菲律宾鹃鸠	*Macropygia tenuirostris*	Philippine Cuckoo Dove	LC		
0143	鸽形目	COLUMBIFORMES	Columbidae	鸠鸽科	小鹃鸠	*Macropygia ruficeps*	Little Cuckoo Dove	LC		
0144	鸽形目	COLUMBIFORMES	Columbidae	鸠鸽科	绿翅金鸠	*Chalcophaps indica*	Emerald Dove	LC		
0145	鸽形目	COLUMBIFORMES	Columbidae	鸠鸽科	橙胸绿鸠	*Treron bicinctus*	Orange-breasted Green Pigeon	NT		
0146	鸽形目	COLUMBIFORMES	Columbidae	鸠鸽科	灰头绿鸠	*Treron pompadora*	Pompadour Green Pigeon	NT		
0147	鸽形目	COLUMBIFORMES	Columbidae	鸠鸽科	厚嘴绿鸠	*Treron curvirostra*	Thick-billed Green Pigeon	NT		
0148	鸽形目	COLUMBIFORMES	Columbidae	鸠鸽科	黄脚绿鸠	*Treron phoenicopterus*	Yellow-footed Green Pigeon	NT		
0149	鸽形目	COLUMBIFORMES	Columbidae	鸠鸽科	针尾绿鸠	*Treron apicauda*	Pin-tailed Green Pigeon	NT		
0150	鸽形目	COLUMBIFORMES	Columbidae	鸠鸽科	楔尾绿鸠	*Treron sphenurus*	Wedge-tailed Green Pigeon	NT		
0151	鸽形目	COLUMBIFORMES	Columbidae	鸠鸽科	红翅绿鸠	*Treron sieboldii*	White-bellied Green Pigeon	LC		
0152	鸽形目	COLUMBIFORMES	Columbidae	鸠鸽科	红顶绿鸠	*Treron formosae*	Whistling Green Pigeon	VU	C1	
0153	鸽形目	COLUMBIFORMES	Columbidae	鸠鸽科	黑颏果鸠	*Ptilinopus leclancheri*	Black-chinned Fruit Dove	LC		
0154	鸽形目	COLUMBIFORMES	Columbidae	鸠鸽科	绿皇鸠	*Ducula aenea*	Green Imperial Pigeon	EN	A4acd, C2a	
0155	鸽形目	COLUMBIFORMES	Columbidae	鸠鸽科	山皇鸠	*Ducula badia*	Mountain Imperial Pigeon	NT		
0156	沙鸡目	PTEROCLIFORMES	Pteroclidae	沙鸡科	西藏毛腿沙鸡	*Syrrhaptes tibetanus*	Tibetan Sandgrouse	LC		
0157	沙鸡目	PTEROCLIFORMES	Pteroclidae	沙鸡科	毛腿沙鸡	*Syrrhaptes paradoxus*	Pallas's Sandgrouse	LC		
0158	沙鸡目	PTEROCLIFORMES	Pteroclidae	沙鸡科	黑腹沙鸡	*Pterocles orientalis*	Black-bellied Sandgrouse	NT		
0159	夜鹰目	CAPRIMULGIFORMES	Podargidae	蛙口夜鹰科	黑顶蛙口夜鹰	*Batrachostomus hodgsoni*	Hodgson's Frogmouth	LC		

续表

	目	科	物种中文名	物种学名	物种英文名	濒危等级	评估依据	是否特有
0160	夜鹰目 CAPRIMULGIFORMES	夜鹰科 Caprimulgidae	毛腿夜鹰	*Lyncornis macrotis*	Great Eared Nightjar	DD		
0161	夜鹰目 CAPRIMULGIFORMES	夜鹰科 Caprimulgidae	普通夜鹰	*Caprimulgus indicus*	Grey Nightjar	LC		
0162	夜鹰目 CAPRIMULGIFORMES	夜鹰科 Caprimulgidae	欧夜鹰	*Caprimulgus europaeus*	European Nightjar	LC		
0163	夜鹰目 CAPRIMULGIFORMES	夜鹰科 Caprimulgidae	埃及夜鹰	*Caprimulgus aegyptius*	Egyptian Nightjar	DD(NE)		
0164	夜鹰目 CAPRIMULGIFORMES	夜鹰科 Caprimulgidae	中亚夜鹰	*Caprimulgus centralasicus*	Vaurie's Nightjar	DD		√
0165	夜鹰目 CAPRIMULGIFORMES	夜鹰科 Caprimulgidae	长尾夜鹰	*Caprimulgus macrurus*	Large-tailed Nightjar	LC		
0166	夜鹰目 CAPRIMULGIFORMES	夜鹰科 Caprimulgidae	林夜鹰	*Caprimulgus affinis*	Savanna Nightjar	LC		
0167	夜鹰目 CAPRIMULGIFORMES	凤头雨燕科 Hemiprocnidae	凤头雨燕	*Hemiprocne coronata*	Crested Treeswift	LC		
0168	夜鹰目 APODIFORMES	雨燕科 Apodidae	短嘴金丝燕	*Aerodramus brevirostris*	Himalayan Swiftlet	NT		
0169	夜鹰目 APODIFORMES	雨燕科 Apodidae	爪哇金丝燕	*Aerodramus fuciphagus*	Edible-nest Swiftlet	CR	B1b(ii, iii), C1+2a(i)	
0170	夜鹰目 APODIFORMES	雨燕科 Apodidae	大金丝燕	*Aerodramus maximus*	Black-nest Swiftlet	DD		
0171	夜鹰目 APODIFORMES	雨燕科 Apodidae	白喉针尾雨燕	*Hirundapus caudacutus*	White-throated Needletail	LC		
0172	夜鹰目 APODIFORMES	雨燕科 Apodidae	灰喉针尾雨燕	*Hirundapus cochinchinensis*	Silver-backed Needletail	NT		
0173	夜鹰目 APODIFORMES	雨燕科 Apodidae	褐背针尾雨燕	*Hirundapus giganteus*	Brown-backed Needletail	LC		
0174	夜鹰目 APODIFORMES	雨燕科 Apodidae	紫针尾雨燕	*Hirundapus celebensis*	Purple Spinetail	NT		
0175	夜鹰目 APODIFORMES	雨燕科 Apodidae	棕雨燕	*Cypsiurus balasiensis*	Asian Palm Swift	LC		
0176	夜鹰目 APODIFORMES	雨燕科 Apodidae	高山雨燕	*Tachymarptis melba*	Alpine Swift	LC		
0177	夜鹰目 APODIFORMES	雨燕科 Apodidae	普通雨燕	*Apus apus*	Common Swift	LC		
0178	夜鹰目 APODIFORMES	雨燕科 Apodidae	白腰雨燕	*Apus pacificus*	Fork-tailed Swift	LC		
0179	夜鹰目 APODIFORMES	雨燕科 Apodidae	暗背雨燕	*Apus acuticauda*	Dark-rumped Swift	DD(NE)		
0180	夜鹰目 APODIFORMES	雨燕科 Apodidae	小白腰雨燕	*Apus nipalensis*	House Swift	LC		
0181	鹃形目 CUCULIFORMES	杜鹃科 Cuculidae	褐翅鸦鹃	*Centropus sinensis*	Greater Coucal	LC		
0182	鹃形目 CUCULIFORMES	杜鹃科 Cuculidae	小鸦鹃	*Centropus bengalensis*	Lesser Coucal	LC		
0183	鹃形目 CUCULIFORMES	杜鹃科 Cuculidae	绿嘴地鹃	*Phaenicophaeus tristis*	Green-billed Malkoha	LC		
0184	鹃形目 CUCULIFORMES	杜鹃科 Cuculidae	红翅凤头鹃	*Clamator coromandus*	Chestnut-winged Cuckoo	LC		
0185	鹃形目 CUCULIFORMES	杜鹃科 Cuculidae	斑翅凤头鹃	*Clamator jacobinus*	Jacobin Cuckoo	LC		

续表

序号	目		科		物种中文名	物种学名	物种英文名	濒危等级	评估依据	是否特有
0186	鹃形目	CUCULIFORMES	杜鹃科	Cuculidae	噪鹃	*Eudynamys scolopaceus*	Common Koel	LC		
0187	鹃形目	CUCULIFORMES	杜鹃科	Cuculidae	翠金鹃	*Chrysococcyx maculatus*	Asian Emerald Cuckoo	NT		
0188	鹃形目	CUCULIFORMES	杜鹃科	Cuculidae	紫金鹃	*Chrysococcyx xanthorhynchus*	Violet Cuckoo	NT		
0189	鹃形目	CUCULIFORMES	杜鹃科	Cuculidae	栗斑杜鹃	*Cacomantis sonneratii*	Banded Bay Cuckoo	LC		
0190	鹃形目	CUCULIFORMES	杜鹃科	Cuculidae	八声杜鹃	*Cacomantis merulinus*	Plaintive Cuckoo	LC		
0191	鹃形目	CUCULIFORMES	杜鹃科	Cuculidae	乌鹃	*Surniculus lugubris*	Drongo Cuckoo	LC		
0192	鹃形目	CUCULIFORMES	杜鹃科	Cuculidae	大鹰鹃	*Hierococcyx sparverioides*	Large Hawk Cuckoo	LC		
0193	鹃形目	CUCULIFORMES	杜鹃科	Cuculidae	普通鹰鹃	*Hierococcyx varius*	Common Hawk Cuckoo	LC		
0194	鹃形目	CUCULIFORMES	杜鹃科	Cuculidae	北棕腹鹰鹃	*Hierococcyx hyperythrus*	Northern Hawk Cuckoo	LC		
0195	鹃形目	CUCULIFORMES	杜鹃科	Cuculidae	棕腹鹰鹃	*Hierococcyx nisicolor*	Whistling Hawk Cuckoo	LC		
0196	鹃形目	CUCULIFORMES	杜鹃科	Cuculidae	小杜鹃	*Cuculus poliocephalus*	Lesser Cuckoo	LC		
0197	鹃形目	CUCULIFORMES	杜鹃科	Cuculidae	四声杜鹃	*Cuculus micropterus*	Indian Cuckoo	LC		
0198	鹃形目	CUCULIFORMES	杜鹃科	Cuculidae	中杜鹃	*Cuculus saturatus*	Himalayan Cuckoo	LC		
0199	鹃形目	CUCULIFORMES	杜鹃科	Cuculidae	东方中杜鹃	*Cuculus optatus*	Oriental Cuckoo	LC		
0200	鹃形目	CUCULIFORMES	杜鹃科	Cuculidae	大杜鹃	*Cuculus canorus*	Common Cuckoo	LC		
0201	鸨形目	OTIDIFORMES	鸨科	Otididae	大鸨	*Otis tarda*	Great Bustard	EN	B2b(iii), C2b	
0202	鸨形目	OTIDIFORMES	鸨科	Otididae	波斑鸨	*Chlamydotis macqueenii*	Macqueen's Bustard	EN	A2cd+3cd, C1	
0203	鸨形目	OTIDIFORMES	鸨科	Otididae	小鸨	*Tetrax tetrax*	Little Bustard	DD		
0204	鹤形目	GRUIFORMES	秧鸡科	Rallidae	花田鸡	*Coturnicops exquisitus*	Swinhoe's Rail	VU	A2cd+3cd, C1	
0205	鹤形目	GRUIFORMES	秧鸡科	Rallidae	红脚斑秧鸡	*Rallina fasciata*	Red-legged Crake	LC		
0206	鹤形目	GRUIFORMES	秧鸡科	Rallidae	白喉斑秧鸡	*Rallina eurizonoides*	Slaty-legged Crake	VU	A1cde	
0207	鹤形目	GRUIFORMES	秧鸡科	Rallidae	灰胸秧鸡	*Lewinia striata*	Slaty-breasted Rail	LC		
0208	鹤形目	GRUIFORMES	秧鸡科	Rallidae	西秧鸡	*Rallus aquaticus*	Water Rail	LC		
0209	鹤形目	GRUIFORMES	秧鸡科	Rallidae	普通秧鸡	*Rallus indicus*	Brown-cheeked Rail	LC		
0210	鹤形目	GRUIFORMES	秧鸡科	Rallidae	长脚秧鸡	*Crex crex*	Corn Crake	VU	A1cde	
0211	鹤形目	GRUIFORMES	秧鸡科	Rallidae	斑胸田鸡	*Porzana porzana*	Spotted Crake	LC		
0212	鹤形目	GRUIFORMES	秧鸡科	Rallidae	红脚田鸡	*Zapornia akool*	Brown Crake	LC		

续表

	目	科	物种中文名	物种学名	物种英文名	濒危等级	评估依据	是否特有
0213	鹤形目 GRUIFORMES	秧鸡科 Rallidae	棕背田鸡	*Zapornia bicolor*	Black-tailed Crake	LC		
0214	鹤形目 GRUIFORMES	秧鸡科 Rallidae	姬田鸡	*Zapornia parva*	Little Crake	LC		
0215	鹤形目 GRUIFORMES	秧鸡科 Rallidae	小田鸡	*Zapornia pusilla*	Baillon's Crake	LC		
0216	鹤形目 GRUIFORMES	秧鸡科 Rallidae	红胸田鸡	*Zapornia fusca*	Ruddy-breasted Crake	NT		
0217	鹤形目 GRUIFORMES	秧鸡科 Rallidae	斑胁田鸡	*Zapornia paykullii*	Band-bellied Crake	VU	A2cd	
0218	鹤形目 GRUIFORMES	秧鸡科 Rallidae	白眉苦恶鸟	*Amaurornis cinerea*	White-browed Crake	LC		
0219	鹤形目 GRUIFORMES	秧鸡科 Rallidae	白胸苦恶鸟	*Amaurornis phoenicurus*	White-breasted Waterhen	LC		
0220	鹤形目 GRUIFORMES	秧鸡科 Rallidae	董鸡	*Gallicrex cinerea*	Watercock	LC		
0221	鹤形目 GRUIFORMES	秧鸡科 Rallidae	紫水鸡	*Porphyrio porphyrio*	Purple Swamphen	VU	A3cd	
0222	鹤形目 GRUIFORMES	秧鸡科 Rallidae	黑水鸡	*Gallinula chloropus*	Common Moorhen	LC		
0223	鹤形目 GRUIFORMES	秧鸡科 Rallidae	白骨顶	*Fulica atra*	Common Coot	LC		
0224	鹤形目 GRUIFORMES	鹤科 Gruidae	白鹤	*Grus leucogeranus*	Siberian Crane	CR	A3bcd+4bcd, B2b	
0225	鹤形目 GRUIFORMES	鹤科 Gruidae	沙丘鹤	*Grus canadensis*	Sandhill Crane	NT		
0226	鹤形目 GRUIFORMES	鹤科 Gruidae	白枕鹤	*Grus vipio*	White-naped Crane	EN	A2ace, C1	
0227	鹤形目 GRUIFORMES	鹤科 Gruidae	赤颈鹤	*Grus antigone*	Sarus Crane	RE		
0228	鹤形目 GRUIFORMES	鹤科 Gruidae	蓑羽鹤	*Grus virgo*	Demoiselle Crane	LC		
0229	鹤形目 GRUIFORMES	鹤科 Gruidae	丹顶鹤	*Grus japonensis*	Red-crowned Crane	EN	C1	
0230	鹤形目 GRUIFORMES	鹤科 Gruidae	灰鹤	*Grus grus*	Common Crane	NT		
0231	鹤形目 GRUIFORMES	鹤科 Gruidae	白头鹤	*Grus monacha*	Hooded Crane	EN	C1+2a(ii)	
0232	鹤形目 GRUIFORMES	鹤科 Gruidae	黑颈鹤	*Grus nigricollis*	Black-necked Crane	VU	C2a(ii)	
0233	鸻形目 CHARADRIIFORMES	石鸻科 Burhinidae	石鸻	*Burhinus oedicnemus*	Eurasian Thick-knee	LC		
0234	鸻形目 CHARADRIIFORMES	石鸻科 Burhinidae	大石鸻	*Esacus recurvirostris*	Great Thick-knee	LC		
0235	鸻形目 CHARADRIIFORMES	蛎鹬科 Haematopodidae	蛎鹬	*Haematopus ostralegus*	Eurasian Oystercatcher	NT		
0236	鸻形目 CHARADRIIFORMES	鹮嘴鹬科 Ibidorhynchidae	鹮嘴鹬	*Ibidorhyncha struthersii*	Ibisbill	LC		
0237	鸻形目 CHARADRIIFORMES	反嘴鹬科 Recurvirostridae	黑翅长脚鹬	*Himantopus himantopus*	Black-winged Stilt	LC		
0238	鸻形目 CHARADRIIFORMES	反嘴鹬科 Recurvirostridae	反嘴鹬	*Recurvirostra avosetta*	Pied Avocet	LC		

附录 中国鸟类濒危等级评估名录

续表

	目	科	科	物种中文名	物种学名	物种英文名	濒危等级	评估依据	是否特有
0239	鸻形目 CHARADRIIFORMES	鸻科	Charadriidae	凤头麦鸡	*Vanellus vanellus*	Northern Lapwing	LC		
0240	鸻形目 CHARADRIIFORMES	鸻科	Charadriidae	距翅麦鸡	*Vanellus duvaucelii*	River Lapwing	NT		
0241	鸻形目 CHARADRIIFORMES	鸻科	Charadriidae	灰头麦鸡	*Vanellus cinereus*	Grey-headed Lapwing	LC		
0242	鸻形目 CHARADRIIFORMES	鸻科	Charadriidae	肉垂麦鸡	*Vanellus indicus*	Red-wattled Lapwing	LC		
0243	鸻形目 CHARADRIIFORMES	鸻科	Charadriidae	黄颊麦鸡	*Vanellus gregarius*	Sociable Lapwing	DD		
0244	鸻形目 CHARADRIIFORMES	鸻科	Charadriidae	白尾麦鸡	*Vanellus leucurus*	White-tailed Lapwing	DD(NE)		
0245	鸻形目 CHARADRIIFORMES	鸻科	Charadriidae	欧金鸻	*Pluvialis apricaria*	European Golden Plover	DD(NE)		
0246	鸻形目 CHARADRIIFORMES	鸻科	Charadriidae	金鸻	*Pluvialis fulva*	Pacific Golden Plover	LC		
0247	鸻形目 CHARADRIIFORMES	鸻科	Charadriidae	灰鸻	*Pluvialis squatarola*	Grey Plover	LC		
0248	鸻形目 CHARADRIIFORMES	鸻科	Charadriidae	剑鸻	*Charadrius hiaticula*	Common Ringed Plover	NT		
0249	鸻形目 CHARADRIIFORMES	鸻科	Charadriidae	长嘴剑鸻	*Charadrius placidus*	Long-billed Plover	LC		
0250	鸻形目 CHARADRIIFORMES	鸻科	Charadriidae	金眶鸻	*Charadrius dubius*	Little Ringed Plover	LC		
0251	鸻形目 CHARADRIIFORMES	鸻科	Charadriidae	环颈鸻	*Charadrius alexandrinus*	Kentish Plover	LC		
0252	鸻形目 CHARADRIIFORMES	鸻科	Charadriidae	蒙古沙鸻	*Charadrius mongolus*	Lesser Sand Plover	LC		
0253	鸻形目 CHARADRIIFORMES	鸻科	Charadriidae	铁嘴沙鸻	*Charadrius leschenaultii*	Greater Sand Plover	LC		
0254	鸻形目 CHARADRIIFORMES	鸻科	Charadriidae	红胸鸻	*Charadrius asiaticus*	Caspian Plover	DD		
0255	鸻形目 CHARADRIIFORMES	鸻科	Charadriidae	东方鸻	*Charadrius veredus*	Oriental Plover	LC		
0256	鸻形目 CHARADRIIFORMES	鸻科	Charadriidae	小嘴鸻	*Eudromias morinellus*	Eurasian Dotterel	NT		
0257	鸻形目 CHARADRIIFORMES	彩鹬科	Rostratulidae	彩鹬	*Rostratula benghalensis*	Greater Painted Snipe	LC		
0258	鸻形目 CHARADRIIFORMES	水雉科	Jacanidae	水雉	*Hydrophasianus chirurgus*	Pheasant-tailed Jacana	NT		
0259	鸻形目 CHARADRIIFORMES	水雉科	Jacanidae	铜翅水雉	*Metopidius indicus*	Bronze-winged Jacana	DD		
0260	鸻形目 CHARADRIIFORMES	鹬科	Scolopacidae	丘鹬	*Scolopax rusticola*	Eurasian Woodcock	LC		
0261	鸻形目 CHARADRIIFORMES	鹬科	Scolopacidae	姬鹬	*Lymnocryptes minimus*	Jack Snipe	LC		
0262	鸻形目 CHARADRIIFORMES	鹬科	Scolopacidae	孤沙锥	*Gallinago solitaria*	Solitary Snipe	LC		
0263	鸻形目 CHARADRIIFORMES	鹬科	Scolopacidae	拉氏沙锥	*Gallinago hardwickii*	Latham's Snipe	DD		
0264	鸻形目 CHARADRIIFORMES	鹬科	Scolopacidae	林沙锥	*Gallinago nemoricola*	Wood Snipe	VU	C2a(i)	
0265	鸻形目 CHARADRIIFORMES	鹬科	Scolopacidae	针尾沙锥	*Gallinago stenura*	Pintail Snipe	LC		

续表

	目	科	物种中文名	物种学名	物种英文名	濒危等级	评估依据	是否特有
0266	鸻形目 CHARADRIIFORMES	鹬科 Scolopacidae	大沙锥	*Gallinago megala*	Swinhoe's Snipe	LC		
0267	鸻形目 CHARADRIIFORMES	鹬科 Scolopacidae	扇尾沙锥	*Gallinago gallinago*	Common Snipe	LC		
0268	鸻形目 CHARADRIIFORMES	鹬科 Scolopacidae	长嘴半蹼鹬	*Limnodromus scolopaceus*	Long-billed Dowitcher	DD		
0269	鸻形目 CHARADRIIFORMES	鹬科 Scolopacidae	半蹼鹬	*Limnodromus semipalmatus*	Asian Dowitcher	NT		
0270	鸻形目 CHARADRIIFORMES	鹬科 Scolopacidae	黑尾塍鹬	*Limosa limosa*	Black-tailed Godwit	LC		
0271	鸻形目 CHARADRIIFORMES	鹬科 Scolopacidae	斑尾塍鹬	*Limosa lapponica*	Bar-tailed Godwit	NT		
0272	鸻形目 CHARADRIIFORMES	鹬科 Scolopacidae	小杓鹬	*Numenius minutus*	Little Curlew	NT		
0273	鸻形目 CHARADRIIFORMES	鹬科 Scolopacidae	中杓鹬	*Numenius phaeopus*	Whimbrel	LC		
0274	鸻形目 CHARADRIIFORMES	鹬科 Scolopacidae	白腰杓鹬	*Numenius arquata*	Eurasian Curlew	NT		
0275	鸻形目 CHARADRIIFORMES	鹬科 Scolopacidae	大杓鹬	*Numenius madagascariensis*	Eastern Curlew	VU	A2cd, C1	
0276	鸻形目 CHARADRIIFORMES	鹬科 Scolopacidae	鹤鹬	*Tringa erythropus*	Spotted Redshank	LC		
0277	鸻形目 CHARADRIIFORMES	鹬科 Scolopacidae	红脚鹬	*Tringa totanus*	Common Redshank	LC		
0278	鸻形目 CHARADRIIFORMES	鹬科 Scolopacidae	泽鹬	*Tringa stagnatilis*	Marsh Sandpiper	LC		
0279	鸻形目 CHARADRIIFORMES	鹬科 Scolopacidae	青脚鹬	*Tringa nebularia*	Common Greenshank	LC		
0280	鸻形目 CHARADRIIFORMES	鹬科 Scolopacidae	小青脚鹬	*Tringa guttifer*	Nordmann's Greenshank	EN	C2a(i)	
0281	鸻形目 CHARADRIIFORMES	鹬科 Scolopacidae	小黄脚鹬	*Tringa flavipes*	Lesser Yellowlegs	DD(NE)		
0282	鸻形目 CHARADRIIFORMES	鹬科 Scolopacidae	白腰草鹬	*Tringa ochropus*	Green Sandpiper	LC		
0283	鸻形目 CHARADRIIFORMES	鹬科 Scolopacidae	林鹬	*Tringa glareola*	Wood Sandpiper	LC		
0284	鸻形目 CHARADRIIFORMES	鹬科 Scolopacidae	灰尾漂鹬	*Tringa brevipes*	Grey-tailed Tattler	LC		
0285	鸻形目 CHARADRIIFORMES	鹬科 Scolopacidae	漂鹬	*Tringa incana*	Wandering Tattler	DD		
0286	鸻形目 CHARADRIIFORMES	鹬科 Scolopacidae	翘嘴鹬	*Xenus cinereus*	Terek Sandpiper	LC		
0287	鸻形目 CHARADRIIFORMES	鹬科 Scolopacidae	矶鹬	*Actitis hypoleucos*	Common Sandpiper	LC		
0288	鸻形目 CHARADRIIFORMES	鹬科 Scolopacidae	翻石鹬	*Arenaria interpres*	Ruddy Turnstone	LC		
0289	鸻形目 CHARADRIIFORMES	鹬科 Scolopacidae	大滨鹬	*Calidris tenuirostris*	Great Knot	VU	A4bcd	
0290	鸻形目 CHARADRIIFORMES	鹬科 Scolopacidae	红腹滨鹬	*Calidris canutus*	Red Knot	VU	A4bcd	
0291	鸻形目 CHARADRIIFORMES	鹬科 Scolopacidae	三趾滨鹬	*Calidris alba*	Sanderling	LC		
0292	鸻形目 CHARADRIIFORMES	鹬科 Scolopacidae	西滨鹬	*Calidris mauri*	Western Sandpiper	DD(NE)		

续表

目	科	物种中文名	物种学名	物种英文名	濒危等级	评估依据	是否特有
0293	鸻形目 CHARADRIIFORMES	鹬科 Scolopacidae	红颈滨鹬	*Calidris ruficollis*	Red-necked Stint	LC		
0294	鸻形目 CHARADRIIFORMES	鹬科 Scolopacidae	勺嘴鹬	*Calidris pygmeus*	Spoon-billed Sandpiper	CR	A2bcd+3bcd+4bcd, C2a(i)	
0295	鸻形目 CHARADRIIFORMES	鹬科 Scolopacidae	小滨鹬	*Calidris minuta*	Little Stint	LC		
0296	鸻形目 CHARADRIIFORMES	鹬科 Scolopacidae	青脚滨鹬	*Calidris temminckii*	Temminck's Stint	LC		
0297	鸻形目 CHARADRIIFORMES	鹬科 Scolopacidae	长趾滨鹬	*Calidris subminuta*	Long-toed Stint	LC		
0298	鸻形目 CHARADRIIFORMES	鹬科 Scolopacidae	白腰滨鹬	*Calidris fuscicollis*	White-rumped Sandpiper	DD(NE)		
0299	鸻形目 CHARADRIIFORMES	鹬科 Scolopacidae	斑胸滨鹬	*Calidris melanotos*	Pectoral Sandpiper	DD		
0300	鸻形目 CHARADRIIFORMES	鹬科 Scolopacidae	黄胸鹬	*Calidris subruficollis*	Buff-breasted Sandpiper	DD(NE)		
0301	鸻形目 CHARADRIIFORMES	鹬科 Scolopacidae	尖尾滨鹬	*Calidris acuminata*	Sharp-tailed Sandpiper	LC		
0302	鸻形目 CHARADRIIFORMES	鹬科 Scolopacidae	阔嘴鹬	*Calidris falcinellus*	Broad-billed Sandpiper	LC		
0303	鸻形目 CHARADRIIFORMES	鹬科 Scolopacidae	流苏鹬	*Calidris pugnax*	Ruff	LC		
0304	鸻形目 CHARADRIIFORMES	鹬科 Scolopacidae	弯嘴滨鹬	*Calidris ferruginea*	Curlew Sandpiper	LC		
0305	鸻形目 CHARADRIIFORMES	鹬科 Scolopacidae	高跷鹬	*Calidris himantopus*	Stilt Sandpiper	DD(NE)		
0306	鸻形目 CHARADRIIFORMES	鹬科 Scolopacidae	岩滨鹬	*Calidris ptilocnemis*	Rock Sandpiper	DD		
0307	鸻形目 CHARADRIIFORMES	鹬科 Scolopacidae	黑腹滨鹬	*Calidris alpina*	Dunlin	LC		
0308	鸻形目 CHARADRIIFORMES	鹬科 Scolopacidae	红颈瓣蹼鹬	*Phalaropus lobatus*	Red-necked Phalarope	LC		
0309	鸻形目 CHARADRIIFORMES	鹬科 Scolopacidae	灰瓣蹼鹬	*Phalaropus fulicarius*	Red Phalarope	LC		
0310	鸻形目 CHARADRIIFORMES	三趾鹑科 Turnicidae	林三趾鹑	*Turnix sylvaticus*	Common Buttonquail	LC		
0311	鸻形目 CHARADRIIFORMES	三趾鹑科 Turnicidae	黄脚三趾鹑	*Turnix tanki*	Yellow-legged Buttonquail	LC		
0312	鸻形目 CHARADRIIFORMES	三趾鹑科 Turnicidae	棕三趾鹑	*Turnix suscitator*	Barred Buttonquail	LC		
0313	鸻形目 CHARADRIIFORMES	燕鸻科 Glareolidae	领燕鸻	*Glareola pratincola*	Collared Pratincole	LC		
0314	鸻形目 CHARADRIIFORMES	燕鸻科 Glareolidae	普通燕鸻	*Glareola maldivarum*	Oriental Pratincole	LC		
0315	鸻形目 CHARADRIIFORMES	燕鸻科 Glareolidae	黑翅燕鸻	*Glareola nordmanni*	Black-winged Pratincole	DD		
0316	鸻形目 CHARADRIIFORMES	燕鸻科 Glareolidae	灰燕鸻	*Glareola lactea*	Little Pratincole	LC		
0317	鸻形目 CHARADRIIFORMES	鸥科 Laridae	白顶玄燕鸥	*Anous stolidus*	Brown Noddy	LC		
0318	鸻形目 CHARADRIIFORMES	鸥科 Laridae	白燕鸥	*Gygis alba*	White Tern	DD(NE)		

续表

	目	科		物种中文名	物种学名	物种英文名	濒危等级	评估依据	是否特有
0319	鸻形目 CHARADRIIFORMES	鸥科	Laridae	三趾鸥	*Rissa tridactyla*	Black-legged Kittiwake	LC		
0320	鸻形目 CHARADRIIFORMES	鸥科	Laridae	叉尾鸥	*Xema sabini*	Sabine's Gull	LC		
0321	鸻形目 CHARADRIIFORMES	鸥科	Laridae	细嘴鸥	*Chroicocephalus genei*	Slender-billed Gull	DD		
0322	鸻形目 CHARADRIIFORMES	鸥科	Laridae	棕头鸥	*Chroicocephalus brunnicephalus*	Brown-headed Gull	LC		
0323	鸻形目 CHARADRIIFORMES	鸥科	Laridae	红嘴鸥	*Chroicocephalus ridibundus*	Black-headed Gull	LC		
0324	鸻形目 CHARADRIIFORMES	鸥科	Laridae	澳洲红嘴鸥	*Chroicocephalus novaehollandiae*	Silver Gull	LC		
0325	鸻形目 CHARADRIIFORMES	鸥科	Laridae	黑嘴鸥	*Saundersilarus saundersi*	Saunders's Gull	VU	A2cde+B1b(iii)+C1	
0326	鸻形目 CHARADRIIFORMES	鸥科	Laridae	小鸥	*Hydrocoloeus minutus*	Little Gull	NT		
0327	鸻形目 CHARADRIIFORMES	鸥科	Laridae	楔尾鸥	*Rhodostethia rosea*	Ross's Gull	DD(NE)		
0328	鸻形目 CHARADRIIFORMES	鸥科	Laridae	弗氏鸥	*Leucophaeus pipixcan*	Franklin's Gull	DD		
0329	鸻形目 CHARADRIIFORMES	鸥科	Laridae	遗鸥	*Ichthyaetus relictus*	Relict Gull	EN	B1b(iii)+1c(iii)	
0330	鸻形目 CHARADRIIFORMES	鸥科	Laridae	渔鸥	*Ichthyaetus ichthyaetus*	Pallas's Gull	LC		
0331	鸻形目 CHARADRIIFORMES	鸥科	Laridae	黑尾鸥	*Larus crassirostris*	Black-tailed Gull	LC		
0332	鸻形目 CHARADRIIFORMES	鸥科	Laridae	普通海鸥	*Larus canus*	Mew Gull	LC		
0333	鸻形目 CHARADRIIFORMES	鸥科	Laridae	灰翅鸥	*Larus glaucescens*	Glaucous-winged Gull	LC		
0334	鸻形目 CHARADRIIFORMES	鸥科	Laridae	北极鸥	*Larus hyperboreus*	Glaucous Gull	LC		
0335	鸻形目 CHARADRIIFORMES	鸥科	Laridae	小黑背银鸥	*Larus fuscus*	Lesser Black-backed Gull	LC		
0336	鸻形目 CHARADRIIFORMES	鸥科	Laridae	西伯利亚银鸥	*Larus smithsonianus*	Siberian Gull	LC		
0337	鸻形目 CHARADRIIFORMES	鸥科	Laridae	黄腿银鸥	*Larus cachinnans*	Caspian Gull	LC		
0338	鸻形目 CHARADRIIFORMES	鸥科	Laridae	灰背鸥	*Larus schistisagus*	Slaty-backed Gull	LC		
0339	鸻形目 CHARADRIIFORMES	鸥科	Laridae	鸥嘴噪鸥	*Gelochelidon nilotica*	Gull-billed Tern	LC		
0340	鸻形目 CHARADRIIFORMES	鸥科	Laridae	红嘴巨燕鸥	*Hydroprogne caspia*	Caspian Tern	LC		
0341	鸻形目 CHARADRIIFORMES	鸥科	Laridae	大凤头燕鸥	*Thalasseus bergii*	Greater Crested Tern	NT		
0342	鸻形目 CHARADRIIFORMES	鸥科	Laridae	小凤头燕鸥	*Thalasseus bengalensis*	Lesser Crested Tern	LC		
0343	鸻形目 CHARADRIIFORMES	鸥科	Laridae	中华凤头燕鸥	*Thalasseus bernsteini*	Chinese Crested Tern	CR	C2a(i)+b, D	
0344	鸻形目 CHARADRIIFORMES	鸥科	Laridae	白嘴端凤头燕鸥	*Thalasseus sandvicensis*	Sandwich Tern	DD(NE)		

续表

	目	科	物种中文名	物种学名	物种英文名	濒危等级	评估依据	是否特有
0345	鸻形目 CHARADRIIFORMES	鸥科 Laridae	白额燕鸥	*Sternula albifrons*	Little Tern	LC		
0346	鸻形目 CHARADRIIFORMES	鸥科 Laridae	白腰燕鸥	*Onychoprion aleuticus*	Aleutian Tern	LC		
0347	鸻形目 CHARADRIIFORMES	鸥科 Laridae	褐翅燕鸥	*Onychoprion anaethetus*	Bridled Tern	LC		
0348	鸻形目 CHARADRIIFORMES	鸥科 Laridae	乌燕鸥	*Onychoprion fuscatus*	Sooty Tern	LC		
0349	鸻形目 CHARADRIIFORMES	鸥科 Laridae	河燕鸥	*Sterna aurantia*	River Tern	NT		
0350	鸻形目 CHARADRIIFORMES	鸥科 Laridae	粉红燕鸥	*Sterna dougallii*	Roseate Tern	LC		
0351	鸻形目 CHARADRIIFORMES	鸥科 Laridae	黑枕燕鸥	*Sterna sumatrana*	Black-naped Tern	LC		
0352	鸻形目 CHARADRIIFORMES	鸥科 Laridae	普通燕鸥	*Sterna hirundo*	Common Tern	LC		
0353	鸻形目 CHARADRIIFORMES	鸥科 Laridae	黑腹燕鸥	*Sterna acuticauda*	Black-bellied Tern	EN	B1, D	
0354	鸻形目 CHARADRIIFORMES	鸥科 Laridae	灰翅浮鸥	*Chlidonias hybrida*	Whiskered Tern	LC		
0355	鸻形目 CHARADRIIFORMES	鸥科 Laridae	白翅浮鸥	*Chlidonias leucopterus*	White-winged Tern	LC		
0356	鸻形目 CHARADRIIFORMES	鸥科 Laridae	黑浮鸥	*Chlidonias niger*	Black Tern	LC		
0357	鸻形目 CHARADRIIFORMES	鸥科 Laridae	剪嘴鸥	*Rynchops albicollis*	Indian Skimmer	DD		
0358	鸻形目 CHARADRIIFORMES	贼鸥科 Stercorariidae	南极贼鸥	*Stercorarius maccormicki*	South Polar Skua	DD		
0359	鸻形目 CHARADRIIFORMES	贼鸥科 Stercorariidae	中贼鸥	*Stercorarius pomarinus*	Pomarine Skua	LC		
0360	鸻形目 CHARADRIIFORMES	贼鸥科 Stercorariidae	短尾贼鸥	*Stercorarius parasiticus*	Parasitic Jaeger	LC		
0361	鸻形目 CHARADRIIFORMES	贼鸥科 Stercorariidae	长尾贼鸥	*Stercorarius longicaudus*	Long-tailed Jaeger	LC		
0362	鸻形目 CHARADRIIFORMES	海雀科 Alcidae	崖海鸦	*Uria aalge*	Common Murre	DD(NE)		
0363	鸻形目 CHARADRIIFORMES	海雀科 Alcidae	长嘴斑海雀	*Brachyramphus perdix*	Long-billed Murrelet	NT		
0364	鸻形目 CHARADRIIFORMES	海雀科 Alcidae	扁嘴海雀	*Synthliboramphus antiquus*	Ancient Murrelet	NT		
0365	鸻形目 CHARADRIIFORMES	海雀科 Alcidae	冠海雀	*Synthliboramphus wumizusume*	Japanese Murrelet	DD(NE)		
0366	鸻形目 CHARADRIIFORMES	海雀科 Alcidae	角嘴海雀	*Cerorhinca monocerata*	Rhinoceros Auklet	DD		
0367	鹲形目 PHAETHONTIFORMES	鹲科 Phaethontidae	红嘴鹲	*Phaethon aethereus*	Red-billed Tropicbird	DD(NE)		
0368	鹲形目 PHAETHONTIFORMES	鹲科 Phaethontidae	红尾鹲	*Phaethon rubricauda*	Red-tailed Tropicbird	DD(NE)		
0369	鹲形目 PHAETHONTIFORMES	鹲科 Phaethontidae	白尾鹲	*Phaethon lepturus*	White-tailed Tropicbird	DD(NE)		
0370	潜鸟目 GAVIIFORMES	潜鸟科 Gaviidae	红喉潜鸟	*Gavia stellata*	Red-throated Diver	LC		
0371	潜鸟目 GAVIIFORMES	潜鸟科 Gaviidae	黑喉潜鸟	*Gavia arctica*	Black-throated Diver	LC		

续表

	目	科	物种中文名	物种学名	物种英文名	濒危等级	评估依据	是否特有
0372	潜鸟目 GAVIIFORMES	潜鸟科 Gaviidae	太平洋潜鸟	*Gavia pacifica*	Pacific Diver	DD		
0373	潜鸟目 GAVIIFORMES	潜鸟科 Gaviidae	黄嘴潜鸟	*Gavia adamsii*	Yellow-billed Diver	DD		
0374	鹱形目 PROCELLARIIFORMES	信天翁科 Diomedeidae	黑背信天翁	*Phoebastria immutabilis*	Laysan Albatross	NT		
0375	鹱形目 PROCELLARIIFORMES	信天翁科 Diomedeidae	黑脚信天翁	*Phoebastria nigripes*	Black-footed Albatross	NT		
0376	鹱形目 PROCELLARIIFORMES	信天翁科 Diomedeidae	短尾信天翁	*Phoebastria albatrus*	Short-tailed Albatross	VU	D2	
0377	鹱形目 PROCELLARIIFORMES	海燕科 Hydrobatidae	黑叉尾海燕	*Hydrobates monorhis*	Swinhoe's Storm Petrel	NT		
0378	鹱形目 PROCELLARIIFORMES	海燕科 Hydrobatidae	白腰叉尾海燕	*Hydrobates leucorhous*	Leach's Storm Petrel	DD(NE)		
0379	鹱形目 PROCELLARIIFORMES	海燕科 Hydrobatidae	褐翅叉尾海燕	*Hydrobates tristrami*	Tristram's Storm Petrel	DD(NE)		
0380	鹱形目 PROCELLARIIFORMES	海燕科 Hydrobatidae	黄蹼洋海燕	*Oceanites oceanicus*	Wilson's Storm Petrel	DD(NE)		
0381	鹱形目 PROCELLARIIFORMES	鹱科 Procellariidae	暴风鹱	*Fulmarus glacialis*	Northern Fulmar	DD(NE)		
0382	鹱形目 PROCELLARIIFORMES	鹱科 Procellariidae	白额圆尾鹱	*Pterodroma hypoleuca*	Bonin Petrel	DD		
0383	鹱形目 PROCELLARIIFORMES	鹱科 Procellariidae	钩嘴圆尾鹱	*Pseudobulweria rostrata*	Tahiti Petrel	DD(NE)		
0384	鹱形目 PROCELLARIIFORMES	鹱科 Procellariidae	白额鹱	*Calonectris leucomelas*	Streaked Shearwater	LC		
0385	鹱形目 PROCELLARIIFORMES	鹱科 Procellariidae	楔尾鹱	*Ardenna pacificus*	Wedge-tailed Shearwater	DD(NE)		
0386	鹱形目 PROCELLARIIFORMES	鹱科 Procellariidae	灰鹱	*Ardenna grisea*	Sooty Shearwater	DD		
0387	鹱形目 PROCELLARIIFORMES	鹱科 Procellariidae	短尾鹱	*Ardenna tenuirostris*	Short-tailed Shearwater	DD(NE)		
0388	鹱形目 PROCELLARIIFORMES	鹱科 Procellariidae	淡足鹱	*Ardenna carneipes*	Flesh-footed Shearwater	DD		
0389	鹱形目 PROCELLARIIFORMES	鹱科 Procellariidae	褐燕鹱	*Bulweria bulwerii*	Bulwer's Petrel	DD		
0390	鹳形目 CICONIIFORMES	鹳科 Ciconiidae	彩鹳	*Mycteria leucocephala*	Painted Stork	DD		
0391	鹳形目 CICONIIFORMES	鹳科 Ciconiidae	钳嘴鹳	*Anastomus oscitans*	Asian Open-bill Stork	LC		
0392	鹳形目 CICONIIFORMES	鹳科 Ciconiidae	黑鹳	*Ciconia nigra*	Black Stork	VU	C2a(i)	
0393	鹳形目 CICONIIFORMES	鹳科 Ciconiidae	白颈鹳	*Ciconia episcopus*	Woolly-necked Stork	DD(NE)		
0394	鹳形目 CICONIIFORMES	鹳科 Ciconiidae	白鹳	*Ciconia ciconia*	White Stork	RE		
0395	鹳形目 CICONIIFORMES	鹳科 Ciconiidae	东方白鹳	*Ciconia boyciana*	Oriental Stork	EN	C1, C2a(i)	
0396	鹳形目 CICONIIFORMES	鹳科 Ciconiidae	秃鹳	*Leptoptilos javanicus*	Lesser Adjutant Stork	DD		
0397	鲣鸟目 SULIFORMES	军舰鸟科 Fregatidae	白腹军舰鸟	*Fregata andrewsi*	Christmas Island Frigatebird	DD		
0398	鲣鸟目 SULIFORMES	军舰鸟科 Fregatidae	黑腹军舰鸟	*Fregata minor*	Great Frigatebird	LC		

附录 中国鸟类濒危等级评估名录

续表

	目		科		物种中文名	物种学名	物种英文名	濒危等级	评估依据	是否特有
0399	鲣鸟目	SULIFORMES	军舰鸟科	Fregatidae	白斑军舰鸟	Fregata ariel	Lesser Frigatebird	DD		
0400	鲣鸟目	SULIFORMES	鲣鸟科	Sulidae	蓝脸鲣鸟	Sula dactylatra	Masked Booby	LC		
0401	鲣鸟目	SULIFORMES	鲣鸟科	Sulidae	红脚鲣鸟	Sula sula	Red-footed Booby	NT		
0402	鲣鸟目	SULIFORMES	鲣鸟科	Sulidae	褐鲣鸟	Sula leucogaster	Brown Booby	LC		
0403	鲣鸟目	SULIFORMES	鸬鹚科	Phalacrocoracidae	黑颈鸬鹚	Microcarbo niger	Little Cormorant	LC		
0404	鲣鸟目	SULIFORMES	鸬鹚科	Phalacrocoracidae	海鸬鹚	Phalacrocorax pelagicus	Pelagic Cormorant	NT		
0405	鲣鸟目	SULIFORMES	鸬鹚科	Phalacrocoracidae	红脸鸬鹚	Phalacrocorax urile	Red-faced Cormorant	LC		
0406	鲣鸟目	SULIFORMES	鸬鹚科	Phalacrocoracidae	普通鸬鹚	Phalacrocorax carbo	Great Cormorant	LC		
0407	鲣鸟目	SULIFORMES	鸬鹚科	Phalacrocoracidae	绿背鸬鹚	Phalacrocorax capillatus	Japanese Cormorant	LC		
0408	鹈形目	PELECANIFORMES	鹮科	Threskiornithidae	黑头白鹮	Threskiornis melanocephalus	Black-headed Ibis	CR	A2bcd, C1+2a(i)	
0409	鹈形目	PELECANIFORMES	鹮科	Threskiornithidae	白肩黑鹮	Pseudibis davisoni	White-shouldered Ibis	DD		
0410	鹈形目	PELECANIFORMES	鹮科	Threskiornithidae	朱鹮	Nipponia nippon	Crested Ibis	EN	B1ab(iii), C1+C2a(ii)	
0411	鹈形目	PELECANIFORMES	鹮科	Threskiornithidae	彩鹮	Plegadis falcinellus	Glossy Ibis	NT		
0412	鹈形目	PELECANIFORMES	鹮科	Threskiornithidae	白琵鹭	Platalea leucorodia	Eurasian Spoonbill	NT		
0413	鹈形目	PELECANIFORMES	鹮科	Threskiornithidae	黑脸琵鹭	Platalea minor	Black-faced Spoonbill	EN	C2a(ii)	
0414	鹈形目	PELECANIFORMES	鹭科	Ardeidae	大麻鳽	Botaurus stellaris	Eurasian Bittern	LC		
0415	鹈形目	PELECANIFORMES	鹭科	Ardeidae	小苇鳽	Ixobrychus minutus	Little Bittern	NT		
0416	鹈形目	PELECANIFORMES	鹭科	Ardeidae	黄斑苇鳽	Ixobrychus sinensis	Yellow Bittern	LC		
0417	鹈形目	PELECANIFORMES	鹭科	Ardeidae	紫背苇鳽	Ixobrychus eurhythmus	Von Schrenck's Bittern	LC		
0418	鹈形目	PELECANIFORMES	鹭科	Ardeidae	栗苇鳽	Ixobrychus cinnamomeus	Cinnamon Bittern	LC		
0419	鹈形目	PELECANIFORMES	鹭科	Ardeidae	黑苇鳽	Ixobrychus flavicollis	Black Bittern	LC		
0420	鹈形目	PELECANIFORMES	鹭科	Ardeidae	海南鳽	Gorsachius magnificus	White-eared Night Heron	EN	C2a(i)	
0421	鹈形目	PELECANIFORMES	鹭科	Ardeidae	栗头鳽	Gorsachius goisagi	Japanese Night Heron	DD		
0422	鹈形目	PELECANIFORMES	鹭科	Ardeidae	黑冠鳽	Gorsachius melanolophus	Malayan Night Heron	NT		
0423	鹈形目	PELECANIFORMES	鹭科	Ardeidae	夜鹭	Nycticorax nycticorax	Black-crowned Night Heron	LC		
0424	鹈形目	PELECANIFORMES	鹭科	Ardeidae	棕夜鹭	Nycticorax caledonicus	Rufous Night Heron	LC		

续表

目	目	科	科	物种中文名	物种学名	物种英文名	濒危等级	评估依据	是否特有
0425	鹈形目	PELECANIFORMES	鹭科	Ardeidae	绿鹭	*Butorides striata*	Striated Heron	LC		
0426	鹈形目	PELECANIFORMES	鹭科	Ardeidae	印度池鹭	*Ardeola grayii*	Indian Pond Heron	LC		
0427	鹈形目	PELECANIFORMES	鹭科	Ardeidae	池鹭	*Ardeola bacchus*	Chinese Pond Heron	LC		
0428	鹈形目	PELECANIFORMES	鹭科	Ardeidae	爪哇池鹭	*Ardeola speciosa*	Javan Pond Heron	LC		
0429	鹈形目	PELECANIFORMES	鹭科	Ardeidae	牛背鹭	*Bubulcus ibis*	Cattle Egret	LC		
0430	鹈形目	PELECANIFORMES	鹭科	Ardeidae	苍鹭	*Ardea cinerea*	Grey Heron	LC		
0431	鹈形目	PELECANIFORMES	鹭科	Ardeidae	白腹鹭	*Ardea insignis*	White-bellied Heron	DD		
0432	鹈形目	PELECANIFORMES	鹭科	Ardeidae	草鹭	*Ardea purpurea*	Purple Heron	LC		
0433	鹈形目	PELECANIFORMES	鹭科	Ardeidae	大白鹭	*Ardea alba*	Great Egret	LC		
0434	鹈形目	PELECANIFORMES	鹭科	Ardeidae	中白鹭	*Ardea intermedia*	Intermediate Egret	LC		
0435	鹈形目	PELECANIFORMES	鹭科	Ardeidae	斑鹭	*Egretta picata*	Pied Heron	DD(NE)		
0436	鹈形目	PELECANIFORMES	鹭科	Ardeidae	白脸鹭	*Egretta novaehollandiae*	White-faced Egret	DD(NE)		
0437	鹈形目	PELECANIFORMES	鹭科	Ardeidae	白鹭	*Egretta garzetta*	Little Egret	LC		
0438	鹈形目	PELECANIFORMES	鹭科	Ardeidae	岩鹭	*Egretta sacra*	Pacific Reef Heron	LC		
0439	鹈形目	PELECANIFORMES	鹭科	Ardeidae	黄嘴白鹭	*Egretta eulophotes*	Chinese Egret	VU	C2a(i)	
0440	鹈形目	PELECANIFORMES	鹈鹕科	Pelecanidae	白鹈鹕	*Pelecanus onocrotalus*	Great White Pelican	EN	B1ab, D	
0441	鹈形目	PELECANIFORMES	鹈鹕科	Pelecanidae	斑嘴鹈鹕	*Pelecanus philippensis*	Spot-billed Pelican	EN	B1ab, D	
0442	鹈形目	PELECANIFORMES	鹈鹕科	Pelecanidae	卷羽鹈鹕	*Pelecanus crispus*	Dalmatian Pelican	EN	A2ce+3ce+4ce	
0443	鹰形目	ACCIPITRIFORMES	鹗科	Pandionidae	鹗	*Pandion haliaetus*	Osprey	NT		
0444	鹰形目	ACCIPITRIFORMES	鹰科	Accipitridae	黑翅鸢	*Elanus caeruleus*	Black-winged Kite	NT		
0445	鹰形目	ACCIPITRIFORMES	鹰科	Accipitridae	胡兀鹫	*Gypaetus barbatus*	Bearded Vulture	NT		
0446	鹰形目	ACCIPITRIFORMES	鹰科	Accipitridae	白兀鹫	*Neophron percnopterus*	Egyptian Vulture	DD(NE)		
0447	鹰形目	ACCIPITRIFORMES	鹰科	Accipitridae	鹃头蜂鹰	*Pernis apivorus*	European Honey Buzzard	DD(NE)		
0448	鹰形目	ACCIPITRIFORMES	鹰科	Accipitridae	凤头蜂鹰	*Pernis ptilorhynchus*	Oriental Honey Buzzard	NT		
0449	鹰形目	ACCIPITRIFORMES	鹰科	Accipitridae	褐冠鹃隼	*Aviceda jerdoni*	Jerdon's Baza	NT		
0450	鹰形目	ACCIPITRIFORMES	鹰科	Accipitridae	黑冠鹃隼	*Aviceda leuphotes*	Black Baza	LC		
0451	鹰形目	ACCIPITRIFORMES	鹰科	Accipitridae	兀鹫	*Gyps fulvus*	Eurasian Griffon	NT		

续表

编号	目	科	科	物种中文名	物种学名	物种英文名	濒危等级	评估依据	是否特有
0452	鹰形目	ACCIPITRIFORMES	Accipitridae	鹰科	长嘴兀鹫	*Gyps indicus*	Indian Vulture	DD		
0453	鹰形目	ACCIPITRIFORMES	Accipitridae	鹰科	白背兀鹫	*Gyps bengalensis*	White-rumped Vulture	DD		
0454	鹰形目	ACCIPITRIFORMES	Accipitridae	鹰科	高山兀鹫	*Gyps himalayensis*	Himalayan Vulture	NT		
0455	鹰形目	ACCIPITRIFORMES	Accipitridae	鹰科	黑兀鹫	*Sarcogyps calvus*	Red-headed Vulture	CR	A2abce+3bce+4abce	
0456	鹰形目	ACCIPITRIFORMES	Accipitridae	鹰科	秃鹫	*Aegypius monachus*	Cinereous Vulture	NT		
0457	鹰形目	ACCIPITRIFORMES	Accipitridae	鹰科	蛇雕	*Spilornis cheela*	Crested Serpent Eagle	NT		
0458	鹰形目	ACCIPITRIFORMES	Accipitridae	鹰科	短趾雕	*Circaetus gallicus*	Short-toed Snake Eagle	NT		
0459	鹰形目	ACCIPITRIFORMES	Accipitridae	鹰科	凤头鹰雕	*Nisaetus cirrhatus*	Changeable Hawk-Eagle	NT		
0460	鹰形目	ACCIPITRIFORMES	Accipitridae	鹰科	鹰雕	*Nisaetus nipalensis*	Mountain Hawk-Eagle	NT		
0461	鹰形目	ACCIPITRIFORMES	Accipitridae	鹰科	棕腹隼雕	*Lophotriorchis kienerii*	Rufous-bellied Hawk-Eagle	NT		
0462	鹰形目	ACCIPITRIFORMES	Accipitridae	鹰科	林雕	*Ictinaetus malaiensis*	Black Eagle	VU	A2cd	
0463	鹰形目	ACCIPITRIFORMES	Accipitridae	鹰科	乌雕	*Clanga clanga*	Greater Spotted Eagle	EN	A2bcde+3cde+4bcde	
0464	鹰形目	ACCIPITRIFORMES	Accipitridae	鹰科	靴隼雕	*Hieraaetus pennatus*	Booted Eagle	VU	A2cd, C1	
0465	鹰形目	ACCIPITRIFORMES	Accipitridae	鹰科	草原雕	*Aquila nipalensis*	Steppe Eagle	VU	A2cd, C1+2b	
0466	鹰形目	ACCIPITRIFORMES	Accipitridae	鹰科	白肩雕	*Aquila heliaca*	Imperial Eagle	EN	A2bcde+3cde+4cbde	
0467	鹰形目	ACCIPITRIFORMES	Accipitridae	鹰科	金雕	*Aquila chrysaetos*	Golden Eagle	VU	A2bcde+3bcde+4bcde, C2a(i)	
0468	鹰形目	ACCIPITRIFORMES	Accipitridae	鹰科	白腹隼雕	*Aquila fasciata*	Bonelli's Eagle	VU	A2cd, C1	
0469	鹰形目	ACCIPITRIFORMES	Accipitridae	鹰科	凤头鹰	*Accipiter trivirgatus*	Crested Goshawk	NT		
0470	鹰形目	ACCIPITRIFORMES	Accipitridae	鹰科	褐耳鹰	*Accipiter badius*	Shikra	NT		
0471	鹰形目	ACCIPITRIFORMES	Accipitridae	鹰科	赤腹鹰	*Accipiter soloensis*	Chinese Sparrowhawk	LC		
0472	鹰形目	ACCIPITRIFORMES	Accipitridae	鹰科	日本松雀鹰	*Accipiter gularis*	Japanese Sparrowhawk	LC		
0473	鹰形目	ACCIPITRIFORMES	Accipitridae	鹰科	松雀鹰	*Accipiter virgatus*	Besra	LC		
0474	鹰形目	ACCIPITRIFORMES	Accipitridae	鹰科	雀鹰	*Accipiter nisus*	Eurasian Sparrowhawk	LC		
0475	鹰形目	ACCIPITRIFORMES	Accipitridae	鹰科	苍鹰	*Accipiter gentilis*	Northern Goshawk	NT		

续表

	目		科		物种中文名	物种学名	物种英文名	濒危等级	评估依据	是否特有
0476	鹰形目	ACCIPITRIFORMES	鹰科	Accipitridae	白头鹞	*Circus aeruginosus*	Western Marsh Harrier	NT		
0477	鹰形目	ACCIPITRIFORMES	鹰科	Accipitridae	白腹鹞	*Circus spilonotus*	Eastern Marsh Harrier	NT		
0478	鹰形目	ACCIPITRIFORMES	鹰科	Accipitridae	白尾鹞	*Circus cyaneus*	Hen Harrier	NT		
0479	鹰形目	ACCIPITRIFORMES	鹰科	Accipitridae	草原鹞	*Circus macrourus*	Pallid Harrier	NT		
0480	鹰形目	ACCIPITRIFORMES	鹰科	Accipitridae	鹊鹞	*Circus melanoleucos*	Pied Harrier	NT		
0481	鹰形目	ACCIPITRIFORMES	鹰科	Accipitridae	乌灰鹞	*Circus pygargus*	Montagu's Harrier	NT		
0482	鹰形目	ACCIPITRIFORMES	鹰科	Accipitridae	黑鸢	*Milvus migrans*	Black Kite	LC		
0483	鹰形目	ACCIPITRIFORMES	鹰科	Accipitridae	栗鸢	*Haliastur indus*	Brahminy Kite	VU	C1	
0484	鹰形目	ACCIPITRIFORMES	鹰科	Accipitridae	白腹海雕	*Haliaeetus leucogaster*	White-bellied Sea Eagle	VU	C2a(ii)	
0485	鹰形目	ACCIPITRIFORMES	鹰科	Accipitridae	玉带海雕	*Haliaeetus leucoryphus*	Pallas's Fish Eagle	EN	A2bdce+3cde+4cbde	
0486	鹰形目	ACCIPITRIFORMES	鹰科	Accipitridae	白尾海雕	*Haliaeetus albicilla*	White-tailed Sea Eagle	VU	C1	
0487	鹰形目	ACCIPITRIFORMES	鹰科	Accipitridae	虎头海雕	*Haliaeetus pelagicus*	Steller's Sea Eagle	EN	A2bcde+3cde+4bcde	
0488	鹰形目	ACCIPITRIFORMES	鹰科	Accipitridae	渔雕	*Ichthyophaga humilis*	Lesser Fish Eagle	NT		
0489	鹰形目	ACCIPITRIFORMES	鹰科	Accipitridae	白眼鵟鹰	*Butastur teesa*	White-eyed Buzzard	DD		
0490	鹰形目	ACCIPITRIFORMES	鹰科	Accipitridae	棕翅鵟鹰	*Butastur liventer*	Rufous-winged Buzzard	DD		
0491	鹰形目	ACCIPITRIFORMES	鹰科	Accipitridae	灰脸鵟鹰	*Butastur indicus*	Grey-faced Buzzard	NT		
0492	鹰形目	ACCIPITRIFORMES	鹰科	Accipitridae	毛脚鵟	*Buteo lagopus*	Rough-legged Buzzard	NT		
0493	鹰形目	ACCIPITRIFORMES	鹰科	Accipitridae	大鵟	*Buteo hemilasius*	Upland Buzzard	VU	A2ac	
0494	鹰形目	ACCIPITRIFORMES	鹰科	Accipitridae	普通鵟	*Buteo japonicus*	Eastern Buzzard	LC		
0495	鹰形目	ACCIPITRIFORMES	鹰科	Accipitridae	喜山鵟	*Buteo refectus*	Himalayan Buzzard	LC		
0496	鹰形目	ACCIPITRIFORMES	鹰科	Accipitridae	欧亚鵟	*Buteo buteo*	Eurasian Buzzard	LC		
0497	鹰形目	ACCIPITRIFORMES	鹰科	Accipitridae	棕尾鵟	*Buteo rufinus*	Long-legged Buzzard	NT		
0498	鸮形目	STRIGIFORMES	鸱鸮科	Strigidae	黄嘴角鸮	*Otus spilocephalus*	Mountain Scops Owl	NT		
0499	鸮形目	STRIGIFORMES	鸱鸮科	Strigidae	领角鸮	*Otus lettia*	Collared Scops Owl	LC		
0500	鸮形目	STRIGIFORMES	鸱鸮科	Strigidae	北领角鸮	*Otus semitorques*	Japanese Scops Owl	LC		
0501	鸮形目	STRIGIFORMES	鸱鸮科	Strigidae	纵纹角鸮	*Otus brucei*	Pallid Scops Owl	LC		

续表

	目	科	物种中文名	物种学名	物种英文名	濒危等级	评估依据	是否特有
0502	STRIGIFORMES 鸮形目	Strigidae 鸱鸮科	西红角鸮	*Otus scops*	Eurasian Scops Owl	LC		
0503	STRIGIFORMES 鸮形目	Strigidae 鸱鸮科	红角鸮	*Otus sunia*	Oriental Scops Owl	LC		
0504	STRIGIFORMES 鸮形目	Strigidae 鸱鸮科	优雅角鸮	*Otus elegans*	Elegant Scops Owl	NT		
0505	STRIGIFORMES 鸮形目	Strigidae 鸱鸮科	雪鸮	*Bubo scandiacus*	Snowy Owl	NT		
0506	STRIGIFORMES 鸮形目	Strigidae 鸱鸮科	雕鸮	*Bubo bubo*	Eurasian Eagle-Owl	NT		
0507	STRIGIFORMES 鸮形目	Strigidae 鸱鸮科	林雕鸮	*Bubo nipalensis*	Spot-bellied Eagle-Owl	NT		
0508	STRIGIFORMES 鸮形目	Strigidae 鸱鸮科	毛腿雕鸮	*Bubo blakistoni*	Blakiston's Fish Owl	CR	A1ac, B1b(ii, iii), C1+2a(i)	
0509	STRIGIFORMES 鸮形目	Strigidae 鸱鸮科	褐渔鸮	*Ketupa zeylonensis*	Brown Fish Owl	EN	A2bcd+3bcd+4bcd	
0510	STRIGIFORMES 鸮形目	Strigidae 鸱鸮科	黄腿渔鸮	*Ketupa flavipes*	Tawny Fish Owl	EN	A2bcd+3bcd+4bcd	
0511	STRIGIFORMES 鸮形目	Strigidae 鸱鸮科	褐林鸮	*Strix leptogrammica*	Brown Wood Owl	NT		
0512	STRIGIFORMES 鸮形目	Strigidae 鸱鸮科	灰林鸮	*Strix aluco*	Tawny Owl	NT		
0513	STRIGIFORMES 鸮形目	Strigidae 鸱鸮科	长尾林鸮	*Strix uralensis*	Ural Owl	NT		
0514	STRIGIFORMES 鸮形目	Strigidae 鸱鸮科	四川林鸮	*Strix davidi*	Sichuan Wood Owl	VU	C2a(i)	√
0515	STRIGIFORMES 鸮形目	Strigidae 鸱鸮科	乌林鸮	*Strix nebulosa*	Great Grey Owl	NT		
0516	STRIGIFORMES 鸮形目	Strigidae 鸱鸮科	猛鸮	*Surnia ulula*	Hawk Owl	NT		
0517	STRIGIFORMES 鸮形目	Strigidae 鸱鸮科	花头鸺鹠	*Glaucidium passerinum*	Eurasian Pygmy Owlet	NT		
0518	STRIGIFORMES 鸮形目	Strigidae 鸱鸮科	领鸺鹠	*Glaucidium brodiei*	Collared Owlet	LC		
0519	STRIGIFORMES 鸮形目	Strigidae 鸱鸮科	斑头鸺鹠	*Glaucidium cuculoides*	Asian Barred Owlet	LC		
0520	STRIGIFORMES 鸮形目	Strigidae 鸱鸮科	纵纹腹小鸮	*Athene noctua*	Little Owl	LC		
0521	STRIGIFORMES 鸮形目	Strigidae 鸱鸮科	横斑腹小鸮	*Athene brama*	Spotted Owlet	NT		
0522	STRIGIFORMES 鸮形目	Strigidae 鸱鸮科	鬼鸮	*Aegolius funereus*	Boreal Owl	VU	C2a(i)	
0523	STRIGIFORMES 鸮形目	Strigidae 鸱鸮科	鹰鸮	*Ninox scutulata*	Brown Boobook	NT		
0524	STRIGIFORMES 鸮形目	Strigidae 鸱鸮科	日本鹰鸮	*Ninox japonica*	Northern Boobook	NT		
0525	STRIGIFORMES 鸮形目	Strigidae 鸱鸮科	长耳鸮	*Asio otus*	Long-eared Owl	LC		
0526	STRIGIFORMES 鸮形目	Strigidae 鸱鸮科	短耳鸮	*Asio flammeus*	Short-eared Owl	NT		

续表

	目	科	物种中文名	物种学名	物种英文名	濒危等级	评估依据	是否特有
0527	鸮形目 STRIGIFORMES	草鸮科 Tytonidae	仓鸮	*Tyto alba*	Barn Owl	NT		
0528	鸮形目 STRIGIFORMES	草鸮科 Tytonidae	草鸮	*Tyto longimembris*	Eastern Grass Owl	NT		
0529	鸮形目 STRIGIFORMES	草鸮科 Tytonidae	栗鸮	*Phodilus badius*	Bay Owl	NT		
0530	咬鹃目 TROGONIFORMES	咬鹃科 Trogonidae	橙胸咬鹃	*Harpactes oreskios*	Orange-breasted Trogon	NT		
0531	咬鹃目 TROGONIFORMES	咬鹃科 Trogonidae	红头咬鹃	*Harpactes erythrocephalus*	Red-headed Trogon	NT		
0532	咬鹃目 TROGONIFORMES	咬鹃科 Trogonidae	红腹咬鹃	*Harpactes wardi*	Ward's Trogon	NT		
0533	犀鸟目 BUCEROTIFORMES	犀鸟科 Bucerotidae	白喉犀鸟	*Anorrhinus austeni*	Austen's Brown Hornbill	VU	C1	
0534	犀鸟目 BUCEROTIFORMES	犀鸟科 Bucerotidae	冠斑犀鸟	*Anthracoceros albirostris*	Oriental Pied Hornbill	CR	B1ab(iii), C1+2(i)	
0535	犀鸟目 BUCEROTIFORMES	犀鸟科 Bucerotidae	双角犀鸟	*Buceros bicornis*	Great Hornbill	CR	B1ab(ii, iii), D	
0536	犀鸟目 BUCEROTIFORMES	犀鸟科 Bucerotidae	棕颈犀鸟	*Aceros nipalensis*	Rufous-necked Hornbill	CR	B1ab(i, ii, iii), D	
0537	犀鸟目 BUCEROTIFORMES	犀鸟科 Bucerotidae	花冠皱盔犀鸟	*Rhyticeros undulatus*	Wreathed Hornbill	EN	B1ab(iii) + 2ab(iii), D	
0538	犀鸟目 BUCEROTIFORMES	戴胜科 Upupidae	戴胜	*Upupa epops*	Common Hoopoe	LC		
0539	佛法僧目 CORACIIFORMES	蜂虎科 Meropidae	赤须蜂虎	*Nyctyornis amictus*	Red-bearded Bee-eater	LC		
0540	佛法僧目 CORACIIFORMES	蜂虎科 Meropidae	蓝须蜂虎	*Nyctyornis athertoni*	Blue-bearded Bee-eater	VU	B1a+b(ii)	
0541	佛法僧目 CORACIIFORMES	蜂虎科 Meropidae	绿喉蜂虎	*Merops orientalis*	Green Bee-eater	LC		
0542	佛法僧目 CORACIIFORMES	蜂虎科 Meropidae	蓝颊蜂虎	*Merops persicus*	Blue-cheeked Bee-eater	LC		
0543	佛法僧目 CORACIIFORMES	蜂虎科 Meropidae	栗喉蜂虎	*Merops philippinus*	Blue-tailed Bee-eater	LC		
0544	佛法僧目 CORACIIFORMES	蜂虎科 Meropidae	彩虹蜂虎	*Merops ornatus*	Rainbow Bee-eater	DD(NE)		
0545	佛法僧目 CORACIIFORMES	蜂虎科 Meropidae	蓝喉蜂虎	*Merops viridis*	Blue-throated Bee-eater	LC		
0546	佛法僧目 CORACIIFORMES	蜂虎科 Meropidae	栗头蜂虎	*Merops leschenaulti*	Chestnut-headed Bee-eater	LC		
0547	佛法僧目 CORACIIFORMES	蜂虎科 Meropidae	黄喉蜂虎	*Merops apiaster*	European Bee-eater	NT		
0548	佛法僧目 CORACIIFORMES	佛法僧科 Coraciidae	棕胸佛法僧	*Coracias benghalensis*	Indian Roller	NT		
0549	佛法僧目 CORACIIFORMES	佛法僧科 Coraciidae	蓝胸佛法僧	*Coracias garrulus*	European Roller	NT		
0550	佛法僧目 CORACIIFORMES	佛法僧科 Coraciidae	三宝鸟	*Eurystomus orientalis*	Dollarbird	LC		
0551	佛法僧目 CORACIIFORMES	翠鸟科 Alcedinidae	鹳嘴翡翠	*Pelargopsis capensis*	Stork-billed Kingfisher	LC		
0552	佛法僧目 CORACIIFORMES	翠鸟科 Alcedinidae	赤翡翠	*Halcyon coromanda*	Ruddy Kingfisher	LC		

续表

	目	科		物种中文名	物种学名	物种英文名	濒危等级	评估依据	是否特有
0553	CORACIIFORMES	佛法僧目	Alcedinidae 翠鸟科	白胸翡翠	*Halcyon smyrnensis*	White-throated Kingfisher	LC		
0554	CORACIIFORMES	佛法僧目	Alcedinidae 翠鸟科	蓝翡翠	*Halcyon pileata*	Black-capped Kingfisher	LC		
0555	CORACIIFORMES	佛法僧目	Alcedinidae 翠鸟科	白领翡翠	*Todiramphus chloris*	Collared Kingfisher	LC		
0556	CORACIIFORMES	佛法僧目	Alcedinidae 翠鸟科	蓝耳翠鸟	*Alcedo meninting*	Blue-eared Kingfisher	LC		
0557	CORACIIFORMES	佛法僧目	Alcedinidae 翠鸟科	普通翠鸟	*Alcedo atthis*	Common Kingfisher	LC		
0558	CORACIIFORMES	佛法僧目	Alcedinidae 翠鸟科	斑头大翠鸟	*Alcedo hercules*	Blyth's Kingfisher	VU	C1+2a(i)	
0559	CORACIIFORMES	佛法僧目	Alcedinidae 翠鸟科	三趾翠鸟	*Ceyx erithaca*	Oriental Dwarf Kingfisher	LC		
0560	CORACIIFORMES	佛法僧目	Alcedinidae 翠鸟科	冠鱼狗	*Megaceryle lugubris*	Crested Kingfisher	LC		
0561	CORACIIFORMES	佛法僧目	Alcedinidae 翠鸟科	斑鱼狗	*Ceryle rudis*	Pied Kingfisher	LC		
0562	PICIFORMES	啄木鸟目	Megalaimidae 拟啄木鸟科	大拟啄木鸟	*Psilopogon virens*	Great Barbet	LC		
0563	PICIFORMES	啄木鸟目	Megalaimidae 拟啄木鸟科	绿拟啄木鸟	*Psilopogon lineatus*	Lineated Barbet	LC		
0564	PICIFORMES	啄木鸟目	Megalaimidae 拟啄木鸟科	黄纹拟啄木鸟	*Psilopogon faiostrictus*	Green-eared Barbet	NT		
0565	PICIFORMES	啄木鸟目	Megalaimidae 拟啄木鸟科	金喉拟啄木鸟	*Psilopogon franklinii*	Golden-throated Barbet	LC		
0566	PICIFORMES	啄木鸟目	Megalaimidae 拟啄木鸟科	黑眉拟啄木鸟	*Psilopogon faber*	Chinese Barbet	LC		
0567	PICIFORMES	啄木鸟目	Megalaimidae 拟啄木鸟科	台湾拟啄木鸟	*Psilopogon nuchalis*	Taiwan Barbet	LC		√
0568	PICIFORMES	啄木鸟目	Megalaimidae 拟啄木鸟科	蓝喉拟啄木鸟	*Psilopogon asiatica*	Blue-throated Barbet	LC		
0569	PICIFORMES	啄木鸟目	Megalaimidae 拟啄木鸟科	蓝耳拟啄木鸟	*Psilopogon australis*	Blue-eared Barbet	LC		
0570	PICIFORMES	啄木鸟目	Megalaimidae 拟啄木鸟科	赤胸拟啄木鸟	*Psilopogon haemacephalus*	Coppersmith Barbet	LC		
0571	PICIFORMES	啄木鸟目	Indicatoridae 响蜜䴕科	黄腰响蜜䴕	*Indicator xanthonotus*	Yellow-rumped Honeyguide	NT		
0572	PICIFORMES	啄木鸟目	Picidae 啄木鸟科	蚁䴕	*Jynx torquilla*	Eurasian Wryneck	LC		
0573	PICIFORMES	啄木鸟目	Picidae 啄木鸟科	斑姬啄木鸟	*Picumnus innominatus*	Speckled Piculet	LC		
0574	PICIFORMES	啄木鸟目	Picidae 啄木鸟科	白眉棕啄木鸟	*Sasia ochracea*	White-browed Piculet	LC		
0575	PICIFORMES	啄木鸟目	Picidae 啄木鸟科	棕腹啄木鸟	*Dendrocopos hyperythrus*	Rufous-bellied Woodpecker	LC		
0576	PICIFORMES	啄木鸟目	Picidae 啄木鸟科	小星头啄木鸟	*Dendrocopos kizuki*	Pygmy Woodpecker	LC		
0577	PICIFORMES	啄木鸟目	Picidae 啄木鸟科	星头啄木鸟	*Dendrocopos canicapillus*	Grey-capped Woodpecker	LC		
0578	PICIFORMES	啄木鸟目	Picidae 啄木鸟科	小斑啄木鸟	*Dendrocopos minor*	Lesser Spotted Woodpecker	LC		
0579	PICIFORMES	啄木鸟目	Picidae 啄木鸟科	纹腹啄木鸟	*Dendrocopos macei*	Fulvous-breasted Woodpecker	DD		

续表

序号	目		科		物种中文名	物种学名	物种英文名	濒危等级	评估依据	是否特有
0580	䴕形目	PICIFORMES	啄木鸟科	Picidae	纹胸啄木鸟	*Dendrocopos atratus*	Stripe-breasted Woodpecker	DD		
0581	䴕形目	PICIFORMES	啄木鸟科	Picidae	褐额啄木鸟	*Dendrocopos auriceps*	Brown-fronted Woodpecker	LC		
0582	䴕形目	PICIFORMES	啄木鸟科	Picidae	赤胸啄木鸟	*Dendrocopos cathpharius*	Crimson-breasted Woodpecker	LC		
0583	䴕形目	PICIFORMES	啄木鸟科	Picidae	黄颈啄木鸟	*Dendrocopos darjellensis*	Darjeeling Woodpecker	LC		
0584	䴕形目	PICIFORMES	啄木鸟科	Picidae	白背啄木鸟	*Dendrocopos leucotos*	White-backed Woodpecker	LC		
0585	䴕形目	PICIFORMES	啄木鸟科	Picidae	白翅啄木鸟	*Dendrocopos leucopterus*	White-winged Woodpecker	NT		
0586	䴕形目	PICIFORMES	啄木鸟科	Picidae	大斑啄木鸟	*Dendrocopos major*	Great Spotted Woodpecker	LC		
0587	䴕形目	PICIFORMES	啄木鸟科	Picidae	三趾啄木鸟	*Picoides tridactylus*	Three-toed Woodpecker	LC		
0588	䴕形目	PICIFORMES	啄木鸟科	Picidae	白腹黑啄木鸟	*Dryocopus javensis*	White-bellied Woodpecker	NT		
0589	䴕形目	PICIFORMES	啄木鸟科	Picidae	黑啄木鸟	*Dryocopus martius*	Black Woodpecker	LC		
0590	䴕形目	PICIFORMES	啄木鸟科	Picidae	大黄冠啄木鸟	*Chrysophlegma flavinucha*	Greater Yellownape Woodpecker	EN	B1ab(iii)	
0591	䴕形目	PICIFORMES	啄木鸟科	Picidae	黄冠啄木鸟	*Picus chlorolophus*	Lesser Yellownape Woodpecker	NT		
0592	䴕形目	PICIFORMES	啄木鸟科	Picidae	花腹绿啄木鸟	*Picus vittatus*	Laced Woodpecker	DD		
0593	䴕形目	PICIFORMES	啄木鸟科	Picidae	纹喉绿啄木鸟	*Picus xanthopygaeus*	Streak-throated Woodpecker	DD		
0594	䴕形目	PICIFORMES	啄木鸟科	Picidae	鳞腹绿啄木鸟	*Picus squamatus*	Scaly-bellied Woodpecker	DD		
0595	䴕形目	PICIFORMES	啄木鸟科	Picidae	红颈绿啄木鸟	*Picus rabieri*	Red-collared Woodpecker	DD		
0596	䴕形目	PICIFORMES	啄木鸟科	Picidae	灰头绿啄木鸟	*Picus canus*	Grey-headed Woodpecker	LC		
0597	䴕形目	PICIFORMES	啄木鸟科	Picidae	金背啄木鸟	*Dinopium javanense*	Common Flamebacked Woodpecker	DD		
0598	䴕形目	PICIFORMES	啄木鸟科	Picidae	喜山金背啄木鸟	*Dinopium shorii*	Himalayan Flamebacked Woodpecker	DD		
0599	䴕形目	PICIFORMES	啄木鸟科	Picidae	小金背啄木鸟	*Dinopium benghalense*	Lesser Golden-backed Flamebacked Woodpecker	DD		
0600	䴕形目	PICIFORMES	啄木鸟科	Picidae	大金背啄木鸟	*Chrysocolaptes lucidus*	Greater Flamebacked Woodpecker	NT		
0601	䴕形目	PICIFORMES	啄木鸟科	Picidae	竹啄木鸟	*Gecinulus grantia*	Pale-headed Woodpecker	LC		
0602	䴕形目	PICIFORMES	啄木鸟科	Picidae	黄嘴栗啄木鸟	*Blythipicus pyrrhotis*	Bay Woodpecker	LC		
0603	䴕形目	PICIFORMES	啄木鸟科	Picidae	栗啄木鸟	*Micropternus brachyurus*	Rufous Woodpecker	LC		
0604	䴕形目	PICIFORMES	啄木鸟科	Picidae	大灰啄木鸟	*Mulleripicus pulverulentus*	Great Slaty Woodpecker	NT		
0605	隼形目	FALCONIFORMES	隼科	Falconidae	红腿小隼	*Microhierax caerulescens*	Collared Falconet	NT		

续表

	目		科		物种中文名	物种学名	物种英文名	濒危等级	评估依据	是否特有
0606	隼形目	FALCONIFORMES	隼科	Falconidae	白腿小隼	*Microhierax melanoleucos*	Pied Falconet	VU	B1a, C2a(i)	
0607	隼形目	FALCONIFORMES	隼科	Falconidae	黄爪隼	*Falco naumanni*	Lesser Kestrel	VU	A2bcde+3bcde+4bcde, C2a(i)	
0608	隼形目	FALCONIFORMES	隼科	Falconidae	红隼	*Falco tinnunculus*	Common Kestrel	LC		
0609	隼形目	FALCONIFORMES	隼科	Falconidae	西红脚隼	*Falco vespertinus*	Red-footed Falcon	NT		
0610	隼形目	FALCONIFORMES	隼科	Falconidae	红脚隼	*Falco amurensis*	Amur Falcon	NT		
0611	隼形目	FALCONIFORMES	隼科	Falconidae	灰背隼	*Falco columbarius*	Merlin	NT		
0612	隼形目	FALCONIFORMES	隼科	Falconidae	燕隼	*Falco subbuteo*	Eurasian Hobby	LC		
0613	隼形目	FALCONIFORMES	隼科	Falconidae	猛隼	*Falco severus*	Oriental Hobby	NT		
0614	隼形目	FALCONIFORMES	隼科	Falconidae	猎隼	*Falco cherrug*	Saker Falcon	EN	A2bcde	
0615	隼形目	FALCONIFORMES	隼科	Falconidae	矛隼	*Falco rusticolus*	Gyrfalcon	NT		
0616	隼形目	FALCONIFORMES	隼科	Falconidae	游隼	*Falco peregrinus*	Peregrine Falcon	NT		
0617	鹦鹉目	PSITTACIFORMES	鹦鹉科	Psittacidae	短尾鹦鹉	*Loriculus vernalis*	Vernal Hanging Parrot	DD		
0618	鹦鹉目	PSITTACIFORMES	鹦鹉科	Psittacidae	蓝腰鹦鹉	*Psittinus cyanurus*	Blue-rumped Parrot	DD	D2	
0619	鹦鹉目	PSITTACIFORMES	鹦鹉科	Psittacidae	亚历山大鹦鹉	*Psittacula eupatria*	Alexandrine Parakeet	LC		
0620	鹦鹉目	PSITTACIFORMES	鹦鹉科	Psittacidae	红领绿鹦鹉	*Psittacula krameri*	Rose-ringed Parakeet	LC		
0621	鹦鹉目	PSITTACIFORMES	鹦鹉科	Psittacidae	青头鹦鹉	*Psittacula himalayana*	Slaty-headed Parakeet	LC		
0622	鹦鹉目	PSITTACIFORMES	鹦鹉科	Psittacidae	灰头鹦鹉	*Psittacula finschii*	Grey-headed Parakeet	LC		
0623	鹦鹉目	PSITTACIFORMES	鹦鹉科	Psittacidae	花头鹦鹉	*Psittacula roseata*	Blossom-headed Parakeet	DD		
0624	鹦鹉目	PSITTACIFORMES	鹦鹉科	Psittacidae	大紫胸鹦鹉	*Psittacula derbiana*	Lord Derby's Parakeet	VU	B2ab(i), C2a(i)	
0625	鹦鹉目	PSITTACIFORMES	鹦鹉科	Psittacidae	绯胸鹦鹉	*Psittacula alexandri*	Red-breasted Parakeet	VU	B1ab(iii)	
0626	雀形目	PASSERIFORMES	八色鸫科	Pittidae	双辫八色鸫	*Pitta phayrei*	Eared Pitta	VU	C1	
0627	雀形目	PASSERIFORMES	八色鸫科	Pittidae	蓝枕八色鸫	*Pitta nipalensis*	Blue-naped Pitta	VU	C1	
0628	雀形目	PASSERIFORMES	八色鸫科	Pittidae	蓝背八色鸫	*Pitta soror*	Blue-rumped Pitta	EN	B1b(ii), C1	
0629	雀形目	PASSERIFORMES	八色鸫科	Pittidae	栗头八色鸫	*Pitta oatesi*	Rusty-naped Pitta	VU	C1	
0630	雀形目	PASSERIFORMES	八色鸫科	Pittidae	蓝八色鸫	*Pitta cyanea*	Blue Pitta	NT		
0631	雀形目	PASSERIFORMES	八色鸫科	Pittidae	绿胸八色鸫	*Pitta sordida*	Hooded Pitta	VU	C1	

续表

	目	科	物种中文名	物种学名	物种英文名	濒危等级	评估依据	是否特有
0632	雀形目 PASSERIFORMES	八色鸫科 Pittidae	仙八色鸫	*Pitta nympha*	Fairy Pitta	VU	A2cd+3cd+4cd	
0633	雀形目 PASSERIFORMES	八色鸫科 Pittidae	蓝翅八色鸫	*Pitta moluccensis*	Blue-winged Pitta	NT		
0634	雀形目 PASSERIFORMES	阔嘴鸟科 Eurylaimidae	长尾阔嘴鸟	*Psarisomus dalhousiae*	Long-tailed Broadbill	NT		
0635	雀形目 PASSERIFORMES	阔嘴鸟科 Eurylaimidae	银胸丝冠鸟	*Serilophus lunatus*	Silver-breasted Broadbill	NT		
0636	雀形目 PASSERIFORMES	黄鹂科 Oriolidae	金黄鹂	*Oriolus oriolus*	Eurasian Golden Oriole	LC		
0637	雀形目 PASSERIFORMES	黄鹂科 Oriolidae	印度金黄鹂	*Oriolus kundoo*	Indian Golden Oriole	LC		
0638	雀形目 PASSERIFORMES	黄鹂科 Oriolidae	细嘴黄鹂	*Oriolus tenuirostris*	Slender-billed Oriole	LC		
0639	雀形目 PASSERIFORMES	黄鹂科 Oriolidae	黑枕黄鹂	*Oriolus chinensis*	Black-naped Oriole	LC		
0640	雀形目 PASSERIFORMES	黄鹂科 Oriolidae	黑头黄鹂	*Oriolus xanthornus*	Black-hooded Oriole	LC		
0641	雀形目 PASSERIFORMES	黄鹂科 Oriolidae	朱鹂	*Oriolus traillii*	Maroon Oriole	NT		
0642	雀形目 PASSERIFORMES	黄鹂科 Oriolidae	鹊鹂	*Oriolus mellianus*	Silver Oriole	EN	C2a(ii)	
0643	雀形目 PASSERIFORMES	莺雀科 Vireonidae	白腹凤鹛	*Erpornis zantholeuca*	White-bellied Erpornis	LC		
0644	雀形目 PASSERIFORMES	莺雀科 Vireonidae	棕腹鹛鹛	*Pteruthius rufiventer*	Black-headed Shrike Babbler	LC		
0645	雀形目 PASSERIFORMES	莺雀科 Vireonidae	红翅鹛鹛	*Pteruthius aeralatus*	Blyth's Shrike Babbler	LC		
0646	雀形目 PASSERIFORMES	莺雀科 Vireonidae	淡绿鹛鹛	*Pteruthius xanthochlorus*	Green Shrike Babbler	NT		
0647	雀形目 PASSERIFORMES	莺雀科 Vireonidae	栗喉鹛鹛	*Pteruthius melanotis*	Black-eared Shrike Babbler	LC		
0648	雀形目 PASSERIFORMES	莺雀科 Vireonidae	栗额鹛鹛	*Pteruthius intermedius*	Clicking Shrike Babbler	LC		
0649	雀形目 PASSERIFORMES	山椒鸟科 Campephagidae	大鹃鵙	*Coracina macei*	Large Cuckooshrike	LC		
0650	雀形目 PASSERIFORMES	山椒鸟科 Campephagidae	暗灰鹃鵙	*Lalage melaschistos*	Black-winged Cuckooshrike	LC		
0651	雀形目 PASSERIFORMES	山椒鸟科 Campephagidae	斑鹃鵙	*Lalage nigra*	Pied Triller	LC		
0652	雀形目 PASSERIFORMES	山椒鸟科 Campephagidae	粉红山椒鸟	*Pericrocotus roseus*	Rosy Minivet	LC		
0653	雀形目 PASSERIFORMES	山椒鸟科 Campephagidae	小灰山椒鸟	*Pericrocotus cantonensis*	Swinhoe's Minivet	LC		
0654	雀形目 PASSERIFORMES	山椒鸟科 Campephagidae	灰山椒鸟	*Pericrocotus divaricatus*	Ashy Minivet	LC		
0655	雀形目 PASSERIFORMES	山椒鸟科 Campephagidae	琉球山椒鸟	*Pericrocotus tegimae*	Ryukyu Minivet	LC		
0656	雀形目 PASSERIFORMES	山椒鸟科 Campephagidae	灰喉山椒鸟	*Pericrocotus solaris*	Grey-chinned Minivet	LC		
0657	雀形目 PASSERIFORMES	山椒鸟科 Campephagidae	长尾山椒鸟	*Pericrocotus ethologus*	Long-tailed Minivet	LC		
0658	雀形目 PASSERIFORMES	山椒鸟科 Campephagidae	短嘴山椒鸟	*Pericrocotus brevirostris*	Short-billed Minivet	LC		

附录　中国鸟类濒危等级评估名录

续表

	目	科	物种中文名	物种学名	物种英文名	濒危等级	评估依据	是否特有
0659	雀形目 PASSERIFORMES	山椒鸟科 Campephagidae	赤红山椒鸟	*Pericrocotus flammeus*	Scarlet Minivet	LC		
0660	雀形目 PASSERIFORMES	燕鵙科 Artamidae	灰燕鵙	*Artamus fuscus*	Ashy Woodswallow	LC		
0661	雀形目 PASSERIFORMES	钩嘴鵙科 Tephrodornithidae	褐背鹟鵙	*Hemipus picatus*	Bar-winged Flycatcher-shrike	LC		
0662	雀形目 PASSERIFORMES	钩嘴鵙科 Tephrodornithidae	钩嘴林鵙	*Tephrodornis virgatus*	Large Woodshrike	LC		
0663	雀形目 PASSERIFORMES	雀鹎科 Aegithinidae	黑翅雀鹎	*Aegithina tiphia*	Common Iora	LC		
0664	雀形目 PASSERIFORMES	雀鹎科 Aegithinidae	大绿雀鹎	*Aegithina lafresnayei*	Great Iora	LC		
0665	雀形目 PASSERIFORMES	扇尾鹟科 Rhipiduridae	白喉扇尾鹟	*Rhipidura albicollis*	White-throated Fantail	LC		
0666	雀形目 PASSERIFORMES	扇尾鹟科 Rhipiduridae	白眉扇尾鹟	*Rhipidura aureola*	White-browed Fantail	LC		
0667	雀形目 PASSERIFORMES	卷尾科 Dicruridae	黑卷尾	*Dicrurus macrocercus*	Black Drongo	LC		
0668	雀形目 PASSERIFORMES	卷尾科 Dicruridae	灰卷尾	*Dicrurus leucophaeus*	Ashy Drongo	LC		
0669	雀形目 PASSERIFORMES	卷尾科 Dicruridae	鸦嘴卷尾	*Dicrurus annectans*	Crow-billed Drongo	LC		
0670	雀形目 PASSERIFORMES	卷尾科 Dicruridae	古铜色卷尾	*Dicrurus aeneus*	Bronzed Drongo	LC		
0671	雀形目 PASSERIFORMES	卷尾科 Dicruridae	发冠卷尾	*Dicrurus hottentottus*	Hair-crested Drongo	NT		
0672	雀形目 PASSERIFORMES	卷尾科 Dicruridae	小盘尾	*Dicrurus remifer*	Lesser Racket-tailed Drongo	VU	A1c, C1	
0673	雀形目 PASSERIFORMES	卷尾科 Dicruridae	大盘尾	*Dicrurus paradiseus*	Greater Racket-tailed Drongo	LC		
0674	雀形目 PASSERIFORMES	王鹟科 Monarchidae	黑枕王鹟	*Hypothymis azurea*	Black-naped Monarch	LC		
0675	雀形目 PASSERIFORMES	王鹟科 Monarchidae	印度寿带	*Terpsiphone paradisi*	Indian Paradise Flycatcher	LC		
0676	雀形目 PASSERIFORMES	王鹟科 Monarchidae	东方寿带	*Terpsiphone affinis*	Oriental Paradise Flycatcher	LC		
0677	雀形目 PASSERIFORMES	王鹟科 Monarchidae	寿带	*Terpsiphone incei*	Amur Paradise Flycatcher	LC		
0678	雀形目 PASSERIFORMES	王鹟科 Monarchidae	紫寿带	*Terpsiphone atrocaudata*	Japanese Paradise-flycatcher	NT		
0679	雀形目 PASSERIFORMES	伯劳科 Laniidae	虎纹伯劳	*Lanius tigrinus*	Tiger Shrike	LC		
0680	雀形目 PASSERIFORMES	伯劳科 Laniidae	牛头伯劳	*Lanius bucephalus*	Bull-headed Shrike	LC		
0681	雀形目 PASSERIFORMES	伯劳科 Laniidae	红尾伯劳	*Lanius cristatus*	Brown Shrike	LC		
0682	雀形目 PASSERIFORMES	伯劳科 Laniidae	红背伯劳	*Lanius collurio*	Red-backed Shrike	LC		
0683	雀形目 PASSERIFORMES	伯劳科 Laniidae	荒漠伯劳	*Lanius isabellinus*	Isabelline Shrike	LC		
0684	雀形目 PASSERIFORMES	伯劳科 Laniidae	棕尾伯劳	*Lanius phoenicuroides*	Rufous-tailed Shrike	LC		
0685	雀形目 PASSERIFORMES	伯劳科 Laniidae	栗背伯劳	*Lanius collurioides*	Burmese Shrike	NT		

续表

	目	科	物种中文名	物种学名	物种英文名	濒危等级	评估依据	是否特有
0686	雀形目 PASSERIFORMES	伯劳科 Laniidae	棕背伯劳	*Lanius schach*	Long-tailed Shrike	LC		
0687	雀形目 PASSERIFORMES	伯劳科 Laniidae	灰背伯劳	*Lanius tephronotus*	Grey-backed Shrike	LC		
0688	雀形目 PASSERIFORMES	伯劳科 Laniidae	黑额伯劳	*Lanius minor*	Lesser Grey Shrike	LC		
0689	雀形目 PASSERIFORMES	伯劳科 Laniidae	灰伯劳	*Lanius excubitor*	Great Grey Shrike	LC		
0690	雀形目 PASSERIFORMES	伯劳科 Laniidae	楔尾伯劳	*Lanius sphenocercus*	Chinese Grey Shrike	LC		
0691	雀形目 PASSERIFORMES	鸦科 Corvidae	北噪鸦	*Perisoreus infaustus*	Siberian Jay	NT		
0692	雀形目 PASSERIFORMES	鸦科 Corvidae	黑头噪鸦	*Perisoreus internigrans*	Sichuan Jay	VU	C2a(i)	✓
0693	雀形目 PASSERIFORMES	鸦科 Corvidae	松鸦	*Garrulus glandarius*	Eurasian Jay	LC		
0694	雀形目 PASSERIFORMES	鸦科 Corvidae	灰喜鹊	*Cyanopica cyanus*	Azure-winged Magpie	LC		
0695	雀形目 PASSERIFORMES	鸦科 Corvidae	台湾蓝鹊	*Urocissa caerulea*	Taiwan Blue Magpie	LC		✓
0696	雀形目 PASSERIFORMES	鸦科 Corvidae	黄嘴蓝鹊	*Urocissa flavirostris*	Yellow-billed Blue Magpie	LC		
0697	雀形目 PASSERIFORMES	鸦科 Corvidae	红嘴蓝鹊	*Urocissa erythroryncha*	Red-billed Blue Magpie	LC		
0698	雀形目 PASSERIFORMES	鸦科 Corvidae	白翅蓝鹊	*Urocissa whiteheadi*	White-winged Magpie	NT		
0699	雀形目 PASSERIFORMES	鸦科 Corvidae	蓝绿鹊	*Cissa chinensis*	Common Green Magpie	LC		
0700	雀形目 PASSERIFORMES	鸦科 Corvidae	黄胸绿鹊	*Cissa hypoleuca*	Indochinese Green Magpie	NT		
0701	雀形目 PASSERIFORMES	鸦科 Corvidae	棕腹树鹊	*Dendrocitta vagabunda*	Rufous Treepie	LC		
0702	雀形目 PASSERIFORMES	鸦科 Corvidae	灰树鹊	*Dendrocitta formosae*	Grey Treepie	LC		
0703	雀形目 PASSERIFORMES	鸦科 Corvidae	黑额树鹊	*Dendrocitta frontalis*	Collared Treepie	LC		
0704	雀形目 PASSERIFORMES	鸦科 Corvidae	塔尾树鹊	*Temnurus temnurus*	Ratchet-tailed Treepie	NT		
0705	雀形目 PASSERIFORMES	鸦科 Corvidae	喜鹊	*Pica pica*	Common Magpie	LC		
0706	雀形目 PASSERIFORMES	鸦科 Corvidae	黑尾地鸦	*Podoces hendersoni*	Mongolian Ground Jay	VU	C2(i), D1	
0707	雀形目 PASSERIFORMES	鸦科 Corvidae	白尾地鸦	*Podoces biddulphi*	Xinjiang Ground Jay	VU	C2(i), D1	✓
0708	雀形目 PASSERIFORMES	鸦科 Corvidae	星鸦	*Nucifraga caryocatactes*	Spotted Nutcracker	LC		
0709	雀形目 PASSERIFORMES	鸦科 Corvidae	红嘴山鸦	*Pyrrhocorax pyrrhocorax*	Red-billed Chough	LC		
0710	雀形目 PASSERIFORMES	鸦科 Corvidae	黄嘴山鸦	*Pyrrhocorax graculus*	Alpine Chough	LC		
0711	雀形目 PASSERIFORMES	鸦科 Corvidae	寒鸦	*Corvus monedula*	Eurasian Jackdaw	LC		
0712	雀形目 PASSERIFORMES	鸦科 Corvidae	达乌里寒鸦	*Corvus dauuricus*	Daurian Jackdaw	LC		

续表

编号	目	科	物种中文名	物种学名	物种英文名	濒危等级	评估依据	是否特有
0713	PASSERIFORMES	Corvidae	家鸦	*Corvus splendens*	House Crow	LC		
0714	PASSERIFORMES	Corvidae	秃鼻乌鸦	*Corvus frugilegus*	Rook	LC		
0715	PASSERIFORMES	Corvidae	小嘴乌鸦	*Corvus corone*	Carrion Crow	LC		
0716	PASSERIFORMES	Corvidae	冠小嘴乌鸦	*Corvus cornix*	Hooded Crow	LC		
0717	PASSERIFORMES	Corvidae	白颈鸦	*Corvus pectoralis*	Collared Crow	NT		
0718	PASSERIFORMES	Corvidae	大嘴乌鸦	*Corvus macrorhynchos*	Large-billed Crow	LC		
0719	PASSERIFORMES	Corvidae	渡鸦	*Corvus corax*	Common Raven	LC		
0720	PASSERIFORMES	Stenostiridae	黄腹扇尾鹟	*Chelidorhynx hypoxanthus*	Citrine Canary-Flycatcher	LC		
0721	PASSERIFORMES	Stenostiridae	方尾鹟	*Culicicapa ceylonensis*	Grey-headed Canary Flycatcher	LC		
0722	PASSERIFORMES	Paridae	火冠雀	*Cephalopyrus flammiceps*	Fire-capped Tit	LC		
0723	PASSERIFORMES	Paridae	黄眉林雀	*Sylviparus modestus*	Yellow-browed Tit	LC		
0724	PASSERIFORMES	Paridae	冕雀	*Melanochlora sultanea*	Sultan Tit	NT		
0725	PASSERIFORMES	Paridae	棕枕山雀	*Periparus rufonuchalis*	Rufous-naped Tit	LC		
0726	PASSERIFORMES	Paridae	黑冠山雀	*Periparus rubidiventris*	Rufous-vented Tit	LC		
0727	PASSERIFORMES	Paridae	煤山雀	*Periparus ater*	Coal Tit	LC		
0728	PASSERIFORMES	Paridae	黄腹山雀	*Pardaliparus venustulus*	Yellow-bellied Tit	LC		√
0729	PASSERIFORMES	Paridae	褐冠山雀	*Lophophanes dichrous*	Grey-crested Tit	LC		
0730	PASSERIFORMES	Paridae	杂色山雀	*Sittiparus varius*	Varied Tit	NT		
0731	PASSERIFORMES	Paridae	台湾杂色山雀	*Sittiparus castaneoventris*	Chestnut-bellied Tit	LC		√
0732	PASSERIFORMES	Paridae	白眉山雀	*Poecile superciliosus*	White-browed Tit	NT		√
0733	PASSERIFORMES	Paridae	红腹山雀	*Poecile davidi*	Rusty-breasted Tit	LC		√
0734	PASSERIFORMES	Paridae	沼泽山雀	*Poecile palustris*	Marsh Tit	LC		
0735	PASSERIFORMES	Paridae	褐头山雀	*Poecile montanus*	Willow Tit	LC		
0736	PASSERIFORMES	Paridae	四川褐头山雀	*Poecile weigoldicus*	Sichuan Tit	NT		√
0737	PASSERIFORMES	Paridae	灰蓝山雀	*Cyanistes cyanus*	Azure Tit	LC		
0738	PASSERIFORMES	Paridae	地山雀	*Pseudopodoces humilis*	Ground Tit	LC		√
0739	PASSERIFORMES	Paridae	欧亚大山雀	*Parus major*	Great Tit	LC		

序号	目	科	物种中文名	物种学名	物种英文名	濒危等级	评估依据	是否特有
0740	PASSERIFORMES 雀形目	Paridae 山雀科	大山雀	*Parus cinereus*	Cinereous Tit	LC		
0741	PASSERIFORMES 雀形目	Paridae 山雀科	绿背山雀	*Parus monticolus*	Green-backed Tit	LC		
0742	PASSERIFORMES 雀形目	Paridae 山雀科	台湾黄山雀	*Machlolophus holsti*	Yellow Tit	LC		√
0743	PASSERIFORMES 雀形目	Paridae 山雀科	眼纹黄山雀	*Machlolophus xanthogenys*	Himalayan Black-lored Tit	LC		
0744	PASSERIFORMES 雀形目	Paridae 山雀科	黄颊山雀	*Machlolophus spilonotus*	Yellow-cheeked Tit	LC		
0745	PASSERIFORMES 雀形目	Paridae 山雀科	黑头攀雀	*Remiz macronyx*	Black-headed Penduline Tit	LC		
0746	PASSERIFORMES 雀形目	Remizidae 攀雀科	白冠攀雀	*Remiz coronatus*	White-crowned Penduline Tit	LC		
0747	PASSERIFORMES 雀形目	Remizidae 攀雀科	中华攀雀	*Remiz consobrinus*	Chinese Penduline Tit	LC		
0748	PASSERIFORMES 雀形目	Alaudidae 百灵科	歌百灵	*Mirafra javanica*	Australasian Bush Lark	VU	A2abcd, B1b(ii,iii)	
0749	PASSERIFORMES 雀形目	Alaudidae 百灵科	草原百灵	*Melanocorypha calandra*	Calandra Lark	NT		
0750	PASSERIFORMES 雀形目	Alaudidae 百灵科	双斑百灵	*Melanocorypha bimaculata*	Bimaculated Lark	LC		
0751	PASSERIFORMES 雀形目	Alaudidae 百灵科	蒙古百灵	*Melanocorypha mongolica*	Mongolian Lark	VU	A2abcd+B1b(ii,iii)	
0752	PASSERIFORMES 雀形目	Alaudidae 百灵科	长嘴百灵	*Melanocorypha maxima*	Tibetan Lark	LC		
0753	PASSERIFORMES 雀形目	Alaudidae 百灵科	黑百灵	*Melanocorypha yeltoniensis*	Black Lark	LC		
0754	PASSERIFORMES 雀形目	Alaudidae 百灵科	大短趾百灵	*Calandrella brachydactyla*	Greater Short-toed Lark	LC		
0755	PASSERIFORMES 雀形目	Alaudidae 百灵科	细嘴短趾百灵	*Calandrella acutirostris*	Hume's Short-toed Lark	LC		
0756	PASSERIFORMES 雀形目	Alaudidae 百灵科	短趾百灵	*Alaudala cheleensis*	Asian Short-toed Lark	LC		
0757	PASSERIFORMES 雀形目	Alaudidae 百灵科	凤头百灵	*Galerida cristata*	Crested Lark	LC		
0758	PASSERIFORMES 雀形目	Alaudidae 百灵科	白翅百灵	*Alauda leucoptera*	White-winged Lark	LC		
0759	PASSERIFORMES 雀形目	Alaudidae 百灵科	云雀	*Alauda arvensis*	Eurasian Skylark	LC		
0760	PASSERIFORMES 雀形目	Alaudidae 百灵科	小云雀	*Alauda gulgula*	Oriental Skylark	LC		
0761	PASSERIFORMES 雀形目	Alaudidae 百灵科	角百灵	*Eremophila alpestris*	Horned Lark	LC		
0762	PASSERIFORMES 雀形目	Panuridae 文须雀科	文须雀	*Panurus biarmicus*	Bearded Reedling	LC		
0763	PASSERIFORMES 雀形目	Cisticolidae 扇尾莺科	棕扇尾莺	*Cisticola juncidis*	Zitting Cisticola	LC		
0764	PASSERIFORMES 雀形目	Cisticolidae 扇尾莺科	金头扇尾莺	*Cisticola exilis*	Golden-headed Cisticola	LC		
0765	PASSERIFORMES 雀形目	Cisticolidae 扇尾莺科	山鹪莺	*Prinia crinigera*	Striated Prinia	LC		

附录 中国鸟类濒危等级评估名录

续表

	目	科	物种中文名	物种学名	物种英文名	濒危等级	评估依据	是否特有
0766	雀形目 PASSERIFORMES	扇尾莺科 Cisticolidae	褐山鹪莺	Prinia polychroa	Brown Prinia	LC		
0767	雀形目 PASSERIFORMES	扇尾莺科 Cisticolidae	黑喉山鹪莺	Prinia atrogularis	Black-throated Prinia	LC		
0768	雀形目 PASSERIFORMES	扇尾莺科 Cisticolidae	暗冕山鹪莺	Prinia rufescens	Rufescent Prinia	LC		
0769	雀形目 PASSERIFORMES	扇尾莺科 Cisticolidae	灰胸山鹪莺	Prinia hodgsonii	Grey-breasted Prinia	LC		
0770	雀形目 PASSERIFORMES	扇尾莺科 Cisticolidae	黄腹山鹪莺	Prinia flaviventris	Yellow-bellied Prinia	LC		
0771	雀形目 PASSERIFORMES	扇尾莺科 Cisticolidae	纯色山鹪莺	Prinia inornata	Plain Prinia	LC		
0772	雀形目 PASSERIFORMES	扇尾莺科 Cisticolidae	长尾缝叶莺	Orthotomus sutorius	Common Tailorbird	LC		
0773	雀形目 PASSERIFORMES	扇尾莺科 Cisticolidae	黑喉缝叶莺	Orthotomus atrogularis	Dark-necked Tailorbird	LC		
0774	雀形目 PASSERIFORMES	苇莺科 Acrocephalidae	大苇莺	Acrocephalus arundinaceus	Great Reed Warbler	LC		
0775	雀形目 PASSERIFORMES	苇莺科 Acrocephalidae	东方大苇莺	Acrocephalus orientalis	Oriental Reed Warbler	LC		
0776	雀形目 PASSERIFORMES	苇莺科 Acrocephalidae	噪苇莺	Acrocephalus stentoreus	Clamorous Reed Warbler	LC		
0777	雀形目 PASSERIFORMES	苇莺科 Acrocephalidae	须苇莺	Acrocephalus melanopogon	Moustached Warbler	LC		
0778	雀形目 PASSERIFORMES	苇莺科 Acrocephalidae	黑眉苇莺	Acrocephalus bistrigiceps	Black-browed Reed Warbler	LC		
0779	雀形目 PASSERIFORMES	苇莺科 Acrocephalidae	蒲苇莺	Acrocephalus schoenobaenus	Sedge Warbler	LC		
0780	雀形目 PASSERIFORMES	苇莺科 Acrocephalidae	细纹苇莺	Acrocephalus sorghophilus	Streaked Reed Warbler	EN	C2a(ii)	
0781	雀形目 PASSERIFORMES	苇莺科 Acrocephalidae	钝翅苇莺	Acrocephalus concinens	Blunt-winged Warbler	LC		
0782	雀形目 PASSERIFORMES	苇莺科 Acrocephalidae	远东苇莺	Acrocephalus tangorum	Manchurian Reed Warbler	VU	C2a(ii)	
0783	雀形目 PASSERIFORMES	苇莺科 Acrocephalidae	稻田苇莺	Acrocephalus agricola	Paddyfield Warbler	LC		
0784	雀形目 PASSERIFORMES	苇莺科 Acrocephalidae	布氏苇莺	Acrocephalus dumetorum	Blyth's Reed Warbler	LC		
0785	雀形目 PASSERIFORMES	苇莺科 Acrocephalidae	芦莺	Acrocephalus scirpaceus	Eurasian Reed Warbler	LC		
0786	雀形目 PASSERIFORMES	苇莺科 Acrocephalidae	厚嘴苇莺	Arundinax aedon	Thick-billed Warbler	LC		
0787	雀形目 PASSERIFORMES	苇莺科 Acrocephalidae	靴篱莺	Iduna caligata	Booted Warbler	LC		
0788	雀形目 PASSERIFORMES	苇莺科 Acrocephalidae	赛氏篱莺	Iduna rama	Sykes's Warbler	LC		
0789	雀形目 PASSERIFORMES	苇莺科 Acrocephalidae	草绿篱莺	Iduna pallida	Eastern Olivaceous Warbler	LC		
0790	雀形目 PASSERIFORMES	鳞胸鹪鹛科 Pnoepygidae	鳞胸鹪鹛	Pnoepyga albiventer	Scaly-breasted Wren-Babbler	LC		
0791	雀形目 PASSERIFORMES	鳞胸鹪鹛科 Pnoepygidae	台湾鹪鹛	Pnoepyga formosana	Taiwan Wren-Babbler	LC		✓
0792	雀形目 PASSERIFORMES	鳞胸鹪鹛科 Pnoepygidae	尼泊尔鹪鹛	Pnoepyga immaculata	Nepal Wren-Babbler	LC		

续表

	目	科	物种中文名	物种学名	物种英文名	濒危等级	评估依据	是否特有
0793	雀形目 PASSERIFORMES	鳞胸鹪鹛科 Pnoepygidae	小鳞胸鹪鹛	Pnoepyga pusilla	Pygmy Wren-Babbler	LC		
0794	雀形目 PASSERIFORMES	蝗莺科 Locustellidae	高山短翅蝗莺	Locustella mandelli	Russet Bush Warbler	LC		
0795	雀形目 PASSERIFORMES	蝗莺科 Locustellidae	台湾短翅蝗莺	Locustella alishanensis	Taiwan Bush Warbler	LC		✓
0796	雀形目 PASSERIFORMES	蝗莺科 Locustellidae	四川短翅蝗莺	Locustella chengi	Sichuan Bush Warbler	NT		✓
0797	雀形目 PASSERIFORMES	蝗莺科 Locustellidae	斑胸短翅蝗莺	Locustella thoracica	Spotted Bush Warbler	LC		
0798	雀形目 PASSERIFORMES	蝗莺科 Locustellidae	北短翅蝗莺	Locustella davidi	Baikal Bush Warbler	LC		
0799	雀形目 PASSERIFORMES	蝗莺科 Locustellidae	巨嘴短翅蝗莺	Locustella major	Long-billed Bush Warbler	NT		
0800	雀形目 PASSERIFORMES	蝗莺科 Locustellidae	中华短翅蝗莺	Locustella tacsanowskia	Chinese Bush Warbler	LC		
0801	雀形目 PASSERIFORMES	蝗莺科 Locustellidae	棕褐短翅蝗莺	Locustella luteoventris	Brown Bush Warbler	LC		
0802	雀形目 PASSERIFORMES	蝗莺科 Locustellidae	黑斑蝗莺	Locustella naevia	Common Grasshopper Warbler	LC		
0803	雀形目 PASSERIFORMES	蝗莺科 Locustellidae	矛斑蝗莺	Locustella lanceolata	Lanceolated Warbler	NT		
0804	雀形目 PASSERIFORMES	蝗莺科 Locustellidae	鸣蝗莺	Locustella luscinioides	Savi's Warbler	LC		
0805	雀形目 PASSERIFORMES	蝗莺科 Locustellidae	北蝗莺	Locustella ochotensis	Middendorff's Grasshopper Warbler	LC		
0806	雀形目 PASSERIFORMES	蝗莺科 Locustellidae	东亚蝗莺	Locustella pleskei	Pleske's Warbler	VU	C2a(i)	
0807	雀形目 PASSERIFORMES	蝗莺科 Locustellidae	小蝗莺	Locustella certhiola	Pallas's Grasshopper Warbler	LC		
0808	雀形目 PASSERIFORMES	蝗莺科 Locustellidae	苍眉蝗莺	Locustella fasciolata	Gray's Grasshopper Warbler	LC		
0809	雀形目 PASSERIFORMES	蝗莺科 Locustellidae	库页岛蝗莺	Locustella amnicola	Sakhalin Grasshopper Warbler	LC		
0810	雀形目 PASSERIFORMES	蝗莺科 Locustellidae	斑背大尾莺	Locustella pryeri	Marsh Grassbird	NT		
0811	雀形目 PASSERIFORMES	蝗莺科 Locustellidae	沼泽大尾莺	Megalurus palustris	Striated Grassbird	LC		
0812	雀形目 PASSERIFORMES	燕科 Hirundinidae	褐喉沙燕	Riparia paludicola	Brown-throated Martin	LC		
0813	雀形目 PASSERIFORMES	燕科 Hirundinidae	崖沙燕	Riparia riparia	Sand Martin	LC		
0814	雀形目 PASSERIFORMES	燕科 Hirundinidae	淡色崖沙燕	Riparia diluta	Pale Martin	LC		
0815	雀形目 PASSERIFORMES	燕科 Hirundinidae	家燕	Hirundo rustica	Barn Swallow	LC		
0816	雀形目 PASSERIFORMES	燕科 Hirundinidae	洋燕	Hirundo tahitica	Pacific Swallow	LC		
0817	雀形目 PASSERIFORMES	燕科 Hirundinidae	线尾燕	Hirundo smithii	Wire-tailed Swallow	LC		
0818	雀形目 PASSERIFORMES	燕科 Hirundinidae	岩燕	Ptyonoprogne rupestris	Eurasian Crag Martin	LC		
0819	雀形目 PASSERIFORMES	燕科 Hirundinidae	纯色岩燕	Ptyonoprogne concolor	Dusky Crag Martin	NT		

续表

序号	目	科	物种中文名	物种学名	物种英文名	濒危等级	评估依据	是否特有
0820	PASSERIFORMES	Hirundinidae	毛脚燕	*Delichon urbicum*	Common House Martin	LC		
0821	PASSERIFORMES	Hirundinidae	烟腹毛脚燕	*Delichon dasypus*	Asian House Martin	LC		
0822	PASSERIFORMES	Hirundinidae	黑喉毛脚燕	*Delichon nipalense*	Nepal House Martin	LC		
0823	PASSERIFORMES	Hirundinidae	金腰燕	*Cecropis daurica*	Red-rumped Swallow	LC		
0824	PASSERIFORMES	Hirundinidae	斑腰燕	*Cecropis striolata*	Striated Swallow	LC		
0825	PASSERIFORMES	Hirundinidae	黄额燕	*Petrochelidon fluvicola*	Streak-throated Swallow	LC		
0826	PASSERIFORMES	Pycnonotidae	凤头雀嘴鹎	*Spizixos canifrons*	Crested Finchbill	LC		
0827	PASSERIFORMES	Pycnonotidae	领雀嘴鹎	*Spizixos semitorques*	Collared Finchbill	LC		
0828	PASSERIFORMES	Pycnonotidae	黑头鹎	*Pycnonotus atriceps*	Black-headed Bulbul	LC		
0829	PASSERIFORMES	Pycnonotidae	纵纹绿鹎	*Pycnonotus striatus*	Striated Bulbul	LC		
0830	PASSERIFORMES	Pycnonotidae	黑冠黄鹎	*Pycnonotus melanicterus*	Black-crested Bulbul	LC		
0831	PASSERIFORMES	Pycnonotidae	红耳鹎	*Pycnonotus jocosus*	Red-whiskered Bulbul	LC		
0832	PASSERIFORMES	Pycnonotidae	黄臀鹎	*Pycnonotus xanthorrhous*	Brown-breasted Bulbul	LC		
0833	PASSERIFORMES	Pycnonotidae	白头鹎	*Pycnonotus sinensis*	Light-vented Bulbul	LC		
0834	PASSERIFORMES	Pycnonotidae	台湾鹎	*Pycnonotus taivanus*	Styan's Bulbul	VU	A2ce+3ce+4ce	√
0835	PASSERIFORMES	Pycnonotidae	白颊鹎	*Pycnonotus leucogenis*	Himalayan Bulbul	LC		
0836	PASSERIFORMES	Pycnonotidae	黑喉红臀鹎	*Pycnonotus cafer*	Red-vented Bulbul	LC		
0837	PASSERIFORMES	Pycnonotidae	白喉红臀鹎	*Pycnonotus aurigaster*	Sooty-headed Bulbul	LC		
0838	PASSERIFORMES	Pycnonotidae	纹喉鹎	*Pycnonotus finlaysoni*	Stripe-throated Bulbul	LC		
0839	PASSERIFORMES	Pycnonotidae	黄绿鹎	*Pycnonotus flavescens*	Flavescent Bulbul	NT		
0840	PASSERIFORMES	Pycnonotidae	黄腹冠鹎	*Alophoixus flaveolus*	White-throated Bulbul	LC		
0841	PASSERIFORMES	Pycnonotidae	白喉冠鹎	*Alophoixus pallidus*	Puff-throated Bulbul	LC		
0842	PASSERIFORMES	Pycnonotidae	灰眼短脚鹎	*Iole propinqua*	Grey-eyed Bulbul	LC		
0843	PASSERIFORMES	Pycnonotidae	绿翅短脚鹎	*Ixos mcclellandii*	Mountain Bulbul	LC		
0844	PASSERIFORMES	Pycnonotidae	灰短脚鹎	*Hemixos flavala*	Ashy Bulbul	LC		
0845	PASSERIFORMES	Pycnonotidae	栗背短脚鹎	*Hemixos castanonotus*	Chestnut Bulbul	LC		
0846	PASSERIFORMES	Pycnonotidae	黑短脚鹎	*Hypsipetes leucocephalus*	Black Bulbul	LC		

续表

	目	科	科	物种中文名	物种学名	物种英文名	濒危等级	评估依据	是否特有
0847	雀形目	PASSERIFORMES	Pycnonotidae	鸭科	栗耳短脚鹎	Hypsipetes amaurotis	Brown-eared Bulbul	LC		
0848	雀形目	PASSERIFORMES	Phylloscopidae	柳莺科	欧柳莺	Phylloscopus trochilus	Willow Warbler	DD(NE)		
0849	雀形目	PASSERIFORMES	Phylloscopidae	柳莺科	叽喳柳莺	Phylloscopus collybita	Common Chiffchaff	LC		
0850	雀形目	PASSERIFORMES	Phylloscopidae	柳莺科	中亚叽喳柳莺	Phylloscopus sindianus	Mountain Chiffchaff	LC		
0851	雀形目	PASSERIFORMES	Phylloscopidae	柳莺科	林柳莺	Phylloscopus sibilatrix	Wood Warbler	LC		
0852	雀形目	PASSERIFORMES	Phylloscopidae	柳莺科	褐柳莺	Phylloscopus fuscatus	Dusky Warbler	LC		
0853	雀形目	PASSERIFORMES	Phylloscopidae	柳莺科	烟柳莺	Phylloscopus fuligiventer	Smoky Warbler	LC		
0854	雀形目	PASSERIFORMES	Phylloscopidae	柳莺科	黄腹柳莺	Phylloscopus affinis	Tickell's Leaf Warbler	LC		
0855	雀形目	PASSERIFORMES	Phylloscopidae	柳莺科	华西柳莺	Phylloscopus occisinensis	Alpine Leaf Warbler	NT		
0856	雀形目	PASSERIFORMES	Phylloscopidae	柳莺科	棕腹柳莺	Phylloscopus subaffinis	Buff-throated Warbler	LC		
0857	雀形目	PASSERIFORMES	Phylloscopidae	柳莺科	灰柳莺	Phylloscopus griseolus	Sulphur-bellied Warbler	LC		
0858	雀形目	PASSERIFORMES	Phylloscopidae	柳莺科	棕眉柳莺	Phylloscopus armandii	Yellow-streaked Warbler	LC		
0859	雀形目	PASSERIFORMES	Phylloscopidae	柳莺科	巨嘴柳莺	Phylloscopus schwarzi	Radde's Warbler	LC		
0860	雀形目	PASSERIFORMES	Phylloscopidae	柳莺科	橙斑翅柳莺	Phylloscopus pulcher	Buff-barred Warbler	LC		
0861	雀形目	PASSERIFORMES	Phylloscopidae	柳莺科	灰喉柳莺	Phylloscopus maculipennis	Ashy-throated Warbler	LC		
0862	雀形目	PASSERIFORMES	Phylloscopidae	柳莺科	甘肃柳莺	Phylloscopus kansuensis	Gansu Leaf Warbler	LC		✓
0863	雀形目	PASSERIFORMES	Phylloscopidae	柳莺科	云南柳莺	Phylloscopus yunnanensis	Chinese Leaf Warbler	LC		
0864	雀形目	PASSERIFORMES	Phylloscopidae	柳莺科	黄腰柳莺	Phylloscopus proregulus	Pallas's Leaf Warbler	LC		
0865	雀形目	PASSERIFORMES	Phylloscopidae	柳莺科	淡黄腰柳莺	Phylloscopus chloronotus	Lemon-rumped Warbler	LC		
0866	雀形目	PASSERIFORMES	Phylloscopidae	柳莺科	四川柳莺	Phylloscopus forresti	Sichuan Leaf Warbler	LC		
0867	雀形目	PASSERIFORMES	Phylloscopidae	柳莺科	黄眉柳莺	Phylloscopus inornatus	Yellow-browed Warbler	LC		
0868	雀形目	PASSERIFORMES	Phylloscopidae	柳莺科	淡眉柳莺	Phylloscopus humei	Hume's Leaf Warbler	LC		
0869	雀形目	PASSERIFORMES	Phylloscopidae	柳莺科	极北柳莺	Phylloscopus borealis	Arctic Warbler	LC		
0870	雀形目	PASSERIFORMES	Phylloscopidae	柳莺科	日本柳莺	Phylloscopus xanthodryas	Japanese Leaf Warbler	LC		
0871	雀形目	PASSERIFORMES	Phylloscopidae	柳莺科	暗绿柳莺	Phylloscopus trochiloides	Greenish Warbler	LC		
0872	雀形目	PASSERIFORMES	Phylloscopidae	柳莺科	双斑绿柳莺	Phylloscopus plumbeitarsus	Two-barred Warbler	LC		
0873	雀形目	PASSERIFORMES	Phylloscopidae	柳莺科	淡脚柳莺	Phylloscopus tenellipes	Pale-legged Warbler	LC		

续表

	目	科	科	物种中文名	物种学名	物种英文名	濒危等级	评估依据	是否特有
0874	雀形目	PASSERIFORMES	Phylloscopidae	柳莺科	萨岛柳莺	*Phylloscopus borealoides*	Sakhalin Leaf Warbler	LC		
0875	雀形目	PASSERIFORMES	Phylloscopidae	柳莺科	乌嘴柳莺	*Phylloscopus magnirostris*	Large-billed Leaf Warbler	LC		
0876	雀形目	PASSERIFORMES	Phylloscopidae	柳莺科	冕柳莺	*Phylloscopus coronatus*	Eastern Crowned Warbler	LC		
0877	雀形目	PASSERIFORMES	Phylloscopidae	柳莺科	日本冕柳莺	*Phylloscopus ijimae*	Ijima's Leaf Warbler	NT		
0878	雀形目	PASSERIFORMES	Phylloscopidae	柳莺科	西南冠纹柳莺	*Phylloscopus reguloides*	Blyth's Leaf Warbler	LC		
0879	雀形目	PASSERIFORMES	Phylloscopidae	柳莺科	冠纹柳莺	*Phylloscopus claudiae*	Claudia's Leaf Warbler	LC		
0880	雀形目	PASSERIFORMES	Phylloscopidae	柳莺科	华南冠纹柳莺	*Phylloscopus goodsoni*	Hartert's Leaf Warbler	LC		
0881	雀形目	PASSERIFORMES	Phylloscopidae	柳莺科	峨眉柳莺	*Phylloscopus emeiensis*	Emei Leaf Warbler	LC		√
0882	雀形目	PASSERIFORMES	Phylloscopidae	柳莺科	云南白斑尾柳莺	*Phylloscopus davisoni*	Davison's Leaf Warbler	LC		
0883	雀形目	PASSERIFORMES	Phylloscopidae	柳莺科	白斑尾柳莺	*Phylloscopus ogilviegranti*	Kloss's Leaf Warbler	LC		
0884	雀形目	PASSERIFORMES	Phylloscopidae	柳莺科	海南柳莺	*Phylloscopus hainanus*	Hainan Leaf Warbler	VU	B1ab(ii, iii, v)	√
0885	雀形目	PASSERIFORMES	Phylloscopidae	柳莺科	黄胸柳莺	*Phylloscopus cantator*	Yellow-vented Warbler	LC		
0886	雀形目	PASSERIFORMES	Phylloscopidae	柳莺科	灰岩柳莺	*Phylloscopus calciatilis*	Limestone Leaf Warbler	NT		
0887	雀形目	PASSERIFORMES	Phylloscopidae	柳莺科	黑眉柳莺	*Phylloscopus ricketti*	Sulphur-breasted Warbler	LC		
0888	雀形目	PASSERIFORMES	Phylloscopidae	柳莺科	灰头柳莺	*Phylloscopus xanthoschistos*	Grey-hooded Warbler	LC		
0889	雀形目	PASSERIFORMES	Phylloscopidae	柳莺科	白眶鹟莺	*Seicercus affinis*	White-spectacled Warbler	LC		
0890	雀形目	PASSERIFORMES	Phylloscopidae	柳莺科	金眶鹟莺	*Seicercus burkii*	Green-crowned Warbler	LC		
0891	雀形目	PASSERIFORMES	Phylloscopidae	柳莺科	灰冠鹟莺	*Seicercus tephrocephalus*	Grey-crowned Warbler	LC		
0892	雀形目	PASSERIFORMES	Phylloscopidae	柳莺科	韦氏鹟莺	*Seicercus whistleri*	Whistler's Warbler	LC		
0893	雀形目	PASSERIFORMES	Phylloscopidae	柳莺科	比氏鹟莺	*Seicercus valentini*	Bianchi's Warbler	LC		
0894	雀形目	PASSERIFORMES	Phylloscopidae	柳莺科	峨眉鹟莺	*Seicercus omeiensis*	Martens's Warbler	LC		
0895	雀形目	PASSERIFORMES	Phylloscopidae	柳莺科	淡尾鹟莺	*Seicercus soror*	Plain-tailed Warbler	LC		
0896	雀形目	PASSERIFORMES	Phylloscopidae	柳莺科	灰脸鹟莺	*Seicercus poliogenys*	Grey-cheeked Warbler	LC		
0897	雀形目	PASSERIFORMES	Phylloscopidae	柳莺科	栗头鹟莺	*Seicercus castaniceps*	Chestnut-crowned Warbler	LC		
0898	雀形目	PASSERIFORMES	Cettiidae	树莺科	黄腹鹟莺	*Abroscopus superciliaris*	Yellow-bellied Warbler	LC		
0899	雀形目	PASSERIFORMES	Cettiidae	树莺科	棕脸鹟莺	*Abroscopus albogularis*	Rufous-faced Warbler	LC		
0900	雀形目	PASSERIFORMES	Cettiidae	树莺科	黑脸鹟莺	*Abroscopus schisticeps*	Black-faced Warbler	LC		

续表

	目	科	物种中文名	物种学名	物种英文名	濒危等级	评估依据	是否特有
0901	雀形目 PASSERIFORMES	树莺科 Cettiidae	栗头织叶莺	*Phyllergates cucullatus*	Mountain Tailorbird	LC		
0902	雀形目 PASSERIFORMES	树莺科 Cettiidae	宽嘴鹟莺	*Tickellia hodgsoni*	Broad-billed Warbler	LC		
0903	雀形目 PASSERIFORMES	树莺科 Cettiidae	短翅树莺	*Horornis diphone*	Japanese Bush Warbler	LC		
0904	雀形目 PASSERIFORMES	树莺科 Cettiidae	远东树莺	*Horornis canturians*	Manchurian Bush Warbler	LC		
0905	雀形目 PASSERIFORMES	树莺科 Cettiidae	强脚树莺	*Horornis fortipes*	Brownish-flanked Bush Warbler	LC		
0906	雀形目 PASSERIFORMES	树莺科 Cettiidae	喜山黄腹树莺	*Horornis brunnescens*	Hume's Bush Warbler	LC		
0907	雀形目 PASSERIFORMES	树莺科 Cettiidae	黄腹树莺	*Horornis acanthizoides*	Yellow-bellied Bush Warbler	LC		
0908	雀形目 PASSERIFORMES	树莺科 Cettiidae	异色树莺	*Horornis flavolivaceus*	Aberrant Bush Warbler	LC		
0909	雀形目 PASSERIFORMES	树莺科 Cettiidae	灰腹地莺	*Tesia cyaniventer*	Grey-bellied Tesia	LC		
0910	雀形目 PASSERIFORMES	树莺科 Cettiidae	金冠地莺	*Tesia olivea*	Slaty-bellied Tesia	LC		
0911	雀形目 PASSERIFORMES	树莺科 Cettiidae	宽尾树莺	*Cettia cetti*	Cetti's Warbler	LC		
0912	雀形目 PASSERIFORMES	树莺科 Cettiidae	大树莺	*Cettia major*	Chestnut-crowned Bush Warbler	LC		
0913	雀形目 PASSERIFORMES	树莺科 Cettiidae	棕顶树莺	*Cettia brunnifrons*	Grey-sided Bush Warbler	LC		
0914	雀形目 PASSERIFORMES	树莺科 Cettiidae	栗头树莺	*Cettia castaneocoronata*	Chestnut-headed Tesia	LC		
0915	雀形目 PASSERIFORMES	树莺科 Cettiidae	鳞头树莺	*Urosphena squameiceps*	Asian Stubtail	LC		
0916	雀形目 PASSERIFORMES	树莺科 Cettiidae	淡脚树莺	*Hemitesia pallidipes*	Pale-footed Bush Warbler	LC		
0917	雀形目 PASSERIFORMES	长尾山雀科 Aegithalidae	北长尾山雀	*Aegithalos caudatus*	Long-tailed Tit	LC		
0918	雀形目 PASSERIFORMES	长尾山雀科 Aegithalidae	银喉长尾山雀	*Aegithalos glaucogularis*	Silver-throated Bushtit	LC		√
0919	雀形目 PASSERIFORMES	长尾山雀科 Aegithalidae	红头长尾山雀	*Aegithalos concinnus*	Black-throated Bushtit	LC		
0920	雀形目 PASSERIFORMES	长尾山雀科 Aegithalidae	棕额长尾山雀	*Aegithalos iouschistos*	Rufous-fronted Bushtit	LC		
0921	雀形目 PASSERIFORMES	长尾山雀科 Aegithalidae	黑眉长尾山雀	*Aegithalos bonvaloti*	Black-browed Bushtit	LC		
0922	雀形目 PASSERIFORMES	长尾山雀科 Aegithalidae	银脸长尾山雀	*Aegithalos fuliginosus*	Sooty Bushtit	LC		√
0923	雀形目 PASSERIFORMES	长尾山雀科 Aegithalidae	花彩雀莺	*Leptopoecile sophiae*	White-browed Tit Warbler	LC		
0924	雀形目 PASSERIFORMES	长尾山雀科 Aegithalidae	凤头雀莺	*Leptopoecile elegans*	Crested Tit Warbler	NT		√
0925	雀形目 PASSERIFORMES	莺鹛科 Sylviidae	火尾绿鹛	*Myzornis pyrrhoura*	Fire-tailed Myzornis	NT		
0926	雀形目 PASSERIFORMES	莺鹛科 Sylviidae	黑顶林莺	*Sylvia atricapilla*	Eurasian Blackcap	LC		
0927	雀形目 PASSERIFORMES	莺鹛科 Sylviidae	横斑林莺	*Sylvia nisoria*	Barred Warbler	LC		

续表

	目	科	物种中文名	物种学名	物种英文名	濒危等级	评估依据	是否特有
0928	雀形目 PASSERIFORMES	莺鹛科 Sylviidae	白喉林莺	*Sylvia curruca*	Lesser Whitethroat	LC		
0929	雀形目 PASSERIFORMES	莺鹛科 Sylviidae	漠白喉林莺	*Sylvia minula*	Desert Whitethroat	LC		
0930	雀形目 PASSERIFORMES	莺鹛科 Sylviidae	休氏白喉林莺	*Sylvia althaea*	Hume's Whitethroat	LC		
0931	雀形目 PASSERIFORMES	莺鹛科 Sylviidae	东歌林莺	*Sylvia crassirostris*	Eastern Orphean Warbler	LC		
0932	雀形目 PASSERIFORMES	莺鹛科 Sylviidae	荒漠林莺	*Sylvia nana*	Asian Desert Warbler	LC		
0933	雀形目 PASSERIFORMES	莺鹛科 Sylviidae	灰白喉林莺	*Sylvia communis*	Common Whitethroat	LC		
0934	雀形目 PASSERIFORMES	莺鹛科 Sylviidae	金胸雀鹛	*Lioparus chrysotis*	Golden-breasted Fulvetta	LC		
0935	雀形目 PASSERIFORMES	莺鹛科 Sylviidae	宝兴鹛雀	*Moupinia poecilotis*	Rufous-tailed Babbler	LC		√
0936	雀形目 PASSERIFORMES	莺鹛科 Sylviidae	白眉雀鹛	*Fulvetta vinipectus*	White-browed Fulvetta	LC		
0937	雀形目 PASSERIFORMES	莺鹛科 Sylviidae	中华雀鹛	*Fulvetta striaticollis*	Chinese Fulvetta	LC		√
0938	雀形目 PASSERIFORMES	莺鹛科 Sylviidae	棕头雀鹛	*Fulvetta ruficapilla*	Spectacled Fulvetta	LC		
0939	雀形目 PASSERIFORMES	莺鹛科 Sylviidae	路氏雀鹛	*Fulvetta ludlowi*	Ludlow's Fulvetta	LC		
0940	雀形目 PASSERIFORMES	莺鹛科 Sylviidae	褐头雀鹛	*Fulvetta cinereiceps*	Streak-throated Fulvetta	LC		
0941	雀形目 PASSERIFORMES	莺鹛科 Sylviidae	金眼鹛雀	*Chrysomma sinense*	Yellow-eyed Babbler	LC		
0942	雀形目 PASSERIFORMES	莺鹛科 Sylviidae	山鹛	*Rhopophilus pekinensis*	Chinese Hill Babbler	LC		
0943	雀形目 PASSERIFORMES	莺鹛科 Sylviidae	红嘴鸦雀	*Conostoma aemodium*	Great Parrotbill	LC		
0944	雀形目 PASSERIFORMES	莺鹛科 Sylviidae	三趾鸦雀	*Cholornis paradoxus*	Three-toed Parrotbill	NT		√
0945	雀形目 PASSERIFORMES	莺鹛科 Sylviidae	褐鸦雀	*Cholornis unicolor*	Brown Parrotbill	LC		
0946	雀形目 PASSERIFORMES	莺鹛科 Sylviidae	白眶鸦雀	*Sinosuthora conspicillata*	Spectacled Parrotbill	NT		√
0947	雀形目 PASSERIFORMES	莺鹛科 Sylviidae	棕头鸦雀	*Sinosuthora webbiana*	Vinous-throated Parrotbill	LC		
0948	雀形目 PASSERIFORMES	莺鹛科 Sylviidae	灰喉鸦雀	*Sinosuthora alphonsiana*	Ashy-throated Parrotbill	LC		
0949	雀形目 PASSERIFORMES	莺鹛科 Sylviidae	褐翅鸦雀	*Sinosuthora brunnea*	Brown-winged Parrotbill	LC		
0950	雀形目 PASSERIFORMES	莺鹛科 Sylviidae	暗色鸦雀	*Sinosuthora zappeyi*	Grey-hooded Parrotbill	VU	B1ab(i, ii, iii), C2a(i)	√
0951	雀形目 PASSERIFORMES	莺鹛科 Sylviidae	灰冠鸦雀	*Sinosuthora przewalskii*	Rusty-throated Parrotbill	EN	C2a(i), D2	√
0952	雀形目 PASSERIFORMES	莺鹛科 Sylviidae	黄额鸦雀	*Suthora fulvifrons*	Fulvous Parrotbill	LC		
0953	雀形目 PASSERIFORMES	莺鹛科 Sylviidae	黑喉鸦雀	*Suthora nipalensis*	Black-throated Parrotbill	DD		

续表

	目	科	物种中文名	物种学名	物种英文名	濒危等级	评估依据	是否特有
0954	PASSERIFORMES	Sylviidae	金色鸦雀	*Suthora verreauxi*	Golden Parrotbill	NT		
0955	PASSERIFORMES	Sylviidae	短尾鸦雀	*Neosuthora davidiana*	Short-tailed Parrotbill	NT		
0956	PASSERIFORMES	Sylviidae	黑眉鸦雀	*Chleuasicus atrosuperciliaris*	Lesser Rufous-headed Parrotbill	LC		
0957	PASSERIFORMES	Sylviidae	红头鸦雀	*Psittiparus ruficeps*	White-breasted Parrotbill	LC		
0958	PASSERIFORMES	Sylviidae	灰头鸦雀	*Psittiparus gularis*	Grey-headed Parrotbill	LC		
0959	PASSERIFORMES	Sylviidae	点胸鸦雀	*Paradoxornis guttaticollis*	Spot-breasted Parrotbill	LC		
0960	PASSERIFORMES	Sylviidae	斑胸鸦雀	*Paradoxornis flavirostris*	Black-breasted Parrotbill	DD		
0961	PASSERIFORMES	Sylviidae	震旦鸦雀	*Paradoxornis heudei*	Reed Parrotbill	NT		
0962	PASSERIFORMES	Zosteropidae	栗耳凤鹛	*Yuhina castaniceps*	Striated Yuhina	LC		
0963	PASSERIFORMES	Zosteropidae	白颈凤鹛	*Yuhina bakeri*	White-naped Yuhina	LC		
0964	PASSERIFORMES	Zosteropidae	黄颈凤鹛	*Yuhina flavicollis*	Whiskered Yuhina	LC		
0965	PASSERIFORMES	Zosteropidae	纹喉凤鹛	*Yuhina gularis*	Stripe-throated Yuhina	LC		
0966	PASSERIFORMES	Zosteropidae	白领凤鹛	*Yuhina diademata*	White-collared Yuhina	LC		
0967	PASSERIFORMES	Zosteropidae	棕臀凤鹛	*Yuhina occipitalis*	Rufous-vented Yuhina	LC		
0968	PASSERIFORMES	Zosteropidae	褐头凤鹛	*Yuhina brunneiceps*	Taiwan Yuhina	LC		√
0969	PASSERIFORMES	Zosteropidae	黑颏凤鹛	*Yuhina nigrimenta*	Black-chinned Yuhina	LC		
0970	PASSERIFORMES	Zosteropidae	红胁绣眼鸟	*Zosterops erythropleurus*	Chestnut-flanked White-eye	LC		
0971	PASSERIFORMES	Zosteropidae	暗绿绣眼鸟	*Zosterops japonicus*	Japanese White-eye	LC		
0972	PASSERIFORMES	Zosteropidae	低地绣眼鸟	*Zosterops meyeni*	Lowland White-eye	LC		
0973	PASSERIFORMES	Zosteropidae	灰腹绣眼鸟	*Zosterops palpebrosus*	Oriental White-eye	LC		
0974	PASSERIFORMES	Timaliidae	长嘴钩嘴鹛	*Erythrogenys hypoleucos*	Large Scimitar Babbler	LC		
0975	PASSERIFORMES	Timaliidae	斑胸钩嘴鹛	*Erythrogenys gravivox*	Black-streaked Scimitar Babbler	LC		
0976	PASSERIFORMES	Timaliidae	华南斑胸钩嘴鹛	*Erythrogenys swinhoei*	Grey-sided Scimitar-Babbler	LC		√
0977	PASSERIFORMES	Timaliidae	台湾斑胸钩嘴鹛	*Erythrogenys erythrocnemis*	Black-necked Scimitar-Babbler	LC		√
0978	PASSERIFORMES	Timaliidae	灰头钩嘴鹛	*Pomatorhinus schisticeps*	White-browed Scimitar Babbler	DD		
0979	PASSERIFORMES	Timaliidae	棕颈钩嘴鹛	*Pomatorhinus ruficollis*	Streak-breasted Scimitar Babbler	LC		
0980	PASSERIFORMES	Timaliidae	台湾棕颈钩嘴鹛	*Pomatorhinus musicus*	Taiwan Scimitar Babbler	LC		√

附录　中国鸟类濒危等级评估名录

续表

	目	科	物种中文名	物种学名	物种英文名	濒危等级	评估依据	是否特有
0981	雀形目 PASSERIFORMES	Timaliidae 林鹛科	棕头钩嘴鹛	*Pomatorhinus ochraceiceps*	Red-billed Scimitar Babbler	LC		
0982	雀形目 PASSERIFORMES	Timaliidae 林鹛科	红嘴钩嘴鹛	*Pomatorhinus ferruginosus*	Coral-billed Scimitar Babbler	LC		
0983	雀形目 PASSERIFORMES	Timaliidae 林鹛科	细钩嘴鹛	*Pomatorhinus superciliaris*	Slender-billed Scimitar Babbler	NT		
0984	雀形目 PASSERIFORMES	Timaliidae 林鹛科	短尾钩嘴鹛	*Jabouilleia danjoui*	Short-tailed Scimitar Babbler	DD(NE)		
0985	雀形目 PASSERIFORMES	Timaliidae 林鹛科	斑翅鹩鹛	*Spelaeornis troglodytoides*	Bar-winged Wren-Babbler	LC		
0986	雀形目 PASSERIFORMES	Timaliidae 林鹛科	长尾鹩鹛	*Spelaeornis chocolatinus*	Long-tailed Wren-Babbler	NT		
0987	雀形目 PASSERIFORMES	Timaliidae 林鹛科	淡喉鹩鹛	*Spelaeornis kinneari*	Pale-throated Wren-Babbler	VU		
0988	雀形目 PASSERIFORMES	Timaliidae 林鹛科	棕喉鹩鹛	*Spelaeornis caudatus*	Rufous-throated Wren-Babbler	NT		
0989	雀形目 PASSERIFORMES	Timaliidae 林鹛科	锈喉鹩鹛	*Spelaeornis badeigularis*	Rusty-throated Wren-Babbler	DD(NE)		
0990	雀形目 PASSERIFORMES	Timaliidae 林鹛科	楔嘴鹩鹛	*Stachyris roberti*	Wedge-billed Wren-Babbler	NT		
0991	雀形目 PASSERIFORMES	Timaliidae 林鹛科	黑胸楔嘴鹩鹛	*Stachyris humei*	Blackish-breasted Babbler			
0992	雀形目 PASSERIFORMES	Timaliidae 林鹛科	弄岗穗鹛	*Stachyris nonggangensis*	Nonggang Babbler	EN	B1a, C2a(ii), D1	√
0993	雀形目 PASSERIFORMES	Timaliidae 林鹛科	黑头穗鹛	*Stachyris nigriceps*	Grey-throated Babbler	LC		
0994	雀形目 PASSERIFORMES	Timaliidae 林鹛科	斑颈穗鹛	*Stachyris striolata*	Spot-necked Babbler	LC		
0995	雀形目 PASSERIFORMES	Timaliidae 林鹛科	黄喉穗鹛	*Cyanoderma ambiguum*	Buff-chested Babbler	LC		
0996	雀形目 PASSERIFORMES	Timaliidae 林鹛科	红头穗鹛	*Cyanoderma ruficeps*	Rufous-capped Babbler	LC		
0997	雀形目 PASSERIFORMES	Timaliidae 林鹛科	黑颏穗鹛	*Cyanoderma pyrrhops*	Black-chinned Babbler	LC		
0998	雀形目 PASSERIFORMES	Timaliidae 林鹛科	金头穗鹛	*Mixornis gularis*	Golden Babbler	LC		
0999	雀形目 PASSERIFORMES	Timaliidae 林鹛科	纹胸鹛	*Timalia pileata*	Striped Tit-Babbler	LC		
1000	雀形目 PASSERIFORMES	Timaliidae 林鹛科	红顶鹛		Chestnut-capped Babbler	LC		
1001	雀形目 PASSERIFORMES	Pellorneidae 幽鹛科	金额雀鹛	*Schoeniparus variegaticeps*	Golden-fronted Fulvetta	VU	B1ab(ii, iv), C2a(i)	√
1002	雀形目 PASSERIFORMES	Pellorneidae 幽鹛科	黄喉雀鹛	*Schoeniparus cinereus*	Yellow-throated Fulvetta	LC		
1003	雀形目 PASSERIFORMES	Pellorneidae 幽鹛科	栗头雀鹛	*Schoeniparus castaneceps*	Rufous-winged Fulvetta	LC		
1004	雀形目 PASSERIFORMES	Pellorneidae 幽鹛科	棕喉雀鹛	*Schoeniparus rufogularis*	Rufous-throated Fulvetta	LC		
1005	雀形目 PASSERIFORMES	Pellorneidae 幽鹛科	褐胁雀鹛	*Schoeniparus dubius*	Rusty-capped Fulvetta	LC		
1006	雀形目 PASSERIFORMES	Pellorneidae 幽鹛科	褐顶雀鹛	*Schoeniparus brunneus*	Dusky Fulvetta	LC		

续表

	目	科	物种中文名	物种学名	物种英文名	濒危等级	评估依据	是否特有
1007	PASSERIFORMES 雀形目	Pellorneidae 幽鹛科	褐脸雀鹛	*Alcippe poioicephala*	Brown-cheeked Fulvetta	LC		
1008	PASSERIFORMES 雀形目	Pellorneidae 幽鹛科	灰眶雀鹛	*Alcippe morrisonia*	Grey-cheeked Fulvetta	LC		
1009	PASSERIFORMES 雀形目	Pellorneidae 幽鹛科	白眶雀鹛	*Alcippe nipalensis*	Nepal Fulvetta	LC		
1010	PASSERIFORMES 雀形目	Pellorneidae 幽鹛科	灰岩鹪鹛	*Turdinus crispifrons*	Limestone Wren-Babbler	LC		
1011	PASSERIFORMES 雀形目	Pellorneidae 幽鹛科	短尾鹪鹛	*Turdinus brevicaudatus*	Streaked Wren-Babbler	LC		
1012	PASSERIFORMES 雀形目	Pellorneidae 幽鹛科	纹胸鹪鹛	*Napothera epilepidota*	Eyebrowed Wren-Babbler	LC		
1013	PASSERIFORMES 雀形目	Pellorneidae 幽鹛科	白头鹛	*Gampsorhynchus rufulus*	White-hooded Babbler	LC		
1014	PASSERIFORMES 雀形目	Pellorneidae 幽鹛科	长嘴鹩鹛	*Rimator malacoptilus*	Long-billed Wren-Babbler	LC		
1015	PASSERIFORMES 雀形目	Pellorneidae 幽鹛科	白腹幽鹛	*Pellorneum albiventre*	Spot-throated Babbler	LC		
1016	PASSERIFORMES 雀形目	Pellorneidae 幽鹛科	棕头幽鹛	*Pellorneum ruficeps*	Puff-throated Babbler	LC		
1017	PASSERIFORMES 雀形目	Pellorneidae 幽鹛科	棕胸雅鹛	*Trichastoma tickelli*	Buff-breasted Babbler	NT		
1018	PASSERIFORMES 雀形目	Pellorneidae 幽鹛科	中华草鹛	*Graminicola striatus*	Chinese Grass-babbler	NT		
1019	PASSERIFORMES 雀形目	Leiothrichidae 噪鹛科	矛纹草鹛	*Babax lanceolatus*	Chinese Babax	LC		
1020	PASSERIFORMES 雀形目	Leiothrichidae 噪鹛科	大草鹛	*Babax waddelli*	Giant Babax	NT		
1021	PASSERIFORMES 雀形目	Leiothrichidae 噪鹛科	棕草鹛	*Babax koslowi*	Tibetan Babax	NT		√
1022	PASSERIFORMES 雀形目	Leiothrichidae 噪鹛科	画眉	*Garrulax canorus*	Hwamei	NT		
1023	PASSERIFORMES 雀形目	Leiothrichidae 噪鹛科	海南画眉	*Garrulax owstoni*	Hainan Hwamei	NT		√
1024	PASSERIFORMES 雀形目	Leiothrichidae 噪鹛科	台湾画眉	*Garrulax taewanus*	Taiwan Hwamei	NT		√
1025	PASSERIFORMES 雀形目	Leiothrichidae 噪鹛科	白冠噪鹛	*Garrulax leucolophus*	White-crested Laughingthrush	LC		
1026	PASSERIFORMES 雀形目	Leiothrichidae 噪鹛科	白颈噪鹛	*Garrulax strepitans*	White-necked Laughingthrush	LC		
1027	PASSERIFORMES 雀形目	Leiothrichidae 噪鹛科	褐胸噪鹛	*Garrulax maesi*	Grey Laughingthrush	LC		
1028	PASSERIFORMES 雀形目	Leiothrichidae 噪鹛科	栗颈噪鹛	*Garrulax castanotis*	Rufous-cheeked Laughingthrush	LC		
1029	PASSERIFORMES 雀形目	Leiothrichidae 噪鹛科	黑额山噪鹛	*Garrulax sukatschewi*	Snowy-cheeked Laughingthrush	VU	B1ab(i, ii, iii), C2a(i)	√
1030	PASSERIFORMES 雀形目	Leiothrichidae 噪鹛科	灰翅噪鹛	*Garrulax cineraceus*	Moustached Laughingthrush	LC		
1031	PASSERIFORMES 雀形目	Leiothrichidae 噪鹛科	棕颏噪鹛	*Garrulax rufogularis*	Rufous-chinned Laughingthrush	LC		
1032	PASSERIFORMES 雀形目	Leiothrichidae 噪鹛科	斑背噪鹛	*Garrulax lunulatus*	Barred Laughingthrush	LC		√

续表

	目	科	物种中文名	物种学名	物种英文名	濒危等级	评估依据	是否特有
1033	PASSERIFORMES 雀形目	Leiothrichidae 噪鹛科	白点噪鹛	*Garrulax bieti*	White-speckled Laughingthrush	VU	C1	√
1034	PASSERIFORMES 雀形目	Leiothrichidae 噪鹛科	大噪鹛	*Garrulax maximus*	Giant Laughingthrush	LC		√
1035	PASSERIFORMES 雀形目	Leiothrichidae 噪鹛科	眼纹噪鹛	*Garrulax ocellatus*	Spotted Laughingthrush	NT		
1036	PASSERIFORMES 雀形目	Leiothrichidae 噪鹛科	黑脸噪鹛	*Garrulax perspicillatus*	Masked Laughingthrush	LC		
1037	PASSERIFORMES 雀形目	Leiothrichidae 噪鹛科	白喉噪鹛	*Garrulax albogularis*	White-throated Laughingthrush	LC		
1038	PASSERIFORMES 雀形目	Leiothrichidae 噪鹛科	台湾白喉噪鹛	*Garrulax ruficeps*	Rufous-crowned Laughingthrush	LC		√
1039	PASSERIFORMES 雀形目	Leiothrichidae 噪鹛科	小黑领噪鹛	*Garrulax monileger*	Lesser Necklaced Laughingthrush	LC		
1040	PASSERIFORMES 雀形目	Leiothrichidae 噪鹛科	黑领噪鹛	*Garrulax pectoralis*	Greater Necklaced Laughingthrush	LC		
1041	PASSERIFORMES 雀形目	Leiothrichidae 噪鹛科	黑喉噪鹛	*Garrulax chinensis*	Black-throated Laughingthrush	LC		
1042	PASSERIFORMES 雀形目	Leiothrichidae 噪鹛科	栗颈噪鹛	*Garrulax ruficollis*	Rufous-necked Laughingthrush	NT		
1043	PASSERIFORMES 雀形目	Leiothrichidae 噪鹛科	蓝冠噪鹛	*Garrulax courtoisi*	Blue-crowned Laughingthrush	CR	B2ab(i-v), C2a(ii)	√
1044	PASSERIFORMES 雀形目	Leiothrichidae 噪鹛科	栗臀噪鹛	*Garrulax gularis*	Rufous-vented Laughingthrush	LC		
1045	PASSERIFORMES 雀形目	Leiothrichidae 噪鹛科	山噪鹛	*Garrulax davidi*	Plain Laughingthrush	LC		√
1046	PASSERIFORMES 雀形目	Leiothrichidae 噪鹛科	灰胁噪鹛	*Garrulax caerulatus*	Grey-sided Laughingthrush	LC		
1047	PASSERIFORMES 雀形目	Leiothrichidae 噪鹛科	棕噪鹛	*Garrulax berthemyi*	Buffy Laughingthrush	LC		√
1048	PASSERIFORMES 雀形目	Leiothrichidae 噪鹛科	台湾棕噪鹛	*Garrulax poecilorhynchus*	Rusty Laughingthrush	LC		
1049	PASSERIFORMES 雀形目	Leiothrichidae 噪鹛科	白颊噪鹛	*Garrulax sannio*	White-browed Laughingthrush	LC		
1050	PASSERIFORMES 雀形目	Leiothrichidae 噪鹛科	斑胸噪鹛	*Garrulax merulinus*	Spot-breasted Laughingthrush	LC		
1051	PASSERIFORMES 雀形目	Leiothrichidae 噪鹛科	条纹噪鹛	*Grammatoptila striata*	Striated Laughingthrush	LC		
1052	PASSERIFORMES 雀形目	Leiothrichidae 噪鹛科	细纹噪鹛	*Trochalopteron lineatum*	Streaked Laughingthrush	LC		
1053	PASSERIFORMES 雀形目	Leiothrichidae 噪鹛科	蓝翅噪鹛	*Trochalopteron squamatum*	Blue-winged Laughingthrush	LC		
1054	PASSERIFORMES 雀形目	Leiothrichidae 噪鹛科	纯色噪鹛	*Trochalopteron subunicolor*	Scaly Laughingthrush	LC		
1055	PASSERIFORMES 雀形目	Leiothrichidae 噪鹛科	橙翅噪鹛	*Trochalopteron elliotii*	Elliot's Laughingthrush	LC		
1056	PASSERIFORMES 雀形目	Leiothrichidae 噪鹛科	灰腹噪鹛	*Trochalopteron henrici*	Brown-cheeked Laughingthrush	LC		√
1057	PASSERIFORMES 雀形目	Leiothrichidae 噪鹛科	黑顶噪鹛	*Trochalopteron affine*	Black-faced Laughingthrush	LC		
1058	PASSERIFORMES 雀形目	Leiothrichidae 噪鹛科	台湾噪鹛	*Trochalopteron morrisonianum*	White-whiskered Laughingthrush	LC		√

续表

	目	科	物种中文名	物种学名	物种英文名	濒危等级	评估依据	是否特有
1059	PASSERIFORMES 雀形目	Leiothrichidae 噪鹛科	杂色噪鹛	*Trochalopteron variegatum*	Variegated Laughingthrush	LC		
1060	PASSERIFORMES 雀形目	Leiothrichidae 噪鹛科	红头噪鹛	*Trochalopteron erythrocephalum*	Chestnut-crowned Laughingthrush	LC		
1061	PASSERIFORMES 雀形目	Leiothrichidae 噪鹛科	红翅噪鹛	*Trochalopteron formosum*	Red-winged Laughingthrush	LC		
1062	PASSERIFORMES 雀形目	Leiothrichidae 噪鹛科	红尾噪鹛	*Trochalopteron milnei*	Red-tailed Laughingthrush	LC		
1063	PASSERIFORMES 雀形目	Leiothrichidae 噪鹛科	斑胁姬鹛	*Cutia nipalensis*	Himalayan Cutia	LC		
1064	PASSERIFORMES 雀形目	Leiothrichidae 噪鹛科	蓝翅希鹛	*Siva cyanouroptera*	Blue-winged Minla	LC		
1065	PASSERIFORMES 雀形目	Leiothrichidae 噪鹛科	斑喉希鹛	*Chrysominla strigula*	Bar-throated Minla	LC		
1066	PASSERIFORMES 雀形目	Leiothrichidae 噪鹛科	红尾希鹛	*Minla ignotincta*	Red-tailed Minla	LC		
1067	PASSERIFORMES 雀形目	Leiothrichidae 噪鹛科	灰头薮鹛	*Liocichla phoenicea*	Red-faced Liocichla	NT		
1068	PASSERIFORMES 雀形目	Leiothrichidae 噪鹛科	红翅薮鹛	*Liocichla ripponi*	Scarlet-faced Liocichla	LC		
1069	PASSERIFORMES 雀形目	Leiothrichidae 噪鹛科	黑冠薮鹛	*Liocichla bugunorum*	Bugun Liocichla	CR	B1ab(i)	
1070	PASSERIFORMES 雀形目	Leiothrichidae 噪鹛科	灰胸薮鹛	*Liocichla omeiensis*	Emei Shan Liocichla	VU	B1ab(i-v), C2a(i, ii)	√
1071	PASSERIFORMES 雀形目	Leiothrichidae 噪鹛科	黄痣薮鹛	*Liocichla steerii*	Steere's Liocichla	LC		√
1072	PASSERIFORMES 雀形目	Leiothrichidae 噪鹛科	栗额斑翅鹛	*Actinodura egertoni*	Rusty-fronted Barwing	LC		
1073	PASSERIFORMES 雀形目	Leiothrichidae 噪鹛科	白眶斑翅鹛	*Actinodura ramsayi*	Spectacled Barwing	LC		
1074	PASSERIFORMES 雀形目	Leiothrichidae 噪鹛科	纹头斑翅鹛	*Sibia nipalensis*	Hoary-throated Barwing	LC		
1075	PASSERIFORMES 雀形目	Leiothrichidae 噪鹛科	纹胸斑翅鹛	*Sibia waldeni*	Streak-throated Barwing	LC		
1076	PASSERIFORMES 雀形目	Leiothrichidae 噪鹛科	灰头斑翅鹛	*Sibia souliei*	Streaked Barwing	LC		
1077	PASSERIFORMES 雀形目	Leiothrichidae 噪鹛科	台湾斑翅鹛	*Sibia morrisoniana*	Taiwan Barwing	LC		√
1078	PASSERIFORMES 雀形目	Leiothrichidae 噪鹛科	银耳相思鸟	*Leiothrix argentauris*	Silver-eared Mesia	NT		
1079	PASSERIFORMES 雀形目	Leiothrichidae 噪鹛科	红嘴相思鸟	*Leiothrix lutea*	Red-billed Leiothrix	LC		
1080	PASSERIFORMES 雀形目	Leiothrichidae 噪鹛科	栗背奇鹛	*Leioptila annectens*	Rufous-backed Sibia	LC		
1081	PASSERIFORMES 雀形目	Leiothrichidae 噪鹛科	黑顶奇鹛	*Heterophasia capistrata*	Rufous Sibia	LC		
1082	PASSERIFORMES 雀形目	Leiothrichidae 噪鹛科	灰奇鹛	*Heterophasia gracilis*	Grey Sibia	LC		
1083	PASSERIFORMES 雀形目	Leiothrichidae 噪鹛科	黑头奇鹛	*Heterophasia desgodinsi*	Black-headed Sibia	LC		
1084	PASSERIFORMES 雀形目	Leiothrichidae 噪鹛科	白耳奇鹛	*Heterophasia auricularis*	White-eared Sibia	LC		√

续表

	目	科	物种中文名	物种学名	物种英文名	濒危等级	评估依据	是否特有
1085	PASSERIFORMES	Leiothrichidae	丽色奇鹛	*Heterophasia pulchella*	Beautiful Sibia	LC		
1086	PASSERIFORMES	Leiothrichidae	长尾奇鹛	*Heterophasia picaoides*	Long-tailed Sibia	LC		
1087	PASSERIFORMES	Certhiidae	欧亚旋木雀	*Certhia familiaris*	Eurasian Treecreeper	LC		
1088	PASSERIFORMES	Certhiidae	霍氏旋木雀	*Certhia hodgsoni*	Hodgson's Treecreeper	LC		
1089	PASSERIFORMES	Certhiidae	高山旋木雀	*Certhia himalayana*	Bar-tailed Treecreeper	LC		
1090	PASSERIFORMES	Certhiidae	红腹旋木雀	*Certhia nipalensis*	Rusty-flanked Treecreeper	LC		
1091	PASSERIFORMES	Certhiidae	褐喉旋木雀	*Certhia discolor*	Brown-throated Treecreeper	LC		
1092	PASSERIFORMES	Certhiidae	休氏旋木雀	*Certhia manipurensis*	Hume's Treecreeper	LC		
1093	PASSERIFORMES	Certhiidae	四川旋木雀	*Certhia tianquanensis*	Sichuan Treecreeper	VU	C2a(i)	√
1094	PASSERIFORMES	Sittidae	普通䴓	*Sitta europaea*	Eurasian Nuthatch	LC		
1095	PASSERIFORMES	Sittidae	栗臀䴓	*Sitta nagaensis*	Chestnut-vented Nuthatch	LC		
1096	PASSERIFORMES	Sittidae	栗腹䴓	*Sitta castanea*	Chestnut-bellied Nuthatch	LC		
1097	PASSERIFORMES	Sittidae	白尾䴓	*Sitta himalayensis*	White-tailed Nuthatch	NT		
1098	PASSERIFORMES	Sittidae	滇䴓	*Sitta yunnanensis*	Yunnan Nuthatch	VU	A3bcd+4bcd, B2b(i, ii, iii)	√
1099	PASSERIFORMES	Sittidae	黑头䴓	*Sitta villosa*	Chinese Nuthatch	NT		
1100	PASSERIFORMES	Sittidae	白脸䴓	*Sitta leucopsis*	White-cheeked Nuthatch	NT		
1101	PASSERIFORMES	Sittidae	绒额䴓	*Sitta frontalis*	Velvet-fronted Nuthatch	NT		
1102	PASSERIFORMES	Sittidae	淡紫䴓	*Sitta solangiae*	Yellow-billed Nuthatch	VU	B1ab(iii)	
1103	PASSERIFORMES	Sittidae	巨䴓	*Sitta magna*	Giant Nuthatch	EN	C1+2a(i)	
1104	PASSERIFORMES	Sittidae	丽䴓	*Sitta formosa*	Beautiful Nuthatch	EN	B1ab(iii) + 2ab(iii), D	
1105	PASSERIFORMES	Sittidae	红翅旋壁雀	*Tichodroma muraria*	Wallcreeper	LC		
1106	PASSERIFORMES	Troglodytidae	鹪鹩	*Troglodytes troglodytes*	Eurasian Wren	LC		
1107	PASSERIFORMES	Cinclidae	河乌	*Cinclus cinclus*	White-throated Dipper	LC		
1108	PASSERIFORMES	Cinclidae	褐河乌	*Cinclus pallasii*	Brown Dipper	LC		
1109	PASSERIFORMES	Sturnidae	亚洲辉椋鸟	*Aplonis panayensis*	Asian Glossy Starling	LC		
1110	PASSERIFORMES	Sturnidae	斑翅椋鸟	*Saroglossa spiloptera*	Spot-winged Starling	LC		

续表

	目	科	物种中文名	物种学名	物种英文名	濒危等级	评估依据	是否特有
1111	雀形目 PASSERIFORMES	Sturnidae 椋鸟科	金冠树八哥	*Ampeliceps coronatus*	Golden-crested Myna	NT		
1112	雀形目 PASSERIFORMES	Sturnidae 椋鸟科	鹩哥	*Gracula religiosa*	Hill Myna	VU	A2acd	
1113	雀形目 PASSERIFORMES	Sturnidae 椋鸟科	林八哥	*Acridotheres grandis*	Great Myna	LC		
1114	雀形目 PASSERIFORMES	Sturnidae 椋鸟科	八哥	*Acridotheres cristatellus*	Crested Myna	LC		
1115	雀形目 PASSERIFORMES	Sturnidae 椋鸟科	爪哇八哥	*Acridotheres javanicus*	Javan Myna	DD		
1116	雀形目 PASSERIFORMES	Sturnidae 椋鸟科	白领八哥	*Acridotheres albocinctus*	Collared Myna	LC		
1117	雀形目 PASSERIFORMES	Sturnidae 椋鸟科	家八哥	*Acridotheres tristis*	Common Myna	LC		
1118	雀形目 PASSERIFORMES	Sturnidae 椋鸟科	红嘴椋鸟	*Acridotheres burmannicus*	Vinous-breasted Starling	LC		
1119	雀形目 PASSERIFORMES	Sturnidae 椋鸟科	丝光椋鸟	*Spodiopsar sericeus*	Red-billed Starling	LC		
1120	雀形目 PASSERIFORMES	Sturnidae 椋鸟科	灰椋鸟	*Spodiopsar cineraceus*	White-cheeked Starling	LC		
1121	雀形目 PASSERIFORMES	Sturnidae 椋鸟科	黑领椋鸟	*Gracupica nigricollis*	Black-collared Starling	LC		
1122	雀形目 PASSERIFORMES	Sturnidae 椋鸟科	斑椋鸟	*Gracupica contra*	Asian Pied Starling	LC		
1123	雀形目 PASSERIFORMES	Sturnidae 椋鸟科	北椋鸟	*Agropsar sturninus*	Daurian Starling	LC		
1124	雀形目 PASSERIFORMES	Sturnidae 椋鸟科	紫背椋鸟	*Agropsar philippensis*	Chestnut-cheeked Starling	LC		
1125	雀形目 PASSERIFORMES	Sturnidae 椋鸟科	灰背椋鸟	*Sturnia sinensis*	White-shouldered Starling	LC		
1126	雀形目 PASSERIFORMES	Sturnidae 椋鸟科	灰头椋鸟	*Sturnia malabarica*	Chestnut-tailed Starling	LC		
1127	雀形目 PASSERIFORMES	Sturnidae 椋鸟科	黑冠椋鸟	*Sturnia pagodarum*	Brahminy Starling	LC		
1128	雀形目 PASSERIFORMES	Sturnidae 椋鸟科	紫翅椋鸟	*Sturnus vulgaris*	Common Starling	LC		
1129	雀形目 PASSERIFORMES	Sturnidae 椋鸟科	粉红椋鸟	*Pastor roseus*	Rosy Starling	LC		
1130	雀形目 PASSERIFORMES	Turdidae 鸫科	橙头地鸫	*Geokichla citrina*	Orange-headed Thrush	LC		
1131	雀形目 PASSERIFORMES	Turdidae 鸫科	白眉地鸫	*Geokichla sibirica*	Siberian Thrush	LC		
1132	雀形目 PASSERIFORMES	Turdidae 鸫科	淡背地鸫	*Zoothera mollissima*	Plain-backed Thrush	LC		
1133	雀形目 PASSERIFORMES	Turdidae 鸫科	四川淡背地鸫	*Zoothera griseiceps*	Sichuan Thrush	NT		
1134	雀形目 PASSERIFORMES	Turdidae 鸫科	喜山淡背地鸫	*Zoothera salimalii*	Himalayan Thrush	NT		
1135	雀形目 PASSERIFORMES	Turdidae 鸫科	长尾地鸫	*Zoothera dixoni*	Long-tailed Thrush	LC		
1136	雀形目 PASSERIFORMES	Turdidae 鸫科	虎斑地鸫	*Zoothera aurea*	White's Thrush	LC		
1137	雀形目 PASSERIFORMES	Turdidae 鸫科	小虎斑地鸫	*Zoothera dauma*	Scaly Thrush	LC		

续表

	目	科	物种中文名	物种学名	物种英文名	濒危等级	评估依据	是否特有
1138	雀形目 PASSERIFORMES	鸫科 Turdidae	大长嘴地鸫	*Zoothera monticola*	Long-billed Thrush	LC		
1139	雀形目 PASSERIFORMES	鸫科 Turdidae	长嘴地鸫	*Zoothera marginata*	Dark-sided Thrush	LC		
1140	雀形目 PASSERIFORMES	鸫科 Turdidae	灰背鸫	*Turdus hortulorum*	Grey-backed Thrush	LC		
1141	雀形目 PASSERIFORMES	鸫科 Turdidae	蒂氏鸫	*Turdus unicolor*	Tickell's Thrush	LC		
1142	雀形目 PASSERIFORMES	鸫科 Turdidae	黑胸鸫	*Turdus dissimilis*	Black-breasted Thrush	NT		
1143	雀形目 PASSERIFORMES	鸫科 Turdidae	乌灰鸫	*Turdus cardis*	Japanese Thrush	LC		
1144	雀形目 PASSERIFORMES	鸫科 Turdidae	白颈鸫	*Turdus albocinctus*	White-collared Blackbird	LC		
1145	雀形目 PASSERIFORMES	鸫科 Turdidae	灰翅鸫	*Turdus boulboul*	Grey-winged Blackbird	LC		
1146	雀形目 PASSERIFORMES	鸫科 Turdidae	欧亚乌鸫	*Turdus merula*	Common Blackbird	LC		
1147	雀形目 PASSERIFORMES	鸫科 Turdidae	乌鸫	*Turdus mandarinus*	Chinese Blackbird	LC		√
1148	雀形目 PASSERIFORMES	鸫科 Turdidae	藏乌鸫	*Turdus maximus*	Tibetan Blackbird	LC		
1149	雀形目 PASSERIFORMES	鸫科 Turdidae	白头鸫	*Turdus niveiceps*	Taiwan Thrush	LC		
1150	雀形目 PASSERIFORMES	鸫科 Turdidae	灰头鸫	*Turdus rubrocanus*	Chestnut Thrush	LC		
1151	雀形目 PASSERIFORMES	鸫科 Turdidae	棕背黑头鸫	*Turdus kessleri*	Kessler's Thrush	LC		
1152	雀形目 PASSERIFORMES	鸫科 Turdidae	褐头鸫	*Turdus feae*	Grey-sided Thrush	VU	C2a(ii)	
1153	雀形目 PASSERIFORMES	鸫科 Turdidae	白眉鸫	*Turdus obscurus*	Eyebrowed Thrush	LC		
1154	雀形目 PASSERIFORMES	鸫科 Turdidae	白腹鸫	*Turdus pallidus*	Pale Thrush	LC		
1155	雀形目 PASSERIFORMES	鸫科 Turdidae	赤胸鸫	*Turdus chrysolaus*	Brown-headed Thrush	LC		
1156	雀形目 PASSERIFORMES	鸫科 Turdidae	黑颈鸫	*Turdus atrogularis*	Black-throated Thrush	LC		
1157	雀形目 PASSERIFORMES	鸫科 Turdidae	赤颈鸫	*Turdus ruficollis*	Red-throated Thrush	LC		
1158	雀形目 PASSERIFORMES	鸫科 Turdidae	红尾斑鸫	*Turdus naumanni*	Naumann's Thrush	LC		
1159	雀形目 PASSERIFORMES	鸫科 Turdidae	斑鸫	*Turdus eunomus*	Dusky Thrush	LC		
1160	雀形目 PASSERIFORMES	鸫科 Turdidae	田鸫	*Turdus pilaris*	Fieldfare	LC		
1161	雀形目 PASSERIFORMES	鸫科 Turdidae	白眉歌鸫	*Turdus iliacus*	Redwing	LC		
1162	雀形目 PASSERIFORMES	鸫科 Turdidae	欧歌鸫	*Turdus philomelos*	Song Thrush	LC		
1163	雀形目 PASSERIFORMES	鸫科 Turdidae	宝兴歌鸫	*Turdus mupinensis*	Chinese Thrush	LC		√
1164	雀形目 PASSERIFORMES	鸫科 Turdidae	槲鸫	*Turdus viscivorus*	Mistle Thrush	LC		

续表

	目	科		物种中文名	物种学名	物种英文名	濒危等级	评估依据	是否特有
1165	雀形目	PASSERIFORMES	鸫科	Turdidae	紫宽嘴鸫	*Cochoa purpurea*	Purple Cochoa	LC		
1166	雀形目	PASSERIFORMES	鸫科	Turdidae	绿宽嘴鸫	*Cochoa viridis*	Green Cochoa	LC		
1167	雀形目	PASSERIFORMES	鹟科	Muscicapidae	欧亚鸲	*Erithacus rubecula*	European Robin	LC		
1168	雀形目	PASSERIFORMES	鹟科	Muscicapidae	日本歌鸲	*Larvivora akahige*	Japanese Robin	LC		
1169	雀形目	PASSERIFORMES	鹟科	Muscicapidae	琉球歌鸲	*Larvivora komadori*	Ryukyu Robin	DD(NE)		
1170	雀形目	PASSERIFORMES	鹟科	Muscicapidae	红尾歌鸲	*Larvivora sibilans*	Rufous-tailed Robin	LC		
1171	雀形目	PASSERIFORMES	鹟科	Muscicapidae	红喉歌鸲	*Calliope calliope*	Siberian Rubythroat	LC		
1172	雀形目	PASSERIFORMES	鹟科	Muscicapidae	黑胸歌鸲	*Calliope pectoralis*	White-tailed Rubythroat	NT		
1173	雀形目	PASSERIFORMES	鹟科	Muscicapidae	白须黑胸歌鸲	*Calliope tschebaiewi*	Chinese Rubythroat	LC		
1174	雀形目	PASSERIFORMES	鹟科	Muscicapidae	黑喉歌鸲	*Calliope obscura*	Blackthroat	EN	C1+2a(i)	
1175	雀形目	PASSERIFORMES	鹟科	Muscicapidae	金胸歌鸲	*Calliope pectardens*	Firethroat	VU	C1	
1176	雀形目	PASSERIFORMES	鹟科	Muscicapidae	白腹短翅鸲	*Luscinia phoenicuroides*	White-bellied Redstart	LC		
1177	雀形目	PASSERIFORMES	鹟科	Muscicapidae	蓝喉歌鸲	*Luscinia svecica*	Bluethroat	LC		
1178	雀形目	PASSERIFORMES	鹟科	Muscicapidae	新疆歌鸲	*Luscinia megarhynchos*	Common Nightingale	LC		
1179	雀形目	PASSERIFORMES	鹟科	Muscicapidae	棕头歌鸲	*Larvivora ruficeps*	Rufous-headed Robin	EN	C2a(ii)	
1180	雀形目	PASSERIFORMES	鹟科	Muscicapidae	栗腹歌鸲	*Larvivora brunnea*	Indian Blue Robin	LC		
1181	雀形目	PASSERIFORMES	鹟科	Muscicapidae	蓝歌鸲	*Larvivora cyane*	Siberian Blue Robin	LC		
1182	雀形目	PASSERIFORMES	鹟科	Muscicapidae	红胁蓝尾鸲	*Tarsiger cyanurus*	Orange-flanked Bluetail	LC		
1183	雀形目	PASSERIFORMES	鹟科	Muscicapidae	蓝眉林鸲	*Tarsiger rufilatus*	Himalayan Bluetail	LC		
1184	雀形目	PASSERIFORMES	鹟科	Muscicapidae	白眉林鸲	*Tarsiger indicus*	White-browed Bush Robin	LC		
1185	雀形目	PASSERIFORMES	鹟科	Muscicapidae	棕腹林鸲	*Tarsiger hyperythrus*	Rufous-breasted Bush Robin	NT		
1186	雀形目	PASSERIFORMES	鹟科	Muscicapidae	台湾林鸲	*Tarsiger johnstoniae*	Collared Bush Robin	LC		√
1187	雀形目	PASSERIFORMES	鹟科	Muscicapidae	金色林鸲	*Tarsiger chrysaeus*	Golden Bush Robin	LC		
1188	雀形目	PASSERIFORMES	鹟科	Muscicapidae	栗背短翅鸫	*Heteroxenicus stellatus*	Gould's Shortwing	LC		
1189	雀形目	PASSERIFORMES	鹟科	Muscicapidae	锈腹短翅鸫	*Brachypteryx hyperythra*	Rusty-bellied Shortwing	NT		
1190	雀形目	PASSERIFORMES	鹟科	Muscicapidae	白喉短翅鸫	*Brachypteryx leucophris*	Lesser Shortwing	LC		
1191	雀形目	PASSERIFORMES	鹟科	Muscicapidae	蓝短翅鸫	*Brachypteryx montana*	White-browed Shortwing	LC		

附录 中国鸟类濒危等级评估名录

续表

	目		科		物种中文名	物种学名	物种英文名	濒危等级	评估依据	是否特有
1192	雀形目	PASSERIFORMES	鹟科	Muscicapidae	棕薮鸲	*Cercotrichas galactotes*	Rufous-tailed Scrub Robin	NT		
1193	雀形目	PASSERIFORMES	鹟科	Muscicapidae	鹊鸲	*Copsychus saularis*	Oriental Magpie Robin	LC		
1194	雀形目	PASSERIFORMES	鹟科	Muscicapidae	白腰鹊鸲	*Kittacincla malabarica*	White-rumped Shama	LC		
1195	雀形目	PASSERIFORMES	鹟科	Muscicapidae	红背红尾鸲	*Phoenicuropsis erythronotus*	Eversmann's Redstart	LC		
1196	雀形目	PASSERIFORMES	鹟科	Muscicapidae	蓝头红尾鸲	*Phoenicuropsis coeruleocephala*	Blue-capped Redstart	LC		
1197	雀形目	PASSERIFORMES	鹟科	Muscicapidae	白喉红尾鸲	*Phoenicuropsis schisticeps*	White-throated Redstart	LC		
1198	雀形目	PASSERIFORMES	鹟科	Muscicapidae	蓝额红尾鸲	*Phoenicuropsis frontalis*	Blue-fronted Redstart	LC		
1199	雀形目	PASSERIFORMES	鹟科	Muscicapidae	贺兰山红尾鸲	*Phoenicurus alaschanicus*	Ala Shan Redstart	EN	B1b(ii, iii); C2a(i, ii)+b	√
1200	雀形目	PASSERIFORMES	鹟科	Muscicapidae	赭红尾鸲	*Phoenicurus ochruros*	Black Redstart	LC		
1201	雀形目	PASSERIFORMES	鹟科	Muscicapidae	欧亚红尾鸲	*Phoenicurus phoenicurus*	Common Redstart	LC		
1202	雀形目	PASSERIFORMES	鹟科	Muscicapidae	黑喉红尾鸲	*Phoenicurus hodgsoni*	Hodgson's Redstart	LC		
1203	雀形目	PASSERIFORMES	鹟科	Muscicapidae	北红尾鸲	*Phoenicurus auroreus*	Daurian Redstart	LC		
1204	雀形目	PASSERIFORMES	鹟科	Muscicapidae	红腹红尾鸲	*Phoenicurus erythrogastrus*	White-winged Redstart	LC		
1205	雀形目	PASSERIFORMES	鹟科	Muscicapidae	红尾水鸲	*Rhyacornis fuliginosa*	Plumbeous Water Redstart	LC		
1206	雀形目	PASSERIFORMES	鹟科	Muscicapidae	白顶溪鸲	*Chaimarrornis leucocephalus*	White-capped Water Redstart	LC		
1207	雀形目	PASSERIFORMES	鹟科	Muscicapidae	白尾地鸲	*Myiomela leucura*	White-tailed Robin	LC		
1208	雀形目	PASSERIFORMES	鹟科	Muscicapidae	蓝额地鸲	*Cinclidium frontale*	Blue-fronted Robin	LC		
1209	雀形目	PASSERIFORMES	鹟科	Muscicapidae	台湾紫啸鸫	*Myophonus insularis*	Taiwan Whistling Thrush	LC		√
1210	雀形目	PASSERIFORMES	鹟科	Muscicapidae	紫啸鸫	*Myophonus caeruleus*	Blue Whistling Thrush	LC		
1211	雀形目	PASSERIFORMES	鹟科	Muscicapidae	蓝大翅鸲	*Grandala coelicolor*	Grandala	LC		
1212	雀形目	PASSERIFORMES	鹟科	Muscicapidae	小燕尾	*Enicurus scouleri*	Little Forktail	LC		
1213	雀形目	PASSERIFORMES	鹟科	Muscicapidae	黑背燕尾	*Enicurus immaculatus*	Black-backed Forktail	LC		
1214	雀形目	PASSERIFORMES	鹟科	Muscicapidae	灰背燕尾	*Enicurus schistaceus*	Slaty-backed Forktail	LC		
1215	雀形目	PASSERIFORMES	鹟科	Muscicapidae	白额燕尾	*Enicurus leschenaulti*	White-crowned Forktail	LC		
1216	雀形目	PASSERIFORMES	鹟科	Muscicapidae	斑背燕尾	*Enicurus maculatus*	Spotted Forktail	LC		
1217	雀形目	PASSERIFORMES	鹟科	Muscicapidae	白喉石䳭	*Saxicola insignis*	White-throated Bushchat	EN	C2a(ii)	

续表

	目	科	物种中文名	物种学名	物种英文名	濒危等级	评估依据	是否特有
1218	雀形目 PASSERIFORMES	鹟科 Muscicapidae	黑喉石鵖	*Saxicola maurus*	Siberian Stonechat	LC		
1219	雀形目 PASSERIFORMES	鹟科 Muscicapidae	白斑黑石鵖	*Saxicola caprata*	Pied Bushchat	LC		
1220	雀形目 PASSERIFORMES	鹟科 Muscicapidae	黑白林鵖	*Saxicola jerdoni*	Jerdon's Bushchat	LC		
1221	雀形目 PASSERIFORMES	鹟科 Muscicapidae	灰林鵖	*Saxicola ferreus*	Grey Bushchat	LC		
1222	雀形目 PASSERIFORMES	鹟科 Muscicapidae	沙鵖	*Oenanthe isabellina*	Isabelline Wheatear	LC		
1223	雀形目 PASSERIFORMES	鹟科 Muscicapidae	穗鵖	*Oenanthe oenanthe*	Northern Wheatear	LC		
1224	雀形目 PASSERIFORMES	鹟科 Muscicapidae	白顶鵖	*Oenanthe pleschanka*	Pied Wheatear	LC		
1225	雀形目 PASSERIFORMES	鹟科 Muscicapidae	漠鵖	*Oenanthe deserti*	Desert Wheatear	LC		
1226	雀形目 PASSERIFORMES	鹟科 Muscicapidae	东方斑鵖	*Oenanthe picata*	Variable Wheatear	LC		
1227	雀形目 PASSERIFORMES	鹟科 Muscicapidae	白背矶鸫	*Monticola saxatilis*	Common Rock Thrush	LC		
1228	雀形目 PASSERIFORMES	鹟科 Muscicapidae	蓝头矶鸫	*Monticola cinclorhyncha*	Blue-capped Rock Thrush	LC		
1229	雀形目 PASSERIFORMES	鹟科 Muscicapidae	蓝矶鸫	*Monticola solitarius*	Blue Rock Thrush	LC		
1230	雀形目 PASSERIFORMES	鹟科 Muscicapidae	栗腹矶鸫	*Monticola rufiventris*	Chestnut-bellied Rock Thrush	LC		
1231	雀形目 PASSERIFORMES	鹟科 Muscicapidae	白喉矶鸫	*Monticola gularis*	White-throated Rock Thrush	LC		
1232	雀形目 PASSERIFORMES	鹟科 Muscicapidae	斑鹟	*Muscicapa striata*	Spotted Flycatcher	LC		
1233	雀形目 PASSERIFORMES	鹟科 Muscicapidae	灰纹鹟	*Muscicapa griseisticta*	Grey-streaked Flycatcher	LC		
1234	雀形目 PASSERIFORMES	鹟科 Muscicapidae	乌鹟	*Muscicapa sibirica*	Dark-sided Flycatcher	LC		
1235	雀形目 PASSERIFORMES	鹟科 Muscicapidae	北灰鹟	*Muscicapa dauurica*	Asian Brown Flycatcher	LC		
1236	雀形目 PASSERIFORMES	鹟科 Muscicapidae	褐胸鹟	*Muscicapa muttui*	Brown-breasted Flycatcher	LC		
1237	雀形目 PASSERIFORMES	鹟科 Muscicapidae	棕尾褐鹟	*Muscicapa ferruginea*	Ferruginous Flycatcher	LC		
1238	雀形目 PASSERIFORMES	鹟科 Muscicapidae	栗尾姬鹟	*Ficedula ruficauda*	Rusty-tailed Flycatcher	LC		
1239	雀形目 PASSERIFORMES	鹟科 Muscicapidae	斑姬鹟	*Ficedula hypoleuca*	European Pied Flycatcher	DD		
1240	雀形目 PASSERIFORMES	鹟科 Muscicapidae	白眉姬鹟	*Ficedula zanthopygia*	Yellow-rumped Flycatcher	LC		
1241	雀形目 PASSERIFORMES	鹟科 Muscicapidae	黄眉姬鹟	*Ficedula narcissina*	Narcissus Flycatcher	LC		
1242	雀形目 PASSERIFORMES	鹟科 Muscicapidae	琉球姬鹟	*Ficedula owstoni*	Ryukyu Flycatcher	LC		
1243	雀形目 PASSERIFORMES	鹟科 Muscicapidae	绿背姬鹟	*Ficedula elisae*	Green-backed Flycatcher	NT		
1244	雀形目 PASSERIFORMES	鹟科 Muscicapidae	侏蓝姬鹟	*Ficedula hodgsoni*	Pygmy Blue Flycatcher	LC		

续表

	目	科	物种中文名	物种学名	物种英文名	濒危等级	评估依据	是否特有
1245	雀形目 PASSERIFORMES	Muscicapidae 鹟科	鸲姬鹟	*Ficedula mugimaki*	Mugimaki Flycatcher	LC		
1246	雀形目 PASSERIFORMES	Muscicapidae 鹟科	锈胸蓝姬鹟	*Ficedula sordida*	Slaty-backed Flycatcher	LC		
1247	雀形目 PASSERIFORMES	Muscicapidae 鹟科	橙胸姬鹟	*Ficedula strophiata*	Rufous-gorgeted Flycatcher	LC		
1248	雀形目 PASSERIFORMES	Muscicapidae 鹟科	红胸姬鹟	*Ficedula parva*	Red-breasted Flycatcher	DD(NE)		
1249	雀形目 PASSERIFORMES	Muscicapidae 鹟科	红喉姬鹟	*Ficedula albicilla*	Taiga Flycatcher	LC		
1250	雀形目 PASSERIFORMES	Muscicapidae 鹟科	棕胸蓝姬鹟	*Ficedula hyperythra*	Snowy-browed Flycatcher	LC		
1251	雀形目 PASSERIFORMES	Muscicapidae 鹟科	小斑姬鹟	*Ficedula westermanni*	Little Pied Flycatcher	LC		
1252	雀形目 PASSERIFORMES	Muscicapidae 鹟科	白眉蓝姬鹟	*Ficedula superciliaris*	Ultramarine Flycatcher	LC		
1253	雀形目 PASSERIFORMES	Muscicapidae 鹟科	灰蓝姬鹟	*Ficedula tricolor*	Slaty-blue Flycatcher	LC		
1254	雀形目 PASSERIFORMES	Muscicapidae 鹟科	玉头姬鹟	*Ficedula sapphira*	Sapphire Flycatcher	LC		
1255	雀形目 PASSERIFORMES	Muscicapidae 鹟科	白腹蓝鹟	*Cyanoptila cyanomelana*	Blue-and-white Flycatcher	LC		
1256	雀形目 PASSERIFORMES	Muscicapidae 鹟科	白腹暗蓝鹟	*Cyanoptila cumatilis*	Zappey's Flycatcher	LC		
1257	雀形目 PASSERIFORMES	Muscicapidae 鹟科	铜蓝鹟	*Eumyias thalassinus*	Verditer Flycatcher	LC		
1258	雀形目 PASSERIFORMES	Muscicapidae 鹟科	白喉林鹟	*Cyornis brunneatus*	Brown-chested Jungle Flycatcher	VU	C2a(ii)	
1259	雀形目 PASSERIFORMES	Muscicapidae 鹟科	海南蓝仙鹟	*Cyornis hainanus*	Hainan Blue Flycatcher	LC		
1260	雀形目 PASSERIFORMES	Muscicapidae 鹟科	纯蓝仙鹟	*Cyornis unicolor*	Pale Blue Flycatcher	LC		
1261	雀形目 PASSERIFORMES	Muscicapidae 鹟科	灰颊仙鹟	*Cyornis poliogenys*	Pale-chinned Flycatcher	LC		
1262	雀形目 PASSERIFORMES	Muscicapidae 鹟科	山蓝仙鹟	*Cyornis banyumas*	Hill Blue Flycatcher	LC		
1263	雀形目 PASSERIFORMES	Muscicapidae 鹟科	蓝喉仙鹟	*Cyornis rubeculoides*	Blue-throated Flycatcher	LC		
1264	雀形目 PASSERIFORMES	Muscicapidae 鹟科	中华仙鹟	*Cyornis glaucicomans*	Chinese Blue Flycatcher	LC		
1265	雀形目 PASSERIFORMES	Muscicapidae 鹟科	白尾蓝仙鹟	*Cyornis concretus*	White-tailed Flycatcher	LC		
1266	雀形目 PASSERIFORMES	Muscicapidae 鹟科	白喉姬鹟	*Anthipes monileger*	White-gorgeted Flycatcher	LC		
1267	雀形目 PASSERIFORMES	Muscicapidae 鹟科	棕腹大仙鹟	*Niltava davidi*	Fujian Niltava	LC		
1268	雀形目 PASSERIFORMES	Muscicapidae 鹟科	棕腹仙鹟	*Niltava sundara*	Rufous-bellied Niltava	LC		
1269	雀形目 PASSERIFORMES	Muscicapidae 鹟科	大仙鹟	*Niltava vivida*	Vivid Niltava	LC		
1270	雀形目 PASSERIFORMES	Muscicapidae 鹟科	大仙鹟	*Niltava grandis*	Large Niltava	LC		
1271	雀形目 PASSERIFORMES	Muscicapidae 鹟科	小仙鹟	*Niltava macgrigoriae*	Small Niltava	LC		

续表

	目	科	科	物种中文名	物种学名	物种英文名	濒危等级	评估依据	是否特有
1272	雀形目	PASSERIFORMES	戴菊科	Regulidae	台湾戴菊	*Regulus goodfellowi*	Flamecrest	LC		√
1273	雀形目	PASSERIFORMES	戴菊科	Regulidae	戴菊	*Regulus regulus*	Goldcrest	LC		
1274	雀形目	PASSERIFORMES	太平鸟科	Bombycillidae	太平鸟	*Bombycilla garrulus*	Bohemian Waxwing	LC		
1275	雀形目	PASSERIFORMES	太平鸟科	Bombycillidae	小太平鸟	*Bombycilla japonica*	Japanese Waxwing	LC		
1276	雀形目	PASSERIFORMES	丽星鹩鹛科	Elachuridae	丽星鹩鹛	*Elachura formosa*	Elachura	NT		
1277	雀形目	PASSERIFORMES	和平鸟科	Irenidae	和平鸟	*Irena puella*	Asian Fairy Bluebird	NT		
1278	雀形目	PASSERIFORMES	叶鹎科	Chloropseidae	蓝翅叶鹎	*Chloropsis cochinchinensis*	Blue-winged Leafbird	LC		
1279	雀形目	PASSERIFORMES	叶鹎科	Chloropseidae	金额叶鹎	*Chloropsis aurifrons*	Golden-fronted Leafbird	NT		
1280	雀形目	PASSERIFORMES	叶鹎科	Chloropseidae	橙腹叶鹎	*Chloropsis hardwickii*	Orange-bellied Leafbird	LC		
1281	雀形目	PASSERIFORMES	啄花鸟科	Dicaeidae	厚嘴啄花鸟	*Dicaeum agile*	Thick-billed Flowerpecker	LC		
1282	雀形目	PASSERIFORMES	啄花鸟科	Dicaeidae	黄臀啄花鸟	*Dicaeum chrysorrheum*	Yellow-vented Flowerpecker	LC		
1283	雀形目	PASSERIFORMES	啄花鸟科	Dicaeidae	黄腹啄花鸟	*Dicaeum melanozanthum*	Yellow-bellied Flowerpecker	LC		
1284	雀形目	PASSERIFORMES	啄花鸟科	Dicaeidae	纯色啄花鸟	*Dicaeum concolor*	Plain Flowerpecker	LC		
1285	雀形目	PASSERIFORMES	啄花鸟科	Dicaeidae	红胸啄花鸟	*Dicaeum ignipectus*	Fire-breasted Flowerpecker	LC		
1286	雀形目	PASSERIFORMES	啄花鸟科	Dicaeidae	朱背啄花鸟	*Dicaeum cruentatum*	Scarlet-backed Flowerpecker	LC		
1287	雀形目	PASSERIFORMES	花蜜鸟科	Nectariniidae	紫颊太阳鸟	*Chalcoparia singalensis*	Ruby-cheeked Sunbird	LC		
1288	雀形目	PASSERIFORMES	花蜜鸟科	Nectariniidae	褐喉食蜜鸟	*Anthreptes malacensis*	Brown-throated Sunbird	LC		
1289	雀形目	PASSERIFORMES	花蜜鸟科	Nectariniidae	蓝枕花蜜鸟	*Hypogramma hypogrammicum*	Purple-naped Sunbird	LC		
1290	雀形目	PASSERIFORMES	花蜜鸟科	Nectariniidae	紫花蜜鸟	*Cinnyris asiaticus*	Purple Sunbird	LC		
1291	雀形目	PASSERIFORMES	花蜜鸟科	Nectariniidae	黄腹花蜜鸟	*Cinnyris jugularis*	Olive-backed Sunbird	LC		
1292	雀形目	PASSERIFORMES	花蜜鸟科	Nectariniidae	蓝喉太阳鸟	*Aethopyga gouldiae*	Mrs Gould's Sunbird	LC		
1293	雀形目	PASSERIFORMES	花蜜鸟科	Nectariniidae	绿喉太阳鸟	*Aethopyga nipalensis*	Green-tailed Sunbird	LC		
1294	雀形目	PASSERIFORMES	花蜜鸟科	Nectariniidae	叉尾太阳鸟	*Aethopyga christinae*	Fork-tailed Sunbird	LC		
1295	雀形目	PASSERIFORMES	花蜜鸟科	Nectariniidae	黑胸太阳鸟	*Aethopyga saturata*	Black-throated Sunbird	LC		
1296	雀形目	PASSERIFORMES	花蜜鸟科	Nectariniidae	黄腰太阳鸟	*Aethopyga siparaja*	Crimson Sunbird	LC		
1297	雀形目	PASSERIFORMES	花蜜鸟科	Nectariniidae	火尾太阳鸟	*Aethopyga ignicauda*	Fire-tailed Sunbird	LC		
1298	雀形目	PASSERIFORMES	花蜜鸟科	Nectariniidae	长嘴捕蛛鸟	*Arachnothera longirostra*	Little Spiderhunter	LC		

续表

	目	科	物种中文名	物种学名	物种英文名	濒危等级	评估依据	是否特有
1299	雀形目 PASSERIFORMES	花蜜鸟科 Nectariniidae	纹背捕蛛鸟	*Arachnothera magna*	Streaked Spiderhunter	LC		
1300	雀形目 PASSERIFORMES	岩鹨科 Prunellidae	领岩鹨	*Prunella collaris*	Alpine Accentor	LC		
1301	雀形目 PASSERIFORMES	岩鹨科 Prunellidae	高原岩鹨	*Prunella himalayana*	Altai Accentor	LC		
1302	雀形目 PASSERIFORMES	岩鹨科 Prunellidae	鸲岩鹨	*Prunella rubeculoides*	Robin Accentor	LC		
1303	雀形目 PASSERIFORMES	岩鹨科 Prunellidae	棕胸岩鹨	*Prunella strophiata*	Rufous-breasted Accentor	LC		
1304	雀形目 PASSERIFORMES	岩鹨科 Prunellidae	棕眉山岩鹨	*Prunella montanella*	Siberian Accentor	LC		
1305	雀形目 PASSERIFORMES	岩鹨科 Prunellidae	褐岩鹨	*Prunella fulvescens*	Brown Accentor	LC		
1306	雀形目 PASSERIFORMES	岩鹨科 Prunellidae	黑喉岩鹨	*Prunella atrogularis*	Black-throated Accentor	LC		
1307	雀形目 PASSERIFORMES	岩鹨科 Prunellidae	贺兰山岩鹨	*Prunella koslowi*	Mongolian Accentor	VU	C2a(i)	
1308	雀形目 PASSERIFORMES	岩鹨科 Prunellidae	栗背岩鹨	*Prunella immaculata*	Maroon-backed Accentor	LC		
1309	雀形目 PASSERIFORMES	朱鹀科 Urocynchramidae	朱鹀	*Urocynchramus pylzowi*	Pink-tailed Rosefinch	NT		√
1310	雀形目 PASSERIFORMES	织雀科 Ploceidae	纹胸织雀	*Ploceus manyar*	Streaked Weaver	LC		
1311	雀形目 PASSERIFORMES	织雀科 Ploceidae	黄胸织雀	*Ploceus philippinus*	Baya Weaver	LC		
1312	雀形目 PASSERIFORMES	梅花雀科 Estrildidae	红梅花雀	*Amandava amandava*	Red Avadavat	LC		
1313	雀形目 PASSERIFORMES	梅花雀科 Estrildidae	长尾鹦雀	*Erythrura prasina*	Pin-tailed Parrotfinch	LC		
1314	雀形目 PASSERIFORMES	梅花雀科 Estrildidae	橙颊梅花雀	*Estrilda melpoda*	Orange-cheeked Waxbill	DD		
1315	雀形目 PASSERIFORMES	梅花雀科 Estrildidae	白喉文鸟	*Euodice malabarica*	White-throated Munia	LC		
1316	雀形目 PASSERIFORMES	梅花雀科 Estrildidae	白腰文鸟	*Lonchura striata*	White-rumped Munia	LC		
1317	雀形目 PASSERIFORMES	梅花雀科 Estrildidae	斑文鸟	*Lonchura punctulata*	Scaly-breasted Munia	LC		
1318	雀形目 PASSERIFORMES	梅花雀科 Estrildidae	栗腹文鸟	*Lonchura atricapilla*	Chestnut Munia	LC		
1319	雀形目 PASSERIFORMES	梅花雀科 Estrildidae	禾雀	*Lonchura oryzivora*	Java Sparrow	VU	A2bde+3bde+4bde	
1320	雀形目 PASSERIFORMES	雀科 Passeridae	黑顶麻雀	*Passer ammodendri*	Saxaul Sparrow	LC		
1321	雀形目 PASSERIFORMES	雀科 Passeridae	家麻雀	*Passer domesticus*	House Sparrow	LC		
1322	雀形目 PASSERIFORMES	雀科 Passeridae	黑胸麻雀	*Passer hispaniolensis*	Spanish Sparrow	LC		
1323	雀形目 PASSERIFORMES	雀科 Passeridae	山麻雀	*Passer cinnamomeus*	Russet Sparrow	LC		
1324	雀形目 PASSERIFORMES	雀科 Passeridae	麻雀	*Passer montanus*	Eurasian Tree Sparrow	LC		

目	科	物种中文名	物种学名	物种英文名	濒危等级	评估依据	是否特有
1325 雀形目 PASSERIFORMES	雀科 Passeridae	石雀	*Petronia petronia*	Rock Sparrow	LC		
1326 雀形目 PASSERIFORMES	雀科 Passeridae	白斑翅雪雀	*Montifringilla nivalis*	White-winged Snowfinch	LC		
1327 雀形目 PASSERIFORMES	雀科 Passeridae	藏雪雀	*Montifringilla henrici*	Henri's Snowfinch	NT		√
1328 雀形目 PASSERIFORMES	雀科 Passeridae	褐翅雪雀	*Montifringilla adamsi*	Tibetan Snowfinch	LC		
1329 雀形目 PASSERIFORMES	雀科 Passeridae	白腰雪雀	*Onychostruthus taczanowskii*	White-rumped Snowfinch	LC		
1330 雀形目 PASSERIFORMES	雀科 Passeridae	黑喉雪雀	*Pyrgilauda davidiana*	Père David's Snowfinch	LC		
1331 雀形目 PASSERIFORMES	雀科 Passeridae	棕颈雪雀	*Pyrgilauda ruficollis*	Rufous-necked Snowfinch	LC		
1332 雀形目 PASSERIFORMES	雀科 Passeridae	棕背雪雀	*Pyrgilauda blanfordi*	Blanford's Snowfinch	LC		
1333 雀形目 PASSERIFORMES	鹡鸰科 Motacillidae	山鹡鸰	*Dendronanthus indicus*	Forest Wagtail	LC		
1334 雀形目 PASSERIFORMES	鹡鸰科 Motacillidae	西黄鹡鸰	*Motacilla flava*	Western Yellow Wagtail	LC		
1335 雀形目 PASSERIFORMES	鹡鸰科 Motacillidae	黄鹡鸰	*Motacilla tschutschensis*	Eastern Yellow Wagtail	LC		
1336 雀形目 PASSERIFORMES	鹡鸰科 Motacillidae	黄头鹡鸰	*Motacilla citreola*	Citrine Wagtail	LC		
1337 雀形目 PASSERIFORMES	鹡鸰科 Motacillidae	灰鹡鸰	*Motacilla cinerea*	Grey Wagtail	LC		
1338 雀形目 PASSERIFORMES	鹡鸰科 Motacillidae	白鹡鸰	*Motacilla alba*	White Wagtail	LC		
1339 雀形目 PASSERIFORMES	鹡鸰科 Motacillidae	日本鹡鸰	*Motacilla grandis*	Japanese Wagtail	LC		
1340 雀形目 PASSERIFORMES	鹡鸰科 Motacillidae	田鹨	*Anthus richardi*	Richard's Pipit	LC		
1341 雀形目 PASSERIFORMES	鹡鸰科 Motacillidae	东方田鹨	*Anthus rufulus*	Paddyfield Pipit	LC		
1342 雀形目 PASSERIFORMES	鹡鸰科 Motacillidae	布氏鹨	*Anthus godlewskii*	Blyth's Pipit	LC		
1343 雀形目 PASSERIFORMES	鹡鸰科 Motacillidae	平原鹨	*Anthus campestris*	Tawny Pipit	LC		
1344 雀形目 PASSERIFORMES	鹡鸰科 Motacillidae	草地鹨	*Anthus pratensis*	Meadow Pipit	LC		
1345 雀形目 PASSERIFORMES	鹡鸰科 Motacillidae	林鹨	*Anthus trivialis*	Tree Pipit	LC		
1346 雀形目 PASSERIFORMES	鹡鸰科 Motacillidae	树鹨	*Anthus hodgsoni*	Olive-backed Pipit	LC		
1347 雀形目 PASSERIFORMES	鹡鸰科 Motacillidae	北鹨	*Anthus gustavi*	Pechora Pipit	LC		
1348 雀形目 PASSERIFORMES	鹡鸰科 Motacillidae	粉红胸鹨	*Anthus roseatus*	Rosy Pipit	LC		
1349 雀形目 PASSERIFORMES	鹡鸰科 Motacillidae	红喉鹨	*Anthus cervinus*	Red-throated Pipit	LC		
1350 雀形目 PASSERIFORMES	鹡鸰科 Motacillidae	黄腹鹨	*Anthus rubescens*	Buff-bellied Pipit	LC		
1351 雀形目 PASSERIFORMES	鹡鸰科 Motacillidae	水鹨	*Anthus spinoletta*	Water Pipit	LC		

续表

	目	科	物种中文名	物种学名	物种英文名	濒危等级	评估依据	是否特有
1352	雀形目 PASSERIFORMES	Motacillidae	山鹨	*Anthus sylvanus*	Upland Pipit	LC		
1353	雀形目 PASSERIFORMES	Fringillidae	苍头燕雀	*Fringilla coelebs*	Common Chaffinch	LC		
1354	雀形目 PASSERIFORMES	Fringillidae	燕雀	*Fringilla montifringilla*	Brambling	LC		
1355	雀形目 PASSERIFORMES	Fringillidae	黄颈拟蜡嘴雀	*Mycerobas affinis*	Collared Grosbeak	LC		
1356	雀形目 PASSERIFORMES	Fringillidae	白点翅拟蜡嘴雀	*Mycerobas melanozanthos*	Spot-winged Grosbeak	LC		
1357	雀形目 PASSERIFORMES	Fringillidae	白斑翅拟蜡嘴雀	*Mycerobas carnipes*	White-winged Grosbeak	LC		
1358	雀形目 PASSERIFORMES	Fringillidae	锡嘴雀	*Coccothraustes coccothraustes*	Hawfinch	LC		
1359	雀形目 PASSERIFORMES	Fringillidae	黑尾蜡嘴雀	*Eophona migratoria*	Chinese Grosbeak	LC		
1360	雀形目 PASSERIFORMES	Fringillidae	黑头蜡嘴雀	*Eophona personata*	Japanese Grosbeak	NT		
1361	雀形目 PASSERIFORMES	Fringillidae	松雀	*Pinicola enucleator*	Pine Grosbeak	LC		
1362	雀形目 PASSERIFORMES	Fringillidae	褐灰雀	*Pyrrhula nipalensis*	Brown Bullfinch	LC		
1363	雀形目 PASSERIFORMES	Fringillidae	红头灰雀	*Pyrrhula erythrocephala*	Red-headed Bullfinch	LC		
1364	雀形目 PASSERIFORMES	Fringillidae	灰头灰雀	*Pyrrhula erythaca*	Grey-headed Bullfinch	LC		
1365	雀形目 PASSERIFORMES	Fringillidae	红腹灰雀	*Pyrrhula pyrrhula*	Eurasian Bullfinch	LC		
1366	雀形目 PASSERIFORMES	Fringillidae	红翅沙雀	*Rhodopechys sanguineus*	Eurasian Crimson-winged Finch	LC		
1367	雀形目 PASSERIFORMES	Fringillidae	蒙古沙雀	*Bucanetes mongolicus*	Mongolian Finch	LC		
1368	雀形目 PASSERIFORMES	Fringillidae	巨嘴沙雀	*Rhodospiza obsoleta*	Desert Finch	LC		
1369	雀形目 PASSERIFORMES	Fringillidae	赤朱雀	*Agraphospiza rubescens*	Blanford's Rosefinch	LC		
1370	雀形目 PASSERIFORMES	Fringillidae	金枕黑雀	*Pyrrhoplectes epauletta*	Gold-naped Finch	LC		
1371	雀形目 PASSERIFORMES	Fringillidae	暗胸朱雀	*Procarduelis nipalensis*	Dark-breasted Rosefinch	LC		
1372	雀形目 PASSERIFORMES	Fringillidae	林岭雀	*Leucosticte nemoricola*	Plain Mountain Finch	LC		
1373	雀形目 PASSERIFORMES	Fringillidae	高山岭雀	*Leucosticte brandti*	Brandt's Mountain Finch	LC		
1374	雀形目 PASSERIFORMES	Fringillidae	粉红腹岭雀	*Leucosticte arctoa*	Asian Rosy Finch	LC		
1375	雀形目 PASSERIFORMES	Fringillidae	普通朱雀	*Carpodacus erythrinus*	Common Rosefinch	LC		
1376	雀形目 PASSERIFORMES	Fringillidae	褐头朱雀	*Carpodacus sillemi*	Sillem's Mountain Finch	DD		✓
1377	雀形目 PASSERIFORMES	Fringillidae	血雀	*Carpodacus sipahi*	Scarlet Finch	LC		
1378	雀形目 PASSERIFORMES	Fringillidae	拟大朱雀	*Carpodacus rubicilloides*	Streaked Rosefinch	NT		

目		科	物种中文名	物种学名	物种英文名	濒危等级	评估依据	是否特有
1379	雀形目 PASSERIFORMES	燕雀科 Fringillidae	大朱雀	*Carpodacus rubicilla*	Spotted Great Rosefinch	LC		
1380	雀形目 PASSERIFORMES	燕雀科 Fringillidae	红腰朱雀	*Carpodacus rhodochlamys*	Red-mantled Rosefinch	LC		
1381	雀形目 PASSERIFORMES	燕雀科 Fringillidae	红眉朱雀	*Carpodacus pulcherrimus*	Himalayan Beautiful Rosefinch	LC		
1382	雀形目 PASSERIFORMES	燕雀科 Fringillidae	中华朱雀	*Carpodacus davidianus*	Chinese Beautiful Rosefinch	LC		√
1383	雀形目 PASSERIFORMES	燕雀科 Fringillidae	曙红朱雀	*Carpodacus waltoni*	Pink-rumped Rosefinch	LC		
1384	雀形目 PASSERIFORMES	燕雀科 Fringillidae	粉红朱雀	*Carpodacus rodochroa*	Pink-browed Rosefinch	LC		
1385	雀形目 PASSERIFORMES	燕雀科 Fringillidae	棕朱雀	*Carpodacus edwardsii*	Dark-rumped Rosefinch	LC		
1386	雀形目 PASSERIFORMES	燕雀科 Fringillidae	点翅朱雀	*Carpodacus rodopeplus*	Spot-winged Rosefinch	LC		
1387	雀形目 PASSERIFORMES	燕雀科 Fringillidae	淡腹点翅朱雀	*Carpodacus verreauxii*	Sharpe's Rosefinch	LC		
1388	雀形目 PASSERIFORMES	燕雀科 Fringillidae	酒红朱雀	*Carpodacus vinaceus*	Vinaceous Rosefinch	LC		
1389	雀形目 PASSERIFORMES	燕雀科 Fringillidae	台湾酒红朱雀	*Carpodacus formosanus*	Taiwan Rosefinch	NT		√
1390	雀形目 PASSERIFORMES	燕雀科 Fringillidae	沙色朱雀	*Carpodacus stoliczkae*	Pale Rosefinch	LC		
1391	雀形目 PASSERIFORMES	燕雀科 Fringillidae	藏雀	*Carpodacus roborowskii*	Tibetan Rosefinch	VU	B2b (i-iv), C2(i, ii), D2.	√
1392	雀形目 PASSERIFORMES	燕雀科 Fringillidae	长尾雀	*Carpodacus sibiricus*	Long-tailed Rosefinch	LC		
1393	雀形目 PASSERIFORMES	燕雀科 Fringillidae	北朱雀	*Carpodacus roseus*	Pallas's Rosefinch	LC		
1394	雀形目 PASSERIFORMES	燕雀科 Fringillidae	斑翅朱雀	*Carpodacus trifasciatus*	Three-banded Rosefinch	LC		√
1395	雀形目 PASSERIFORMES	燕雀科 Fringillidae	喜山白眉朱雀	*Carpodacus thura*	Himalayan White-browed Rosefinch	LC		
1396	雀形目 PASSERIFORMES	燕雀科 Fringillidae	白眉朱雀	*Carpodacus dubius*	Chinese White-browed Rosefinch	LC		
1397	雀形目 PASSERIFORMES	燕雀科 Fringillidae	红胸朱雀	*Carpodacus puniceus*	Red-fronted Rosefinch	LC		
1398	雀形目 PASSERIFORMES	燕雀科 Fringillidae	红眉松雀	*Carpodacus subhimachalus*	Crimson-browed Finch	LC		
1399	雀形目 PASSERIFORMES	燕雀科 Fringillidae	红眉金翅雀	*Callacanthis burtoni*	Spactacled Finch	LC		
1400	雀形目 PASSERIFORMES	燕雀科 Fringillidae	欧金翅雀	*Chloris chloris*	European Greenfinch	LC		
1401	雀形目 PASSERIFORMES	燕雀科 Fringillidae	金翅雀	*Chloris sinica*	Grey-capped Greenfinch	LC		
1402	雀形目 PASSERIFORMES	燕雀科 Fringillidae	高山金翅雀	*Chloris spinoides*	Yellow-breasted Greenfinch	LC		
1403	雀形目 PASSERIFORMES	燕雀科 Fringillidae	黑头金翅雀	*Chloris ambigua*	Black-headed Greenfinch	LC		
1404	雀形目 PASSERIFORMES	燕雀科 Fringillidae	黄嘴朱顶雀	*Linaria flavirostris*	Twite	LC		

续表

	目	科	物种中文名	物种学名	物种英文名	濒危等级	评估依据	是否特有
1405	雀形目 PASSERIFORMES	Fringillidae	赤胸朱顶雀	*Linaria cannabina*	Common Linnet	LC		
1406	雀形目 PASSERIFORMES	Fringillidae	白腰朱顶雀	*Acanthis flammea*	Common Redpoll	LC		
1407	雀形目 PASSERIFORMES	Fringillidae	极北朱顶雀	*Acanthis hornemanni*	Arctic Redpoll	LC		
1408	雀形目 PASSERIFORMES	Fringillidae	红交嘴雀	*Loxia curvirostra*	Red Crossbill	LC		
1409	雀形目 PASSERIFORMES	Fringillidae	白翅交嘴雀	*Loxia leucoptera*	White-winged Crossbill	LC		
1410	雀形目 PASSERIFORMES	Fringillidae	红额金翅雀	*Carduelis carduelis*	European Goldfinch	LC		
1411	雀形目 PASSERIFORMES	Fringillidae	金额丝雀	*Serinus pusillus*	Red-fronted Serin	LC		
1412	雀形目 PASSERIFORMES	Fringillidae	藏黄雀	*Spinus thibetana*	Tibetan Serin	NT		
1413	雀形目 PASSERIFORMES	Fringillidae	黄雀	*Spinus spinus*	Eurasian Siskin	NT		
1414	雀形目 PASSERIFORMES	Calcariidae	铁爪鹀	*Calcarius lapponicus*	Lapland Longspur	NT		
1415	雀形目 PASSERIFORMES	Calcariidae	雪鹀	*Plectrophenax nivalis*	Snow Bunting	LC		
1416	雀形目 PASSERIFORMES	Emberizidae	凤头鹀	*Melophus lathami*	Crested Bunting	LC		
1417	雀形目 PASSERIFORMES	Emberizidae	蓝鹀	*Emberiza siemsseni*	Slaty Bunting	LC		✓
1418	雀形目 PASSERIFORMES	Emberizidae	黍鹀	*Emberiza calandra*	Corn Bunting	LC		
1419	雀形目 PASSERIFORMES	Emberizidae	黄鹀	*Emberiza citrinella*	Yellowhammer	NT		
1420	雀形目 PASSERIFORMES	Emberizidae	白头鹀	*Emberiza leucocephalos*	Pine Bunting	LC		
1421	雀形目 PASSERIFORMES	Emberizidae	淡灰眉岩鹀	*Emberiza cia*	Rock Bunting	LC		
1422	雀形目 PASSERIFORMES	Emberizidae	灰眉岩鹀	*Emberiza godlewskii*	Godlewski's Bunting	LC		
1423	雀形目 PASSERIFORMES	Emberizidae	三道眉草鹀	*Emberiza cioides*	Meadow Bunting	LC		
1424	雀形目 PASSERIFORMES	Emberizidae	白顶鹀	*Emberiza stewarti*	White-capped Bunting	DD(NE)		
1425	雀形目 PASSERIFORMES	Emberizidae	栗斑腹鹀	*Emberiza jankowskii*	Jankowski's Bunting	EN	A2abc+3bc+4bc; B2a+c(iii)	
1426	雀形目 PASSERIFORMES	Emberizidae	灰颈鹀	*Emberiza buchanani*	Grey-necked Bunting	LC		
1427	雀形目 PASSERIFORMES	Emberizidae	圃鹀	*Emberiza hortulana*	Ortolan Bunting	LC		
1428	雀形目 PASSERIFORMES	Emberizidae	白眉鹀	*Emberiza tristrami*	Tristram's Bunting	NT		
1429	雀形目 PASSERIFORMES	Emberizidae	栗耳鹀	*Emberiza fucata*	Chestnut-eared Bunting	LC		
1430	雀形目 PASSERIFORMES	Emberizidae	小鹀	*Emberiza pusilla*	Little Bunting	LC		

续表

	目	科	物种中文名	物种学名	物种英文名	濒危等级	评估依据	是否特有
1431	PASSERIFORMES	Emberizidae	黄眉鹀	*Emberiza chrysophrys*	Yellow-browed Bunting	LC		
1432	PASSERIFORMES	Emberizidae	田鹀	*Emberiza rustica*	Rustic Bunting	NT		
1433	PASSERIFORMES	Emberizidae	黄喉鹀	*Emberiza elegans*	Yellow-throated Bunting	LC		
1434	PASSERIFORMES	Emberizidae	黄胸鹀	*Emberiza aureola*	Yellow-breasted Bunting	CR	A1acd	
1435	PASSERIFORMES	Emberizidae	栗鹀	*Emberiza rutila*	Chestnut Bunting	NT		
1436	PASSERIFORMES	Emberizidae	藏鹀	*Emberiza koslowi*	Tibetan Bunting	VU	B1b(i-iv)+C1	√
1437	PASSERIFORMES	Emberizidae	黑头鹀	*Emberiza melanocephala*	Black-headed Bunting	LC		
1438	PASSERIFORMES	Emberizidae	褐头鹀	*Emberiza bruniceps*	Red-headed Bunting	LC		
1439	PASSERIFORMES	Emberizidae	硫黄鹀	*Emberiza sulphurata*	Yellow Bunting	VU	C2a(ii)	
1440	PASSERIFORMES	Emberizidae	灰头鹀	*Emberiza spodocephala*	Black-faced Bunting	LC		
1441	PASSERIFORMES	Emberizidae	灰鹀	*Emberiza variabilis*	Grey Bunting	LC		
1442	PASSERIFORMES	Emberizidae	苇鹀	*Emberiza pallasi*	Pallas's Bunting	LC		
1443	PASSERIFORMES	Emberizidae	红颈苇鹀	*Emberiza yessoensis*	Ochre-rumped Bunting	NT		
1444	PASSERIFORMES	Emberizidae	芦鹀	*Emberiza schoeniclus*	Reed Bunting	LC		
1445	PASSERIFORMES	Emberizidae	白冠带鹀	*Zonotrichia leucophrys*	White-crowned Sparrow	LC		

物种中文名索引
Index of Chinese Names of Species

A

阿尔泰雪鸡　196, 372
埃及夜鹰　378
鹌鹑　373
暗背雨燕　378
暗腹雪鸡　372
暗灰鹃鵙　396
暗绿柳莺　404
暗绿绣眼鸟　408
暗冕山鹪莺　401
暗色鸦雀　312, 407
暗胸朱雀　423
澳洲红嘴鸥　384

B

八哥　414
八声杜鹃　379
白斑翅拟蜡嘴雀　423
白斑翅雪雀　422
白斑黑石䳭　418
白斑军舰鸟　387
白斑尾柳莺　405
白背矶鸫　418
白背兀鹫　389
白背啄木鸟　394
白翅百灵　400
白翅浮鸥　385
白翅交嘴雀　425
白翅蓝鹊　398
白翅啄木鸟　394
白点翅拟蜡嘴雀　423
白点噪鹛　318, 411
白顶鵖　418
白顶鹀　425

白顶溪鸲　417
白顶玄燕鸥　383
白额鹱　386
白额雁　374
白额燕鸥　385
白额燕尾　417
白额圆尾鹱　386
白耳奇鹛　412
白腹暗蓝鹟　419
白腹鸫　415
白腹短翅鸲　416
白腹凤鹛　396
白腹海雕　256, 390
白腹黑啄木鸟　394
白腹锦鸡　374
白腹军舰鸟　386
白腹蓝鹟　419
白腹鹭　388
白腹隼雕　252, 389
白腹鸫　390
白腹幽鹛　410
白骨顶　380
白冠长尾雉　104, 374
白冠带鹀　426
白冠攀雀　400
白冠噪鹛　410
白鹳　52, 386
白鹤　66, 380
白喉斑秧鸡　218, 379
白喉短翅鸫　416
白喉冠鹎　403
白喉红臀鹎　403
白喉红尾鸲　417
白喉矶鸫　418
白喉姬鹟　419
白喉林鹟　334, 419

白喉林莺　407
白喉扇尾莺　397
白喉石䳭　182, 417
白喉文鸟　421
白喉犀鸟　266, 392
白喉噪鹛　411
白喉针尾雨燕　378
白鹛鸰　422
白颊鸭　403
白颊黑雁　374
白颊山鹧鸪　372
白颊噪鹛　411
白肩雕　146, 389
白肩黑鹮　387
白颈长尾雉　204, 374
白颈鸫　415
白颈凤鹛　408
白颈鹳　386
白颈鸦　399
白颈噪鹛　410
白眶斑翅鹛　412
白眶雀鹛　410
白眶鹟莺　405
白眶鸦雀　407
白脸鸭　413
白脸鹭　388
白领八哥　414
白领翡翠　393
白领凤鹛　408
白鹭　388
白马鸡　373
白眉地鸫　414
白眉鸫　415
白眉歌鸫　415
白眉姬鹟　418
白眉苦恶鸟　380
白眉蓝姬鹟　419
白眉林鸲　416
白眉雀鹛　407
白眉山雀　399
白眉山鹧鸪　186, 372
白眉扇尾莺　397
白眉鸦　425
白眉鸭　375
白眉朱雀　424
白眉棕啄木鸟　393
白琵鹭　387
白鹈鹕　138, 388
白头鸭　403

白头鸫　415
白头鹤　124, 380
白头鹀鹛　410
白头鸦　425
白头鹎　390
白头硬尾鸭　62, 376
白腿小隼　272, 395
白尾鸭　413
白尾地鸲　298, 417
白尾地鸦　398
白尾海雕　258, 390
白尾蓝仙鹟　419
白尾麦鸡　381
白尾鹲　385
白尾梢虹雉　100, 373
白尾鹞　390
白兀鹫　388
白鹇　373
白胸翡翠　393
白胸苦恶鸟　380
白须黑胸歌鸲　416
白眼鵟鹰　390
白眼潜鸭　375
白燕鸥　383
白腰杓鹬　382
白腰滨鹬　383
白腰草鹬　382
白腰叉尾海燕　386
白腰鹊鸲　417
白腰文鸟　421
白腰雪雀　422
白腰燕鸥　385
白腰雨燕　378
白腰朱顶雀　425
白枕鹤　120, 380
白嘴端凤头燕鸥　384
斑背大尾莺　402
斑背潜鸭　376
斑背燕尾　417
斑背噪鹛　410
斑翅凤头鹃　378
斑翅椋鸟　413
斑翅鹩鹛　409
斑翅山鹑　373
斑翅朱雀　424
斑鸫　415
斑喉希鹛　412
斑姬鹟　418
斑姬啄木鸟　393

物种中文名索引

斑颈穗鹛　409
斑鹃鸠　396
斑脸海番鸭　376
斑椋鸟　414
斑林鸽　376
斑鹭　388
斑头大翠鸟　270, 393
斑头秋沙鸭　376
斑头鸺鹠　391
斑头雁　374
斑尾膝鹬　382
斑尾鹃鸠　377
斑尾林鸽　376
斑尾榛鸡　372
斑文鸟　421
斑鹟　418
斑胁姬鹛　412
斑胁田鸡　222, 380
斑胸滨鹬　383
斑胸短翅蝗莺　402
斑胸钩嘴鹛　408
斑胸田鸡　379
斑胸鸦雀　408
斑胸噪鹛　411
斑腰燕　403
斑鱼狗　393
斑嘴鹈鹕　140, 388
斑嘴鸭　375
半蹼鹬　382
宝兴歌鸫　415
宝兴鹛雀　407
暴风鹱　386
北长尾山雀　406
北短翅蝗莺　402
北红尾鸲　417
北蝗莺　402
北灰鹟　418
北极鸥　384
北椋鸟　414
北领角鸮　390
北鹨　422
北噪鸦　398
北朱雀　424
北棕腹鹰鹃　379
比氏鹟莺　405
扁嘴海雀　385
波斑鸨　118, 379
布氏鹨　422
布氏苇莺　401

C

彩鹳　386
彩虹蜂虎　392
彩鹬　387
彩鹮　381
仓鸮　392
苍鹭　388
苍眉蝗莺　402
苍头燕雀　423
苍鹰　389
草地鹨　422
草鹭　388
草绿篱莺　401
草鸮　392
草原百灵　400
草原雕　248, 389
草原鹞　390
叉尾鸥　384
叉尾太阳鸟　420
长耳鸮　391
长脚秧鸡　220, 379
长尾地鸫　414
长尾缝叶莺　401
长尾阔嘴鸟　396
长尾鹩鹛　409
长尾林鸮　391
长尾奇鹛　413
长尾雀　424
长尾山椒鸟　396
长尾鸭　60, 376
长尾夜鹰　378
长尾鹦雀　421
长尾贼鸥　385
长趾滨鹬　383
长嘴百灵　400
长嘴斑海雀　385
长嘴半蹼鹬　382
长嘴捕蛛鸟　420
长嘴地鸫　415
长嘴钩嘴鹛　408
长嘴剑鸻　381
长嘴鹩鹛　410
长嘴兀鹫　389
橙斑翅柳莺　404
橙翅噪鹛　411
橙腹叶鹎　420
橙颊梅花雀　421
橙头地鸫　414

橙胸姬鹟 419
橙胸绿鸠 377
橙胸咬鹃 392
池鹭 388
赤膀鸭 375
赤翡翠 392
赤腹鹰 389
赤红山椒鸟 397
赤颈䴘 376
赤颈鸫 415
赤颈鹤 50, 380
赤颈鸭 375
赤麻鸭 375
赤胸鸫 415
赤胸拟啄木鸟 393
赤胸朱顶雀 425
赤胸啄木鸟 394
赤须蜂虎 392
赤朱雀 423
赤嘴潜鸭 375
丑鸭 376
纯蓝仙鹟 419
纯色山鹪莺 401
纯色岩燕 402
纯色噪鹛 411
纯色啄花鸟 420
翠金鹃 379

D

达乌里寒鸦 398
大白鹭 388
大斑啄木鸟 394
大鸨 116, 379
大杓鹬 230, 382
大滨鹬 232, 382
大草鹛 410
大长嘴地鸫 415
大杜鹃 379
大短趾百灵 400
大凤头燕鸥 384
大红鹳 376
大黄冠啄木鸟 158, 394
大灰啄木鸟 394
大金背啄木鸟 394
大金丝燕 378
大鹃鵙 396
大鵟 260, 390
大绿雀鹎 397

大麻鳽 387
大拟啄木鸟 393
大盘尾 292, 397
大沙锥 382
大山雀 400
大石鸻 380
大石鸡 372
大树莺 406
大天鹅 375
大苇莺 401
大仙鹟 419
大鹰鹃 379
大噪鹛 411
大朱雀 424
大紫胸鹦鹉 278, 395
大嘴乌鸦 399
戴菊 420
戴胜 392
丹顶鹤 122, 380
淡背地鸫 414
淡腹点翅朱雀 424
淡喉鹩鹛 409
淡黄腰柳莺 404
淡灰眉岩鹀 425
淡脚柳莺 404
淡脚树莺 406
淡绿鵙鹛 396
淡眉柳莺 404
淡色崖沙燕 402
淡尾鹟莺 405
淡紫䴓 326, 413
淡足鹱 386
稻田苇莺 401
低地绣眼鸟 408
地山雀 399
蒂氏鸫 415
滇䴓 324, 413
点翅朱雀 424
点胸鸦雀 408
雕鸮 391
东方白鹳 130, 386
东方斑鸭 418
东方大苇莺 401
东方鸻 381
东方寿带 397
东方田鹨 422
东方中杜鹃 379
东歌林莺 407
东亚蝗莺 306, 402

董鸡　380
豆雁　374
渡鸦　399
短翅树莺　406
短耳鸮　391
短尾钩嘴鹛　409
短尾鹱　386
短尾鹪鹛　410
短尾信天翁　238, 386
短尾鸦雀　408
短尾鹦鹉　395
短尾贼鸥　385
短趾百灵　400
短趾雕　389
短嘴豆雁　374
短嘴金丝燕　378
短嘴山椒鸟　396
钝翅苇莺　401

E

峨眉柳莺　405
峨眉鹟莺　405
鹗　388

F

发冠卷尾　397
帆背潜鸭　375
翻石鹬　382
反嘴鹬　380
方尾鹟　399
绯胸鹦鹉　280, 395
菲律宾鹃鸠　377
粉红腹岭雀　423
粉红椋鸟　414
粉红山椒鸟　396
粉红胸鹨　422
粉红燕鸥　385
粉眉朱雀　424
凤头百灵　400
凤头蜂鹰　388
凤头麦鸡　381
凤头鹀鹛　376
凤头潜鸭　375
凤头雀莺　406
凤头雀嘴鹎　403
凤头鸦　425
凤头鹰　389

凤头鹰雕　389
凤头雨燕　378
弗氏鸥　384

G

甘肃柳莺　404
高跷鹬　383
高山短翅蝗莺　402
高山金翅雀　424
高山岭雀　423
高山兀鹫　389
高山旋木雀　413
高山雨燕　378
高原山鹑　373
高原岩鹨　421
歌百灵　300, 400
钩嘴林鵙　397
钩嘴圆尾鹱　386
孤沙锥　381
古铜色卷尾　397
冠斑犀鸟　78, 392
冠海雀　385
冠纹柳莺　405
冠小嘴乌鸦　399
冠鱼狗　393
鹳嘴翡翠　392
鬼鸮　264, 391

H

海鸬鹚　387
海南画眉　410
海南鳽　136, 387
海南孔雀雉　54, 374
海南蓝仙鹟　419
海南柳莺　310, 405
海南山鹧鸪　92, 372
寒鸦　398
禾雀　338, 421
和平鸟　420
河乌　413
河燕鸥　385
贺兰山红尾鸲　180, 417
贺兰山岩鹨　336, 421
褐背鹟鵙　397
褐背针尾雨燕　378
褐翅叉尾海燕　386
褐翅雪雀　422

褐翅鸦鹃　378	黑顶噪鹛　411
褐翅鸦雀　407	黑短脚鹎　403
褐翅燕鸥　385	黑额伯劳　398
褐顶雀鹛　409	黑额山噪鹛　316, 410
褐额啄木鸟　394	黑额树鹊　398
褐耳鹰　389	黑浮鸥　385
褐冠鹃隼　388	黑腹滨鹬　383
褐冠山雀　399	黑腹军舰鸟　386
褐河乌　413	黑腹沙鸡　377
褐喉沙燕　402	黑腹燕鸥　385
褐喉食蜜鸟　420	黑冠黄鹎　403
褐喉旋木雀　413	黑冠鸦　387
褐灰雀　423	黑冠鹃隼　388
褐鲣鸟　387	黑冠椋鸟　414
褐脸雀鹛　410	黑冠山雀　399
褐林鸮　391	黑冠薮鹛　86, 412
褐柳莺　404	黑鹳　240, 386
褐马鸡　200, 373	黑海番鸭　376
褐山鹪莺　401	黑喉鸫　415
褐头鸫　330, 415	黑喉缝叶莺　401
褐头凤鹛　408	黑喉歌鸲　176, 416
褐头雀鹛　407	黑喉红臀鹎　403
褐头山雀　399	黑喉红尾鸲　417
褐头鹀　426	黑喉毛脚燕　403
褐头朱雀　423	黑喉潜鸟　385
褐胁雀鹛　409	黑喉山鹪莺　401
褐胸山鹧鸪　372	黑喉石䳭　418
褐胸鹟　418	黑喉雪雀　422
褐胸噪鹛　410	黑喉鸦雀　407
褐鸦雀　407	黑喉岩鹨　421
褐岩鹨　421	黑喉噪鹛　411
褐燕䴉　386	黑脚信天翁　386
褐渔鸮　152, 391	黑颈长尾雉　206, 374
鹤鹬　382	黑颈鹤　226, 380
黑白林鵙　418	黑颈䴙䴘　387
黑百灵　400	黑颈鸬鹚　376
黑斑蝗莺　402	黑卷尾　397
黑背信天翁　386	黑颏凤鹛　408
黑背燕尾　417	黑颏果鸠　377
黑叉尾海燕　386	黑颏穗鹛　409
黑长尾雉　374	黑脸琵鹭　134, 387
黑翅长脚鹬　380	黑脸鹟莺　405
黑翅雀鹎　397	黑脸噪鹛　411
黑翅燕鸻　383	黑林鸽　377
黑翅鸢　388	黑领椋鸟　414
黑顶林莺　406	黑领噪鹛　411
黑顶麻雀　421	黑眉长尾山雀　406
黑顶奇鹛　412	黑眉柳莺　405
黑顶蛙口夜鹰　377	黑眉拟啄木鸟　393

黑眉苇莺　401
黑眉鸦雀　408
黑琴鸡　372
黑水鸡　380
黑头白鹮　72, 387
黑头鸭　403
黑头黄鹂　396
黑头角雉　373
黑头金翅雀　424
黑头蜡嘴雀　423
黑头攀雀　400
黑头奇鹛　412
黑头䴓　413
黑头穗鹛　409
黑头鸫　426
黑头噪鸦　294, 398
黑尾塍鹬　382
黑尾地鸦　296, 398
黑尾蜡嘴雀　423
黑尾鸥　384
黑苇鳽　387
黑兀鹫　74, 389
黑鹇　373
黑胸鸫　415
黑胸歌鸲　416
黑胸麻雀　421
黑胸太阳鸟　420
黑胸楔嘴穗鹛　409
黑雁　374
黑鸢　390
黑枕黄鹂　396
黑枕王鹟　397
黑枕燕鸥　385
黑啄木鸟　394
黑嘴鸥　236, 384
黑嘴松鸡　96, 372
横斑腹小鸮　391
横斑林莺　406
红背伯劳　397
红背红尾鸲　417
红翅凤头鹃　378
红翅鵙鹛　396
红翅绿鸠　377
红翅沙雀　423
红翅薮鹛　412
红翅旋壁雀　413
红翅噪鹛　412
红顶绿鸠　214, 377
红顶鹛　409

红额金翅雀　425
红耳鹎　403
红腹滨鹬　234, 382
红腹红尾鸲　417
红腹灰雀　423
红腹角雉　373
红腹锦鸡　374
红腹山雀　399
红腹旋木雀　413
红腹咬鹃　392
红喉歌鸲　416
红喉姬鹟　419
红喉鹨　422
红喉潜鸟　385
红喉山鹧鸪　372
红喉雉鹑　192, 372
红交嘴雀　425
红角鸮　391
红脚斑秧鸡　379
红脚鲣鸟　387
红脚隼　395
红脚田鸡　379
红脚鹬　382
红颈瓣蹼鹬　383
红颈滨鹬　383
红颈绿啄木鸟　394
红颈苇鹀　426
红脸鸬鹚　387
红领绿鹦鹉　395
红眉金翅雀　424
红眉松雀　424
红眉朱雀　424
红梅花雀　421
红隼　395
红头长尾山雀　406
红头灰雀　423
红头潜鸭　375
红头穗鹛　409
红头鸦雀　408
红头咬鹃　392
红头噪鹛　412
红腿小隼　394
红尾斑鸫　415
红尾伯劳　397
红尾歌鸲　416
红尾鹲　385
红尾水鸲　417
红尾希鹛　412
红尾噪鹛　412

红胁蓝尾鸲　416
红胁绣眼鸟　408
红胸黑雁　374
红胸鸻　381
红胸姬鹟　419
红胸角雉　198, 373
红胸秋沙鸭　376
红胸山鹧鸪　188, 372
红胸田鸡　380
红胸朱雀　424
红胸啄花鸟　420
红腰朱雀　424
红原鸡　373
红嘴钩嘴鹛　409
红嘴巨燕鸥　384
红嘴蓝鹊　398
红嘴椋鸟　414
红嘴鹲　385
红嘴鸥　384
红嘴山鸦　398
红嘴相思鸟　412
红嘴鸦雀　407
鸿雁　210, 374
厚嘴绿鸠　377
厚嘴苇莺　401
厚嘴啄花鸟　420
胡兀鹫　388
槲鸫　415
虎斑地鸫　414
虎头海雕　150, 390
虎纹伯劳　397
花彩雀莺　406
花腹绿啄木鸟　394
花冠皱盔犀鸟　156, 392
花脸鸭　375
花田鸡　216, 379
花头㶉鹈　391
花头鹦鹉　395
花尾榛鸡　372
华南斑胸钩嘴鹛　408
华南冠纹柳莺　405
华西柳莺　404
画眉　410
环颈鸻　381
环颈山鹧鸪　372
环颈雉　374
䴉嘴鹬　380
荒漠伯劳　397
荒漠林莺　407

黄斑苇鸦　387
黄额鸦雀　407
黄额燕　403
黄腹冠鹎　403
黄腹花蜜鸟　420
黄腹角雉　98, 373
黄腹柳莺　404
黄腹鹨　422
黄腹山鹪莺　401
黄腹山雀　399
黄腹扇尾鹟　399
黄腹树莺　406
黄腹鹟莺　405
黄腹啄花鸟　420
黄冠啄木鸟　394
黄喉蜂虎　392
黄喉雀鹛　409
黄喉穗鹛　409
黄喉鹀　426
黄喉雉鹑　194, 372
黄鹡鸰　422
黄颊麦鸡　381
黄颊山雀　400
黄脚绿鸠　377
黄脚三趾鹑　383
黄颈凤鹛　408
黄颈拟蜡嘴雀　423
黄颈啄木鸟　394
黄绿鸭　403
黄眉姬鹟　418
黄眉林雀　399
黄眉柳莺　404
黄眉鹀　426
黄蹼洋海燕　386
黄雀　425
黄头鹡鸰　422
黄腿银鸥　384
黄腿渔鸮　154, 391
黄臀鹎　403
黄臀啄花鸟　420
黄纹拟啄木鸟　393
黄鹀　425
黄胸柳莺　405
黄胸绿鹊　398
黄胸鹀　88, 426
黄胸鹟　383
黄胸织雀　421
黄腰柳莺　404
黄腰太阳鸟　420

物种中文名索引

黄腰响蜜䴕　393
黄痣薮鹛　412
黄爪隼　274, 395
黄嘴白鹭　242, 388
黄嘴角鸮　390
黄嘴蓝鹊　398
黄嘴栗啄木鸟　394
黄嘴潜鸟　386
黄嘴山鸦　398
黄嘴朱顶雀　424
灰白喉林莺　407
灰斑鸠　377
灰瓣蹼鹬　383
灰背伯劳　398
灰背鸫　415
灰背椋鸟　414
灰背鸥　384
灰背隼　395
灰背燕尾　417
灰伯劳　398
灰翅鸫　415
灰翅浮鸥　385
灰翅鸥　384
灰翅噪鹛　410
灰短脚鹎　403
灰腹地莺　406
灰腹角雉　373
灰腹绣眼鸟　408
灰腹噪鹛　411
灰冠鹟莺　405
灰冠鸦雀　168, 407
灰鹤　380
灰鸻　381
灰喉柳莺　404
灰喉山椒鸟　396
灰喉鸦雀　407
灰喉针尾雨燕　378
灰鹮　386
灰鹡鸰　422
灰颊仙鸫　419
灰颈鹀　425
灰卷尾　397
灰孔雀雉　106, 374
灰眶雀鹛　410
灰蓝姬鹟　419
灰蓝山雀　399
灰脸鵟鹰　390
灰脸鹟莺　405
灰椋鸟　414

灰林鸽　376
灰林鹎　418
灰林鸮　391
灰柳莺　404
灰眉岩鹀　425
灰奇鹛　412
灰山鹑　373
灰山椒鸟　396
灰树鹊　398
灰头斑翅鹛　412
灰头鸫　415
灰头钩嘴鹛　408
灰头灰雀　423
灰头椋鸟　414
灰头柳莺　405
灰头绿鸠　377
灰头绿啄木鸟　394
灰头麦鸡　381
灰头薮鹛　412
灰头鸦　426
灰头鸦雀　408
灰头鹦鹉　395
灰尾漂鹬　382
灰纹鹟　418
灰鸦　426
灰喜鹊　398
灰胁噪鹛　411
灰胸山鹪莺　401
灰胸薮鹛　320, 412
灰胸秧鸡　379
灰胸竹鸡　373
灰岩鹪鹛　410
灰岩柳莺　405
灰眼短脚鹎　403
灰雁　374
灰燕鸻　383
灰燕鹀　397
火斑鸠　377
火冠雀　399
火尾绿鹛　406
火尾太阳鸟　420
霍氏旋木雀　413

J

叽喳柳莺　404
矶鹬　382
姬田鸡　380
姬鹬　381

极北柳莺　404

极北朱顶雀　425

加拿大雁　374

家八哥　414

家麻雀　421

家鸦　399

家燕　402

尖尾滨鹬　383

剪嘴鸥　385

剑鸻　381

鹪鹩　413

角䴙䴘　376

角百灵　400

角嘴海雀　385

金背啄木鸟　394

金翅雀　424

金雕　250, 389

金额雀鹛　314, 409

金额丝雀　425

金额叶鹎　420

金冠地莺　406

金冠树八哥　414

金鸻　381

金喉拟啄木鸟　393

金黄鹂　396

金眶鸻　381

金眶鹟莺　405

金色林鸲　416

金色鸦雀　408

金头扇尾莺　400

金头穗鹛　409

金胸歌鸲　332, 416

金胸雀鹛　407

金眼鹛雀　407

金腰燕　403

金枕黑雀　423

酒红朱雀　424

巨䴓　172, 413

巨嘴短翅蝗莺　402

巨嘴柳莺　404

巨嘴沙雀　423

距翅麦鸡　381

鹃头蜂鹰　388

卷羽鹈鹕　142, 388

K

库页岛蝗莺　402

宽尾树莺　406

宽嘴鹟莺　406

阔嘴鹬　383

L

拉氏沙锥　381

蓝八色鸫　395

蓝背八色鸫　162, 395

蓝翅八色鸫　396

蓝翅希鹛　412

蓝翅叶鹎　420

蓝翅噪鹛　411

蓝大翅鸲　417

蓝短翅鸫　416

蓝额地鸲　417

蓝额红尾鸲　417

蓝耳翠鸟　393

蓝耳拟啄木鸟　393

蓝翡翠　393

蓝腹鹇　373

蓝歌鸲　416

蓝冠噪鹛　84, 411

蓝喉蜂虎　392

蓝喉歌鸲　416

蓝喉拟啄木鸟　393

蓝喉太阳鸟　420

蓝喉仙鹟　419

蓝矶鸫　418

蓝颊蜂虎　392

蓝脸鲣鸟　387

蓝绿鹊　398

蓝马鸡　374

蓝眉林鸲　416

蓝头红尾鸲　417

蓝头矶鸫　418

蓝鹀　425

蓝胸鹑　373

蓝胸佛法僧　392

蓝须蜂虎　268, 392

蓝腰鹦鹉　395

蓝枕八色鸫　284, 395

蓝枕花蜜鸟　420

丽色奇鹛　413

丽䴓　174, 413

丽星鹩鹛　420

栗斑杜鹃　379

栗斑腹鹀　184, 425

栗背伯劳　397

栗背短翅鸫　416

栗背短脚鹎　403
栗背奇鹛　412
栗背岩鹨　421
栗额斑翅鹛　412
栗额鵙鹛　396
栗耳短脚鹎　404
栗耳凤鹛　408
栗耳鹀　425
栗腹歌鸲　416
栗腹矶鸫　418
栗腹鳾　413
栗腹文鸟　421
栗喉蜂虎　392
栗喉鵙鹛　396
栗颊噪鹛　410
栗颈噪鹛　411
栗树鸭　208, 374
栗头八色鸫　286, 395
栗头蜂虎　392
栗头鹛　387
栗头雀鹛　409
栗头树莺　406
栗头䴖莺　405
栗头织叶莺　406
栗臀鳾　413
栗臀噪鹛　411
栗尾姬鹟　418
栗苇鳽　387
栗鹀　426
栗鸮　392
栗鸢　254, 390
栗啄木鸟　394
蛎鹬　380
镰翅鸡　48, 372
鹩哥　328, 414
猎隼　160, 395
林八哥　414
林雕　244, 389
林雕鸮　391
林岭雀　423
林柳莺　404
林鹨　422
林三趾鹑　383
林沙锥　228, 381
林夜鹰　378
林鹛　382
鳞腹绿啄木鸟　394
鳞头树莺　406
鳞胸鹪鹛　401

领角鸮　390
领雀嘴鹎　403
领䴗鹟　391
领岩鹨　421
领燕鸻　383
流苏鹬　383
琉球歌鸲　416
琉球姬鹟　418
琉球山椒鸟　396
硫黄鹀　344, 426
瘤鸭　375
柳雷鸟　190, 372
芦鹀　426
芦莺　401
路氏雀鹛　407
绿背姬鹟　418
绿背鸫鹟　387
绿背山雀　400
绿翅短脚鹎　403
绿翅金鸠　377
绿翅鸭　375
绿喉蜂虎　392
绿喉太阳鸟　420
绿皇鸠　114, 377
绿脚树鹧鸪　372
绿孔雀　56, 374
绿宽嘴鸫　416
绿鹭　388
绿眉鸭　375
绿拟啄木鸟　393
绿头鸭　375
绿尾虹雉　102, 373
绿胸八色鸫　288, 395
绿嘴地鹃　378
罗纹鸭　375

M

麻雀　421
毛脚鵟　390
毛脚燕　403
毛腿雕鸮　76, 391
毛腿沙鸡　377
毛腿夜鹰　378
矛斑蝗莺　402
矛隼　395
矛纹草鹛　410
煤山雀　399
猛隼　395

猛鸮　391
蒙古百灵　302, 400
蒙古沙鸻　381
蒙古沙雀　423
棉凫　108, 375
冕柳莺　405
冕雀　399
漠白喉林莺　407
漠䳭　418

N

南极贼鸥　385
尼泊尔鹪鹛　401
拟大朱雀　423
牛背鹭　388
牛头伯劳　397
弄岗穗鹛　170, 409

O

欧斑鸠　377
欧鸽　376
欧歌鸫　415
欧金翅雀　424
欧金鸻　381
欧柳莺　404
欧亚大山雀　399
欧亚红尾鸲　417
欧亚鵟　390
欧亚鸲　416
欧亚乌鸫　415
欧亚旋木雀　413
欧夜鹰　378
鸥嘴噪鸥　384

P

琵嘴鸭　375
漂鹬　382
平原鹨　422
蒲苇莺　401
䴙鷉　425
普通翠鸟　393
普通海鸥　384
普通鵟　390
普通鸬鹚　387
普通秋沙鸭　376
普通鸭　413

普通燕鸻　383
普通燕鸥　385
普通秧鸡　379
普通夜鹰　378
普通鹰鹃　379
普通雨燕　378
普通朱雀　423

Q

钳嘴鹳　386
强脚树莺　406
翘鼻麻鸭　375
翘嘴鹬　382
青脚滨鹬　383
青脚鹬　382
青头潜鸭　58, 375
青头鹦鹉　395
丘鹬　381
鸲蝗莺　402
鸲姬鹟　419
鸲岩鹨　421
雀鹰　389
鹊鹂　164, 396
鹊鸲　417
鹊鸭　376
鹊鹞　390

R

日本歌鸲　416
日本鹡鸰　422
日本柳莺　404
日本冕柳莺　405
日本松雀鹰　389
日本鹰鸮　391
绒额䴓　413
肉垂麦鸡　381

S

萨岛柳莺　405
赛氏篱莺　401
三宝鸟　392
三道眉草鹀　425
三趾滨鹬　382
三趾翠鸟　393
三趾鸥　384
三趾鸦雀　407

物种中文名索引

三趾啄木鸟　394
沙䴖　418
沙丘鹤　380
沙色朱雀　424
山斑鸠　377
山皇鸠　377
山鹃鸽　422
山鹪莺　400
山蓝仙鹟　419
山鹨　423
山麻雀　421
山鹛　407
山噪鹛　411
扇尾沙锥　382
勺鸡　373
勺嘴鹬　68, 383
蛇雕　389
石鸻　380
石鸡　372
石雀　422
寿带　397
黍鹀　425
曙红朱雀　424
树鹨　422
双斑百灵　400
双斑绿柳莺　404
双辫八色鸫　282, 395
双角犀鸟　80, 392
水鹨　422
水雉　381
丝光椋鸟　414
四川淡背地鸫　414
四川短翅蝗莺　402
四川褐头山雀　399
四川林鸮　262, 391
四川柳莺　404
四川山鹧鸪　90, 372
四川旋木雀　322, 413
四声杜鹃　379
松鸡　94, 372
松雀　423
松雀鹰　389
松鸦　398
穗䳭　418
蓑羽鹤　380

T

塔尾树鹊　398

台湾白喉噪鹛　411
台湾斑翅鹛　412
台湾斑胸钩嘴鹛　408
台湾鹎　308, 403
台湾戴菊　420
台湾短翅蝗莺　402
台湾画眉　410
台湾黄山雀　400
台湾鹪鹛　401
台湾酒红朱雀　424
台湾蓝鹊　398
台湾林鸲　416
台湾拟啄木鸟　393
台湾山鹧鸪　372
台湾杂色山雀　399
台湾噪鹛　411
台湾竹鸡　373
台湾紫啸鸫　417
台湾棕颈钩嘴鹛　408
台湾棕噪鹛　411
太平鸟　420
太平洋潜鸟　386
田鹨　415
田鹀　422
田鸫　426
条纹噪鹛　411
铁爪鹀　425
铁嘴沙鸻　381
铜翅水雉　381
铜蓝鹟　419
秃鼻乌鸦　399
秃鹳　386
秃鹫　389

W

弯嘴滨鹬　383
韦氏鹟莺　405
苇鹀　426
文须雀　400
纹背捕蛛鸟　421
纹腹啄木鸟　393
纹喉鹎　403
纹喉凤鹛　408
纹喉绿啄木鸟　394
纹头斑翅鹛　412
纹胸斑翅鹛　412
纹胸鹪鹛　410
纹胸鹛　409

纹胸织雀　421
纹胸啄木鸟　394
乌雕　144, 389
乌鸫　415
乌灰鸫　415
乌灰鹞　390
乌鹃　379
乌林鸮　391
乌鹟　418
乌燕鸥　385
乌嘴柳莺　405
兀鹫　388

X

西鹌鹑　373
西滨鹬　382
西伯利亚银鸥　384
西红角鸮　391
西红脚隼　395
西黄鹡鸰　422
西南冠纹柳莺　405
西秧鸡　379
西藏毛腿沙鸡　377
锡嘴雀　423
喜鹊　398
喜山白眉朱雀　424
喜山淡背地鸫　414
喜山黄腹树莺　406
喜山金背啄木鸟　394
喜山鵟　390
细钩嘴鹛　409
细纹苇莺　166, 401
细纹噪鹛　411
细嘴短趾百灵　400
细嘴黄鹂　396
细嘴鸥　384
仙八色鸫　290, 396
线尾燕　402
小鹀鹎　376
小白额雁　212, 374
小白腰雨燕　378
小斑姬鹟　419
小斑啄木鸟　393
小䴉　379
小杓鹬　382
小滨鹬　383
小杜鹃　379
小凤头燕鸥　384

小黑背银鸥　384
小黑领噪鹛　411
小虎斑地鸫　414
小黄脚鹬　382
小蝗莺　402
小灰山椒鸟　396
小金背啄木鸟　394
小鹃鸠　377
小鳞胸鹪鹛　402
小美洲黑雁　374
小鸥　384
小盘尾　397
小青脚鹬　126, 382
小绒鸭　376
小太平鸟　420
小天鹅　375
小田鸡　380
小苇鳽　387
小鸦　425
小仙鹟　419
小星头啄木鸟　393
小鸦鹃　378
小燕尾　417
小云雀　400
小嘴鸻　381
小嘴乌鸦　399
楔尾伯劳　398
楔尾鹱　386
楔尾绿鸠　377
楔尾鸥　384
楔嘴穗鹛　409
新疆歌鸲　416
星头啄木鸟　393
星鸦　398
休氏白喉林莺　407
休氏旋木雀　413
锈腹短翅鸫　416
锈喉鹩鹛　409
锈胸蓝姬鹟　419
须苇莺　401
靴篱莺　401
靴隼雕　246, 389
雪鹑　372
雪鸽　376
雪鸦　425
雪鹀　391
雪雁　374
血雀　423
血雉　373

物种中文名索引

Y

鸦嘴卷尾　397
崖海鸦　385
崖沙燕　402
亚历山大鹦鹉　395
亚洲辉椋鸟　413
烟腹毛脚燕　403
烟柳莺　404
岩滨鹬　383
岩鸽　376
岩雷鸟　372
岩鹭　388
岩燕　402
眼纹黄山雀　400
眼纹噪鹛　411
燕雀　423
燕隼　395
洋燕　402
夜鹭　387
遗鸥　128, 384
蚁䴕　393
异色树莺　406
银耳相思鸟　412
银喉长尾山雀　406
银脸长尾山雀　406
银胸丝冠鸟　396
印度斑嘴鸭　375
印度池鹭　388
印度金黄鹂　396
印度寿带　397
鹰雕　389
鹰鸮　391
优雅角鸮　391
疣鼻天鹅　374
游隼　395
渔雕　390
渔鸥　384
玉带海雕　148, 390
玉头姬鹟　419
鸳鸯　375
原鸽　376
远东树莺　406
远东苇莺　304, 401
云南白斑尾柳莺　405
云南柳莺　404
云雀　400
云石斑鸭　375

Z

杂色山雀　399
杂色噪鹛　412
噪鹃　379
噪苇莺　401
藏黄雀　425
藏马鸡　373
藏雀　340, 424
藏乌鸫　415
藏鹀　342, 426
藏雪鸡　372
藏雪雀　422
泽鹬　382
沼泽大尾莺　402
沼泽山雀　399
赭红尾鸲　417
针尾绿鸠　377
针尾沙锥　381
针尾鸭　375
震旦鸦雀　408
中白鹭　388
中杓鹬　382
中杜鹃　379
中华草鹛　410
中华短翅蝗莺　402
中华凤头燕鸥　70, 384
中华攀雀　400
中华秋沙鸭　110, 376
中华雀鹛　407
中华仙鹟　419
中华鹧鸪　373
中华朱雀　424
中亚鸽　376
中亚叽喳柳莺　404
中亚夜鹰　378
中贼鸥　385
朱背啄花鸟　420
朱鹮　132, 387
朱鹂　396
朱鸦　421
侏蓝姬鹟　418
珠颈斑鸠　377
竹啄木鸟　394
爪哇八哥　414
爪哇池鹭　388
爪哇金丝燕　64, 378
紫背椋鸟　414

441

紫背苇鳽　387
紫翅椋鸟　414
紫花蜜鸟　420
紫颊太阳鸟　420
紫金鹃　379
紫宽嘴鸫　416
紫林鸽　112, 377
紫寿带　397
紫水鸡　224, 380
紫啸鸫　417
紫针尾雨燕　378
棕斑鸠　377
棕背伯劳　398
棕背黑头鸫　415
棕背田鸡　380
棕背雪雀　422
棕草鹛　410
棕翅鵟鹰　390
棕顶树莺　406
棕额长尾山雀　406
棕腹大仙鹟　419
棕腹䴗鹛　396
棕腹蓝仙鹟　419
棕腹林鸽　416
棕腹柳莺　404
棕腹树鹊　398
棕腹隼雕　389
棕腹仙鹟　419
棕腹鹰鹃　379
棕腹啄木鸟　393
棕褐短翅蝗莺　402
棕喉鹩鹛　409
棕喉雀鹛　409
棕颈钩嘴鹛　408

棕颈犀鸟　82, 392
棕颈雪雀　422
棕颈鸭　375
棕颏噪鹛　410
棕脸鹟莺　405
棕眉柳莺　404
棕眉山岩鹨　421
棕三趾鹑　383
棕扇尾莺　400
棕薮鸲　417
棕头歌鸲　178, 416
棕头钩嘴鹛　409
棕头鸥　384
棕头雀鹛　407
棕头鸦雀　407
棕头幽鹛　410
棕臀凤鹛　408
棕尾伯劳　397
棕尾褐鹟　418
棕尾虹雉　373
棕尾鵟　390
棕胸佛法僧　392
棕胸蓝姬鹟　419
棕胸雅鹛　410
棕胸岩鹨　421
棕胸竹鸡　373
棕夜鹭　387
棕雨燕　378
棕噪鹛　411
棕枕山雀　399
棕朱雀　424
纵纹腹小鸮　391
纵纹角鸮　390
纵纹绿鹎　403

物种学名索引
Index of Scientific Names of Species

A

Abroscopus albogularis 405
Abroscopus schisticeps 405
Abroscopus superciliaris 405
Acanthis flammea 425
Acanthis hornemanni 425
Accipiter badius 389
Accipiter gentilis 389
Accipiter gularis 389
Accipiter nisus 389
Accipiter soloensis 389
Accipiter trivirgatus 389
Accipiter virgatus 389
Aceros nipalensis 82, 392
Acridotheres albocinctus 414
Acridotheres burmannicus 414
Acridotheres cristatellus 414
Acridotheres grandis 414
Acridotheres javanicus 414
Acridotheres tristis 414
Acrocephalus agricola 401
Acrocephalus arundinaceus 401
Acrocephalus bistrigiceps 401
Acrocephalus concinens 401
Acrocephalus dumetorum 401
Acrocephalus melanopogon 401
Acrocephalus orientalis 401
Acrocephalus schoenobaenus 401
Acrocephalus scirpaceus 401
Acrocephalus sorghophilus 166, 401
Acrocephalus stentoreus 401
Acrocephalus tangorum 304, 401
Actinodura egertoni 412
Actinodura ramsayi 412

Actitis hypoleucos 382
Aegithalos bonvaloti 406
Aegithalos caudatus 406
Aegithalos concinnus 406
Aegithalos fuliginosus 406
Aegithalos glaucogularis 406
Aegithalos iouschistos 406
Aegithina lafresnayei 397
Aegithina tiphia 397
Aegolius funereus 264, 391
Aegypius monachus 389
Aerodramus brevirostris 378
Aerodramus fuciphagus 64, 378
Aerodramus maximus 378
Aethopyga christinae 420
Aethopyga gouldiae 420
Aethopyga ignicauda 420
Aethopyga nipalensis 420
Aethopyga saturata 420
Aethopyga siparaja 420
Agraphospiza rubescens 423
Agropsar philippensis 414
Agropsar sturninus 414
Aix galericulata 375
Alauda arvensis 400
Alauda gulgula 400
Alauda leucoptera 400
Alaudala cheleensis 400
Alcedo atthis 393
Alcedo hercules 270, 393
Alcedo meninting 393
Alcippe morrisonia 410
Alcippe nipalensis 410
Alcippe poioicephala 410
Alectoris chukar 372

Alectoris magna 372
Alophoixus flaveolus 403
Alophoixus pallidus 403
Amandava amandava 421
Amaurornis cinerea 380
Amaurornis phoenicurus 380
Ampeliceps coronatus 414
Anas acuta 375
Anas crecca 375
Anas luzonica 375
Anas platyrhynchos 375
Anas poecilorhyncha 375
Anas zonorhyncha 375
Anastomus oscitans 386
Anorrhinus austeni 266, 392
Anous stolidus 383
Anser albifrons 374
Anser anser 374
Anser caerulescens 374
Anser cygnoid 210, 374
Anser erythropus 212, 374
Anser fabalis 374
Anser indicus 374
Anser serrirostris 374
Anthipes monileger 419
Anthracoceros albirostris 78, 392
Anthreptes malacensis 420
Anthus campestris 422
Anthus cervinus 422
Anthus godlewskii 422
Anthus gustavi 422
Anthus hodgsoni 422
Anthus pratensis 422
Anthus richardi 422
Anthus roseatus 422
Anthus rubescens 422
Anthus rufulus 422
Anthus spinoletta 422
Anthus sylvanus 423
Anthus trivialis 422
Aplonis panayensis 413
Apus acuticauda 378
Apus apus 378
Apus nipalensis 378
Apus pacificus 378
Aquila chrysaetos 250, 389
Aquila fasciata 252, 389
Aquila heliaca 146, 389
Aquila nipalensis 248, 389

Arachnothera longirostra 420
Arachnothera magna 421
Arborophila ardens 92, 372
Arborophila atrogularis 372
Arborophila brunneopectus 372
Arborophila crudigularis 372
Arborophila gingica 186, 372
Arborophila mandellii 188, 372
Arborophila rufipectus 90, 372
Arborophila rufogularis 372
Arborophila torqueola 372
Ardea alba 388
Ardea cinerea 388
Ardea insignis 388
Ardea intermedia 388
Ardea purpurea 388
Ardenna carneipes 386
Ardenna grisea 386
Ardenna pacificus 386
Ardenna tenuirostris 386
Ardeola bacchus 388
Ardeola grayii 388
Ardeola speciosa 388
Arenaria interpres 382
Artamus fuscus 397
Arundinax aedon 401
Asio flammeus 391
Asio otus 391
Athene brama 391
Athene noctua 391
Aviceda jerdoni 388
Aviceda leuphotes 388
Aythya baeri 58, 375
Aythya ferina 375
Aythya fuligula 375
Aythya marila 376
Aythya nyroca 375
Aythya valisineria 375

B

Babax koslowi 410
Babax lanceolatus 410
Babax waddelli 410
Bambusicola fytchii 373
Bambusicola sonorivox 373
Bambusicola thoracicus 373
Batrachostomus hodgsoni 377
Blythipicus pyrrhotis 394

Bombycilla garrulus　420
Bombycilla japonica　420
Botaurus stellaris　387
Brachypteryx hyperythra　416
Brachypteryx leucophris　416
Brachypteryx montana　416
Brachyramphus perdix　385
Branta bernicla　374
Branta canadensis　374
Branta hutchinsii　374
Branta leucopsis　374
Branta ruficollis　374
Bubo blakistoni　76, 391
Bubo bubo　391
Bubo nipalensis　391
Bubo scandiacus　391
Bubulcus ibis　388
Bucanetes mongolicus　423
Bucephala clangula　376
Buceros bicornis　80, 392
Bulweria bulwerii　386
Burhinus oedicnemus　380
Butastur indicus　390
Butastur liventer　390
Butastur teesa　390
Buteo buteo　390
Buteo hemilasius　260, 390
Buteo japonicus　390
Buteo lagopus　390
Buteo refectus　390
Buteo rufinus　390
Butorides striata　388

C

Cacomantis merulinus　379
Cacomantis sonneratii　379
Calandrella acutirostris　400
Calandrella brachydactyla　400
Calcarius lapponicus　425
Calidris acuminata　383
Calidris alba　382
Calidris alpina　383
Calidris canutus　234, 382
Calidris falcinellus　383
Calidris ferruginea　383
Calidris fuscicollis　383
Calidris himantopus　383
Calidris mauri　382

Calidris melanotos　383
Calidris minuta　383
Calidris ptilocnemis　383
Calidris pugnax　383
Calidris pygmeus　68, 383
Calidris ruficollis　383
Calidris subminuta　383
Calidris subruficollis　383
Calidris temminckii　383
Calidris tenuirostris　232, 382
Callacanthis burtoni　424
Calliope calliope　416
Calliope obscura　176, 416
Calliope pectardens　332, 416
Calliope pectoralis　416
Calliope tschebaiewi　416
Calonectris leucomelas　386
Caprimulgus aegyptius　378
Caprimulgus affinis　378
Caprimulgus centralasicus　378
Caprimulgus europaeus　378
Caprimulgus indicus　378
Caprimulgus macrurus　378
Carduelis carduelis　425
Carpodacus davidianus　424
Carpodacus dubius　424
Carpodacus edwardsii　424
Carpodacus erythrinus　423
Carpodacus formosanus　424
Carpodacus pulcherrimus　424
Carpodacus puniceus　424
Carpodacus rhodochlamys　424
Carpodacus roborowskii　340, 424
Carpodacus rodochroa　424
Carpodacus rodopeplus　424
Carpodacus roseus　424
Carpodacus rubicilla　424
Carpodacus rubicilloides　423
Carpodacus sibiricus　424
Carpodacus sillemi　423
Carpodacus sipahi　423
Carpodacus stoliczkae　424
Carpodacus subhimachalus　424
Carpodacus thura　424
Carpodacus trifasciatus　424
Carpodacus verreauxii　424
Carpodacus vinaceus　424
Carpodacus waltoni　424
Cecropis daurica　403

Cecropis striolata　403
Centropus bengalensis　378
Centropus sinensis　378
Cephalopyrus flammiceps　399
Cercotrichas galactotes　417
Cerorhinca monocerata　385
Certhia discolor　413
Certhia familiaris　413
Certhia himalayana　413
Certhia hodgsoni　413
Certhia manipurensis　413
Certhia nipalensis　413
Certhia tianquanensis　322, 413
Ceryle rudis　393
Cettia brunnifrons　406
Cettia castaneocoronata　406
Cettia cetti　406
Cettia major　406
Ceyx erithaca　393
Chaimarrornis leucocephalus　417
Chalcoparia singalensis　420
Chalcophaps indica　377
Charadrius alexandrinus　381
Charadrius asiaticus　381
Charadrius dubius　381
Charadrius hiaticula　381
Charadrius leschenaultii　381
Charadrius mongolus　381
Charadrius placidus　381
Charadrius veredus　381
Chelidorhynx hypoxanthus　399
Chlamydotis macqueenii　118, 379
Chleuasicus atrosuperciliaris　408
Chlidonias hybrida　385
Chlidonias leucopterus　385
Chlidonias niger　385
Chloris ambigua　424
Chloris chloris　424
Chloris sinica　424
Chloris spinoides　424
Chloropsis aurifrons　420
Chloropsis cochinchinensis　420
Chloropsis hardwickii　420
Cholornis paradoxus　407
Cholornis unicolor　407
Chroicocephalus brunnicephalus　384
Chroicocephalus genei　384
Chroicocephalus novaehollandiae　384
Chroicocephalus ridibundus　384

Chrysococcyx maculatus　379
Chrysococcyx xanthorhynchus　379
Chrysocolaptes lucidus　394
Chrysolophus amherstiae　374
Chrysolophus pictus　374
Chrysominla strigula　412
Chrysomma sinense　407
Chrysophlegma flavinucha　158, 394
Ciconia boyciana　130, 386
Ciconia ciconia　52, 386
Ciconia episcopus　386
Ciconia nigra　240, 386
Cinclidium frontale　417
Cinclus cinclus　413
Cinclus pallasii　413
Cinnyris asiaticus　420
Cinnyris jugularis　420
Circaetus gallicus　389
Circus aeruginosus　390
Circus cyaneus　390
Circus macrourus　390
Circus melanoleucos　390
Circus pygargus　390
Circus spilonotus　390
Cissa chinensis　398
Cissa hypoleuca　398
Cisticola exilis　400
Cisticola juncidis　400
Clamator coromandus　378
Clamator jacobinus　378
Clanga clanga　144, 389
Clangula hyemalis　60, 376
Coccothraustes coccothraustes　423
Cochoa purpurea　416
Cochoa viridis　416
Columba eversmanni　376
Columba hodgsonii　376
Columba janthina　377
Columba leuconota　376
Columba livia　376
Columba oenas　376
Columba palumbus　376
Columba pulchricollis　376
Columba punicea　112, 377
Columba rupestris　376
Conostoma aemodium　407
Copsychus saularis　417
Coracias benghalensis　392
Coracias garrulus　392

Coracina macei 396
Corvus corax 399
Corvus cornix 399
Corvus corone 399
Corvus dauuricus 398
Corvus frugilegus 399
Corvus macrorhynchos 399
Corvus monedula 398
Corvus pectoralis 399
Corvus splendens 399
Coturnicops exquisitus 216, 379
Coturnix coturnix 373
Coturnix japonica 373
Crex crex 220, 379
Crossoptilon auritum 374
Crossoptilon crossoptilon 373
Crossoptilon harmani 373
Crossoptilon mantchuricum 200, 373
Cuculus canorus 379
Cuculus micropterus 379
Cuculus optatus 379
Cuculus poliocephalus 379
Cuculus saturatus 379
Culicicapa ceylonensis 399
Cutia nipalensis 412
Cyanistes cyanus 399
Cyanoderma ambiguum 409
Cyanoderma chrysaeum 409
Cyanoderma pyrrhops 409
Cyanoderma ruficeps 409
Cyanopica cyanus 398
Cyanoptila cumatilis 419
Cyanoptila cyanomelana 419
Cygnus columbianus 375
Cygnus cygnus 375
Cygnus olor 374
Cyornis banyumas 419
Cyornis brunneatus 334, 419
Cyornis concretus 419
Cyornis glaucicomans 419
Cyornis hainanus 419
Cyornis poliogenys 419
Cyornis rubeculoides 419
Cyornis unicolor 419
Cypsiurus balasiensis 378

D

Delichon dasypus 403

Delichon nipalense 403
Delichon urbicum 403
Dendrocitta formosae 398
Dendrocitta frontalis 398
Dendrocitta vagabunda 398
Dendrocopos atratus 394
Dendrocopos auriceps 394
Dendrocopos canicapillus 393
Dendrocopos cathpharius 394
Dendrocopos darjellensis 394
Dendrocopos hyperythrus 393
Dendrocopos kizuki 393
Dendrocopos leucopterus 394
Dendrocopos leucotos 394
Dendrocopos macei 393
Dendrocopos major 394
Dendrocopos minor 393
Dendrocygna javanica 208, 374
Dendronanthus indicus 422
Dicaeum agile 420
Dicaeum chrysorrheum 420
Dicaeum concolor 420
Dicaeum cruentatum 420
Dicaeum ignipectus 420
Dicaeum melanozanthum 420
Dicrurus aeneus 397
Dicrurus annectans 397
Dicrurus hottentottus 397
Dicrurus leucophaeus 397
Dicrurus macrocercus 397
Dicrurus paradiseus 292, 397
Dicrurus remifer 397
Dinopium benghalense 394
Dinopium javanense 394
Dinopium shorii 394
Dryocopus javensis 394
Dryocopus martius 394
Ducula aenea 114, 377
Ducula badia 377

E

Egretta eulophotes 242, 388
Egretta garzetta 388
Egretta novaehollandiae 388
Egretta picata 388
Egretta sacra 388
Elachura formosa 420
Elanus caeruleus 388

Emberiza aureola　88, 426
Emberiza bruniceps　426
Emberiza buchanani　425
Emberiza calandra　425
Emberiza chrysophrys　426
Emberiza cia　425
Emberiza cioides　425
Emberiza citrinella　425
Emberiza elegans　426
Emberiza fucata　425
Emberiza godlewskii　425
Emberiza hortulana　425
Emberiza jankowskii　184, 425
Emberiza koslowi　342, 426
Emberiza leucocephalos　425
Emberiza melanocephala　426
Emberiza pallasi　426
Emberiza pusilla　425
Emberiza rustica　426
Emberiza rutila　426
Emberiza schoeniclus　426
Emberiza siemsseni　425
Emberiza spodocephala　426
Emberiza stewarti　425
Emberiza sulphurata　344, 426
Emberiza tristrami　425
Emberiza variabilis　426
Emberiza yessoensis　426
Enicurus immaculatus　417
Enicurus leschenaulti　417
Enicurus maculatus　417
Enicurus schistaceus　417
Enicurus scouleri　417
Eophona migratoria　423
Eophona personata　423
Eremophila alpestris　400
Erithacus rubecula　416
Erpornis zantholeuca　396
Erythrogenys erythrocnemis　408
Erythrogenys gravivox　408
Erythrogenys hypoleucos　408
Erythrogenys swinhoei　408
Erythrura prasina　421
Esacus recurvirostris　380
Estrilda melpoda　421
Eudromias morinellus　381
Eudynamys scolopaceus　379
Eumyias thalassinus　419
Euodice malabarica　421

Eurystomus orientalis　392

F

Falcipennis falcipennis　48, 372
Falco amurensis　395
Falco cherrug　160, 395
Falco columbarius　395
Falco naumanni　274, 395
Falco peregrinus　395
Falco rusticolus　395
Falco severus　395
Falco subbuteo　395
Falco tinnunculus　395
Falco vespertinus　395
Ficedula albicilla　419
Ficedula elisae　418
Ficedula hodgsoni　418
Ficedula hyperythra　419
Ficedula hypoleuca　418
Ficedula mugimaki　419
Ficedula narcissina　418
Ficedula owstoni　418
Ficedula parva　419
Ficedula ruficauda　418
Ficedula sapphira　419
Ficedula sordida　419
Ficedula strophiata　419
Ficedula superciliaris　419
Ficedula tricolor　419
Ficedula westermanni　419
Ficedula zanthopygia　418
Francolinus pintadeanus　373
Fregata andrewsi　386
Fregata ariel　387
Fregata minor　386
Fringilla coelebs　423
Fringilla montifringilla　423
Fulica atra　380
Fulmarus glacialis　386
Fulvetta cinereiceps　407
Fulvetta ludlowi　407
Fulvetta ruficapilla　407
Fulvetta striaticollis　407
Fulvetta vinipectus　407

G

Galerida cristata　400

Gallicrex cinerea　380
Gallinago gallinago　382
Gallinago hardwickii　381
Gallinago megala　382
Gallinago nemoricola　228, 381
Gallinago solitaria　381
Gallinago stenura　381
Gallinula chloropus　380
Gallus gallus　373
Gampsorhynchus rufulus　410
Garrulax albogularis　411
Garrulax berthemyi　411
Garrulax bieti　318, 411
Garrulax caerulatus　411
Garrulax canorus　410
Garrulax castanotis　410
Garrulax chinensis　411
Garrulax cineraceus　410
Garrulax courtoisi　84, 411
Garrulax davidi　411
Garrulax gularis　411
Garrulax leucolophus　410
Garrulax lunulatus　410
Garrulax maesi　410
Garrulax maximus　411
Garrulax merulinus　411
Garrulax monileger　411
Garrulax ocellatus　411
Garrulax owstoni　410
Garrulax pectoralis　411
Garrulax perspicillatus　411
Garrulax poecilorhynchus　411
Garrulax ruficeps　411
Garrulax ruficollis　411
Garrulax rufogularis　410
Garrulax sannio　411
Garrulax strepitans　410
Garrulax sukatschewi　316, 410
Garrulax taewanus　410
Garrulus glandarius　398
Gavia adamsii　386
Gavia arctica　385
Gavia pacifica　386
Gavia stellata　385
Gecinulus grantia　394
Gelochelidon nilotica　384
Geokichla citrina　414
Geokichla sibirica　414
Glareola lactea　383

Glareola maldivarum　383
Glareola nordmanni　383
Glareola pratincola　383
Glaucidium brodiei　391
Glaucidium cuculoides　391
Glaucidium passerinum　391
Gorsachius goisagi　387
Gorsachius magnificus　136, 387
Gorsachius melanolophus　387
Gracula religiosa　328, 414
Gracupica contra　414
Gracupica nigricollis　414
Graminicola striatus　410
Grammatoptila striata　411
Grandala coelicolor　417
Grus antigone　50, 380
Grus canadensis　380
Grus grus　380
Grus japonensis　122, 380
Grus leucogeranus　66, 380
Grus monacha　124, 380
Grus nigricollis　226, 380
Grus vipio　120, 380
Grus virgo　380
Gygis alba　383
Gypaetus barbatus　388
Gyps bengalensis　389
Gyps fulvus　388
Gyps himalayensis　389
Gyps indicus　389

H

Haematopus ostralegus　380
Halcyon coromanda　392
Halcyon pileata　393
Halcyon smyrnensis　393
Haliaeetus albicilla　258, 390
Haliaeetus leucogaster　256, 390
Haliaeetus leucoryphus　148, 390
Haliaeetus pelagicus　150, 390
Haliastur indus　254, 390
Harpactes erythrocephalus　392
Harpactes oreskios　392
Harpactes wardi　392
Hemiprocne coronata　378
Hemipus picatus　397
Hemitesia pallidipes　406
Hemixos castanonotus　403

Hemixos flavala   403
Heterophasia auricularis   412
Heterophasia capistrata   412
Heterophasia desgodinsi   412
Heterophasia gracilis   412
Heterophasia picaoides   413
Heterophasia pulchella   413
Heteroxenicus stellatus   416
Hieraaetus pennatus   246, 389
Hierococcyx hyperythrus   379
Hierococcyx nisicolor   379
Hierococcyx sparverioides   379
Hierococcyx varius   379
Himantopus himantopus   380
Hirundapus caudacutus   378
Hirundapus celebensis   378
Hirundapus cochinchinensis   378
Hirundapus giganteus   378
Hirundo rustica   402
Hirundo smithii   402
Hirundo tahitica   402
Histrionicus histrionicus   376
Horornis acanthizoides   406
Horornis brunnescens   406
Horornis canturians   406
Horornis diphone   406
Horornis flavolivaceus   406
Horornis fortipes   406
Hydrobates leucorhous   386
Hydrobates monorhis   386
Hydrobates tristrami   386
Hydrocoloeus minutus   384
Hydrophasianus chirurgus   381
Hydroprogne caspia   384
Hypogramma hypogrammicum   420
Hypothymis azurea   397
Hypsipetes amaurotis   404
Hypsipetes leucocephalus   403

I

Ibidorhyncha struthersii   380
Ichthyaetus ichthyaetus   384
Ichthyaetus relictus   128, 384
Ichthyophaga humilis   390
Ictinaetus malaiensis   244, 389
Iduna caligata   401
Iduna pallida   401
Iduna rama   401

Indicator xanthonotus   393
Iole propinqua   403
Irena puella   420
Ithaginis cruentus   373
Ixobrychus cinnamomeus   387
Ixobrychus eurhythmus   387
Ixobrychus flavicollis   387
Ixobrychus minutus   387
Ixobrychus sinensis   387
Ixos mcclellandii   403

J

Jabouilleia danjoui   409
Jynx torquilla   393

K

Ketupa flavipes   154, 391
Ketupa zeylonensis   152, 391
Kittacincla malabarica   417

L

Lagopus lagopus   190, 372
Lagopus muta   372
Lalage melaschistos   396
Lalage nigra   396
Lanius bucephalus   397
Lanius collurio   397
Lanius collurioides   397
Lanius cristatus   397
Lanius excubitor   398
Lanius isabellinus   397
Lanius minor   398
Lanius phoenicuroides   397
Lanius schach   398
Lanius sphenocercus   398
Lanius tephronotus   398
Lanius tigrinus   397
Larus cachinnans   384
Larus canus   384
Larus crassirostris   384
Larus fuscus   384
Larus glaucescens   384
Larus hyperboreus   384
Larus schistisagus   384
Larus smithsonianus   384
Larvivora akahige   416
Larvivora brunnea   416
Larvivora cyane   416

Larvivora komadori　416
Larvivora ruficeps　178, 416
Larvivora sibilans　416
Leioptila annectens　412
Leiothrix argentauris　412
Leiothrix lutea　412
Leptopoecile elegans　406
Leptopoecile sophiae　406
Leptoptilos javanicus　386
Lerwa lerwa　372
Leucophaeus pipixcan　384
Leucosticte arctoa　423
Leucosticte brandti　423
Leucosticte nemoricola　423
Lewinia striata　379
Limnodromus scolopaceus　382
Limnodromus semipalmatus　382
Limosa lapponica　382
Limosa limosa　382
Linaria cannabina　425
Linaria flavirostris　424
Liocichla bugunorum　86, 412
Liocichla omeiensis　320, 412
Liocichla phoenicea　412
Liocichla ripponi　412
Liocichla steerii　412
Lioparus chrysotis　407
Locustella alishanensis　402
Locustella amnicola　402
Locustella certhiola　402
Locustella chengi　402
Locustella davidi　402
Locustella fasciolata　402
Locustella lanceolata　402
Locustella luscinioides　402
Locustella luteoventris　402
Locustella major　402
Locustella mandelli　402
Locustella naevia　402
Locustella ochotensis　402
Locustella pleskei　306, 402
Locustella pryeri　402
Locustella tacsanowskia　402
Locustella thoracica　402
Lonchura atricapilla　421
Lonchura oryzivora　338, 421
Lonchura punctulata　421
Lonchura striata　421
Lophophanes dichrous　399

Lophophorus impejanus　373
Lophophorus lhuysii　102, 373
Lophophorus sclateri　100, 373
Lophotriorchis kienerii　389
Lophura leucomelanos　373
Lophura nycthemera　373
Lophura swinhoii　373
Loriculus vernalis　395
Loxia curvirostra　425
Loxia leucoptera　425
Luscinia megarhynchos　416
Luscinia phoenicuroides　416
Luscinia svecica　416
Lymnocryptes minimus　381
Lyncornis macrotis　378
Lyrurus tetrix　372

M

Machlolophus holsti　400
Machlolophus spilonotus　400
Machlolophus xanthogenys　400
Macropygia ruficeps　377
Macropygia tenuirostris　377
Macropygia unchall　377
Mareca americana　375
Mareca falcata　375
Mareca penelope　375
Mareca strepera　375
Marmaronetta angustirostris　375
Megaceryle lugubris　393
Megalurus palustris　402
Melanitta americana　376
Melanitta fusca　376
Melanochlora sultanea　399
Melanocorypha bimaculata　400
Melanocorypha calandra　400
Melanocorypha maxima　400
Melanocorypha mongolica　302, 400
Melanocorypha yeltoniensis　400
Melophus lathami　425
Mergellus albellus　376
Mergus merganser　376
Mergus serrator　376
Mergus squamatus　110, 376
Merops apiaster　392
Merops leschenaulti　392
Merops orientalis　392
Merops ornatus　392

Merops persicus　392
Merops philippinus　392
Merops viridis　392
Metopidius indicus　381
Microcarbo niger　387
Microhierax caerulescens　394
Microhierax melanoleucos　272, 395
Micropternus brachyurus　394
Milvus migrans　390
Minla ignotincta　412
Mirafra javanica　300, 400
Mixornis gularis　409
Monticola cinclorhyncha　418
Monticola gularis　418
Monticola rufiventris　418
Monticola saxatilis　418
Monticola solitarius　418
Montifringilla adamsi　422
Montifringilla henrici　422
Montifringilla nivalis　422
Motacilla alba　422
Motacilla cinerea　422
Motacilla citreola　422
Motacilla flava　422
Motacilla grandis　422
Motacilla tschutschensis　422
Moupinia poecilotis　407
Mulleripicus pulverulentus　394
Muscicapa dauurica　418
Muscicapa ferruginea　418
Muscicapa griseisticta　418
Muscicapa muttui　418
Muscicapa sibirica　418
Muscicapa striata　418
Mycerobas affinis　423
Mycerobas carnipes　423
Mycerobas melanozanthos　423
Mycteria leucocephala　386
Myiomela leucura　417
Myophonus caeruleus　417
Myophonus insularis　417
Myzornis pyrrhoura　406

N

Napothera epilepidota　410
Neophron percnopterus　388
Neosuthora davidiana　408
Netta rufina　375

Nettapus coromandelianus　108, 375
Niltava davidi　419
Niltava grandis　419
Niltava macgrigoriae　419
Niltava sundara　419
Niltava vivida　419
Ninox japonica　391
Ninox scutulata　391
Nipponia nippon　132, 387
Nisaetus cirrhatus　389
Nisaetus nipalensis　389
Nucifraga caryocatactes　398
Numenius arquata　382
Numenius madagascariensis　230, 382
Numenius minutus　382
Numenius phaeopus　382
Nycticorax caledonicus　387
Nycticorax nycticorax　387
Nyctyornis amictus　392
Nyctyornis athertoni　268, 392

O

Oceanites oceanicus　386
Oenanthe deserti　418
Oenanthe isabellina　418
Oenanthe oenanthe　418
Oenanthe picata　418
Oenanthe pleschanka　418
Onychoprion aleuticus　385
Onychoprion anaethetus　385
Onychoprion fuscatus　385
Onychostruthus taczanowskii　422
Oriolus chinensis　396
Oriolus kundoo　396
Oriolus mellianus　164, 396
Oriolus oriolus　396
Oriolus tenuirostris　396
Oriolus traillii　396
Oriolus xanthornus　396
Orthotomus atrogularis　401
Orthotomus sutorius　401
Otis tarda　116, 379
Otus brucei　390
Otus elegans　391
Otus lettia　390
Otus scops　391
Otus semitorques　390
Otus spilocephalus　390

Otus sunia　391
Oxyura leucocephala　62, 376

P

Pandion haliaetus　388
Panurus biarmicus　400
Paradoxornis flavirostris　408
Paradoxornis guttaticollis　408
Paradoxornis heudei　408
Pardaliparus venustulus　399
Parus cinereus　400
Parus major　399
Parus monticolus　400
Passer ammodendri　421
Passer cinnamomeus　421
Passer domesticus　421
Passer hispaniolensis　421
Passer montanus　421
Pastor roseus　414
Pavo muticus　56, 374
Pelargopsis capensis　392
Pelecanus crispus　142, 388
Pelecanus onocrotalus　138, 388
Pelecanus philippensis　140, 388
Pellorneum albiventre　410
Pellorneum ruficeps　410
Perdix dauurica　373
Perdix hodgsoniae　373
Perdix perdix　373
Pericrocotus brevirostris　396
Pericrocotus cantonensis　396
Pericrocotus divaricatus　396
Pericrocotus ethologus　396
Pericrocotus flammeus　397
Pericrocotus roseus　396
Pericrocotus solaris　396
Pericrocotus tegimae　396
Periparus ater　399
Periparus rubidiventris　399
Periparus rufonuchalis　399
Perisoreus infaustus　398
Perisoreus internigrans　294, 398
Pernis apivorus　388
Pernis ptilorhynchus　388
Petrochelidon fluvicola　403
Petronia petronia　422
Phaenicophaeus tristis　378
Phaethon aethereus　385

Phaethon lepturus　385
Phaethon rubricauda　385
Phalacrocorax capillatus　387
Phalacrocorax carbo　387
Phalacrocorax pelagicus　387
Phalacrocorax urile　387
Phalaropus fulicarius　383
Phalaropus lobatus　383
Phasianus colchicus　374
Phodilus badius　392
Phoebastria albatrus　238, 386
Phoebastria immutabilis　386
Phoebastria nigripes　386
Phoenicopterus roseus　376
Phoenicuropsis coeruleocephala　417
Phoenicuropsis erythronotus　417
Phoenicuropsis frontalis　417
Phoenicuropsis schisticeps　417
Phoenicurus alaschanicus　180, 417
Phoenicurus auroreus　417
Phoenicurus erythrogastrus　417
Phoenicurus hodgsoni　417
Phoenicurus ochruros　417
Phoenicurus phoenicurus　417
Phyllergates cucullatus　406
Phylloscopus affinis　404
Phylloscopus armandii　404
Phylloscopus borealis　404
Phylloscopus borealoides　405
Phylloscopus calciatilis　405
Phylloscopus cantator　405
Phylloscopus chloronotus　404
Phylloscopus claudiae　405
Phylloscopus collybita　404
Phylloscopus coronatus　405
Phylloscopus davisoni　405
Phylloscopus emeiensis　405
Phylloscopus forresti　404
Phylloscopus fuligiventer　404
Phylloscopus fuscatus　404
Phylloscopus goodsoni　405
Phylloscopus griseolus　404
Phylloscopus hainanus　310, 405
Phylloscopus humei　404
Phylloscopus ijimae　405
Phylloscopus inornatus　404
Phylloscopus kansuensis　404
Phylloscopus maculipennis　404
Phylloscopus magnirostris　405

Phylloscopus occisinensis 404
Phylloscopus ogilviegranti 405
Phylloscopus plumbeitarsus 404
Phylloscopus proregulus 404
Phylloscopus pulcher 404
Phylloscopus reguloides 405
Phylloscopus ricketti 405
Phylloscopus schwarzi 404
Phylloscopus sibilatrix 404
Phylloscopus sindianus 404
Phylloscopus subaffinis 404
Phylloscopus tenellipes 404
Phylloscopus trochiloides 404
Phylloscopus trochilus 404
Phylloscopus xanthodryas 404
Phylloscopus xanthoschistos 405
Phylloscopus yunnanensis 404
Pica pica 398
Picoides tridactylus 394
Picumnus innominatus 393
Picus canus 394
Picus chlorolophus 394
Picus rabieri 394
Picus squamatus 394
Picus vittatus 394
Picus xanthopygaeus 394
Pinicola enucleator 423
Pitta cyanea 395
Pitta moluccensis 396
Pitta nipalensis 284, 395
Pitta nympha 290, 396
Pitta oatesi 286, 395
Pitta phayrei 282, 395
Pitta sordida 288, 395
Pitta soror 162, 395
Platalea leucorodia 387
Platalea minor 134, 387
Plectrophenax nivalis 425
Plegadis falcinellus 387
Ploceus manyar 421
Ploceus philippinus 421
Pluvialis apricaria 381
Pluvialis fulva 381
Pluvialis squatarola 381
Pnoepyga albiventer 401
Pnoepyga formosana 401
Pnoepyga immaculata 401
Pnoepyga pusilla 402
Podiceps auritus 376

Podiceps cristatus 376
Podiceps grisegena 376
Podiceps nigricollis 376
Podoces biddulphi 298, 398
Podoces hendersoni 296, 398
Poecile davidi 399
Poecile montanus 399
Poecile palustris 399
Poecile superciliosus 399
Poecile weigoldicus 399
Polyplectron bicalcaratum 106, 374
Polyplectron katsumatae 54, 374
Polysticta stelleri 376
Pomatorhinus ferruginosus 409
Pomatorhinus musicus 408
Pomatorhinus ochraceiceps 409
Pomatorhinus ruficollis 408
Pomatorhinus schisticeps 408
Pomatorhinus superciliaris 409
Porphyrio porphyrio 224, 380
Porzana porzana 379
Prinia atrogularis 401
Prinia crinigera 400
Prinia flaviventris 401
Prinia hodgsonii 401
Prinia inornata 401
Prinia polychroa 401
Prinia rufescens 401
Procarduelis nipalensis 423
Prunella atrogularis 421
Prunella collaris 421
Prunella fulvescens 421
Prunella himalayana 421
Prunella immaculata 421
Prunella koslowi 336, 421
Prunella montanella 421
Prunella rubeculoides 421
Prunella strophiata 421
Psarisomus dalhousiae 396
Pseudibis davisoni 387
Pseudobulweria rostrata 386
Pseudopodoces humilis 399
Psilopogon asiatica 393
Psilopogon australis 393
Psilopogon faber 393
Psilopogon faiostrictus 393
Psilopogon franklinii 393
Psilopogon haemacephalus 393
Psilopogon lineatus 393

Psilopogon nuchalis　393
Psilopogon virens　393
Psittacula alexandri　280, 395
Psittacula derbiana　278, 395
Psittacula eupatria　395
Psittacula finschii　395
Psittacula himalayana　395
Psittacula krameri　395
Psittacula roseata　395
Psittinus cyanurus　395
Psittiparus gularis　408
Psittiparus ruficeps　408
Pterocles orientalis　377
Pterodroma hypoleuca　386
Pteruthius aeralatus　396
Pteruthius intermedius　396
Pteruthius melanotis　396
Pteruthius rufiventer　396
Pteruthius xanthochlorus　396
Ptilinopus leclancheri　377
Ptyonoprogne concolor　402
Ptyonoprogne rupestris　402
Pucrasia macrolopha　373
Pycnonotus atriceps　403
Pycnonotus aurigaster　403
Pycnonotus cafer　403
Pycnonotus finlaysoni　403
Pycnonotus flavescens　403
Pycnonotus jocosus　403
Pycnonotus leucogenis　403
Pycnonotus melanicterus　403
Pycnonotus sinensis　403
Pycnonotus striatus　403
Pycnonotus taivanus　308, 403
Pycnonotus xanthorrhous　403
Pyrgilauda blanfordi　422
Pyrgilauda davidiana　422
Pyrgilauda ruficollis　422
Pyrrhocorax graculus　398
Pyrrhocorax pyrrhocorax　398
Pyrrhoplectes epauletta　423
Pyrrhula erythaca　423
Pyrrhula erythrocephala　423
Pyrrhula nipalensis　423
Pyrrhula pyrrhula　423

R

Rallina eurizonoides　218, 379

Rallina fasciata　379
Rallus aquaticus　379
Rallus indicus　379
Recurvirostra avosetta　380
Regulus goodfellowi　420
Regulus regulus　420
Remiz consobrinus　400
Remiz coronatus　400
Remiz macronyx　400
Rhipidura albicollis　397
Rhipidura aureola　397
Rhodopechys sanguineus　423
Rhodospiza obsoleta　423
Rhodostethia rosea　384
Rhopophilus pekinensis　407
Rhyacornis fuliginosa　417
Rhyticeros undulatus　156, 392
Rimator malacoptilus　410
Riparia diluta　402
Riparia paludicola　402
Riparia riparia　402
Rissa tridactyla　384
Rostratula benghalensis　381
Rynchops albicollis　385

S

Sarcogyps calvus　74, 389
Sarkidiornis melanotos　375
Saroglossa spiloptera　413
Sasia ochracea　393
Saundersilarus saundersi　236, 384
Saxicola caprata　418
Saxicola ferreus　418
Saxicola insignis　182, 417
Saxicola jerdoni　418
Saxicola maurus　418
Schoeniparus brunneus　409
Schoeniparus castaneceps　409
Schoeniparus cinereus　409
Schoeniparus dubius　409
Schoeniparus rufogularis　409
Schoeniparus variegaticeps　314, 409
Scolopax rusticola　381
Seicercus affinis　405
Seicercus burkii　405
Seicercus castaniceps　405
Seicercus omeiensis　405
Seicercus poliogenys　405

Seicercus soror　405
Seicercus tephrocephalus　405
Seicercus valentini　405
Seicercus whistleri　405
Serilophus lunatus　396
Serinus pusillus　425
Sibia morrisoniana　412
Sibia nipalensis　412
Sibia souliei　412
Sibia waldeni　412
Sibirionetta formosa　375
Sinosuthora alphonsiana　407
Sinosuthora brunnea　407
Sinosuthora conspicillata　407
Sinosuthora przewalskii　168, 407
Sinosuthora webbiana　407
Sinosuthora zappeyi　312, 407
Sitta castanea　413
Sitta europaea　413
Sitta formosa　174, 413
Sitta frontalis　413
Sitta himalayensis　413
Sitta leucopsis　413
Sitta magna　172, 413
Sitta nagaensis　413
Sitta solangiae　326, 413
Sitta villosa　413
Sitta yunnanensis　324, 413
Sittiparus castaneoventris　399
Sittiparus varius　399
Siva cyanouroptera　412
Spatula clypeata　375
Spatula querquedula　375
Spelaeornis badeigularis　409
Spelaeornis caudatus　409
Spelaeornis chocolatinus　409
Spelaeornis kinneari　409
Spelaeornis troglodytoides　409
Spilopelia chinensis　377
Spilornis cheela　389
Spinus spinus　425
Spinus thibetana　425
Spizixos canifrons　403
Spizixos semitorques　403
Spodiopsar cineraceus　414
Spodiopsar sericeus　414
Stachyris humei　409
Stachyris nigriceps　409
Stachyris nonggangensis　170, 409

Stachyris roberti　409
Stachyris strialata　409
Stercorarius longicaudus　385
Stercorarius maccormicki　385
Stercorarius parasiticus　385
Stercorarius pomarinus　385
Sterna acuticauda　385
Sterna aurantia　385
Sterna dougallii　385
Sterna hirundo　385
Sterna sumatrana　385
Sternula albifrons　385
Streptopelia decaocto　377
Streptopelia orientalis　377
Streptopelia senegalensis　377
Streptopelia tranquebarica　377
Streptopelia turtur　377
Strix aluco　391
Strix davidi　262, 391
Strix leptogrammica　391
Strix nebulosa　391
Strix uralensis　391
Sturnia malabarica　414
Sturnia pagodarum　414
Sturnia sinensis　414
Sturnus vulgaris　414
Sula dactylatra　387
Sula leucogaster　387
Sula sula　387
Surnia ulula　391
Surniculus lugubris　379
Suthora fulvifrons　407
Suthora nipalensis　407
Suthora verreauxi　408
Sylvia althaea　407
Sylvia atricapilla　406
Sylvia communis　407
Sylvia crassirostris　407
Sylvia curruca　407
Sylvia minula　407
Sylvia nana　407
Sylvia nisoria　406
Sylviparus modestus　399
Synoicus chinensis　373
Synthliboramphus antiquus　385
Synthliboramphus wumizusume　385
Syrmaticus ellioti　204, 374
Syrmaticus humiae　206, 374
Syrmaticus mikado　374

Syrmaticus reevesii　104, 374
Syrrhaptes paradoxus　377
Syrrhaptes tibetanus　377

T

Tachybaptus ruficollis　376
Tachymarptis melba　378
Tadorna ferruginea　375
Tadorna tadorna　375
Tarsiger chrysaeus　416
Tarsiger cyanurus　416
Tarsiger hyperythrus　416
Tarsiger indicus　416
Tarsiger johnstoniae　416
Tarsiger rufilatus　416
Temnurus temnurus　398
Tephrodornis virgatus　397
Terpsiphone affinis　397
Terpsiphone atrocaudata　397
Terpsiphone incei　397
Terpsiphone paradisi　397
Tesia cyaniventer　406
Tesia olivea　406
Tetrao urogalloides　96, 372
Tetrao urogallus　94, 372
Tetraogallus altaicus　196, 372
Tetraogallus himalayensis　372
Tetraogallus tibetanus　372
Tetraophasis obscurus　192, 372
Tetraophasis szechenyii　194, 372
Tetrastes bonasia　372
Tetrastes sewerzowi　372
Tetrax tetrax　379
Thalasseus bengalensis　384
Thalasseus bergii　384
Thalasseus bernsteini　70, 384
Thalasseus sandvicensis　384
Threskiornis melanocephalus　72, 387
Tichodroma muraria　413
Tickellia hodgsoni　406
Timalia pileata　409
Todiramphus chloris　393
Tragopan blythii　373
Tragopan caboti　98, 373
Tragopan melanocephalus　373
Tragopan satyra　198, 373
Tragopan temminckii　373
Treron apicauda　377

Treron bicinctus　377
Treron curvirostra　377
Treron formosae　214, 377
Treron phoenicopterus　377
Treron pompadora　377
Treron sieboldii　377
Treron sphenurus　377
Trichastoma tickelli　410
Tringa brevipes　382
Tringa erythropus　382
Tringa flavipes　382
Tringa glareola　382
Tringa guttifer　126, 382
Tringa incana　382
Tringa nebularia　382
Tringa ochropus　382
Tringa stagnatilis　382
Tringa totanus　382
Trochalopteron affine　411
Trochalopteron elliotii　411
Trochalopteron erythrocephalum　412
Trochalopteron formosum　412
Trochalopteron henrici　411
Trochalopteron lineatum　411
Trochalopteron milnei　412
Trochalopteron morrisonianum　411
Trochalopteron squamatum　411
Trochalopteron subunicolor　411
Trochalopteron variegatum　412
Troglodytes troglodytes　413
Tropicoperdix chloropus　372
Turdinus brevicaudatus　410
Turdinus crispifrons　410
Turdus albocinctus　415
Turdus atrogularis　415
Turdus boulboul　415
Turdus cardis　415
Turdus chrysolaus　415
Turdus dissimilis　415
Turdus eunomus　415
Turdus feae　330, 415
Turdus hortulorum　415
Turdus iliacus　415
Turdus kessleri　415
Turdus mandarinus　415
Turdus maximus　415
Turdus merula　415
Turdus mupinensis　415
Turdus naumanni　415

Turdus niveiceps　415
Turdus obscurus　415
Turdus pallidus　415
Turdus philomelos　415
Turdus pilaris　415
Turdus rubrocanus　415
Turdus ruficollis　415
Turdus unicolor　415
Turdus viscivorus　415
Turnix suscitator　383
Turnix sylvaticus　383
Turnix tanki　383
Tyto alba　392
Tyto longimembris　392

U

Upupa epops　392
Uria aalge　385
Urocissa caerulea　398
Urocissa erythroryncha　398
Urocissa flavirostris　398
Urocissa whiteheadi　398
Urocynchramus pylzowi　421
Urosphena squameiceps　406

V

Vanellus cinereus　381
Vanellus duvaucelii　381
Vanellus gregarius　381
Vanellus indicus　381
Vanellus leucurus　381
Vanellus vanellus　381

X

Xema sabini　384

Xenus cinereus　382

Y

Yuhina bakeri　408
Yuhina brunneiceps　408
Yuhina castaniceps　408
Yuhina diademata　408
Yuhina flavicollis　408
Yuhina gularis　408
Yuhina nigrimenta　408
Yuhina occipitalis　408

Z

Zapornia akool　379
Zapornia bicolor　380
Zapornia fusca　380
Zapornia parva　380
Zapornia paykullii　222, 380
Zapornia pusilla　380
Zonotrichia leucophrys　426
Zoothera aurea　414
Zoothera dauma　414
Zoothera dixoni　414
Zoothera griseiceps　414
Zoothera marginata　415
Zoothera mollissima　414
Zoothera monticola　415
Zoothera salimalii　414
Zosterops erythropleurus　408
Zosterops japonicus　408
Zosterops meyeni　408
Zosterops palpebrosus　408